GEOENVIRONMENTAL RECLAMATION

International Symposium
20–22 November, Nagpur, India

Editors

A.G. Paithankar
P.K. Jha
R.K. Agarwal

A.A. BALKEMA/ROTTERDAM/BROOKFIELD/2001

ISBN 90 5809 219 4

A.A. Balkema, P.O. Box 1675, 3000 BR Rotterdam, Netherlands
Fax: +31.10.4135947; e-mail: balkema@balkema.nl
Internet site: http://www.balkema.nl

Distributed in USA and Canada by
A.A. Balkema Publishers, Old Post Road, Brookfield, Vermont 05036, USA
Fax: 802.276.3837; e-mail: Info@ashgate.com

निसर्गो रक्षति रक्षितः

IF YOU PROTECT THE NATURE, THE NATURE WILL PROTECT YOU

INTERNATIONAL SYMPOSIUM
ON
GEOENVIRONMENTAL RECLAMATION

(SHRI RAMDEO BABA KAMLA NEHRU ENGG. COLLEGE, NAGPUR. NOVEMBER 20 – 22, 2000)

PATRON

Shri B. L. Purohit
(Chairman, R.S.S.S, Nagpur)

Dr. Arun. S. Satputaley,
(Vice Chancellor, Nagpur
University)

CHAIRMAN

Shri Govindlalji L. Agrawal,
(Secretary, R.S.S.S, Nagpur)

**CHAIRMAN OF EXE.
COMMITTEE**

Prof. Dr. N. T. Khobragade
(Principal, RKNEC)

CONVENORS

Prof. Dr. A.G.Paithankar
(Chairman, Earth Resources
Mngt. Centre (P) Ltd., Nagpur)

Prof. Dr. Keith Atkinson
(Dy. Vice.Chancellor,
Uni. Of Exeter, U.K.)

CO-CONVENORS

Prof. P.K. Jha
(HOD, Mining, RKNEC)

Dr. Mike Heath
(Camborne School of Mines,
United Kingdom)

CO-ORDINATORS

Prof. J.S.de Omenaca,
(Universidad de Cantabria Spain)

Prof. Dr. R.T. Jadhav
(HOD, Chemistry Dept, RKNEC)

SECRETARY

A.K. Agarwal
(Asst. Prof. in Mining, RKNEC)

JOINT SECRETARIES

Prof. Dr. R.R.Lakhe
(H.O.D. Industrial Engg. RKNEC)

M.S.Tiwari
(Lecturer, Mining Engg. RKNEC)

FROM THE CONVENOR'S DESK

An International Symposium on Geo-Environmental Reclamation is the most appropriate theme in the present scenario when global population is exploding and man is in mad pursuit of physical comforts and is out to destroy the nature. Man seems to have forgotten that this earth is like a flying plane and he has to behave like a disciplined passenger or else the plane will crash and along with it all the passengers will perish.

International Society for Geo-Environmental Reclamation (ISGR) along with Ramdeobaba Kamla Nehru Engineering College (RKNEC) has taken a lead in organizing this first global event.

Today the condition of this earth, the only one of its kind in the Universe, is in a state of total shamble. Every country is behaving like an independent variable, some are burning fossil fuels indiscriminately thus polluting the atmosphere and giving rise to green house gases, some are running after indiscriminate mining of nonrenewable mineral resources thus inflicting permanent wounds on the mother earth, some are after stock piling and detonation of nuclear devices, some are firing guided missiles and with success in very such "Physical" activity champagne corks pop up

The organizers want to emphasize that they are not against industrialization are against measures for self-defence. That is the fundamental right of every government. But all activities should have a certain limit for which self discipline and restraint is required. We

iii

COMMUNICATION ADDRESS
CONVENOR, INTERNATIONAL SYMPOSIUM,
SHREE R.K.N. ENGG. COLLEGE, KATOL ROAD, NAGPUR (INDIA), 440013.
TEL.: (91) - (712) - 582844, 580177 FAX: (91)-(712) – 583237, E – MAIL < rknintsy@webginn.com >
For more details please visit us at our web site www.rknec.edu

PATRON

Shri B. L. Purohit
(Chairman, R.S.S.S, Nagpur)

Dr. Arun. S. Satputaley,
(Vice Chancellor, Nagpur University)

CHAIRMAN

Shri Govindlalji L. Agrawal,
(Secretary, R.S.S.S, Nagpur)

CHAIRMAN OF EXE. COMMITTEE

Prof. Dr. N. T. Khobragade
(Principal, RKNEC)

CONVENORS

Prof. Dr. A.G.Paithankar
(Chairman, Earth Resources Mngt. Centre (P) Ltd., Nagpur)

Prof. Dr. Keith Atkinson
(Dy. Vice.Chancellor, Uni. Of Exeter, U.K.)

CO-CONVENORS

Prof. P.K. Jha
(HOD, Mining, RKNEC)

Dr. Mike Heath
(Camborne School of Mines, United Kingdom)

CO-ORDINATORS

Prof. J.S.de Omenaca,
(Universidad de Cantabria Spain)

Prof. Dr. R.T. Jadhav
(HOD, Chemistry Dept, RKNEC)

SECRETARY

A.K. Agarwal
(Asst. Prof. In Mining, RKNEC)

JOINT SECRETARIES

Prof. Dr. R.R.Lakhe
(H.O.D. Industrial Engg. RKNEC)

M.S.Tiwari
(Lecturer, Mining Engg. RKNEC)

should not forget that, when we pollute the atmosphere or river water or cause land degradation it is our bounden duly to undertake reclamation to restore nature to its pristine condition.

We should introspect and find out what are our real needs for existence on this earth. If these are satisfied we should get out of this mad race of "Physical" Pursuits and devote some time of our limited life to take care of the earth. It should never be forgotten that when we inherited this earth from our forefathers it was certainly nearer to nature with proper ecological balance. Is it not then our primary duty to hand over this earth to our children in a condition covered with green forests, with streams dancing with sparklingly clean water, with birds chirping on branches, with flowers blossoming is gardens with sky clean where in the night the moon can be seen clearly by our children and they enjoy its peaceful beauty?

This is not the time to remain silent spectators to the destruction that is taking place on this earth. Let us unite on a common global platform and shout with a chorus what everybody wants i.e. "to protect nature in order to protect ourselves"

(Prof. Dr. Anand Paithankar)
Convenor

iv

COMMUNICATION ADDRESS
CONVENOR, INTERNATIONAL SYMPOSIUM,
SHREE R.K.N. ENGG. COLLEGE, KATOL ROAD, NAGPUR (INDIA), 440013.
TEL.: (91) - (712) - 582844, 580177 FAX: (91)-(712) – 583237, E – MAIL < rknintsy@webginn.com >
For more details please visit us at our web site www.rknec.edu

निसर्गो रक्षति रक्षितः

IF YOU PROTECT THE NATURE, THE NATURE WILL PROTECT YOU

INTERNATIONAL SYMPOSIUM
ON
GEOENVIRONMENTAL RECLAMATION

(SHRI RAMDEO BABA KAMLA NEHRU ENGG. COLLEGE, NAGPUR. NOVEMBER 20 – 22, 2000)

PATRON

Shri B. L. Purohit
(Chairman, R.S.S.S, Nagpur)

Dr. Arun. S. Satputaley,
(Vice Chancellor, Nagpur
University)

CHAIRMAN

Shri Govindlalji L. Agrawal,
(Secretary, R.S.S.S, Nagpur)

**CHAIRMAN OF EXE.
COMMITTEE**

Prof. Dr. N. T. Khobragade
(Principal, RKNEC)

CONVENORS

Prof. Dr. A.G.Paithankar
(Chairman, Earth Resources
Mngt. Centre (P) Ltd., Nagpur)

Prof. Dr. Keith Atkinson
(Dy. Vice.Chancellor,
Uni. Of Exeter, U.K.)

CO-CONVENORS

Prof. P.K. Jha
(HOD, Mining, RKNEC)

Dr. Mike Heath
(Camborne School of Mines,
United Kingdom)

CO-ORDINATORS

Prof. J.S.de Omenaca,
(Universidad de Cantabria Spain)

Prof. Dr. R.T. Jadhav
(HOD, Chemistry Dept, RKNEC)

SECRETARY

A.K. Agarwal
(Asst. Prof. in Mining, RKNEC)

JOINT SECRETARIES

Prof. Dr. R.R.Lakhe
(H.O.D. Industrial Engg. RKNEC)

M.S.Tiwari
(Lecturer, Mining Engg. RKNEC)

FROM THE EDTIOR'S DESK

The scope of the Symposium is indeed wide and encompasses all human activities and also natural forces, which cause land degradation and water and air pollution. Valuable papers have been contributed on almost all themes by experts of international repute and from different countries. They provide a view of the problem from different angles obviously because the perception is different. When considering the whole lot of the papers included in the proceeding it makes an extremely interesting reading. We are sure the proceeding will be a sparkling ornament in any library in the world and pride possession for every human being interested in Geo-Environmental Reclamation.

(Prof. Dr. Anand Paithankar)

v

COMMUNICATION ADDRESS
CONVENOR, INTERNATIONAL SYMPOSIUM,
SHREE R.K.N. ENGG. COLLEGE, KATOL ROAD, NAGPUR (INDIA), 440013.
TEL.: (91) - (712) - 582844, 580177 FAX: (91)-(712) – 583237, E – MAIL < rknintsy@webginn.com >
For more details please visit us at our web site www.rknec.edu

Contents

Theme 1 : Land Degradation and Reclamation

Theme 2 : Waste Disposal and Eco-friendly Working

Theme 6 : General on Environment, Population Control etc.

THEME 1 :
Land Degradation and Reclamation

Environmental Assessment of the Cierro De San Pantaleón Reclamation (Cantabria, N. Spain)

J.A. Saiz de Omeñaca[1], A. Prieto Rodríguez[1], R. Fernández Suárez[2] and J. Sáiz de Omeñaca[3]

[1]*Escuela T.S.I. de Montes, Universidad Politécnica de Madrid, Ciudad Universitaria, 28040 Madrid (Spain)*

[2]*Escuela U.I.T. Forestal, Universidad Politécnica de Madrid, Ciudad Universitaria, 28040 Madrid (Spain)*

[3]*DCITYMAC, Universidad de Cantabria, Edificio de Ciencias Económicas, Av. De Ios Castros s/n., 39005 Santander, Cantabria (Spain)*

INTRODUCTION

The Cierro Grande de San Pantaleón, today a land area of around 80 hectares covered by arboreal vegetation, was, in the seventies, a rocky area, dry and unproductive. This paper outlines the main characteristics of the area, the initial situation and the actions undertaken to make the area productive again, both in an economic and an environmental sense.

CHARACTERISTICS OF THE LAND

The most significant characteristics are described below (for further details, Cendrero & Saiz de Omeñaca, 1975; DG, 1976, 1978, 1980; Cendrero & al., 1980b; Rodríguez Barreal & al., 1990; Prieto et al., 2000):

Relief: Hills and ridges alternate with closed depressions of up to 650 ft. in diameter. Heights are within a range of 240 and 650 ft. above sea level. Slope gradients vary greatly, generally being between 12 and 15°, though it is not unusual to find gradients of over 30°. Flat extensions are scarce and small. The surface roughness is high or even very high and the predominant exposure is to the north-east.

Climate

This is a warm maritime climate, with a mesothermal thermal domain – supermaritime and with a subhumid humidity domain. The average annual temperature is 55°F, with moderate thermal oscillations (January average 45°F and August 65°F). The number of hours of sunshine is low (1250/year), the insulation rate is low and there is a predominance of cloudy or overcast days. Rainfall reaches 43 in. per year, generally with maximums in autumn and with drought periods, occasionally lasting for months, occurring irregularly, most often in spring and summer.

The prevailing winds, from The Atlantic (north-east) are ususally accompanied by rainfall of varying intensity. Continental winds (south and south-east), warm and dry and of the foehn type, are also very frequent and often reach considerable speeds, with gusts of around 65 to 80 mi/h, which can cause considerable damage to the vegetation.

Lithology

The subsoil is composed essentially of limestone, occasionally marly limestone or dolostones, except for a thin stretch in the south-east, where there is only marly limestone. At the bottom of the great depressions, regoliths of essentially residual origins have been formed.

Geological Structure

The limestone is arranged in poorly defined strata, to such a point that these often appear to be massive. The dolomotisation is irregular, but the marliest feature is linked to the existence of beds. The marly limestone of the south-east has a better stratification, arranged in thin, clearly defined layers, generally of a thickness of less than one meter. The limestone and marly limestone have a rather irregular dip of between 15° and 20° towards the north-east throughout the subsoil of the Cierro.

Morphology
Due to the climatic conditions to which the area is and has been exposed, its karstification has been very intense and has given rise to a truly complex topography, with numerous holes, concavities, entrances, grooves, rifts, striations, pinnacles, needles, crests, etc. In places where the corrosion has been most intense, closed depressions and sinkholes of varying diameters have been formed.

Soils
These are orthic luvisoils or lithosoils of a largely residual nature, often very fragile. The former, with an average continuity and variable thickness, poor in organic matter and not very developed, often tend towards lithosoils. The latter, with irregular distribution and very variable thickness, have suffered substantial deterioration. Only the orthic luvisoils that stretch over the marly limestone areas of the south-east have an acceptable continuity and thickness.

Not only are rock-beds abundant throughout most of the Cierro, but also the thickness of the soil varies greatly from one area to another. Thus, it is not unusual for the thickness of the regoliths found next to a crest which rises from a few centimeters to more than a metre above the surface to reach several metres.

Waters
The water infiltrates easily into the rock mass and flows underground through a complex network of cracks and conducts. Thus, despite the abundant rainfall in the area, there are no surface water courses, either seasonal or permanent. The vegetation only disposes in practice of the waters retained in the soils. Although there is a subterranean aquifer, the water table is, at its most favourable point, around 165 ft below the surface.

The aquifer, which stretches out under at least 90% of the Cierro, has been included in the inventories of the Geomineral Technology Institute of Spain. Both because of the volume of water and because it is located close to potential centres of consumption, the conservation of the quality of these waters is of great socio-economic interest.

Landscape and other considerations
With the exception of the highest peaks, which are visible from far away, most of the Cierro can only be seen from close up. There is no human occupation and, apart from a minor road along the northern edge, there no other communication lines across it. There are no known historical, scientific, educational or historic resources in the area.

INITIAL SITUATION
The area was, then, one without any use for economic profit, with a complex relief, great roughness, high aridity (surface and subsurface) and with little developed soils, poor in organic matter, with an uneven development, fragile and of varying thickness. Moreover, there were limitations on use, such as those imposed by the need to protect the deep aquifer. Thus, any human action needed to be planned and executed with the utmost care.

The potential vegetation of the area corresponds to that named by Guinea (1943, 1953, 1954) as "Cantabrian evergreen oak forest" characterised by the oak-tree (*Quercus ilex* L.), accompanied by other more or less xerophytic species, a peculiar association which is more commonly found in Mediterranean rather than Atlantic climates. It could also be said to have certain characteristics of a mixed forest.

In foregone times, human actions (to obtain firewood and charcoal) deforested the area, leaving barren lands with few elements of the natural biomes. The area, not being suitable for fields or pastures, the traditional uses of land in the region, was abandoned. Many of the soils were eroded and their physical elements were carried off towards the various karstic depressions, where they accumulated. In the higher parts of the inner area of the Cierro, the rock was left with no soil or with little, impoverished soil, reduced to small traces in cracks, recesses and hollows.

Throughout most of the Cierro, a vegetation of very little value developed, dominated by a few species several of which were highly ulcerating. In the vales, with a comparatively rich and thicker soil, as well as conditions of greater humidity and less exposure to the winds, an autochthonous germoplasm was partially preserved. In the higher areas, the development of the little existing vegetation was conditioned by the greater surface aridity and by their exposure to high winds, but this development did attain a certain biological value. Before the reclamation, the only "use" made of the area was a small illegal wild dump on the northern edge.

4

ACTIONS

An attempt to return to conditions similar to the initial conditions would have been condemned to failure. Firstly, because the evergreen oak in Cantabria is a relict species, belonging to previous eras with a different climate from today's, and would be very difficult and slow to reintroduce. The use of other species of oak would run into even greater difficulties, due to both the lack of the appropriate environmental conditions and the long periods of time required. Moreover, the socio-economic conditions of the region and the fact that the lands were private property made it advisable to find an economically productive use for the area, so that the long-term continuity of the actions would be guaranteed.

A partial reforestation was carried out with a non-native (and controversial in the region) species of rapid growth, the *Eucaliptus globulus* Labill., fairly resistant to the winds. The mild temperatures throughout the year and the fact that most of the land was in shade, combined with the relative regularity of the rainfall during the period of growth of the vegetation, were all favourable factors in the growth of this species. The trees were planted irregularly, rather than in squares or in lines, in the areas with the greatest soil-depth. It was also hoped that the perspective of monetary profit might encourage the land-owners to exploit the forest in an appropriate way, something which, for various reasons, does not occur too frequently in Cantabria (Saiz de Omeñaca & al., 1990; Rodríguez Barreal & al. 1990; Reimat, 1993; Saiz de Omeñaca & al., 1993a y 1993b).

Both the karstic hollows and the high areas were respected, as well as any spontaneous vegetation of a certain quality, which remained protected by the planting. Since the second felling, the wood has been felled in sections, as, by acting on only a relatively small section of the forest area, the negative environmental effects derived from the felling are reduced. Other precautions are also taken to reduce the negative effects on the soils of erosion, excessive leaching, alkalinisation or mineralisation.

RESULTS

With the exception of some high rock areas, the surface of the Cierro is covered by woodland, the soils have been preserved and in most areas have improved, the protection of the aquifer has increased, the risk of flooding in the nearby depressed areas has been reduced, a contribution has been made to a greater atmospheric quality (in the neighbouring areas, mining activities have been initiated which produce large amounts of dust), the quality of the landscape has improved, even inside the repopulated forest mass, as an artificial look has been avoided (by not planting in lines and due to the different rates of growth resulting from the differences in soil quality).

Above all, a space for the spontaneous growth of flora and fauna has been recovered (Sánchez y Valdeolivas, 1995, p. 15-16; Prieto & al., 2000). Indeed, under the woodland, spontaneous undergrowth covers practically 100% of the soils and most of the rock area. Although the system cannot yet be classified as mature, it is made up of a rich and varied set of species, since, though no attempt has yet been made at an exhaustive analysis, 156 vegetal species have been classified, apart from the *E. Globulus*, of which 29 are bush or tree species. None of these are present in all of the areas of the forest, though 5 of them are very numerous and at least another 7 (4 woody, 2 climbers, and one herbaceous) are scattered over many areas. There are also species which, while abundant in some areas, are not found in others (often even in most of the Cierro), in some cases due to competitive regression. The edges and boundaries often present peculiar features and sometimes include "opportunist" species, more characteristic of meadows and open spaces. On the slopes, species common in undergrowth or meadows coexist with others normally found in rocky clearings or spoilt and even open spaces. This data, which reflects the wide range of microhabitats in the Cierro, is all the more surprising when one bears in mind that the lack or poverty of the soil in most of the area and the relatively high surface aridity might hinder the development of the undergrowth.

Special mention should be made of the vales. In them, protected against the wind and other aggressions by the woodland, and with better humidity conditions, a rich and varied vegetation has developed which includes 12 woody species of average or great height (7 spontaneous and 5 test remains) and another 34 frequent species. These vales, have, as to a lesser extent have the high areas and the rock lands, played an interesting role not only in the conservation of species in the Cierro, but also in the expansion of these species.

As for the fauna, it is equally rich. As an example, the list of vertebrate species located in the Cierro includes 4 amphibians, 6 reptiles, 32 birds and 15 mammals; and it is possible that the number of these which inhabit the Cierro or which come there for food, shelter or a place to reproduce is far higher than this. In any case, the predatory nature of many of these leaves no doubt as to the abundance of prey in the habitat.

Moreover, the conservation of the area seems to be assured, since monetary profits are regularly obtained. In fact, although in some areas, the trees do not grow very high and in others there are many tree trunks with fissures caused by the wind (some more than two metres long; trees more than 30m high have been seen to be bent by the wind until the highest parts of the tree-top whip the ground), the profits are, in general, comparable to eucalyptus forests planted for exclusively economic purposes.

CONCLUSIONS

The reclamation of the Cierro Grande de San Pantaleón, based on: respect towards the remaining natural vegetation, adaptation to the environmental conditions and on the reforestation of a large part of the Cierro with a rapid-growth species has been carried out successfully. At present, the Cierro presents characteristics which, though they do not make it a forest in the strictest sense, do give rise to a forest microclimate and although its environmental value may be assumed to be less than that which a climatic forest would have, it is clearly greater than that of plains or pasture. Moreover, with respect to the future, the continuity of the initiative seems to be assured due to the economic benefits.

REFERENCES

Cendrero, A. & Saiz de Omeñaca, J. (1975). Criterios de definición y valoración de unidades ambientales en una zona costera y su apllicación a la estimación de impactos ambientales. *Ac. I Cong. Iberoamericano de Medio Ambiente*, 4: 1813-1833.

Cendrero, A.; Díaz de Terán, J.R.; Saiz de Omeñaca, J.; Loriente, E.; Cifuentes, P. & al. (1980). Environmental survey along the Santander-Unquera coastal strip (Northern Spain) and assessment of its capability for development. *Landscape Planning*, 7: 23-56.

DG (1976). Clasificación y valoración del medio natural en la franja costera Santander-Unquera. Departamento de Geología, Universidad de Santander, Spain.

DG (1978). Estudio integral de definición de unidades ambientales y ordenación del territorio de Santander. Departamento de Geología, Universidad de Santander, Spain.

DG (1980). Análisis territorial integrado e inventariación de recursos del territorio en la provincia de Santander. Departamento de Geología, Universidad de Santander, Spain.

Guinea López, E. (1949). *Vizcaya y su paisaje vegetal*. Junta de Cultura de Vizcaya,204 pp.

Guinea López, E. (1953). *Geografía botánica de Santander*. Diputación Provincial de Santander, 450 pp.

Guinea López, E. (1954). El susector cantábrico del N. de España. *Anal. Inst. Bot. A.J. Cavanilles*, 12, 1, pp. 509-521.

Prieto, A.; Saiz de Omeñaca, J.A; Saiz de Omeñaca, J. & al. (2000). Proyecto de ordenación del Cierro Grande de San Pantaleón.

Reimat Burgués, M.J. (1993). Estudio de opinión sobre la influencia de las plantaciones de eucalipto en el medio físico y en el entorno social de la Cantabria rural. Dr. Th., ETSI de Montes, Universidad Politécnica de Madrid.

Rodríguez Barreal, J.A.; Saiz de Omeñaca, J.A.; Saiz de Omeñaca, J. & Laza, B.R. (1990). Atributos y peculiaridades biogeofísicas del territorio en Peñas Negras (Camargo, Cantabria).

Saiz de Omeñaca, J.; Martín Hernández, A.; Saiz de Omeñaca, J.A.; Reimat, M.J.; Solar, M. & Gómez Martínez, D. (1990). Análisis de opinión sobre influencia de los eucaliptos en el medio biogeofísico y en entorno socioeconómico de la Cantabria rural.

Saiz de Omeñaca, J.; Gómez Martínez, D.; Reimat, M.J.; Saiz de Omeñaca, M.G.; Solar, M.; Martín Hernández, A. & Saiz de Omeñaca, J.A. (1993a). Análisis sociológico de la opinión entre la población rural de Cantabria sobre el impacto de las plantaciones de eucalipto. *Investigación agraria. Sistemas y recursos forestales*, vol. 2 (1), pp. 71-88.

Saiz de Omeñaca, J.A; Martín, A.; Saiz de Omeñaca, J.; Saiz de Omeñaca, M.G. & Reimat, M.J. (1993b). Orígenes y entorno social de los incendios forestales: estudio de opinión en la Cantabria rural. *Congreso Forestal Español, Lourizán. Ponencias y Comunicaciones*, t. 3, 251-256.

Sánchez Martínez, C. & Valdeolivas Bartolomé, G. (1995). *Guía de la fauna y flora de un municipio cantábrico: Camargo*. Elabra, 283 pp.

Design and Management of Spoil Dumps

Y.V. Rao, M. Aruna and Harsha Vardhan

Department of Mining Engineering, Karnataka Regional Engineering College, Surathkal 574 157 (D.K.), India

ABSTRACT: *Mining all over the world a definitely a much required process for advancement of technology and it is one of the most pollution causing operation. India has unique blend of big and small, manual and mechanised, opencast and underground mines. One critical point in mining industry is over burdening, requiring large volumes of materials to be handled and extensive requirement for dumping. During monsoons the rain water carries the washed out material from the mine waste dumps to the adjoining agricultural fields and water streams. There is thus the need to increase the base knowledge on the existing practices of spoil dump design and rehabilitation. An efficient engineering design of spoil dumps aims to ensure the long term stability of the dumps at the end of the construction. This paper discusses some of the design aspects and* management of waste dumps.

1 INTRODUCTION

For developed as well as developing countries, mining of minerals to sustain national development is of paramount importance. Over 75% of the global mineral output is coming out from the surface mining industry and there has been increasing share of surface mining industry vis-a-vis the underground sector world over. In India 60% of the total mineral production is from surface mines. The current production trend which will be doubled in the near future, will be contributed by surface mining only. Over 80% of the production of non coal minerals are from surface mines and coal production contributes more than 60%.In process of winning, huge quantity of waste are generated and these waste are piling up in and around the mining pits covering a vast stretches of land. The mining wastes are dumped either on mineralised section beyond the final pit limits in the lease hold area or are carried over a distance to suitable non-mineralised areas, for dumping. Engineering design of these spoil dumps is very important. The success or failure of a land reclamation project is to a considerable extent determined by the management of surface runoff and drainage.

Reclaimed overburden tends to allow little infiltration because of compaction during reclamation (1). Slope is a major factor controlling the rate of erosion during rainy period in any mining area. As slope increases, soil and water loss also increases (2). In one of the study, it was found that the rate of sediment loss from steep waste dump slopes was 10-12 times more that in non-mining/agricultural/forest lands in the same drainage area during rainy period (3). Intensity and duration of rainfall are important factors of soil erosion (4). The extent of rain water erosion, is largely a function of the slope, but the length of the slope is also equally important (5). Efficient spoil dump design and rehabilitation practices are acknowledged as important phases in mine design and planning (6).

2 IMPORTANCE OF SPOIL DUMP DESIGN

Whether it is strip or opencast system the damage to the environment and the ecosystem is the same in the long run. In most of the time, because of non availability of dumping space at a stretch is posing problem in systematic planning of the dumping and waste disposal. Waste dump erosion is a very crucial phenomena causing degradation of loose, non-cohesive material which are dumped during mining operations. Engineered slopes of mine spoils may be stable at the end of the construction, but they can deteriorate over time. During monsoons the rain water carries the washed out material from the mine waste dumps to the adjoining agricultural fields and water streams. The damage to the agricultural fields has been maximum which are located close to the mining areas and the reject dumps.

The fast running water along the slopes, removes the fine newly dispersed soil particles, leaving behind the hard pebbles on the surface, which makes a poor seed bed and provide little or no anchorage for the plant roots. The huge quantities of rejects flow down from both active and dead reject dumps to the river bed causing considerable silting downstream which effects the required draught standards for navigation (7). While this by itself is a harmful consequence of the mining industry, such heavy silting will affect the course of the river and result in the flooding of the adjacent fields. The rate of erosion /sediment loss varies from place to place depending upon various factors. There is thus the need to increase the base knowledge on the existing practices of spoil dump design and management.

3 SPOIL DUMP CHARECTERSTICS

Waste dump generation during extraction of valuable ore/mineral is unavoidable. These overburden having quite different properties from the natural material. They may vary in type, thickness, extent and stratigraphic relationships of individual formations, chemical composition and physical properties. In most of the cases the land being in short supply, dumps are typically steep, which are more susceptible to erosion. The dumps do not provide suitable substrate for vegetation and are, usually susceptible to slumps, slides and erosion during rains, especially in heavy monsoon. Toy and Hadley (8) refer to various studies which have shown erosion to be greatest on uniform hill slopes especially on easily erodible material. Haigh (9) finds that it is the upper convexity of complex slope profiles which suffers most erosion. Godmann and Haigh (10) show that younger sites suffer more erosion than older sites and that the zone of maximal erosion is found nearer to the midslope than the slope crest. These factors should be considered when planning post mining use.

Several constituents of the dump material go down the slopes to lower valleys and plains, either in mechanical suspension or in solution to affect the depositional areas in various ways. The stabilisation of a slope is the function of a many characteristics like: type of debris material, its angle of repose (wet and dry), particle size distribution, slope, age of dump, length, water retention capacity etc. Small heaps composed of large fine particulate material, mild slopes and smaller lengths are susceptible to become stable much sooner than large heaps of steeper angles.

4 SPOIL DUMP DESIGN

An efficient engineering design of spoil dumps aims to ensure the long term stability of the dumps at the end of the construction. It also seeks to reduce the erosion potential through correct selection of an initial stable landform scheme. Higher runoff will have more the sediment discharge during rainy periods. One of the field survey shows that the factors used in the design of stable spoil dumps are: climate, rate of erosion, height of dumps, slope gradient, slope length, equipment, soil characteristics, legislation and costs involved (Figure.1).

Spoil dumps were either reshaped with benches, terraces or reduced to that of undulating landscape prior to revegetation for improved erosional and mass stability. Terraces may be of single or multistage, depending on height of dump. Prunty and Kirkham (11) found that the total seepage rate for the four-terrace topography is ten times greater than for the uniform slope topography . Field survey (6) shows that the track type bulldozer is the most suited equipment for the reshaping of spoil dumps to legislative requirements. It must also be remembered that due to accelerated weathering artificial slopes will evolve relatively quickly after contouring (12).

8

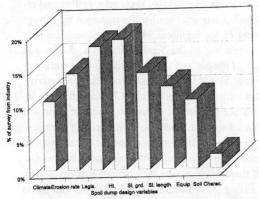

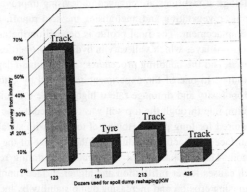

Fig. 1 Industrial responses of variables used for spoil dump design (after Goh et al.)

Fig. 2 Power of dozers used for spoil dump reshaping (track/tyre mounted)

5 DUMP MANAGEMENT AND REHABILITATION

The dumps are generally categorised as dead dumps and active dumps. Dead dumps are old dumps where no further dumping is being done. Active dumps are running dumps where dumping of materials is continuing. Control of erosion is given importance in the rehabilitation programme. The erosion mitigation measures popularly used in the industry include surface ripping, contour banks, sediment traps, diversion banks, runoff velocity control, lined waterways, grassed waterways and hydromulching. The rehabilitation of dead dumps is mostly being done through some bio-engineering measures. Priority shall be given to utilise the top soil which is a very valuable material and also waste materials like fly ash, sewage sludge etc. can be used suitably for improving the soil properties (13)

To reduce the wind velocity and to give protection to newly planted areas against wind erosion, a wild belt of trees, planted in rows all along the vertical boundary. It was found that such shelter belt can help in reducing the wind velocity by 20% to 30% (14). To make the dumps hydrologically compatible, terraces are made at definite intervals with proper width. Terraced land form with short steep slopes (angle of repose) and gently sloping terraces are stable and less obstrusive. The terrace are generally sloping inward to allow rain water to concentrate at the centre and infilter instead of running down the slope. The slopes can be stabilised by:

- plantation of trees on the slopes
- growing shrubs and cactus
- and by grass turfing

Until an adequate cover of vegetation has been established, it becomes imperative to make provision to control erosion from these slopes. This is normally achieved by covering the sloping sides of the dumps with pebbles of assorted sizes thereby splashing of particles by rainfall and their dislodgement are avoided. In case of active dumps, series of parapet walls are constructed at the toe of dump plots to arrest water flowing down the slope. Sometimes a trench is also made parallel to the parapet wall, a few meters away. In case of hilly terrain, a retention wall shall be constructed at the lower side of the dump site. Horizontal roughing filters should be constructed along the natural water courses to control the siltation due to flow of mining rejects along the monsoon water.

Jensen et al. (15) suggested surface manipulation as a method of reshaping the spoils in an attempt to reduce the rate of surface runoff by intercepting and storing runoff and permitting infiltration. It must also remembered that due to accelerated weathering artificial slopes will evolve relatively quickly after contouring (12). Rippers attached to bulldozers assist in reducing the compaction resulting from heavy machinery traffic during spoil dump reshaping operations.

6 CONCLUSIONS

Reclamation planning is an integral parts of surface mining industries in many parts of the world. Indeed, new approaches such as 'concurrent reclamation' are being led by the industry itself. Many technological institutes are engaged in devising improved methods of drainage, reseeding, soiling and re-shaping, revegetating and nuetralising the acid runoff. Applied top soils tend to be damaged by storage and emplacement. The final profile is depend on the nature of the replacement operation, and secondly the top soil layer which will vary in hydraulic properties depending on the depth of the layer, the top soil material, and the handling procedures. The optimum depth of the top soil can be obtained by experimental studies. In one of the experiment conducted by Chong et al. (16) shows that hydraulic conductivity, macroporosity and drainage rate is higher in the top soiled minespoil. Ward et al. (17) found that the infiltration rate through the top soil was controlled by the surface conditions. If preservation of the top soil is required for long period ripping of top soil with suitable subsoil machine is to be carried out to relieve the surface compaction produced by the passage of the scrapper machine and to improve the aeration of the soil. The compaction of the overburden during reclamation often renders it almost impermeable, which causes lower infiltration and higher surface runoff thereby increasing the erosion. Ripping aids in minimising erosion and increasing dump stability by increasing water infiltration and reducing runoffs on the spoil dump surfaces.

The track type equipment is most suited for steep slope reshaping and grading of the overburden spoil dumps. An industrial survey carried out by the authors in Goa region shows that, track mounted equipments are widely used for grading spoil dumps (Figure.2). Wherever possible, urban garbage should be mixed with mining rejects, to improve the quality of the soil to develop high value plantation. It also help in the disposing of garbage. Owing to the costs involved in the construction of stable slopes and regulatory requirements aimed at minimising off-site effects, the efficient design of the spoil dump is regarded as a significant engineering task. Careful planning and optimal design of spoil dumps can assist in saving money during the construction and rehabilitation of spoil dumps.

REFERENCES

1. Marianne P. Kilmartin (1989). Hydrology of reclaimed opencast coal-mined land: A review. International Journal of Surface Mining 3, pp. 71-82.

2. Tejwani K.G. & Bharadwaj S. P. (1982). Soil and water conservation in research, In Review of Soil Research in India Part-II, 12TH Int. Congr. Of Soil Sci., New Delhi, India, pp. 608-621.

3. Ashutosh Dubey & Rath, R. (1997). Effects of slope and rainfall in waste dump/overburden erosion in opencast mining area - A case study. The Indian Mining Journal: 19-34

4. Singh N.T. and Verma K.S (1975). Effect of rainfall intensity and surface condition on water erosion in foothill soils of the Punjab, Jl. Ind. Soc. Soil Sci., Vol. 23 (1), pp. 27-30.

5. Hussain A. (1989). Biological plant colonization on lateritic waste dumps of bauxite mines - a case study, Journal of Mines, Metals & Fuels, December: 391-395.

6. Goh, E.K.H., Aspinall, T.O. & Kuszmaul, J.S.(1998). Spoil dump design and rehabilitation managament practices (Australia). International Journal of Surface Mining, Reclamation and Environment 12: 57-60.

7. Poduval A. M. K (1997). Pollution in mining with special reference to Goa, Proceedings of AICTE-ISTE Winter School on Environmental Pollution Control and Reprecussios, December 1-13, 1997. Karnataka Regional Engineering College, Surathkal, pp. 10-20.

8. Toy T.J & R.F.Hadley (1987). Geomorphology and reclamation of disturbed lands, Washington D.C. Academic Press Inc. Ltd.

9. Haigh M.J (1988). Slope evolution on coal-mine disturbed land, In A.S. Balasubramaniam, S. Chandra, D. T. Bergado & Prinya Nutallyl (eds.), Environmental geotechnics and problematic soils and rocks, pp. 3-13.

10. Godman J. & M.J.Haigh (1981). Evolution of slopes on abandoned spoil banks in Eastern Oklahoma, Physical Geography 2(2), pp. 160-173.

11. Prunty L. & D. Kirkham (1980). Seepage vs. Terrace density in reclaimed mineland soil, J. Environ. Qual, 9. (2): 273-278.

12. Haigh M.J., (1978). Evolution of slopes on artificial landforms, Blaenavon, U. K. University of Chicago, Department of Geography, Research Paper 183, pp. 1-293.

13. Velan. M & P. Sivagnanam (1997). Conservation and utilization of top soil for reclamation, Proceedings of National Seminar on Opencast Mining, Neyveli, Tamilnadu, July 21-22, 1997, pp. V-6-1 to V-6-7.

14. Hussain, A. (1990). Results of land reclamation schemes adopted in the worked out areas of BALCO's mines. The Indian Mining and Engineering Journal, March: 15-21.

15. Jensen I.B., R.L.Hodder & D.J.Dollhopf (1978). Effects of surface manipulation on the hydrologic balance of surface mined land, In M.K.Wali (eds.), Ecology and Coal Resource Development, New York, Pergamon Press. pp. 754-761.

16. Chong S. K., M. A. Becker, S. M. Moore & G. T. Weaver (1986). Characterization of reclaimed land with and without top soil, Journal Environs. Qual., 15. (2): 157-160.

17. Ward A.D, A.G.Wells & R.E.Phillips (1983). Infiltration through rconstructed surface mined spoils and soils, American Society of Agricultural Engineers, Transactions 5: 821-829

Physical, Chemical and Biological Processes of Geo-Environmental Reclamation

N.K. Sharma

Director (Technical), Coal India Limited, 10, N.S. Road, Calcutta 700 001, India

BACKGROUND

The vast reserve of coal, the major and only dependable source of energy at present in India, etimated at about 212 billion tonnes is sufficient to cater for projected demand of the nation for a very long period.The growth in production from the opencast mines in our country during the recent years has been remarkable and quantitatively India is at present in the forefront of world coal scene. Opencast mining operation through which nearly 80% of country's coal is produced, results in removal of huge quantities of the overburden materials as waste and later on placing it either outside the quarry area or within the created void as backfill. In recent years with the increasing trend in production from opencast mines, number of waste dumps has been on the rise. This is of major environmental concern as it leads to degradation of land, possibility of slope failures of different magnitude along with chances of pollution of air and water including silting of nearby watercourses.

As more and more areas come under opencast mining, corresponding, land becomes biologically unproductive. The requirement of land increases to accommodate additionally created external overburden dumps. India accommodates more than 16% of global population with its share of land as 2.3% only. So far, availability of land is concerned, the situation is thus quite serious in the country and land as a scarce non-renewable natural resource is well recognised. Thus, impact on land due to winning coal and mining in general deserves special attention for making good the damages.

RECLAMATION OF LAND

Reclamation is the process of reforming and regrading the waste land approximately to the original topography and bring about permanent self sustaining vegetation. Thus, reclamation includes,

1.) Restoration of the approximate contour and recreating conditions suitable for the previous use of land /area.
2.) Achieving a landscape that is visually acceptable and fits into the surrounding.
3.) Protection of failure of slopes and protection of erosion of surface and prevention / reduction of silting arising from that downward.
4.) Excavation of topsoil, its storage, preservation and reuse by spreading on filled deposit in order to support vegetation.
5.) Establishment of vegetation in stages comparable to pre-mining conditions.

OB DUMP STABILITY

In opencast mining operation, formation of overburden (OB) dumps , external or internal with surfaces having slopes is an associated phenomenon.

Loss of land productivity, erosion of surface soil, instability of slopes are the problems associated with OB dumps. These problems will intensify if the dump is left unattended and no attempt is made to stabilise the dumps by taking suitable measures.

The most tragic incident of spoil heap failure which brought the problem of slope stability to public attention occurred at Aberfan, South Wales, UK in 1966 in which a school was buried under colliery waste causing death of 144 people including 119 children aged between 7 to 10 years. It is, therefore, essential to take up some engineering measures for slope stabilisation including soil conservation and arrangement of drainage of water.

It is important that the original mine design must take account of all factors, not merely of the mining itself but those applying to final restoration and rehabilitation of the land. Coal India has incorporated this in its environmental policy.

SAFETY OF DUMP SLOPES

Safety of slope of soil, rock or their mixture is coventionally assessed by determining its factor of safety (FOS). The ratio of summation of all the resisting forces to the cumulative forces intending to disturb the stability is the FOS for a surface within the mass. The resisting forces are dependent on a few properties of the constituent materials and extent of presence of water in it. These contribute to the shearing strength of the mass under the influence of the forces. Properties like grain size and its distribution, angle of internal friction, cohesion etc. are evaluated in laboratories.

RISK RATING OF DUMP

The other approach for assessment of status of safety / hazard related to the dumps is method of 'hazard ranking'. Under this, hazard - ranking for the dump is done based on the dump geometry i.e. its height , area etc. and other general characteristics like dump material, ground slope, nature of foundation, degree of restoration work done on the dump and existence of infrastructure and related features near the toe of the dump. Hazard ranking may be done as "high", "medium" & "low".

Having assessed the ranking, the dumps are taken up for restoration. Priority and extent of technical reclamation depend on the assessed ranking.

The potential hazard in a dump is assessed by adding the score for each case in the following table:

CASE	SCORE		
Average Natural Ground Slope	> 1 in 6 (2)	>1 in 12 <1 in 6 (1)	< 1 in 12 (0)
Height (m)	>50 (4)	>25 <50 (2)	<25 (0)
Adjacent Active Dumping	Contiguous (0.5)	Contiguous and higher (1)	None (0)
Technical Restoration	None (3)	Partial (1.5)	Full (0)
Biological Restoration	None (1)	Only Top Surface (2)	Top & Flanks 0
Material	Soil (2)	Soil and Rock (1)	Rock (0)
Nearby water course	Alluvial bed / bank (high velocity) (3)	Alluvial bed/ bank (medium velocity) (1)	Bed / bank of rock (0)
Foundation	Alluvium (1)	Black cotton / expansive soil (3)	Hard rock (0)

Using the above table each dump can be given a mark adding all the scores. For example, a 65 m high active dump without any contiguous active dumping adjacent to a stream of high velocity with alluvium bed / bank on ground sloping at 1 in 10, made predominantly of soil, founded on hard rock with partial technical restoration and biologically restored on the top only would have a score of 4+0+3+1+2+0+ 1.5 + 2 = 13.5. The score can then be compared with the following table for ranking:

Ranking High   12 or more
Medium   Greater than 8 but less than 12
Low   8 or less

The above site with a score of 13.5 would be ranked high for hazard and will attract priority for restoration; technical and biological.

TECHNICAL RESTORATION:

Reprofiling, drainage, use of geotextiles, controlling gullies & routine inspection are the different aspects of technical restoration for improvement of stability of a dump.

While reprofiling is aimed to give a flatter slope, drainage will eliminate or reduce the detrimental effect of water. Application of jute/ coir / synthetic geotextiles protects against erosion, allowing growth of vegetation quickly by holding the seed, sapling and adjoining soil in position. Controlling gullies formed due to surface erosion is done by providing check dams in series / steps in combination with brushwood filling for arresting silt or laying locally available boulders as preventive measures against erosion.

VEGETATION:
Growth of vegetation assists to a great extent in technical restoration by reducing the erodability of the surface. Biological restoration of the flanks can increase stability by varying amount depending on heights.

TOPSOIL
Due to the loss of organic matters and micro-organisms and changed structure, the supporting capacity of the soil mass to vegetation is nearly destroyed.

On account of this , careful management of topsoil including its separation, storage, amendment if necessary to suit vegetation and respreading over surface of the dump assumes significance. CIL has recognised this and an exercise has been carried out to determine the quality of top soil of a number of mines for finalising the reclamation strategy.

BIOLOGICAL PROCESS
The method of restoration currently used is to place a loose layer of topsoil before planting six-month old saplings in pits at about 2m centres. Where topsoil is scarce, pits are created and filled with good earth mixed with manure and pesticide. The results are generally good with a high quality of restoration being achieved.

Approach for plantation on the slopes should be done after proper benching / terracing with suitable width and height depending on local condition. Normally, terracing at contour levels of 3m with bund all along should be considered a good practice.

As the vegetation on the top becomes more established, the percentage of run-off seems to increase, causing further problems with surface erosion. It is essential that planting of the slopes is also taken up closely after and if possible concurrently with the planting of the top surface.

Plantation has to be of locally available species including grasses, fruit and medicinal plants. The areas for plantation may be taken up in phases. The activities associated with reclamation processes will also offer an opportunity to the tribal, scheduled castes and other backward people, who were original inhabitants of the area to supplement resources out of vegetation thus created.

CASE STUDY
Let us now discuss a case of reclamation of OB dump of Bina opencast mine in Northern Coalfields, one of the subsidiaries of CIIL .
A part of the internal dump at Bina opencast mine was taken for restoring the relatively steep slope of 46^0 with provision of geosynthetic net / mat .
Relevant informations on the dump under consideration are:
- The dump is located on foundation strata of coarse grained sand stone.
- The dump material mainly consists of clay and weathered sand stone.
- Backfilling in the portion was done by the draglines from the year 1990 to 1995.
- Period of restoration work : 8 months

The steps involved in reclamation of the dump were;

PHYSICAL RESTORATION

Configuration Of The Dump And Physical Properties Of The Dump Materials
Being cast on strong sandstone of the Turra Seam floor dipping at $2-3^0$ having high bearing capacity, stability of the foundation under the loads of the internal OB dump is ensured with high safety factor.

The strength properties of dumped materials were tested at different times through reputed agencies. Based on these test results and the experience of the experts in researching the stability of dumps for years, numerical figures for limiting dump heights related to flat profiled sloping angle have been obtained by calculations. Relevant figures are given here.
The following shear strength specifications were adopted as initial data.
Internal frictional angle $\Phi=33.4^0$

15

Cohesion $c = 3.6 \text{ t/m}^2$
Bulk density $\gamma = 1.8 \text{ t / m}^3$

Total flat profiled dump slope angle (degree)	Dump Height with due account of various oozing height "h" at factor of safety, n=1.2 (in Metres)			
	h=0 metre	h= 2 metres	h= 4 metres	h= 8 metres
50	35.5	33.8	31.0	15
45	47.5	44.5	41	31.5
42	58.7	54.5	50.5	40.0
40	69.4	64.0	59.0	47.5
38	84.6	77.0	71.2	58.0
36	108.3	98.0	90.0	73.0
35	125.9	112.5	104.0	85.0

Based on this and as the height of dump was about 53 metres, a profile was designed with four metres wide berms at about eleven metres vertical intervals to bring down the effective slope to less than 39^0. From above table it would appear that at an angle of 40^0 with oozing height at 4 metres the calculated safe height of dump is 59 metres. There is however no recorded observation for the water logging status at the site. But, in view of the location of the dump, face and dip of the sandstone foundation towards the area of mining operation it was reasonably concluded that the status would resemble mostly to dry / drained condition. Yet , to be more safe, an oozing height of 4 metres was considered while designing the profile.

The designed profile was flatter and less likely to be affected by water than the condition of oozing at 4 metre height.

Accordingly the designed profile of the dump to be treated with a height of about 53 metres and an effective slope of 38.8^0 was in order for physical restoration / global stability of the dump.

Excavation And Making The Profile:
Mechanical excavator (Poclain) was engaged for the purpose in combination with dumpers. A uniform grade and four metre wide berms were provided at regular intervals of about eleven metres. Intimate contact of the mat with the dump surface is essential for avoiding local undermining caused by flow of water beneath the material. Accordingly, surface is prepared with sufficient care to avoid unevenness. Excavated soil from the trench was placed as bunds at the edge of the shoulder. Simultaneously gullies formed due to erosion were treated by filling cavities with stony materials topped with compacted soil.

Grading The Top Of The Dump

The top of the treated part of the slope was treated by dressing and grading the top surface of the dump so that a gentle slope towards the edge of the dump was formed to enable the surface runoff flow easily with a non erosive velocity. The surface at top was covered with plantation in course of time to reduce the loss of ingredient particles by action of running water from precipitation.

Providing Benches Of Convenient Width For Reduction Of The Effective Angle
Bench refers to the levelled surface or terrace which may be used for plantation also. The difference of height between successive levels depends on angle of slope, extent of space available, nature of material dumped, moisture content etc. This varies from dump to dump. In this case the angle was steep with ingredients characteristically resembling mostly sand. The vertical distances between successive benches were kept around 11 (eleven) metres.

The terraces were 4 (four) metres wide and slightly dipping towards the centre of dump to facilitate automatic flow of water to the drain which is provided at the inner side of the terrace.

Plantation was aimed at the terrace levels and was subsequently taken up during monsoon, 1999.
With the provision of these terraces / benches, the inclination with the horizontal of the imaginary line joining the lowest toe and highest shoulder of the sloped parts of the dump is about 38.8 degrees. Thus the effective slope of the dump was reduced by more than 7 degrees.

A small mound of earth built with the excavated material from trenching was created with a purpose to prevent flow of water down the slope of the dump. Surface of this bund was also later covered with grass. (Fig – 1, Annex 1)

Laying Geosynthetic Matting To Act As A Separator Between Soil Surface And Atmosphere
Jute nets or geo-textiles (synthetic) are used to facilitate growth of vegetation. In the instant case since the slope was quite steep, geosynthetic netting was chosen. To keep the matting in place it is required to

be anchored and accordingly anchor trenches at the toe and shoulder of the individual slope were excavated.

The geo-matting is a flexible, three dimensional matting produced from Nylon filaments having large free space. They serve multiple functions. While the primary function is to arrest erosion, secondary function is to resist downward and outward movement of the soil / dump mass owing to anchorage and support.

The geo composites placed on the sloped surface separates the ground surface from the prevailing atmospheric condition and high precipitation in particular. Though water is predominantly responsible for dislodging soil particles or seed from its in-situ position, wind is also a potential medium for similar kind of damage.

Trenches of size 0.5M x 0.5M were provided at every toe and shoulder of the individual slope for the purpose of anchorage of the mat. Matting was placed in the trench adjacent to the shoulder of each slope and then laid flat down the slope and into the anchor trench at each toe. For prevention of displacement, the mat is pinned with 6 mm dia MS pins at I metre centres.

The anchor trenches were 500 mm x 500 mm in size and these were filled with stone materials (50-63 mm). Within the stone filled trench perforated PVC pipes @ 150 mm dia were placed centrally to collect and convey water for drainage. The entire filling material was wrapped in non- woven geosynthetic that protect the trench from being clogged.

Protection of the Toe

Gabions were provided along the toe of the dump. These are units square in section made of rock / boulder of size from 90 - 300 mm wrapped up in geo synthetic nets of suitable mesh forming massive flexible blocks 1M x 1M X 1M to resist movement of materials from the toe of the dump. (Fig. – 2, Annex 2)

Network Of Drains

The anchor trenches as described above, provided at and along the toe and top of individual slope were to serve as french drains in those alignment. These were further connected by drains of similar construction running along the slope spaced at interval of fifty metres. Through this network of drains water reaches the bottom of the dump and discharges into an open catch water drain leading to its ultimate removal from the site.

BIOLOGICAL RESTORATION

The Overburden Materials Generally Lacks Of Nutrients

The mine spoils of which the dump was formed was nitrogen deficient. This was due to absence of soil organic matter normally made available by decay of plant materials. Besides, soil organic matter was not decomposed due to lack of soil microflora.

Such deficiencies were made good by use of top soil obtained by stripping in course of mining. Farm yard manure (FYM) was applied for conditioning of the soil and increasing its plant life supporting capacity.

Amendment Of The Topsoil Before Spreading Over The Top , Benches And Surface Of The Dump

The soil being slightly acidic in nature was treated with lime. Dressed slope was covered with thick layer of topsoil / good agricultural soil mixed with manure / nutrients and mulched with straw / hay @ 5 kg / sqm. Seed of grass was spread over the top surface. Thus the soil was amended and requisite nutrients added to support the vegetation over it.

Pit Plantation And Irrigation

Pits were dug on the benches and saplings planted after refilling the pits with good agricultural soil mixed with manure / nutrients and pesticides. Simple system of irrigating the area was arranged by providing water to the plants and treated surface through temporary pipeline connected with water tanker placed at a higher bench. Gradually grasses came up. The matting structure reinforces the grass stems and resistance was exhibited against being washed away.

MAINTENANCE

In the monsoon however out of the area of two hectares a few small patches of about 20 to 30 sqm in size at the upper part of the slope came down with grasses. These patches were attended immediately by mending the surface by re-grading and providing good earth as before. Grass turf was placed in those places and further damage has not been reported.

The job was done contractually in a period of 8 months from October 1997 to June 1998 with a maintenance period of 6 months in-built in the contract. Cost involved in the job was Rs. 1.5 crore.

From the photograph (Fig. – 3, Annex 2) it will appear that the slope after treatment offers a smooth green surface aesthetically acceptable in contrast with dumps that are not reclaimed. The individual slopes are inclined at around 45^0 and the resultant slope due to provision of the intermediate benches is around 39^0.

Inspection pegs were installed at different places to indicate any sign of movement including settlement. Top levels of the pegs were connected to a benchmark. Any variation in level will give signal of settlement and tilt of the peg tells about movement of the soil mass.

Having done all these as narrated above, the potential hazard in the dump has been evaluated by adding the score for each case pertaining to the status of the dump before and after restoration.

CASE	Before treatment		After treatment	
	Status	SCORE	Status	SCORE
Average Natural Ground Slope	< 1 in 12	0	< 1 in 12	0
Height (m)	>50	4	>50	4
Adjacent Active Dumping	Contiguous	0.5	None	0
Technical Restoration	None	3	Full	0
Biological Restoration	Only Top Surface	2	Top & Flanks	0
Material	Soil	2	Soil	2
Nearby water course	NA	0	NA	0
Foundation	Alluvium Hard rock	0	Hard rock	0
	TOTAL	11.5		6

From the above exercise it was seen that the hazard ranking of the dump before treatment was 11.5 i.e. very close to those grouped as high hazard rank. After treatment the ranking figure came down to 6 indicating its group as low hazard dump. This indicated that the degree of restoration had been satisfactory.

To maintain this status of safety undisturbed, inspection team has been formed in the mine. The team inspects the dump at regular intervals and on finding any sign of potential damage will bring the matter to the notice of the CGM of the project for decision as required. They however submit their inspection report regularly as a routine as well.

Though the mode of treatment is considered to be effective, in view of the fact that the internal OB dump slopes are significant to the mining operation and social life near about for a comparatively short period, alternative methodology is under consideration as trial to make restoration works less cost intensive

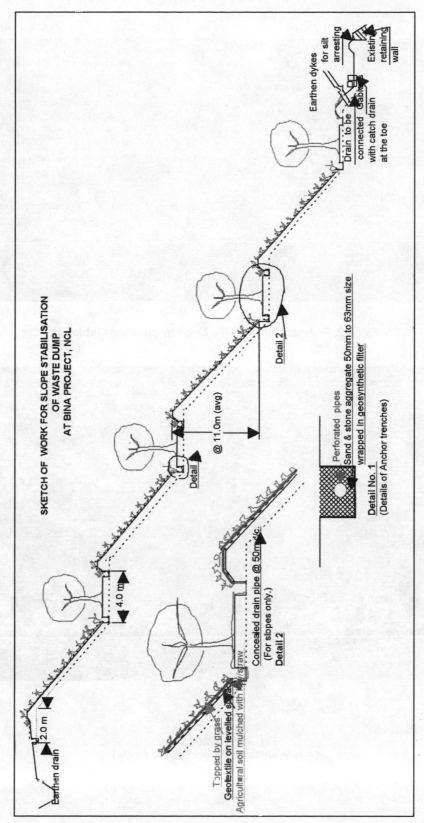

SKETCH OF WORK FOR SLOPE STABILISATION
OF WASTE DUMP
AT BINA PROJECT, NCL

Earthen drain

2.0 m

4.0 m

Detail

@ 11.0m (avg)

Detail 2

Topped by grass

Geotextile on levelled surface

Agricultural soil mulched with hay/straw

Concealed drain pipe @ 50mm dic.
(For slopes only.)
Detail 2

Perforated pipes

Sand & stone aggregate 50mm to 63mm size
wrapped in geosynthetic filter

Detail No. 1
(Details of Anchor trenches)

Earthen dykes
for silt
arresting

Drain to be
connected
with catch drain
at the toe

Gabions

Existing
retaining
wall

Fig. - 1

19

Fig. – 2: Protection of toe of the Slope by providing **Gabions**

Fig. – 3: Treated slope
(A safe aesthetically acceptable profile)

20

Environmental Problems due to Red Mud from Alumina Plant and Possible Solutions

V.P. Deshpandey* and C.K. Kale**

* Scientist & Sr. Asst. Director, ** Rt. De. Director,
NEERI, Nagpur 440 020, India

1. PREAMBLE

The aluminium industry in India is currently in a low profile; the per capita consumption of this metal in India is only 0.5 kg, as compared with 24 Kg in U.S., 20 kg in Japan & 3 kg each in Brazil and Egypt[1]. More than half of the aluminium consumption in India is in the electrical sector; demand for the metal is rising from other sectors such as housing, packaging and automobiles[2], in addition to many other utility items [3]

The Bharat Aluminium Co. Ltd. (Balco) has been the first public Sector Company to enter the primary aluminium industry in 60s. Balco produced about 16400 tons of the aluminium metal by the end of 1993 out of 62800 tons of total production [4]

Bauxite is the aluminium ore used all over the world. Balco has taken over a bauxite mine at Mainpat in Sarguja district of Madhya Pradesh (India), 160 kms from Korba, where Balco has commissioned the aluminium smelter. This bauxite deposit at Mainpat is estimated to have reserves[1] of 6.5 million tons of a high-grade bauxite.

During the production of alumina from bauxite, two environmental problems are encountered; the first is the generation of fluorine in the atmosphere and the second is the generation of a by-product-red mud slurry. Fluorine is generally dispersed in the atmosphere; its concentration rapidly decreases with distance from the alumina plant. The fate of fluorine and its impact on the environment have been adequately covered by the research workers[5-8]. The second environmental problem, that of the red mud slurry, needs to be studied properly for identifying related environmental issues and formulating appropriate strategies therefor.

As far as the production of alumina from bauxite is concerned, a conceptual presentation is covered up to the stage of the generation of the by-product red mud slurry. Subsequently, a brief account is furnished about the red mud slurry composition and its disposal and the associated environmental problems. The main theme of this paper is the proposition of a scientific plantation programme over and around the red mud ponds.

2. GENERATION, COMPOSITION AND DISPOSAL OF THE RED MUD SLURRY

The production of alumina from bauxite is based on Bayer-process. The main raw material, bauxite, transported by railway wagons and trucks from the mines, is stored in a storage yard and the ore is taken to the crusher by conveyer belt. During production of alumina from bauxite, various unit operations such as crushing, grinding, digestion and settling are carried out. The product after digestion is the alumina in slurry from wherein impurities mainly red mud and caustic soda are produced. The settled red mud is causticized with lime to recover

NaOH. The red mud is taken to specially designed polythene lined red mud ponds, whereas the recovered NaOH is recycled back to the process through washing system of red mud.

The major constituents of the red mud are Fe_2O_3 TiO_2 Al_2O_3, SiO_2, Na_2O and CaO (Table 1). The minor constituents are V_2O_5, P_2O_5, Ga_2O_3, ZrO_2, MnO_2, Zno, FeO, NaF etc. The red-mud, at times, is treated with calcium hydroxide to recover the bound soda from sodium-aluminium- silicate or sodalite.   The sodium oxide substitution by calcium oxide is not instantaneous. The liberation of soda is spontaneous because of delayed reaction kinetics as given below :

$3(Na_2O.Al_2O_3.2SiO_2) Na_2O. x n H_2O + Ca(OH)_2$
Sodalite   Calcium Hydroxide
$--3(Na_2O.Al_2O_3.2SiO_2) CaO x. n H_2O + 2 NaOH$
Sodalite   Calcium Hydroxide

After recovery of soda by filtration and washing, the alkalinity of the red mud is not lost completely, because of above, the disposal site or red mud pond (RMP) operates as a part of water and caustic recovery system.

The red mud is transported to the impoundment area – the RMPs-, with the help of positive displacement pumps for long distance or by centrifugal pumps for shorter distance.

The most important ingredient of Indian bauxite residue is the titanium dioxide (TiO_2) which constitutes about 20% of the red mud. The mud as such is a mixture of coarse and fine particles. The coarse particles are separated hydrometallurgically and attempts are made to reduce it as Fe-Ti alloy by alumina thermic method. But due to very high bond energy of TiO2 success was very limited. The fine fractions are treated with mineral acids and the TiO2 could

Table 1

Chemical Composition of Red Mud

Sr.No.	Constituents	Values(%)
1	Al_2O_3	15-20
2	Fe_2O_3	25-35
3	TiO_2	12-18
4	SiO_2	6-9
5	Na_2O	4-6
6	CaO	8-10
7	Loss on ignition (LOI)	8-12

be separated from the red mud. The titanium oxide, thus recovered still contains traces of iron, silicates and alumina and can be used as such as starting material for production of pigment grade TiO_2.

3.   THE RED MUD PONDS & THE ENVIRONMENTAL ISSUES

Earlier the aluminium industries used to discharge their red-mud in the surroundings of their plants. The Bharat Aluminium Company Ltd., in order to protect the environment from harmful effects of entrained liquors of red mud; initiated the construction of red mud ponds in isolated areas. The ponds so constructed gradually get filled up and necessitate the construction of expensive new ponds, one after another. These ponds because of hydraulic pressure of the slurry on earthen dykes may leak, especially, during rainy season. The base of the pond notwithstanding its impervious layers seep and contaminate the surface and ground water. The Bayer technology based Alumina plant generate one tonne of red-mud for each tonne of alumina produced. Due to strict enforcement of pollution control regulations, during the eighties, and also because of restriction on clearance of land for construction of new ponds by the regional regulatory agencies, lead to a realization that there is a risk of holding enormous quantity of red mud in mud ponds for all the time, and new land for construction of red mud pond is difficult to acquire.

The red mud ponds would cause certain apprehensions due to generation of fine dust particles, especially during a dry stormy weather. This apprehension can be reduced/eliminated by creating a screen of suitable tree species around each pond to arrest the dust particles.

At present in the aluminium industry where studies were conducted, red mud is discharged in the red mud ponds provided with impervious synthetic lining. Some of the ponds have already been filled up. During the rainy season, due to erosion of slopes, leakage and seepage might take place and in the process harmful constituents may pollute the surrounding area. Therefore, proper repair and maintenance of red mud ponds is necessary.

4.   PROPOSED USE OF PLANT SPECIES FOR MITIGATION OF ENVIRONMENTAL PROBLEMS

Red mud ponds represent elevated areas along the pond edges sloping down outward. Inadvertently, though, these areas simulate small hilly lateritic slopes and offer an opportunity for plantation of many of the plant species that naturally exist on areas of similar physiography in nearby forests. Also, as these ponds accommodate alkaline residues generated at the aluminium plant, the suggested vegetation around these areas should include some of the plant species, that can withstand certain level of salinity/alkalinity in these lateritic soils. General list of plant species is presented in Table 2. Plantations on these areas should be carried out on a 40-meter wide strip all around the ponds. RMPs –2 and –3 have been covered by a common greenbelt; similarly RMPs, 4 and 5 have a common greenbelt strip (Figure 1), with an additional small strip of (270 m x 20 m). The sizes of these strips and details about the selected species, spacings, number of plants, irrigation water requirement etc. are reported in Table 3. However, a conceptual presentation about the placement of each plant species in each of the five rows of plants in a representative section of the greenbelt strip around RMP-1 has been shown in Figure-2.

On these RMPs, before the surface dries up and becomes hard, it is suggested to transplant pre-seeded and raised grass strips of the following species of the family gramineae; Brachiaria mutica – Paragrass; Chloris gayana – Rhodesgrass ; Panicum Antidotale – Gunara ; Saccharum bengalense – Munj.

If possible, all the above four species should be planted in each pond, one year's experience would further guide to select two good species,if not 3 or all the 4 to maintain on regular basis. After two years, some trees of the species of Acacia, Dalbergia, Prosopis, Cassia and Sesbania can be effectively raised on dried red mud in the pond that would be adequately opened up by the proliferating grass roots. It should be noted that, whether grass, or tree species, clippings of the green material should be taken up at suitable intervals so that the useful organic nutrients are effectively removed. These plant species naturally

Table 2 List of Plant Species for Plantation around the Red Mud Ponds

Plant Categor y Sr.No. (1)	Botanical Name Common Name, Family (2)	RMP No. (3)	Characteristics and usage					
			*D Or E (4)	Wind Break (5)	Aerosol Caputre (6)	Soil Improv ement (7)	Odour Maskin g (8)	Any other (9)
M₁	Acacia catechu, Khair, Mimosaceae	1	E		*	*		Can withatsan d drought and salinity
M₂	Acacia leucophloea; Hiwar, Mimosaceae	2&3	D			*		Fuel wood
L₁	Adina cordifolra; Haldu, Rubiaceae	2&3	D		*			Wood

23

L₂	Ailanthus excelsa; Maharukh, Simoubaceae	5	D			*		Shade & anti – erosion
M₃	Anogeissus latifolia; Dhoara, Combrataceae	2 & 3	D			*		Timber
M₄	Bopswellia serrata; Salai, Bursaraceae	2&3	D					Can grow on poor shallow soils
M₅	Bridelia retusa; Kasai, Euphorbiaceae	2&3	D		*			Fuel
S₁	Buchanania lanzen; Achar, Anacardiaceae	4 & 5	E			*		For dry slopes
M₆	Butea monosperma; Palas, Legumiosae	5	D					Fuel and in greenbelts
L₃	Cassia siamea; Buikhakhasa, Legumiosae	4 & 5	D		*	*		Birds
M₇	Casuarina equisetifolia; Jangli Suru, Casurinaceae	5	D	*	*			Noise Attenuation
L₄	Chloroxylon Swietenia; Bhirra, Flindersiaceae	4 & 5	D		*			Fuel
S₂	Cleistanthus Collinus; Garari, Euphorbiaceae	1,4 & 5	D			*		Wood for Posts
L₇	Eucalyptus Globulus; The Blue Gum, Myrtaceae	1 & 5	E					Abstracts water from Depths, Gum
M₈	Cochlospermum; Religiosum, Galgal Cochlospermaceae	4 & 5	D			*		Afforestation Of barren areas
L₅	Dalberg a sissoo; Shisham, leguminosac	5	D			*		Timber, birds
S₃	Dendrocalamus strictus; Bans, Gramineae	1-5 , in gaps	E	*	*	*		Birds
L₆	Diosphyres Malanoxylon; Tendu, Ebenaceae	4 & 5	E			*		Ebony yielding tree

24

L_8	Ficus benghalensis; Bargad, Moraceae	2, 3 & 5	D			*			Birds
L_9	Ficus glomerata; Umber, Moraceae	1	D			*			Birds
L_{10}	Ficus religiosa; peapal, moraceae	4 & 5	D			*			Birds
S_4	Gardenia turgida; Gadra Rubiaceae	2 & 3	D				*		Ornamental Value
S_5	Holarrhena Antidysentrica; Kurchi, Apocyanaceae	4 & 5	D				*		Ornamental value
M_9	Lannea caromandclica; Jhingan, Anacardiaceae	1	D			*			Ornamental Value
S_6	Mallotus philippensis; Kamalagundi Euphorbiaceac	1	E				*		Supports growth of sal trees
S_7	Ougeinia oojeinensis; Tinsa, Leguminosae	2 & 3	D						Aesthetic Value
L_{11}	Petrocar marsupium; Bija, Leguminosae	2 & 3	D			*			Firewood and Timber
S_8	Randia dumetorum; Mainphal, Rubiaceae	2 & 3	D					*	Fragrant Flowers
S_9	Randia uliginosa; Phentra, Rubiaceae	2 & 3	D					*	Fragrant Flowers
L_{12}	Saraca indica; Asoka, Leguminosae	5	E			*	*		Noise Attention
M_{10}	Schrebera Swietenioides; Mokha, Sapindaceae	2 &3	D					*	Ornamental; fragrant flowers
L_{13}	Shorea robusta; Sal, Dipterocarpeae	4 & 5	D			*			Wood, fuel
M_{11}	Saymida febrifuga; Rohan, Meliaceae	4 & 5	D						Wood, tanning and aesthetics
M_{12}	Sterculia urens; Kulu, Sterculiaceae	1	D			*			Ecological complement on RMP areas

L_{14}	Stereospermum suaveolens; Padar, Bignoniaceae	1	D		*			Fruits; birds
L_{15}	Tectona grandis; Teak, Verbenaceae	1	D		*			Wood
L_{16}	Teraminalia bellerica; Behera, Combretaceae	1	D		*			Fuel wood
S_{10}	Wendlardia exserta; Tilwa Rubiaceae	1	E		*			Leaves used as fodder

L = Large Trees   M = Medium Size Trees
S = Shrubs/Small Trees   D = Deciduous Trees
E = Evergreen Trees

Table 3   Design Details of Plantations on Areas Around Red Mud Ponds (RMPs)

Sr. No.	Site	Dimensions of the greenbelt; total running length x width (m^2) (area in hectares)	Sr.No. of Plant Row	Total no of plants	No.of plants of each species	Quantity (liters) of irrigation water required per day	
	(1)	(2)	(3)*	(4)	(5)**	(6)***	Total L/d
1	RMP – 1	1310 x 40 (5.24)	1	132	66	M_1, S_6	
			2	132	66	M_{12}, L_{15}	
			3	132	66	L_{14}, L_{16}	
			4	132	66	M_7, M_9	
			5	132	132	L_7	6600
2	RMPs – 2 & 3	2330 x 40 (9.32)	1	234	177	M_2, S_7	
			2	234	177	L_1, S_8	
			3	234	177	M_4, M_{10}	
			4	234	78	M_5, L_{11}, S_8	
			5	234		L_7	11700
3	RMPs 4 &5	1370 x 40 (5.48) and 270 x 20 (0.54)	1	138	46	L_4, M_8, L_{13}	
			2	138	69	S_2, M_{11}	
			3	138	46	L_3, S_5, L_2	
			4	138	46	S_1, L_6, S_{10}	
			5	138	46	M_6, L_5, L_9	
					138	L_7	
			AND				
			1	28	14	L_2, M_6	
			2	28	14	L_{11}, M_7	
			3	28	28	L_7	7740
	TOTAL			2604			26040

recycle the atmospheric carbon and nitrogen through their regulated growth and periodic harvest.
*   The first row, in each case, is 3-4 meter from the edge of the pond

26

** Within each row, the species are alternatingly placed in a representative
Sequence with 10 meters **(Fig.-2)** plant to plant spacing and row to row spacing also 10
meters; for names of species refer **Table-2**

*** 50 liters per plant at a time at an interval of 5 days

These grass species can also be planted on areas between the tree species in the greenbelt strips around the RMPs for slope stabilization

Even with all these measures, it is suggested that, additionally, to collect surface run-off from the slopes, a collection drain, should be made at the base of each of these greenbelt strips, traversing across the slope, such drains from all the RMPs should open into a settling pond from where a relatively clear supernatent would be released to the nearby nullah.

5   EPILOGUE

The Balco has contemplated improvement and modernization of the entire aluminium production and red mud disposal processes in phases. Balco has signed a contract with Kaiser Aluminium and Technical Services Inc. of the US, the Alusuisse – a Swiss Company and Aluminium Pechiney of France for the first phase, which includes revamping and retrofitting of smelter plant to improve energy and material efficiency and saving of carbon and aluminium fluoride2.

Concentration of fluorine gas in the atmosphere and fluoride in soils will reduce further due to the proposed changes in the process. Due to process changes in the aluminium complex, there will be no further increase in quantity of solid waste proportionately.

The red mud ponds with barren surroundings do require to be subjected to an appropriate EMP as suggested. This would reduce any chance of fine dust moving beyond the area confined by the proposed greenbelt barriers around these ponds and grass growth on the pond surfaces.

Additionally, it is hoped that proposed management plan will go a long way in mitigating most of the environmental problems, e.g. by way at capturing of the alkaline dust by various tree species, loosening the red mud surface thereby encouraging root proliferation, accelerating percolation of metals, sodium and dissolved aluminium out of the root zone10 and abstraction of the pond leachate/percolates by the roots of the Eucalyptus plants suggested at the lowest level of the diked soil surface.

REFERENCES

Vinayak, M (1993). Aluminium levies bring down growth rate. The Hindu Survey of Industry. Pp.271-271. Kasturi & Sons Ltd. Chennai (India)

Vithoba. (1995). Aluminium Geared for the global market. The Hindu Survey of Indian Industry pp. 357-358 Kasturi & Sons Ltd. Chennai (India)

Ralvi, S. W.M. (1999). Aluminium Rolling – Adequate capacity. The Hindu Survey of Industry pp. 357-358 Kasturi & Sons Ltd. Chennai (India)

Special Correspondent. (1994). Alluminium Challenges of a free market. The Hindu Survey of Industry pp. 357-358 Kasturi & Sons Ltd. Chennai (India)

Pandya G.H.: Phadke, K.M. : Rao, K.SM. and Swamwarth, R. (1988). Fluoride concentration in and around an aluminium (Reduction) plant. Indian J. Environ. Hlth. 30 (3) : 222-227

Chamblee, J.W: arey, F, ;Powell, J.W. and Heckel, E. (1980). Fluoride in ambient air over the Pamlico estuary in North Carolina. Journal of the Air Pollution Control Association 30(3) : 397-400

Pande, R.N. and Vaidya S.D. (1975). Plants andpollution. Scavenger, Jan. Pp. 8-14

Maclean, DC. (1982). Air quality standards for fluoride to protect vegetation : Regional seasonal and other considerations. Journal of the Air Pollution Control Association 32(2),; pp. 82-84

NEERI, (1994). Comprehensive EIA of Aluminium Complex Korba Balco, National Environmental Engineering Research Institute, Nagpur

Fuller R.D. ; Nelson, E.D.P. and Richardson, C.J. (1982). Reclamation of Red Mud (Bauxite Residue_ using alkaline tolerant grasses with organic amendments J. Environ. Qual. 11 (3); 533-539

Solid Waste Management in Indian Mines

T.N. Singh[1], A. Dubey[1] and S.K. Singh[2]

[1]Department of Mining Engineering, Institute of Technology, Banaras Hindu University, Varanasi 221 005, U.P., India
[2]Department of Geography, Banaras Hindu University, Varanasi 221 005, U.P., India

INTRODUCTION

Generation of waste material is an important part of human activities from time immemorial. Disposal of wastes generated by human under the care of atmosphere, over land and water bodies has become a normal practice since the very begining. The capacity to cleanse the wastes by natural process is limited. The generation of waste isa direct of human activities and technology choice. The design of mining and mineral industries has been driven historically more by technological issues relating to resourceextaction and mineral beneficiation than it has been by waste management and disposal considerations(Stewart and Petrie, 2000). Increasing population and pressure on land has drawn attention towards solid was waste disposal management. Today, it has become a matter of concern because of the systamatic approach of viewing the earth as the total system and man as one of its many subsystems. Many opportunities may be lost if urgent resource mangement and regeneration plans are not out into action. Mining, one of the important sector in India, is the second largest sector after agriculture (Trivedi, 1990). In India, the total area occupied by mining leases (excluding atomic mineral, minor mineral and petroleum and natural gas) is about 0.22 percent of the land area of the country (Anon, 1997). Surface mining causes spreading of waste material over large areas. In mining equal or larger area than the excavated one is needed for external overburden dumping and a further area is needed for storage of wastes and tailings, beneficiary plants, washing and other plants and infrastructural facilities like roads, colonies, powerlines, etc.(Banerjee, 1987). It is estimated that on an average about four hectare of land surface area is disturbed for every million tonne of coal extracted by opencast mining and a similar area is also damaged by creation of external overburden dump. Environmental degradation starts in the mining sector from exploration and drilling stage. As an integral part of mining operations, forests are cleared at first and land is excavated causing serious land degradation in the form of vast tracts of dug quarries and heaps of overburden and waste piles. ·In addition, ancillary activities and residential colonies also generate wastes.

 The mining industry needs to be able to ameliorate the environmental impacts of all its activities. The volume of waste produced is expected to increase in future, this problem of waste management will increase manifolds. The main problems associated with waste management include the scaring of landscape, instability of waste dumps, chances of spontaneous combustion occurring in spoil heaps, water and air pollution and land degradation.

The waste products generasted at various stages from mineral inventory and exploration and extraction through concentration smelting, refining, fabricated and manufacturing, and finally decommissioning, and their ill environmental effects to which they give rise are given in fig. 1.(Warhust and Noronha, 2000).

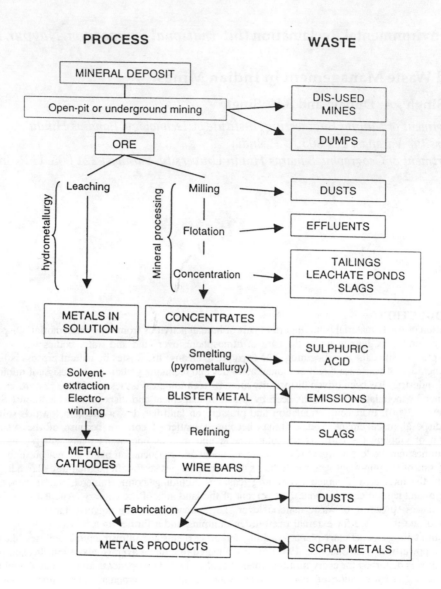

Fig. 1a : Generation of waste due to different activities (after Warhurst and Noronha, 2000)

1 show the production of coal and generation of waste materials in various mines of Singrauli Coalfields Ltd, Sidhi Dist. (M.P.).

MEDIA AND HAZARD

EFFECTS

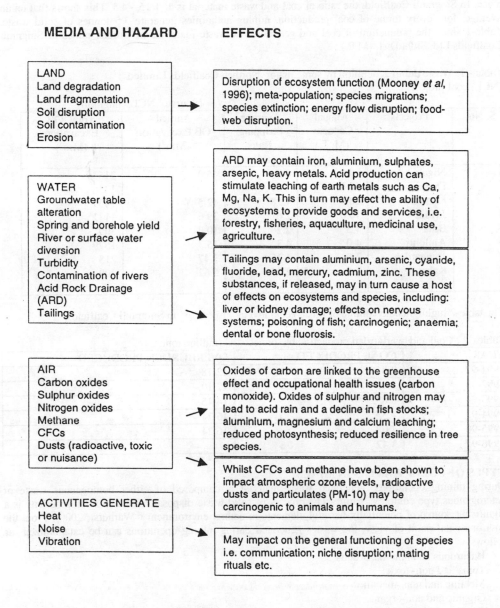

LAND
Land degradation
Land fragmentation
Soil disruption
Soil contamination
Erosion

→ Disruption of ecosystem function (Mooney *et al*, 1996); meta-population; species migrations; species extinction; energy flow disruption; food-web disruption.

WATER
Groundwater table alteration
Spring and borehole yield
River or surface water diversion
Turbidity
Contamination of rivers
Acid Rock Drainage (ARD)
Tailings

→ ARD may contain iron, aluminium, sulphates, arsenic, heavy metals. Acid production can stimulate leaching of earth metals such as Ca, Mg, Na, K. This in turn may effect the ability of ecosystems to provide goods and services, i.e. forestry, fisheries, aquaculture, medicinal use, agriculture.

→ Tailings may contain aluminium, arsenic, cyanide, fluoride, lead, mercury, cadmium, zinc. These substances, if released, may in turn cause a host of effects on ecosystems and species, including: liver or kidney damage; effects on nervous systems; poisoning of fish; carcinogenic; anaemia; dental or bone fluorosis.

AIR
Carbon oxides
Sulphur oxides
Nitrogen oxides
Methane
CFCs
Dusts (radioactive, toxic or nuisance)

→ Oxides of carbon are linked to the greenhouse effect and occupational health issues (carbon monoxide). Oxides of sulphur and nitrogen may lead to acid rain and a decline in fish stocks; aluminium, magnesium and calcium leaching; reduced photosynthesis; reduced resilience in tree species.

→ Whilst CFCs and methane have been shown to impact atmospheric ozone levels, radioactive dusts and particulates (PM-10) may be carcinogenic to animals and humans.

ACTIVITIES GENERATE
Heat
Noise
Vibration

→ May impact on the general functioning of species i.e. communication; niche disruption; mating rituals etc.

Fig. 1b : Different environmental hazards and its effect on eco-system (Warhurst and Noronha, 2000)

WASTE MATERIAL GENERATION IN MINES

Waste material generation is important and integral part of every mining activity. During mining activity, valuable economic ores are excavated and unwanted material is left as waste. The ratio of ore and waste materials varies from mine to mine. The generated waste material is disposed at a certain place. In Singrauli Coalfield, the ratio of coal and waste material is 1: 1.15 - 4.5. This shows that on an average, for every tonne of coal production, mining authorities generate 3.5 tonnes of solid waste. Table-1 show the production of coal and generation of waste materials in various mines of Singrauli Coalfields Ltd, Sidhi Dist. (M.P.).

Table - 1: Waste dump generation by some major Northen Coalfields Limited (NCL) coal projects.

(Source : NCL, Singrauli)

S. No.	Projects	Annual Capacity (M. T.)	Average Stripping Rate (m³/t)	Annual OB Excavation (Mm³)	Total OB Excavation (Mm³)
1.	Nigahi	14.0	4.3	45.0	2163
2.	Dudhichua	10.0	3.29	36.0	1356
3.	Khadia	10.0	4.28	42.5	1457
4.	Jayant	10.0	2.6	29.6	1132
5.	Bina	4.5	2.2	10.0	309
6.	Amlohri	4.0	4.3	15.3	1579
7.	Jhingardah	3.0	1.15	3.47	215
8.	Kakri	2.5	2.25	5.62	209

The table-2 highlights the volume of waste generated by Project 'A' in Singrauli Coalfields.

Table : 2 : Coal and overburden production by project 'A' (million tonnes)

YEAR	COAL PRODUCED	OVERBURDEN PRODUCED
1991-92	2.81	12.18
1992-93	3.45	12.10
1993-94	3.83	14.45
1994-95	4.33	16.24
1995-96	4.08	16.24
1996-97	3.42	--

TYPES OF SOLID WASTES IN MINES

During mining operations, generated waste material is composed of either homogenous waste or heterogenous type complex nature. The process of solid waste disposal related to mining activity is a significant source of potentially harmful elements in natural environment (Warhurst, 2000). From the environmental point of view, the wastes generated during mining operations can be catagoraized as follows:

1. Hazardous and non- hazardous;
2. Toxic and non- toxic;
3. Metallic and non -metallic;
4. Organic and non-organic;
5. Combustible and non-combustible;
6. Radio-active and non- radio-active;
7. Soluble and insoluble; and
8. Usable and unusable waste.

Each of the above mentioned wastes may have certain criteria for their further categorization. Absence or presence of a specific element/mineral can be used for categorization, such as presence of sulfur in coal waste dumps may be responsible for combustion in coal waste dumps and causes acid mine drainage problem. These waste materials are responsible for air, water and land pollution and cause ecological problems in the area in which they are generated. Other effects such as social and aesthetic problems are also due to these wastes in and around mines.

SEASONAL PROBLEMS IN SOLID WASTE MANAGEMENT IN INDIAN MINES

Action of natural agencies alongwith faulty action of waste disposal can cause adverse impact on surrounding environment. It is realized that waste management problems vary from season to season. The seasonal variation of problems can be categorized as -

1. Pre- Monsoon problems;
2. Monsoon problems;
3. Post monsoon problems;

The process of solid waste disposal related to mining activity is a significant source of potentially harmful elements in natural environemntal(Warhurst,2000).It is well known fact that rocks are composed of various elements occurring in nature. These elements constitute solid waste which are generated by human economic activity. Solid wastes are aggregate of unwanted materials containing a number trace element (i.e. elements in a trace amount). When concentration of these elements exceeds the threshold limit value they become hazardous and cause severe environmental problem on local and regional scale (Dubey, 1992) (fig. 2). For example, leaching of hazardous trace elements with rainwater and its contamination may cause health hazard in local inhabitants of the area in which water is used for drinking and other potable purpose. Management of these trace elements needs proper care and treatment during their disposal.

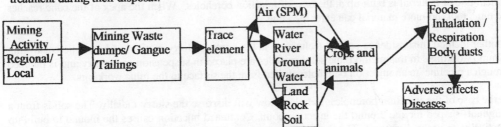

Fig. 2 : Representation of pathway of trace elements to human body

Aesthetics:
Efforts have to be under taken to improve the landscape. Although it is not possible to restore the landscape to the original state, it is very much possible to improve the aesthetic impact of present and past activities. Reclamation of the mined area has to be carried out such that revegetation becomes possible.

Stability:
Waste dumps comprising of uncohesive and uncompacted materials can be unstable when saturated and unvegetated. Tailing dams become unstable mounds of waste may lead to catastrophic events causing loss to life and property.

Spontaneous combustion:
Spoil heaps containing combustible material are susceptible to spontaneous combustion as a result oxidation of these materials. Proper compaction of waste is required to prevent air from entering the heaps and causing oxidation.

Water quality:
The disposal of solid waste should be such that the impact on surface and ground water is minimized. The leaching of heavy metals from the waste dumps has a toxic effect on the soil biota and vegetation.

Dust:
Fine particulate matter can be blown from waste dumps and lagoon deposits. This problem gets compounded if dust contains heavy metals or other contaminants. Thus, suitable methods of dust suppression need to be adopted whenever necessary.

The most obvious solution for the disposal of solid waste is the filling of underground mine openings after excavation is complete. Hydraulic back filling of the voids is an effective means of protection the environment form the hazards of mining operations. Another possibility consists of filling in the pits left by derelict quarries to restore the landscape to the original state.

Disposal on spoil tips is another alternative that can be used to tackle the problem of waste disposal. Tailing should be so stored that effects of overheating and the noxious effect are minimized. Construction of refuse cell is also advantageous.

HYDRAULIC BACKFILLING

Backfilling of existing mine voids by slurring crushed coal preparation waste or other solid wastes and conveying it to the portions of a mine where support is needed serves the dual purpose of not only providing support but also helps in the disposal of waste. In subsidence control projects, waste rock from refuse banks may be used as fill material. It is imperative to carry out compressibility tests on the crushed samples to evaluate the load bearing capacity of the fill. The source of mine waste should be some nearby mine so to make it available at a low coast of transportation.

The utilization of refuse from the mine for filling helps in the removal of unsightly piles of waste. The damage to the environment, both on air and water can be considerably checked by such utilization of waste. The problem of fire in the waste pits due to spontaneous combustion (especially for those having high sulphur content) also gets solved.

For mines where workers can gain safe access, 'Controlled Flushing' can be effectively used for hydraulic backfilling. Herein, bulkheads are built in mine workings to confine the backfill. From the surface, the fill material is sluiced down injection boreholes into connecting horizontal pipelines and slurry is directed to the selected area of the mine.

The method of 'Blind Flushing' can be adopted for mines that are, due to some reasons, inaccessible. A conical pile of material is built up at the base of injection boreholes. When the apex of the cone reaches the mine roof, no more material can enter the mine.

Another method in subsidence control techniques is the pumped slurry method of hydraulic backfilling of mine workings. In this method, the solid particles are placed in suspension as a slurry and is pumped through a pipeline to an injection bore holes drilled from the surface to the mine workings.

At the base of the injection boreholes, turbulent flow will disperse the slurry radially. The solids from a doughnut- shaped mound around the injection point. Continued injection causes the mound to build up practically up to the mine roof. The mound first grows upwards and gradually as pressure increases due to constriction of channel space, it starts growing horizontally.

REFUSE HANDLING

The relatively new concept of refuse cell is based on the idea of encapsulating refuse material to isolate it from the environment. The cells are placed out in the mined out areas. Limekiln dust is mixed with the refuse to increase alkalinity, inhibit bacterial growth.

Construction of cells begins with the making of a narrow berm around the cell area. An inner dyke is then built to contain the refuse. The floor of the cell is covered with a thick layer of fragmented rock. A three to five feet layer of shale is placed over this to make an impervious layer. The placement of refuse is then initiated and when a layer has been placed, the next higher level is constructed using second inner dyke structure. This process is continued in cycle till the desired height is reached. The site and top of the cell are covered with shale placed in a dome like shape. The shale, soil and dome shape assists in diverting rainfall to outside the berm, thereby reducing infiltration of water. The refuse should be mixed with limestone dust not only because of its neutralizing potential but also because it acts as a drying agent, forming calcium hydroxide when it comes in contact with water. The rock bridge placed between layers of refuse is required to support the weight of fully loaded trucks on top of the refuse. The bridge allows Turks to move over the refuse, though it consumes considerable volume of the cell.

With the use of limekiln dust, the requirement of rock bridges is not necessary. Thus, the use of limestone dust must be made for its neutralization aspects and for its stabilizing properties. The stabilization process is driven by the pH, pozzolan nature of calcium hydroxide, fly ash and clay.

USE OF REFUSE FOR HIGH WAY CONSTRUCTIONS

The refuse produced as a fall out of mining can be used for the construction of highways. It is both an environmental clean up procedure and a high way construction process. The mining industry through this method can not only get rid of the large volumes of solid waste produced but also realize monetary gains out of it.

Raw coal refuse can be used for a variety of construction purposes. Adequate compaction of the refuse will ensure against the material causing air or water pollution. The performance of this method is comparable to that of naturally occurring rock and engineering soils. Due to sufficient compaction, the susceptibility of waste material (mostly in case of coal mines) to spontaneous combustion will be greatly reduced. It also checks the penetration of water into and through a properly constructed embankment and thus the problem of leaching is eliminated. The use of unburnt coal refuse for high way constructions is being perused in several countries under the approval from the environment safe guarding agencies. It is proper to test the spontaneous combustion potentiality, leaching and swelling indices, and porosity of the material to be used. One of the assets of coal refuse compared to many engineering materials is the fact that it is workable during wet whether. Cement stabilization is also being considered. By using a small amount of cement, coal refuse can also be used as cheap concrete. A mixture of cement (about 5%) and fly ash along with the refuse will give a satisfactory performance.

PROBLEM REALIZATION

During past several years, little attention was paid for proper solid waste disposal in mining industry in our country. Till today, there is no accurate law which governs the safe solid waste disposal in mines. It is only a part of environmental management plan. Waste materials are disposed without proper consideration. Therefore, they cause environmental and associated problems in nearby areas. The danger of contamination of solid wastes becomes more through environmental degradation. Therefore, it is demand of time to formulate and implement an accurate solid waste disposal and management programme even after mining has to be ceased.

SOLID WASTE MANAGEMENT PLANNING

There needs an acccurate and separate solid waste management planning for each and eery mine. This should be categorized on the basis of types of wastes generated during mining activity. The focal areas in solid waste disposal planning should be considered as:

1. Geology - The planners should know detail about geological factors such as dip and strike of strata, fault planes, shear zones, lithological characteristics, etc.
2. Geomorphology - This covers terrain morphology drainage pattern, slopes, etc.
3. Climatic variables - include temperature, wind velocity and direction, humidity, rain intensity, amount and duration, seasonal changes, etc.
4. Hydrological parameters such as drainage characteristics, river course and flow velocity and other attributes should be studied in detail.
5. Soil characteristics including physical, chemical and physico-mechanical characteristics.
6. Hydrogeology and hydrogeochemistry such as ground water conditions, recharge and discharge, leaching of minerals, contamination pattern, potability, etc. should be investigated.

SOLID WASTE MANAGEMENT IN INDIAN MINES

Environmental management is one of the important aspects of every mining project to combat the problem of environmental deterioration by mining activity. To avoid the environmental problems, Ministry of Environment and Forest, Govt. of India, has set up Environmental Appraisal Committee to examine mining projects from environmental angle. For the purpose, each mining project can be opened or worked unless it has on approved Environmental Management Plan (EMP). This also covers management including solid wastes.

Proper design of solid waste management system involves planning and designing of methods of collection, handling, transportation, treatment and ultimate disposal of solid waste. Physical and chemical characteristics of solid waste must be investigated before a suitable plan has to be drawn up for its handling, transportation, recycling and disposal. In the mining areas, waste is generally dumped on the outskirts of the mining areas giving rise to a lot of problems. The contamination of water, soil and air, the proliferation of reptiles, insects and worms and consequently emergence of various diseases among the town's population, particularly those living in the vicinity of the dumps are major problems arising on account of improper disposal of wastes. Majority of dweller's throw their daily wastes along the roads and streets.

The number of sweepers and vehicles are insufficient in the mining areas for waste collection. There is need to increase the number of sweepers and transport vehicles in almost every mining area and adjacent towns. Governmental and Non-Governmental Organizations must be involved in awareness programmes for the local inhabitants towards problems generated due to improper disposal of waste. It is the responsibility of every citizen to keep the environment clean. Sanitary land filling should be made available in all the mining areas for waste disposal. It will minimize contamination of land and water from toxic substances.

The Ministry of Environment and Forest has come out with the draft for Management of Municipal Solid Waste (Management and Handling) Rules, 1998. The Ministry of Urban Development has approved it. Thereafter rules comprise collection of waste, its storage, transportation, processing and disposal with the overall objective to ensure proper management of municipal solid waste to maintain good hygienic and sanitary conditions in mining areas and towns for the protection of health of human population literacy of municipal solid waste should be prohibited in mines and towns.

CONCLUSION

It is clear from the above discussion that solid waste disposal is a burning problem in mining areas because it not only poses environmental problems but also social and economic problems. Therefore proper back filling, appropriate waste disposal plans, effective tailing disposal, phased reclamation programs, and rehabilitation programs should be taken into consideration. In this connection, availability of technical resources, including appropriate institutional arrangements, staff, analytical facilities, equipments, and finance alongwith pre-mining base line data and post mining estimation should be done. In mining sector, government should implement a proper solid waste management Programme and may raise a cess fund with appropriate share which will not only be helpful in controlling the solid waste disposal problem but also for environmental management.

REFERENCES

1. Anon. (1997) : Ind. Mining & Engineering Jl. Vol.36 (5). Pp.15-18.
2. Banerjee,S.P. (1987) : Land reclamation in mined areas. in Nat. Workshop on Env. Management in India - A status paper (ed.) B.B.Dhar. Dept. of Mining Engg., I.T., B.H.U., Varanasi. pp. 63-71.
3. Dubey. A. (1992) : Impact of trace element air pollution on human health. Ind. Jl. Env. Prot. Vol. 12(7). pp. 512-514.
4. Stewart, M. And Petrie, J. (2000) : Plannning for waste management and disposal in mineral processing . in Environemntal Policy in Mining-Corporate Strategy and Planning for Closure. (ed.) Alyson Warhurst and Ligia Noronha. Lewis Publications London. pp. 145-174.
5. Trivedy,R.K. (1990) : Mining and environment : an overview. in Impact of Mining on Environment (ed.) R.K.Trivedy and M.P.Sinha. Ashish Publ. House, New Delhi. pp. 1-18.
6. Warhurst,A. (2000) : Mining, mineral processing, and extractive metallurgy : an overview of the technologies and their impact on the physical environment. in Environemntal Policy in Mining-Corporate Strategy and Planning for Closure (ed.) Alyson Warhurst and Ligia Noronha. Lewis Publications London. pp.33-56.
7. Warhurst, A. And Noronha, L. (2000) : Integrated environmental mangement through planning for closure from the outset. in Environemntal Policy in Mining-Corporate Strategy and Planning for Closure (ed.) Alyson Warhurst and Ligia Noronha. Lewis Publications London. pp. 13-32.

Assessment of Geo-Environmental Impacts of River Flooding

Aabha Sargaonkar* and Chitra Gowda**

*Scientist, **Project Fellow
National Environmental Engineering Research Institute, Nagpur, India

INTRODUCTION

Natural disasters have been threatening humankind since time memorial. Damage, destruction and loss of life are characteristic of such events. A disaster is defined as an event – natural or man-made, sudden or progressive – that seriously disrupts the functioning of society causing human, material or environmental losses of such severity that exceptional measures are required to cope. River flooding is one such disaster, which affects severely most of the developing countries as judged by deaths per million population. The geo-environmental effects of flooding are many, such as soil erosion, sedimentation, sand deposition, water logging, and inundation.

The paper presents the causes and impacts of river flooding and mathematical techniques used to evaluate the effects of two major impacts viz., soil erosion and inundation. Some measures to alleviate or prevent flooding and its effects are also discussed.

GEO-ENVIRONMENTAL IMPACTS OF RIVER FLOODING

River flooding may be caused by either excessive rainfall in a season or by unusually heavy storms. Thus, rainfall which is otherwise used for irrigation, energy production, service of livelihoods and the like, becomes a major cause of death and destruction of property as well as changes in the geo-environment.

River flooding may be classified into overbank spill and inundation. The latter is caused primarily by local run-off and an inefficient natural or constructed drainage system. Inundation therefore dictates land-use patterns, whereas flooding is more a measure of occasional damage and loss. The features that characterize and quantify flooding include timing of flood, depth of flood, duration of flood, sediment concentration, sediment size, velocity of flood water, and rate of water rise during flooding. The impact of these flood characteristics is measured in terms of the extent of soil erosion, sedimentation, sand deposition and the demarcation of areas prone to water logging or inundation.

Soil Erosion and Inundation

Soil erosion is the result of removal of surface layers of soil effected due to a process of both particle detachment and transport by the agencies of wind, water and ice. Stream bank erosion occurs where the velocity of the flowing water is high and bank material resistivity, often low. Erosion of the riverbed is caused by inundation for continuous or prolonged periods of time, along with other degradation processes.

Flooding and consequently overspill into the river flood plains cause sheet and rill erosion besides stream channel erosion. Sheet erosion is the removal in thin layers or sheets of soil from the surface, whereas rill erosion is the removal of soil by water from small, well-defined channels where there is overflow at areas of depression. The consequences of sheet and rill erosion are:
- declining soil productivity on-site
- pollution and sedimentation off-site
- unsightly scarring of the landscape

Although climate, soil, landform, and vegetation cover all interact to determine how much erosion takes place, it is the vegetation cover which ultimately controls how much protection is afforded to the land. Therefore soil conservation can be provided based on the evaluation of the land with respect to its suitability for different uses and information available from a geo-morphologist regarding the mechanics of water erosion (Ref 1).

Inundation, although responsible for loss of human life and property due to settlements and cultivation in the floodplains, is important for the following environmental issues that river floodplains must receive water every year even for a low discharge of water:

- the floodplains with 'green' buffer zones connected by ditches, floodways, wetlands or grasslands support the environmental quality in the river flood plain
- wide buffer zones of diverse biota are a part of nature conservation, and they considerably reduce water quality deterioration by filtering, hence reducing the input pollution load
- aquatic flora on the margins of these channels and vegetation along the banks provide cover for trapping and hence reduction of fine solid material besides reduce chances of chemical pollution of main channels

ASSESSMENT OF IMPACTS OF RIVER FLOODING

Impacts of river flooding can be assessed by using mathematical models and site specific data. For the case study presented herein, the *Universal Soil Loss Equation (USLE)*, a semi-empirical equation developed by the USDA Agricultural Research Service in the early 1960s and modified in subsequent years is used for soil loss prediction, and *HEC-2* model (Version 4.6) developed at Hydrologic Engineering Center by the U.S. Army Corps of Engineers is utilized to determine inundation of the river floodplain.

Study Area and Vulnerability to Floods

River Yamuna originates from Yamunotri at a height of 3320 m and traverses a distance of 1400 km before meeting river Ganga at Allahabad. The length of river Yamuna in National Capital Region of Delhi is about 50 km with half this length within the present urban sector and the other half in rural sector. Width of the river varies from 1.5 km to 3 km in this stretch. The topography of the territory is such that it recieves drainage water from the adjoining states of Haryana and Rajasthan. The problem of floods due to the spill of Yamuna has been reduced by constructing embankments on both sides of the river. However, intense floods occur every year in River Yamuna and the complete floodplain gets inundated. Gauge records of the last thirty five years indicate that the 1978 flood was a very severe flood in the history of the Yamuna River. A discharge of over 20,000 cumecs was estimated to have passed at Tajewala Headworks (about 200 km upstream of Delhi) in Uttar Pradesh on 3rd September, 1978 and the flood peak was moderated to 7022 cumecs (2.5 lakh cusecs) at National Capital Territory (NCT), Delhi. Flood discharges and levels measured at different gauging stations and recorded for studying the return period of a flood and its impact, indicate that the 1995 flood is comparable to the 1978 flood by means of discharge.

Soil Loss Prediction

The USLE equation considers various factors like climate, soil, topography and vegetative cover. The annual soil loss from a site is given by (Ref 2):

$$A = R \times K \times LS \times C \times P$$

where A = computed soil loss, tons per acre per year
   R = rainfall erosion index, tons per acre $\times$ in/hr
   K = soil erodibility factor, tons per acre per unit of R
   L = slope length factor, dimensionless
   S = steepness factor, dimensionless
   C = vegetation factor, dimensionless
   P = erosion control practice factor, dimensionless

In order to calculate the soil erosion in the floodplains, the above parameters are estimated based on site-specific conditions of soil type, rainfall intensity, vegetative cover and topography of the study region. The details of mathematical computations are given below.

The rainfall erosion index, R, is actually a measure of the erosive force and intensity of rain in a year. Erosion, E, for water plus sandy loam type of soil in the study area is around 0.003 tons per acre and for water plus sandy clay, 0.705 tons per acre (Ref 3). The other factor, rainfall intensity I, for water-logged area due to heavy rains is estimated using the runoff equation,

$$Q = (C \times A \times I) / 360$$

where   Q = runoff rate, m3/s
     C = runoff coefficient

I = precipitation intensity, mm/hr

A = watershed area, hectares

Knowing Q, C and A for the study region **(Ref 4)**, rainfall intensity is estimated to be 51.19 mm/hr. finally, the product of E and I gives R, for sandy loam 0.00603 and for loamy sand, 1.41705.

The susceptibility of soil to erosion is known as erodibility. Some soils are inherently more erodible than others. Estimation of soil erodibility factor K, requires analysis of soil samples of the study area, since erodibility depends upon parameters such as soil, texture and antecedent soil moisture content as well as organic content and clay size fraction of a soil.

Site specific data of soil analysis indicate that the contiguous region of the river stretch has alluvium-derived soils. There are deep loamy sand (85% sand, 15% clay) and sandy loam (30% silt, 60 % sand and 10% clay) soils. The moisture content of various sites along the river and in the floodplain area ranges from 0.0434% to 0.1164 % at surface, and 0.0581% to 0.1419% at a depth of 6 inches and organic matter ranges from 3.40% to 8.86%. The soil erodibility factor K in the USLE equation is determined by using the Nomogram method, knowing the percentages of silt, sand and clay in a soil sample.

Adjustments must be made to account for the organic matter content in the soil sample, as the nomograph assumes 2% organic matter. The correction factor of –0.05 for sandy loam and –0.10 for loamy sand gives K values to be 0.27 and 0.54 respectively.

The topographic factor, LS, for a slope gradient of 2%, and a slope length of 1000m, is 0.40; the vegetation factor C i.e. the ratio of soil loss from land under specified crop or mulch conditions to the corresponding loss from tilled, bare soil is 0.1 for 90% cover annual grass; and the erosion control factor P i.e. the ratio of soil loss with a given surface condition to soil loss with up-and-down plowing is 1.2, for vegetated cover. **(Ref. 2)**

Finally, the soil erosion using USLE equation is found to be 0.0000781488 tons / acre / year for sandy loam soils, and for loamy sand soils 0.036729936 tons / acre / year.

Prediction of Water Spread Area: Inundation

The river stretch downstream of IP Barrage, considered for modeling, has one rail and one road bridge viz. the Nizamuddin Railway Line and the National Highway-24. The waterway of these structures is confined to a width of around 600 m. In case of major floods, flood waters pass over the road due to breaches in the embankments, flooding the vast area in the surrounding region. Mathematical model HEC-2 **(Ref. 6)** has been used to compute water surface profiles for steady, gradually varying flow under both subcritical and supercritical flow profiles. The model simulation for computing water surface profiles in the study stretch considers the effects of various obstructions such as bridges, culverts, and weirs and with irregularly shaped cross sections located at intervals along the stream to characterize the flow capacity of the channel and its adjacent overbank areas. Manning's 'n' for left and right overbank regions and the channel characterize the bed roughness. River flow, initial water surface elevation, topography of the channel and adjacent flood plains also form input data for the model. Data characteristic of the bridges include pier shape coefficient, number of piers, total loss coefficient, coefficient of discharge, bridge span, bottom width of bridge opening, total width of obstruction and elevation of bridge invert at upstream and downstream sides of the bridge. These inputs were provided with due consideration to bridge data and river topography of that cross section.

Model Calibration

The 3 km river stretch from I.P. Barrage to the proposed ILFS-NOIDA Link in NCT-Delhi was segmented into two reaches, with the Nizamuddin Rail Bridge in between. Ground level elevations along cross sections in the river channel and flood plains, at 4 stations in the first reach and 16 stations in the second, define the topography of the region. A schematic diagram of the study stretch and the model settings is shown in the **Fig. 1**. To examine the effect of flood on the wetlands of the study area, model simulations were run for peak flood discharge scenarios of 6360 cumecs (once in 100 years phenomenon), 4182 cumecs (once in 50 years phenomenon) and 2132 cumecs (25th percentile discharge). Known water surface elevations corresponding to these flows also form the model input. Manning's n ranging from 0.023 to 0.035, the *total loss coefficient, K* at bridge cross sections as 1.5 and the *coefficient of discharge, C*, ranging from 1.39 to 1.72 are the important parameters in simulation.

Model simulations for the once in hundred years scenario were compared with remotely sensed data of the satellite LANDSAT 5 TM which was acquired in the form of false color composite (FCC) generated from bands 4 and 5. The satellite image was geo-referenced with respect to Survey of India topographic maps and then interpreted using visual interpretation techniques taking into consideration the ground truth of the area. Critical cross sections, corresponding to those considered during the model run, were digitized on the image and classification for levels of waterlogged and land areas was done, as high

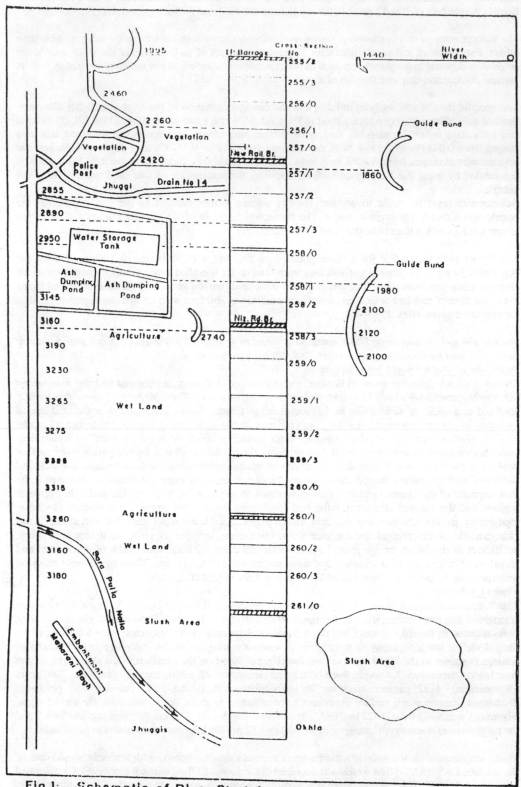

Fig. 1   Schematic of River Stretch used in the Model HEC-2

40

waterlogged (greater than 1m), waterlogged (between 0.5m and 1m), low waterlogged (less than 0.5m) and dry land by superimposing the water surface elevations on the ground elevation levels (in terms of MSL). Statistical analysis, by correlating the model simulation and satellite data, showed a correlation coefficient (R^2) of 0.59. *Model calibration for Manning's roughness coefficient (n) for left and right overbank regions and for the channel as 0.04, 0.03 and 0.025 respectively resulted in a Pearson's correlation coefficient (R^2) of 0.77.*

The water spread areas in the study region, for discharge scenario of once in 50 years, which is used as the guideline for developmental activities in the river bed area is depicted in **Figure 2**.

RESULTS AND DISCUSSION

Soil erosion as estimated using the USLE equation, is found to be of moderate order in the loamy sand soils of the study area, whereas it is very little in the sandy loam soils. In order to ensure that the erosion does not escalate, as the river floodplains are often subject to anthropogenic activities, erosion control measures on the banks of the river may be employed by decreasing the tractive forces which would slow down the velocity of water flowing over the surface or impoundment and also by increasing the erosion resistance of the surface by means of a suitable cover.

Vegetation as a method of erosion control is preferred due to its multifold advantages such as self-regeneration, minimal maintenance costs and inherent engineering properties. Vegetation reduces erosion by absorbing rainfall impact, reducing runoff velocity, increasing percolation into soil, binding soil with roots and protecting soil from wind. Plants planted for the sole purpose of erosion control must suit the local environment, and quickly adapt to the soil and climate. Classification of crops as soil building crops, which improve soil quality by increasing the organic matter in the soil chiefly by root growth, for example permanent sod; and soil conserving crops, for example non-cultivated crops such as alfalfa, clover, lespedezas, kudsu, grass hay and pasture, conserve the organic matter in soil and being closely-growing, prevent erosion and hence loss of organic matter and minerals.

Using geotextiles is another method of erosion control where vegetation is difficult to establish because of a hostile soil and climatic environment. Geotextile is any permeable textile material used with foundation, soil, rock, earth or any geotechnical engineering related material in the form of a mat, sheet, grid or web of natural or artificial fibre and is either placed on the slope or buried in the topsoil to reduce erosion.

Also, runoff velocity on the banks can be kept low by minimizing flow path lengths, constructing channels with gentle gradients and lining channels with rough surfaces. A permanent waterway lined with concrete, asphalt, gravel or grass may be strategically located in order to shorten long slopes and thereby reduce runoff velocity and soil loss besides diverting runoff to a storm drain system.

In order to prevent channel erosion, channel linings that slow velocity and reduce erosion, are used. Permeable linings are preferred as they allow infiltration of water. Various linings are gravel or rock, grass, combination of both, and jute mesh, excelsior mat (composed of dried, shredded wood), or other fabrics.

A long-term and cost-effective solution for reclamation and management of waterlogged areas is planting. Removing the unwanted weeds and other nuisance plants from the area and planting of trees with a high transpiration rate, like Acacia nilotica, Bombax ceiba, Eucalyptus robusta, Terminalia arjuna, are the important steps involved.

Tall plants nursed for two to three years prior to actual planting ensure their sustenance even in a peak flood. Results of the model simulation indicate that most of the Yamuna River floodplain gets inundated even for a 25th percentile discharge. Therefore the zoning of areas according to the risk or frequency of inundation should be done using mathematical techniques to minimize the likelihood of loss of life or damage.

A comprehensive disaster management plan should be prepared for possible flooding, which includes all aspects of planning and responding to the disaster. The two main approaches for reducing or averting the damage and loss of life caused by flooding are:
- Structural, which includes engineering works to protect flood-prone areas
- Non-structural, which involves forecasting and warning of impeding floods so that lives and movable property can be saved

Another approach to alleviating flood damage is to install flood forecasting and warning systems. These systems are based on the telemetry of rainfall and river flow data and the use of models that run in real-

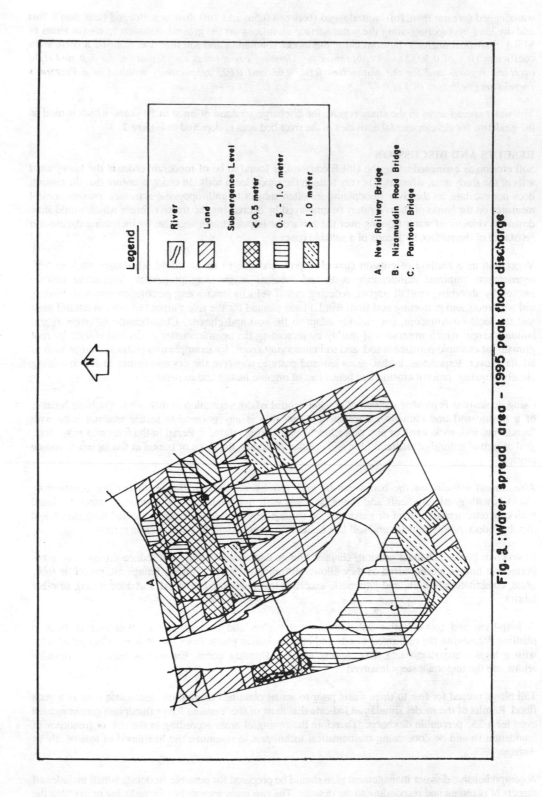

Fig. 2 : Water spread area – 1995 peak flood discharge

Legend

River
Land

Submergence Level

< 0.5 meter
0.5 – 1.0 meter
> 1.0 meter

A. New Railway Bridge
B. Nizamuddin Road Bridge
C. Pantoon Bridge

time. They provide forecasts of flows and river levels over a short period in the future, from a few hours to a matter of days ahead.

CONCLUSIONS

- The increasing population and civilization is seen encroaching in the river floodplains for developmental activities. Flood protection measures are offered to reduce the risk of inundation and to alleviate the loss due to floods. However the importance of the characteristics of the floodplains i.e. riparian zones composed of shifting mosaics of ecological units, various landforms i.e. cut-off channels, alluvial levees (embankments), flats and islands, and various stages of successions that occur on each type of landform, terrestrial vegetation, marshy and swampy areas, plantations, cultivated fields and habitations which support both the aquatic life and contiguous urban areas, must be realized.

- The USLE equation is a quick and convenient way of estimating soil erosion. In the study area, loamy sand type of soil is eroded and increased anthropogenic activities worsen the situation; hence it is imperative that soil erosion control measures are inculcated in the area and more so if there is possibility of extensive development.

- To undertake proposed development in the river bed area, the plan should ensure that the pristine qualities of the flood plain sustain even under the most frequently occurring floods.

- The inundation of the floodplain is essential for supporting ecosystems (such as reed-swamp associations, herbaceous and wooded communities) in the floodplains, which provide interaction mechanism governing the land-water interface phenomena.

- The model HEC-2 is a good hydrodynamic model to determine flood prone areas and levels of inundation. The model output can be used for flood zoning, which is of utmost importance to the town planners and for undertaking disaster mitigation measures.

REFERENCES

1. J. M. Hooke, John Wiley and Sons, (1988). *Geomorphology in Environmental Planning*.
2. Steven J. Goldman, Katherine Jackson, Taras A. Bursztynsky, P.E., McGraw-Hill Inc., (1986) *Erosion and Sediment Control Handbook*.
3. Dr. S.S.Negi, (1991). *Principals of Land Management and Soil Conservation*.
4. CBPCWP. Central Board for the Prevention and Control of Water Pollution, New Delhi (1980 – 1981). Basin Sub Basin Inventory of Water Pollution – The Ganga Basin Part-I, The Yamuna Sub Basin.
5. Daniel H. Hoggan(1989). *Computer Assisted Floodplain Hydrology and Hydraulics*, McGraw Hill Publishing Company.

Reclamation and Rehabilitation of Bauxite Mined Out Areas in India

C.S. Singh

M/S Hindalco Industries Limited, At & P.O. Lohardaga, Bihar 835 302, India

In Les Baux, France, in 1821, the French Chemist P. Bertier discovered a red material, thought to be clay, in fact contained no silica and was not a clay at all. It is composed primarily of one or more aluminium hydroxide minerals and impurities of silica, iron oxide and titanium. The mineral was given the name "Bauxite" after the name of village in which it was discovered.

The abundant reserves of bauxite is found in India and stands on fifth rank in the world. It is only after the discovery of East Coast Bauxite in mid seventies that the reserve position of bauxite in India changed dramatically. The east coast bauxite hold more than two-third of all India bauxite reserves. It occurs on high landforms rising to an altitude of 1000m – 1400m above mean sea level. Bauxite is found in four types of deposits: blanket, pockety & lensoid, interlayered and detrital. In India, large blanket deposits occur. The deposits occur as flat layers lying near the surface and may extend over an area covering many square kilometers. Thickness may vary from a meter or less to 40 meters in exceptional cases, although 4-6 m is average. The topographical areas vary in which blanket deposits are found; some are in low elevations where bauxite is covered by shallow tides, while others occur on gently inclined tops of hills, ridges and plateaux.

Mining

Bauxite mining is carried out in opencast The mining method depends mainly on the nature of deposit rather than the quantum of production. The east coast bauxite deposits are thick and consistent in nature which is amenable to a high degree of mechanisation. The west coast bauxite deposit and plateau region of Bihar and M.P. is pockety and lensoid, which can be better worked by semi-mechanised method employing large number of workers in breaking & sorting of bauxite from blasted material.

The bauxite deposits ends into an uneven corrugated type of floor, so, trench mining is carried out in two phases, after removal of overburden. In first phase, there is mass mining of most of bauxite layer, and in second phase, selective mining of remaining bauxite in contact with wall is carried out.

In case of pockety and lensoid deposit, the soft soil overburden is removed by using backhoe hydraulic excavator and stacked separately at an identified place. The drilling and blasting is carried out in hard strata with jackhammer or small diameter wagon drill holes. Ammonium nitrate mixed with fuel oil (ANFO) alongwith booster is used for blasting. The usable bauxite is sorted out from blasted material and sized manually for transport to alumina plant. The rejects and interburden forms about 40% of the blasted material and are disposed off in the worked out areas for reclamation. The front end loaders in conjunction with 10 tonnes capacity dumpers are used for handling of rejects. In small mines, the entire mining operations are carried out manually, supplemented with drilling and blasting of some shallow holes.

Reclamation and Rehabilitation

According to Indian Mining laws, the reclamation and rehabilitation of mined out area is a statutory requirement. **Quote** " Every holder of prospecting licence or mining lease shall undertake the phased restoration, reclamation and rehabilitation of lands affected by prospecting or mining operations and shall complete this work before the completion of such operations and the abandonment of prospect or mine" **Unquote**.

The moto of any reclamation and rehabilitation programme is to mitigate the adverse effects of mining on land and to establish a self sustaining eco-system for sustainable development. The method of reclamation depends mainly on the nature of deposit and local conditions. The reclamation method of pockety and lensoidal type of deposit differs from east coast bauxite deposit. The east coast bauxite deposit is thick with a small quantity of overburden without much of soil. The material for reclamation has to be imported from the valley portion. The profile can not be maintained as per original position. The pockety and lensoid type of deposit are comparatively flatter with thick soil cover, the proportion of rejects is also higher for use in reclamation. In case of east coast bauxite deposit, there is delayed rehabilitation whereas in case of pockety and lensoid deposit there is progressive rehabilitation.

The reclamation of pockety and lensoid type of deposit is carried out in a cyclic manner. The hard overburden and rejects during mining are dumped in the worked out pit starting from one side and levlled with dozer. Over this levelled surface, the soil stacked earlier is brought and spread over it. Then, reclaimed area is levelled.

After spreading and levelling of soil, shallow holes are dug in a grid pattern for plantation. In these pits, plants are planted on the onset of monsoon. The seedlings of plant are prepared in a nursery. The slopes of dumps are contoured and planted with shrubs, legumes and some fast growing species of plants so as to protect them from erosion. After the monsoon period, small quantity of fertiliser is used and watering is done in dry season. After third year of plantation, plants become self-supporting and does not require manuring or watering. The recommended plant species for afforestation, pasture and avenue are given in table 1 & 2.

Rehabilitation is monitored for seedling survival. The dead plants are replaced by new ones. The planted area is protected from the grazing cattles by erecting pitch wall of laterite around the planted area or barbed wire fencing. Green belt is also developed around mine working, haul roads, dumps etc. to act as a screen. Some fast growing plants are also planted to meet out fuel wood need of local population. In recent years, an experiment has been done to use the reclaimed area for potato cultivation. The level of reclaimed area is kept below the pit level so that, it can retain moisture. Social forestory is also encouraged by providing free seedlings amongst the residents around mining areas.

In case of east coast bauxite deposit, it is very difficult to restore the area to its original topography because of non-availability of waste and soil in sufficient quantity. The initial rehabilitation work in pits are floor ripping and bulldozing, pit sides and floor to create a gentle rolling, well drained topography. In steeper sections, 4m wide berm at a maximum interval of 10m are made on footwall.

Hard laterite material is brought from stock pile, spread in the pit and levelled. The imported soil from valley area is also spread and levelled over it. Soil is also spread on the berm in case of steep slopes. The edge of berm is protected by erecting parapet wall of dry stone masonary in Gabion structure. At the end of slope, along the scarp face, a barrier of insitu bauxite is left to check the flow of water and wash off from the pit during monsoon. A filter bed is also created against this barrier to drain out water free from suspended particles.

After spreading soil, shallow holes are dug for the plantation. On the onset of monsoon, fast growing plants are planted. Rehabilitation is monitored for seedling survival and growth. The planted tree is fenced to check grazing cattles and manured after the monsoon period. There is separate manpower and budget allocation for rehabilitation work.

Conclusion
The reclamation and rehabilitation efforts are made to mitigate the adverse impacts of mining on environment. In general, mining is blamed for creating ugly scar on the land and its degradation. But, it can be proved wrong if rehabilitation programme is implemented in right way from initial development phase. It can be said that mining is the intermediate use of land which can be restored not only to its original form but for better use. In India, the experience of bauxite mine is that the plateau which were originally devoid of any vegetation have now been covered with thick vegetation and have become productive after reclamation of mined out area. There is a great thrust on reclamation and rehabilitation programme. We believe on **"MINING IN HARMONY WITH THE ENVIRONMENT"** as well as **"MAN LIVES WHEN LAND LIVES"**.

Table 1: Recommended Plant Species For Afforestation.

Botanical Name	Common Name	Height (M)	Flower Color	Flowering Time
Acacia arabica	Babool	8-10	Yellow	Mar.-May
Aegle marmelos	Bel	8-10	Whitegreen	Mar.-Apl.
Albizia lebbek	Siris	10-15	Greenyellow	Apl.-May
Anogeossus latifolia	Dhaura	10-15	Greenish	Sep.-Jan.
Anacardium occidentale	Cashew	3-4	Cream	Jan.-Feb.
Anthocephalus Cadamba	Kadam	10-15	Yellow	Oct.-Dec.
Azadiracsta indica	Neem	12-20	Creamwhite	Mar.-Apl.
Cassia Javanica	Java Cassia	12-14	Deep Pink	May.-Jun.
Cassia Fisyula	Amaltas	9-13	Yellow	Mar.-May
Cassia Nodosa	Pinkmohar	12-14	Rosepink	Apl.-May
Cassia Siamea	Minjri	10-13	Yellow	Aug.-Nov.
Bauhinia variegata	Kanchan	7-8	Kanchan	Feb.-Apl.
Delonix regia	Gulmohar	8-12	Orange red	Apl.-Jun.
Eucalyptus hybrid	Blue gum tree	20-25	Cream	May-Jun.
Nyctanthes arbortristis	Siuli	2-3	Creamwhite	Oct.-Nov.
Largerstromea floreginae	Jarul	7-10	Liliae	May-Jun.
Magnifera indica	Amala	7-8	White	Feb-Mar.
Leucaena leucophloea	Subabul	8-10	White	Jul.-Aug.
Mimosops elengi	Bakul	4-5	Creamwhite	Nov.-Dec.
Nerium Odorum	Karabi	2-4	Redwhite	Feb.-Mar.
Peltophorum ferruginium	Radhachura	10-14	Goldenyellow	May.-Oct.
Pithecolobium dulce	Jungle Jelebi	8-10	White	Mar.-Apl.
Putranjiva roxburgsii	Indian Amulet	8-10	Foliage	Mar.-Apl.
Polyalthea longofolia	Devdaru	20-25	White	Mar.-Jun.

Table 2: Recommended Plant Species For Pasture & Avenue.

Plant Species	Flowering Season	Fruiting Season	Growth (Years)	Sowing Season	Planting
SLOW GROWING PLANTS					
Dalbergia sissoo	Mar., Apl.	Dec., Mar.	5	Nov.	Jun.
Acacia aculiformis	Apl., Jun.	Nov., Dec.	4	Nov.	Jul.
Cassia siamea	Nov.	Mar., Apl.	3-5	Dec.	Jun.
Casuarina sp.	Jun.	Sep.	5	Dec.	Jun.
Pongiamia sp.	Feb.,Mar.	May,Jun.	10-12	Nov.	Jul.
Basia latifolia	Mar.,Aug.	Jun.	10	May, Jun.	Jul.
Azardiracta indica	Mar.,Apl.	Jul.	10	Feb.	Jun.
Acacia mangium	Jun.	Apl.	5	Nov.	Jul.
Leicenia lucocephala	Dec.	Jan.	3-4	Nov.	Jun.
FAST GROWING PLANTS					
Bauhinia recemosa	Nov.	Apl.	3	Nov.	Jun.
Samania Saman	Dec.	Apl.	3	Nov.	Jun.
Peltophprum sp.	Mar.	Dec.	3	Nov.	Jul.
Eucalyptus sp.	Nov., May	Mar.	3	Nov.	Jun.
FLOWERING PLANTS					
Cassia Javanica	Nov.	Mar.,Apl.	3	Nov.	Jul.
C. Spectabilis	Dec.	Mar.	2	Nov.	Jun.
C. Fistula	Mar., Apl.	Jul., Aug.	3	Dec.	Jul.
Tecoma sp.	Apl.	Jun.	2	Nov.	Jun.
Delonix sp.	Apl.	May	3	Dec.	Jul.
Spathodea companulata	Apl	Aug	4	Nov	Jun
Millingtonia artensis	Apl.	Aug.	3	Nov.	Jul.
Cassia biflora	Nov.	Apl.	3	Dec.	Jun.
Calvellia recemosa	Mar.	Aug.	3	Nov.	Jul.

Water requirement : 25 lit. x 4 / month (summer)
25 lit. x 2 / month (winter)

Ecological Principles in Restoration of Derelict Mined Ecosystems

M.P. Singh and M. Mahto

Department of Forest Sciences, Birsa Agricultural University,
Kanke, Ranchi 834 006, Bihar, India

ABSTRACT

Mining drastically disturbs the physical, Chemical, biological and socioeconomical features of the area. Reclamation of mine disturbed lands is thus a challenging task. Natural restoration is an extremely slow process but at the same time is the only way for sustainable development of the disturbed sites, and here comes the role of ecological principles in restoration of disturbed ecosystems. Present paper describes the technologies developed for ecological restoration of coal mined lands and replicated in other mined areas.

Findings of studies at Kujju have shown that in these areas diversity index for grasses and herbs increased from 0-1.6 and trees and shrubs to 2.8 (Shanon-weener diversity index) as compared to 1.4 to 1.6 in the adjoining natural forests; organic carbon percent doubled form 0.1.16 and nitrogen percentage increased from .003 to .009 Weathering of rocky portions resulted in an increase of fine fragments (204 mesh) form 16% to 28%.

Besides favourable changes in the soil and plant components of system, the areas after 8 years of restoration were capable to produce a sustainable amount of fuel (14 to nnes/ha), fodder (24 tonnes/ha), litter (882 kg/ha), and that ching helped the local population to meet their day to day requirements as well as the restored systems in improving the nutrient and organic status.

INTRODUCTION

In India 5,510 mining leases are spread all over the country covering an area of 800,000 hectares. Out of these 516 mines are in the coal sector whereas 4994 mines are in non-coal sector for minerals like copper, gold, manganese, iron, zinc, limestone, dolomite and rock phosphate etc. Whereas all these minerals are valuable resources for the industrial development of the country and are constantly required to keep our industries running but at the same time mineral excavation leaves aside vast strethches of derelict lands which are completely devoid of fertile soil, flora and fauna, associated with deteriorated air and water quality. Mining and mineral processing adversely effects the ecology of the area by disturbing the land mass, the water systems, the plant and animal populations and these in turn effect the quality of human life (fig .1)

Restoration of these mined derelict lands is thus a challenging task. Mother nature builds up an inch thickness of soil in hundreds of years and then propagules of plants received through wind, water, birds and other insects growing thereby, a cover of vegetation but at the same time in nature growing on barren rocky area also which are depleted in moisture, nutrients and organic matter and this is where application of ecological basis becomes important. Although number of techniques have been tried extensively to revegetate such derelict/depleted lands all over the world (Davis, 1989; Chadwick and Goodman, 1973; Bradshaw and Chadwick, 1986;). However, only a limited number of quantitative reports are available where emphasis has been laid on plant community development parallel in manner to natural ecosystem development (Jones, 1979; Brenner, 1985, Soni & Vasistha, 1986 and Soni et al. 1989).

Restoration of mined areas is an ecological problem and it is infact essential to study the structure and function of the premining ecosystem before taking up the rehabilitation (Soni et al, 1992). Wali & Freeman (1973) and Fisser & Ries (1975) have stressed the importance of understanding species diversity, community structure, microenvironmental features, edaphic conditions and site specific soil-plant-animal interrelationships of a given ecogeographic region before the initiation of any mining activity and only these informations will enable for a reasonable forecasting of the potential as well as the expected productivity of mined areas. Therefore, in the present paper, role of application of basic principles of ecology in achieving effectively restored ecosystem has been stressed.

Case study of Kujju mined area

Till 1994 when the ecological restoration was taken up by the authors with the collaboration of Central Coal Limited the area directly disturbed by mining was 300 hectares. Top soil and the layer of silt coal was dumped on the sides of this explored area. This area was planted by forest Department with Melia azedarach seedlings in 1985.

Rest of the 300 hectares area contained overburden dumps, exposed areas containing coal and a water course in which overburden was dumped during mining. These mined areas were highly prone to erosion and responsible for high sediment concentration even in Damodar river which is a source of drinking water to parts of Kujju.

Main objectives of this ecological restoration project were to

-   To arrest the erosion of dumps and in turn sedimentation of ricer
-   To develop a diverse and self sustaining ecosystem
-   To develop an ecosystem which can cater to the social and economic requirements of local population besides ecological amelioration.

To achieve these objectives it was a prerequisite to have baseline information of the plant soil-animal relationship in the premined ecosystem. However, since no such records were available adjoining unmined areas were surveyed to select plant species that are :

- fast growing, and of primary colonizing nature

- can fix atmosphetic nitrogen and ameliorate the soil through leaf litter decomposition and enrich soil faunal population

- can attract birds, butterflies and other forms of wildlife

- are of social and economic value for local population.

Table 1 details the selected species, their utilization potential, mode of propagation and survival %. As in nature primary colonizers are mostly grass and shrub species and trees appear at a later stage, similarly a phased programme of re-establishment of vegetation was taken up.

Table 1:   **List of plant species selected for stablilization of mined area**

	Mode of propagation	Utility	Survival %
(A) Trees			
Acacia Catechu	Ds, Si, Sp	Fo, Fu. Ti. Fer	98
Leucaena leucocephala	Ds. Si	Fo,Fu,Fer	95
Dalbergia Sissoo	Ds, Si, Sp,Rs	Fo, Fu, Ti. Fer	90
Cassia nodosa	SI	Ti	90
Tectona grandis	Cu, Si	Fo, Fi, Ss	80
Pongamia pimota	Si	Ti	10
Melia azedarach	Si	Fo, Fu, Ti	25
Gmelina arborea	Cu, Si	Fo, Fu, Ed	1
Moringa pterygosperma	Cu. Si	Fo. Fu. Ed	Nil
(B) Shrubs			
Rumex hastatus	Ds	Sc	90
Wendlandia exserta	Ds	Fu, Fo, Fi	75-80
Ipomea carnea	Cu	Fu, Sc	95
Vitex negundo	Du	Sc, Fu	100
Agave sislana	Bb	Fi. Sc	60
Buddieja assiatica	Ds	Fu, Med, Sc	60-70
Mimosa himalayana	Ds	Fer. Fu	60-70
Eriophorum comosum	Ds, Tu	Sc	90
Crotolaria sericea	Ds	Ser	90
Trema politoria	Ds	Fu, Fo, Sc	60
(C) Grasses			
Stylosanthes hurnilis	Tu	Sc, Fi, Th	100
Pannisetum pedicellatum	Cu, Tu	Sc, Th	90
Tridex procumbens	Cu, Tu	Fo, Sc	100

51

*   Ds-Direct sowing; SI-Seedling; Sp-Stump planting; Cu- Cutting; Rs-Root sucker; Tu- Tussock; Bb-Bulbils.

**   Fo-Fodder; Fu-Fuel; Ti-Timber; Fer-Fertilizer; Fi- Fibre; Ss- Streambank Stabilization ; Med-Medicine ; Th-Thatching; Ed- Edible; Sc- Soil Conserving.

1.   In the first phase, seeds of pioneering shrubs, viz, Rumex hastatus, Mimosa himalayana, Wendlandia exserta, Trema politoria, Buddleja asiatica and Eriophorum Comosum were got collected from the adjoining forests and broad cast seeded over the area.

2.   Contour trenching at a vertical interval of 2 meter was done and planting of cuttings of vitex negundo, Ipomoea carnea, pennisetum purpureum (root stock), Agave sislana, Tectona grandis and tufts of Saccharum spontaneum were planted along the water courses to stabilize the water course.

3.   Small gullies were plugged with thick cuttings of Moringa pterygosperma and Lannea Coromandelica

4.   Within a year the area was covered with shrub-grass cover and became favourable for plantation of tree species. Since the grass and shrub cover gave partial stability to earlier loose strata, tree planting was done in 30x30x30cm pits in between trenches.

However, in chert areas no tree planting could be done and area was left with shrub-grass community only. Since the entire area falls in khair-sissoo zone of tropical dry mixed deciduous forest, large scale trial of these two species viz; Dalbergia Sisoo and Acacia Catechu was undertaken besides a number of other species.

Ecological impact Assessment

Ecological impact of restoration has been assessed by monitoring the vegetation, soil and soil fauna. Observations show that survival of local species like Rumex hastatus, saccharum spontaneum, Tectona grandis, Dalbergia sissoo and Acacia catechu besides some other species viz; Leuceana leucocephala and Pennisetum purpureum (Soni et al., 1991). Floristic composition shows that density of herbs and grasses increased significantly in the soils after reclamation as compared to shrubs. Data collected further supports the fact that raising of mixed forms (Grass shrubs-tree mixture) favoured the invasion of a number of natural invaders and hence the diversity index also increased form zero to 2.14 for grasses and herbs and to 1.4 trees and shrubs.

Periodic monitoring of soil characters indicate that percentage of soil 200 mesh is maximum after 5-6 years i.e., 27% and the amount of leaf litter added to favour nutrient cycling is to the tune of 775 kg/ha resulted in improved soil faunal occurrence of different classes of fauna as well as nitrogen and organic matter status of substrate (Table 2 & 3).

Table 2 : **Physical properties of mined areas and natural forest**

Site	%Soil 200 mesh	Sand %	Silt%	Clay %	Textural classes
Unreclaimed dumps	15.73	52.0	21.1	25.4	Sandy clay loam
Reclaimed areas (10 yrs)	27.30	66.0	19.3	13.4	Sandy loam
Natural forest (unmined)	17.60	49.1	28.6	22.16	Sandy clay loam

Table 3: Litter production and chemical analysis of mined areas and natural forest

Site	Litter production kg-ha	pH	% Nitrogen	% Organic carbon
Unreclaimed dumps		7.08	.0.26	.052
Reclaimed areas (10 Yrs)	775	8.0	0.081	0.76
Natural forest (unmined)	443	7.0	0.165	1.10

Monitoring of socio-economic returns form these areas reveal that areas are capable of producing.

-   22 tons of fodder (mainly leaf fodder)
-   12 tons of fuel wood
-   thatching grass for local people

Detail socio-economic studies as well as total biomass and nutrient studies are in progress.

After developing the basic technology for surface mined phosphate areas, following the same ecological principles, another area of abandoned coal mines on Rajrappa Coal Mines has been taken up with the active collaboration Central Coal Limited. Monitoring of vegetation after three years of restoration shows that besides the raised species viz ; Rumex hastatus, Agave sislana, Eriophorum comosum, Dalbergia sissoo and Cassia nodosa more than 20 species of shrubs, grasses and trees have started invading these areas, thus initiating

53

the process of natural succession which is the first favourable process leading to sustainable development of the area.

It is therefore concluded that while approaching restoration of derelict mined lands principles of ecology need to be taken into account for achieving the goal effectively. This will not only ensure speedy recovery from dereliction but also in ecologically and socioeconomically sustainable way.

References :

Bradshaw, A.D. and Chadwick, J. (1986). *The Restoration of Land: The Ecology and Reclamation of Derelict and Degraded land,* Oxford, Blackwell Sci. Pub. pp. 317.

Brenner, F.J. (1985). Land Reclamation after strip coal mining in the United states. *Mining Mag.* Sept. 211-216.

Chadwick, M.J. and Goodman, G.T. (Eds.) (1975). *The Ecology of Resource Degradation and Renewal,* John Wiley & Sons, New York, U.S.A.

Fisser, H.G. and Ries, R.E. (1975). Predisturbance Ecological studies improve and define potential for surface mine reclamation. In *3rd Symp. Surface Mine Reclamation.* Vol. I National Coal *Association / Bituminous Coal Research Conference & Expo II,* Liwsville K.Y. : 128-139

Hutnik, R.J. and Davis, G. (Ed.) (1973). *Ecology and reclamation of Devasted Land.* Vol. I & II Gordon & Breach New York.

Jones, H. (1979). *Iron ore Mining effects and rehabilitation in Management of Land Effected by Mining* (Eds. R.A. Rummery & K. M. W. Howes). Perth, C.S.I.R.O., 15-158.

Soni, P. and Vasistha, H.B. (1986). Reclamation of Mine Spoils for Environmental amelioration *India Forester* 42 (7) : 621-623.

Soni, P., Vasistha, H.B. and Kumar Om (1987). Ecological Approach Towards Reclaiming Mined Ecosystem. *Indian Forester, Vol (512) : 875-883.*

Soni, P., Vasistha, H.B. and kumar Om (1992). Surface Mined Lands-problem and prospects (A report on Eco-restoration of phosphate Mined Areas) I.C.F.R.E.-18, New Forest, Dehra Dun.

Thomas, J.I. (Eds) (1977). Reclamation and use of disturbed lands in the south West Tueson, Arizona, Univ. Of Arizona press.

Wall, M.K. and Freeman, P.G. (1973). Ecology of some surface Mined Areas in North Dakota (Ed. M.K. Wali). *North Dakota Geological Survey Education Series 5,* Grand Forks, H.D. pp. 27-45.

Coastal Geological Alterations: A Case Study

Rekha Thakre* and B.K. Sarangi**

* Scientist & Head,** Scientist
Land Environment Management Division, National Environmental Engineering
Research Institute, Nagpur 440 020, India

INTRODUCTION:

The coastline is a dynamic environment, which gives rise to specific natural hazards that may affect development. Coastal hazards can be divided into those that are essentially "quiet" and progressive, such as coastal erosion and those which are " dramatic" events such as coastal flooding (1).

In this paper a case study of Gujarat coastline has been high lightened which stretches to about 1600 km. It is quite distinct from the rest of the West Coast of India. The coastal zone of Gujarat is characterized by a variety of geomorphic forms and geological features evolved under different structural controls during the Quaternary period (2). The present coastline shows much variation in its trend, shoreline features and the near shore and offshore characters.

The paper elaborates the problems of coastal erosion, and salinity ingress in the coastal part, mechanism of seawater intrusion and mitigation strategy for the protection of coastal seawater intrusion.

STUDY AREA:

The coastline of Gujarat, facing the Arabian Sea, is quite distinctive from the rest of the West Coast of India. The coastal tract has a varying width about 5 80 kms, (3). It covers about 27,000 sq.km. Stretch sharing about 14% area of the entire Gujarat State. Its onshore boundary extends approximately up to 20 m. Contour line and the area below this level shows strong influence of present sea. Its geomorphic evolution to a great extent has also taken place dominantly under the marine conditions of Pliestocene (1.5 MY) and Holocene (10 KY) sea.

The ecological characteristics of Gujarat coast line are not constant all along the coast **(Table 1).** The coastal zone can broadly be divided into three major geographical parts viz. Katchch, Saurashtra and Main land Gujarat, each one characterized by its own distinctive characteristics and diverse geomorphological features **(4)**. Saurashtra coast is divisible into three segments - (i) the northern coast of Navalakhi-Dwarka (ii) southern coast of Dwarka Division and (iii) south southern coast of Division Bhavnagar. Main land Gujarat Coast is divisible in two segments (i) Bhavnagar to Bharuch flanking the Gulf of Khambhat (ii) Bharuch to Umbergaon in South Gujarat on Arabian sea coast.

Coastal Landforms

Coastal Rocky Plain: The study of coastal area has revealed the occurrence of wasteland which is by and large, comprising outcrops of limestone exposed intermittently all along the coast. These outcrops of miliolitic limestone form subdued topography and are most prominent landforms of the coastal area.

Table 1. Geomorphologic Features of Gujarat Coast Line

Parameter	Kachch	Saurashtra			Main Land Gujarat	
		Navalakhi - Dwarka Segment	Bhavnagar to Bharuch segment	Hansot - Umbergaon Segment	Dwarka - Div Segment	Div-Bhavnagar Segment
Area (Km) Stretch	300	250	250	200 width - 5-20 km	300	250, width 20-25 km.
Climate	Arid	Semi arid	Semiarid to subhumid	Sub humid to humid Aridity index 20% to 10%	Semi arid	Semiarid
Rainfall (mm)	250-400	350-400	600-800	800 - 1800	350-550	500-600
Soil Type	Sandy & Silty	Halaquept, & Ustocrepts	Haloquept - Salorthid, Ustocrept & Chromustat.	Chromusterts and Ustocrepts	Halaquept, Salorthid & Ustocrept	Halaquept & Salorthid
Status	Narrow beaches	rocky platforms with narrow beaches	low level saline flat lands, drowned alluvial coast characterized by step diffy river mouths.	Highly indented and sandy at estuarine mouth of Narmada, Tapi,		rocky coastline with small narrow non-calcareous sandy beaches.
Cinfiguration	Irregular & much dissicted muddy with extensive tidal flats	landforms like channels, shoals, submerged islands bars. Hard trappean tertiary sediments.	Highly intented characterised by mudflats and alluvial cliff, paleomudflats and alluvial plains.	Variety of landforms of alluvial plain, midflats(prominent estuarine river mouths).	Smooth and straight well developed sandy beaches.	Indented, mudflats, Cliffs, rocky platforms, beaches and dune ridges
Littoral zone	<2Km,.. Swampy forests, Mangroves.	variable with ranging from 5-10 km,	3-5 Km silty and muddy sediments highly turbid and hyposaline	2-8 km. formed of sandy and silty deposits. Water turbid and hyposaline to saline..	0.5-1.5, growth gradient, water clean and saline	1.5 -2.5, low gradient formed off temigenous sturdy shingle and mud. Water saline to hyposaline and turbid

56

Two sets of current bedding could be seen in this limestone. **(5)** At places, on account of physico-chemical weathering, solution cavities have formed because of which the surface has become rough and corrugated. The miliolitic limestone exposures are distinctly confined to coast wherever it is dissected. The geological mapping has revealed that the dissection of the coast is on account of faulting (6). At places e.g. near Jagri Island, Gopnath etc. folding in miliolitic limestone is observed. The structural disturbances noticed in these areas appear to be due to the re-activation of the faults during the Pleistocene period.

b. Sand dunes: Though the unconsolidated sand dunes have not been observed in the area but study has clearly revealed the occurrences of fossilized aeolian sand dunes comprising miliolitic limestone. Typical linear and Parabolic dunes comprising tests of milioli-fossils and occurring as consolidated to semi consolidated deposits could be observed off the coast of Una, Kodinar, Veraval etc. The linear dunes as well as the axes of parabolic dune are oriented in NE-SW direction. This clearly points to the prevalence of southwesterly wind direction during the pleistocene period. Further, these aeolian deposits of miliolitic limestone are found to be isolated from the main miliolitic limestone deposit noticed on the coast.

Alterations in Geo-morphological features in Gujarat Coastline

The Gujarat Coast line is experiencing the alterations due to various natural as well as anthropogenic consequences (**Figure 1**). The causes and effects are illustrated for Gulf of Kachchh and Gulf of Khambhat as well as the changes in geo- morphological patterns is highlighted in **Table 2.**

Table 2. CHANGES OBSERVED IN GEO-MORPHOLOGICAL FEATURES OF THE GUJARAT COAST LINE

Sr. No.	Area	Alterations/Implications	Comments
1.	GULF OF KACHCH	* Silting up * Migration and joining of different creeks * Re-orientation of tidal current * Ridges and regression of sea * Variation in tidal range * High sedimentation	* Tectonically activated process * Anthropogenic activities
2.	GULF OF KHAMBHAT	* Shoreline changes due to erosional and depositional process * Coastal degradation * Discharge of large volume of sediments from river drainage **Changes in land forms** : * Shoreline * Estuaries * Mudflats * Islands * Cliffs * Flood-plains * Creeks * Dunes * Mangroves	* Anthropogenic activities * Rapid industrialization * Soil erosion * High tidal range * Construction of dams * Destruction of mangroves

Tides, waves and currents provide energy that is constantly working to change landforms such as shoreline, estuaries, mudflats, islands, mangroves, relict alluvium, cliffs, dunes, flood-plains, paleochannels, paleomeanders, oxbow lakes etc. Shoreline around the Gulf of Khambhat is inundated by estuaries and creeks, which are characterized by islands and extreme mudflats and absence of sandy beaches (7). All these features of landforms have evolved due to high tidal range in this region.

Shoreline changes in the Mahi and the Narmada estuaries are significant (8). The erosional processes are predominant in the Mahi estuary while depositional activities are dominant in the Narmada estuary. It is observed that the maximum deposition occurs at the mouth and the gradual expansion of Aliabet Island is an indication of the same. As a result, the channel south of the island has been blocked and filled up consequently.

Salinity ingress

The pivotal and intense ecological problem in Gujarat Coast line is intrusion of salt water in inland ground water table **(9).**

The Mechanism of Sea water Intrusion:

In their natural state, coastal aquifers contain a characteristic vertical distribution of both fresh and saline groundwater. With the volume of recharge from rainfall exceeding any abstraction by pumping, a hydraulic 'gradient' should be established towards the coast and the excess recharge will seep out to sea. **Figure 2 (A).** However, because of the different densities of fresh groundwater and seawater, the system tends toward a hydrostatic equilibrium, with a wedge of saline water exceeding inland from the coast deep into the aquifer (Figure 2 (B).). The depth of the interface between the two types of water at any point is proportional to the elevation of the water table above sea level at that point. If onshore extraction lowers the water table, the interface rises proportionately nearer the surface and the wedge of saline water protrudes further inland. As the water table declines to sea level the entire aquifer fills with seawater (10).

Ground Water Salinity Status in Gujarat Coast Line

The area has three different types of salinity-

i)   Inherent or the geological formation salinity
ii)   Sea water ingress
iii)   Coastal inundation

The causes of salinity can be attributed to meteorological characteristics, geological formations and man made activities. Scanty and irregular rainfall, quick run off and less recharge, high tides with high intensity and force, winds blowing from Rann of Kachch bring salt and silty particles adding to salinity problem. The cavernous and fractured rocks in contact with coastline help easy ingress. Inherent salinity caused due to deposition of Gaj clays and sandy and gravelly formations around Bhavnagar and Jodiya areas. As if not sufficient, the population pressure forces the excess withdrawal of water causing imbalance in hydro static pressures of fresh and saline water interface. Gujarat Government constituted High level committees in 1972, 1976, 1978 and 1984 and studied the salinity ingress problem with the help of various agencies viz. Kharland Development Board, Satellite Application Center, National Bureau of Soil Survey and Land Use Planning, National Remote Sensing Agencies, Central Saline Soil Research Institute and Gujarat Ecological Research Institute from time to time

Salinity ingress levels are ascertained by monitoring the total dissolved solids (TDS) of the ground water. This is because, the sea water characteristic of high salineness. High salinity in ground water is observed in Kachch, Dwarka to Bhavnagar , Dwarka to Madhabpur and Madhabpur to Malia segment (Table 3). It is observed that the salinity has increased throughout the coastline very significantly from 1960 to 1990 especially from 1975 onwards. (Figure 3).

Control of Salinity Ingress

The salinity ingress can be checked to minimum by -
*   Management techniques such as change in cropping patterns, regulating of ground water extraction
*   Recharge techniques by constructing check dams, recharge tanks, wells, spreading channels, afforestation and nala plugging
*   Salinity control techniques - tidal regulator, bandhara, fresh water barrier, extraction barrier, and static barrier and restriction in excessive with drawal of water
*   Reclamation of coastal lands

58

Salinity control schemes can survive only if they are continuously recharged by rain water and regular and continuous filling of structures and strict enforcement of ground water regulation can only save the area (11).

Table 3. Salinity Ingress: Level of TDS in the Gujarat Coast

Sr. No.	Area	Taluk	TDS Level in ppm
1.	KACHCHH		upto 1,02,000
		Lakhpat	660-1,00,000
		Narayan Saravor	1,02,000
		Abdasa	1180-7600
		Mundra	2000-5000
		Bachau, Mandvi	2130-15,000
2.	DWARKA – BHAVNAGAR SEGMENT	Bhavnagar	3000-5000
		Ghogha	
		Talaja	upto 11,910
		Mahuva	1880 – 4660
		Una	2100 – 4110
		Jafrabad	
		Rajula	
3.	DWARKA TO MADHABPUR	Okha mandal	1000-6000
		Gaj beds	
4.	MADHABPUR TO MALIA	Porbandar	1000-6000
		Bhadar	
		Ozat	
		Gosa	
		Bara	
		Harshad	

ACKNOWLEDGEMENT

Authors are grateful to Dr.R.N.Singh, Director National Environmental Engineering Research Institute, Nagpur for permission to present the paper

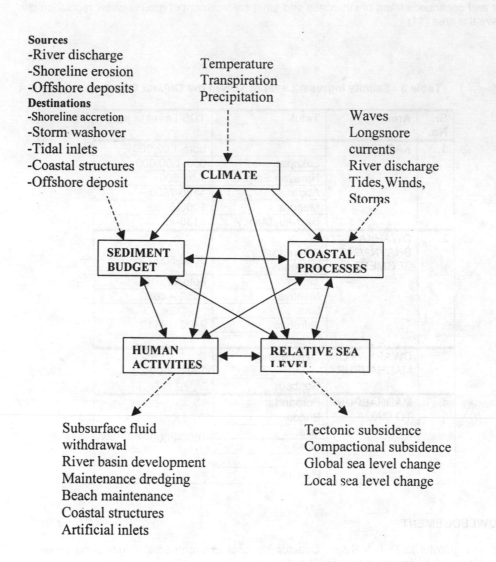

Sources
-River discharge
-Shoreline erosion
-Offshore deposits
Destinations
-Shoreline accretion
-Storm washover
-Tidal inlets
-Coastal structures
-Offshore deposit

Temperature
Transpiration
Precipitation

Waves
Longsnore
currents
River discharge
Tides,Winds,
Storms

CLIMATE

SEDIMENT BUDGET

COASTAL PROCESSES

HUMAN ACTIVITIES

RELATIVE SEA LEVEL

Subsurface fluid withdrawal
River basin development
Maintenance dredging
Beach maintenance
Coastal structures
Artificial inlets

Tectonic subsidence
Compactional subsidence
Global sea level change
Local sea level change

**Figure 1. Interactive Factors Affecting Developmental
Planning in Coastal Zones**

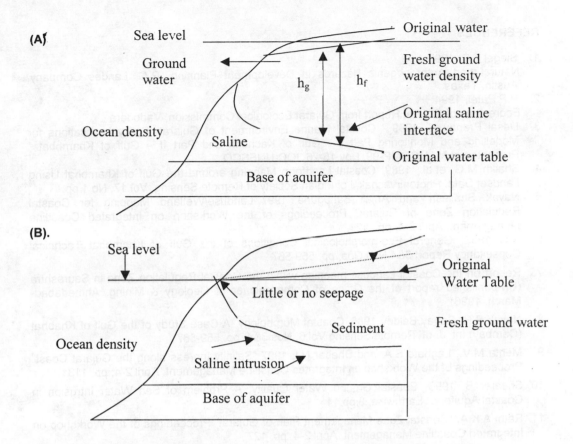

Figure 2. Principe of Saline Intrusion into a Coastal Aquifer (A). Before Pumping (B). After Pumping

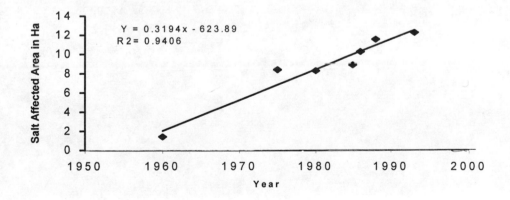

Figure 3. Trends of Salinity Ingress in Gujarat Coast Line 1960-1995

REFERENCES

1. Siegel F.R. 1996
   Natural and Anthropogenic Hazards in Development Planning. R.G. Landes Company, Austin, Texas

2. P.P.Patel, 1994
   Ecoregions of Gujarat Report from Gujarat Ecological Commission, Vadodara

3. Desai Pranav N. 1999 " Coastal Marine Environment of Gujarat. Policy Implications for Modelling and Monitoring Part I – Gulf of Kachchh and Part II – Gulf of Khambhat " Presented at MAMCOMP-99, Nov.15-26, IOC-UNESCO.

4. Shaikh M.G. et.al., 1989 "Coastal Landform Mapping around the Gulf of Khambhat Using Landsat Data. Photonirvachak J of Indian Society of Remote Sensing, Vol.17, No.1 pp.

5. Nayak Shailesh and Anjali Bahuguna 1997."Landuse/Wetland Mapping for Coastal Regulation Zone of Gujarat" Proceedings of the Workshop on Integrated Coastline Managament. April 2-4, pp. 121.

6. NIO 1992. General Eco-morphological Conditions of the Gulf of Khambhat Technical Consultancy Report 3&4 NIO,Goa, pp. 559-567

7. Report on the Coastal Landuse mapping within the Coastal Regulation Zone in Saurashtra (Un-published report of the Office of Commissioner of Geology & Mining, Ahmedabad-March, 1996).

8. Nayak S.R. Sahay Baldev 1985."Coastal Morphology :A Case Study of the Gulf of Khabhat (Cambay) Int. J. of Remote Sensing Vol.6, Nos.3&4, pp. 559-567

9. Mehta M.V., Laghate S.A. and Dhebar N.I. 1997 "Salinity Ingress along the Gujarat Coast" Proceedings of the Workshop on Integrated Coastline Management. April 2-4, pp. 113.

10. Shearer B. 1996 "Coastal Ground Water Quality. A Problem of Sea Water Intrusion in Coastal Aquifers." Earthwise, 9,pp.11

11. Rathi A.K.A. "Coastal Zone Management Plan of Gujarat" Proceedings of the Workshop on Integrated Coastline Management. April 2-4, pp. 127.

Environmental Impact of Mining Operations—Problems and Remedial Measures

R.S. Varshney* and I.B. Chhibber**

*Chairman, **Chief Consultant
Shubhani Engineering Consultants (P) Ltd., India

ABSTRACT

Minerals are the national wealth of the country. India is one of the major mineral producing countries of the world. Indian Bureau of Mines, IBM, is the national organization to control the mining development and law enforcing authority, preparation of mine plans etc. However the extraction of the minerals when not properly cared for has posed various environmental hazards. MOIL, COAL INDIA LTD, NMDC, Zinc Corporation, Copper Corporation, ONGC and States Public Health Deptts have conducted some selective environmental impact assessment studies. These studies have pointed towards environmental hazards. However a large portion of the mining industry is still not taking the remedial measures.

The industrial development is based on mineral industry. India exports minerals to the tune of Rs 3200 crores every year. Mines are source of raw materials for industry. However, mining waste pollutes streams and rivers. Environmental impact of mining depends upon the process of extraction, which if not properly done, can affect adversely the land, water, forests, air as well as agricultural land.

Environmental impacts of coal, iron and limestone mining operations have been highlighted and remedial measures have been suggested to control the ill effects.

Solutions are outlined like establishment of Pollution Alert Centers to monitor the adverse impact of mining and give warnings in time. Involvement of local people in the mine area in controlling the adverse damage done by the mining is desirable and essential to mitigate and to some extent eliminate the possible damages.

THE MINING SCENARIO

Mining is second only to agriculture as the world's oldest and most important industry. India is one of the richest and largest minerals producing country in the world. It has a large number of mines both open cast and underground. As per estimate of 1997-98, there are 2963 mines, out of which about 800 produce metallic minerals and the remaining nonmetallic minerals. Amongst the metallic mineral mines iron ore mines are the largest, followed by bauxite with 174 and manganese ore 150. There are 2168 nonmetallic mineral producing mines of which 576 mines produce limestone and other calcareous material for cement industry, followed by silica sand with 199 mines, steatite 174, kaolin 166, dolomite 145, fire clay 121, quartz 99, mica crude 158, feldspar 57, barites 20 etc.

India's ranking in world's mineral production and exploitation in 1996-97 was; coal production 5th, iron ore 5th, chromite 2nd; bauxite 6th etc. Indian Bureau of Mines, IBM, has updated the National Mineral Inventory, NMI, as on 1.04.2000. The NMI covers about 6000 free hold deposits and 7000 leases mines pertaining to 61 minerals.

To discuss the environmental impact of all the mining operations would encompass a very large space;

only coal, iron ore and limestone mining have been discussed herein.

2.0 ENVIRONMENTAL IMPACT OF MINING

Most of the mining activity in India is being carried out in forested regions. The obvious result is the deforestation and erosion. Underground mining also denudes significantly the forests because timber is used for supporting mine galleries. In Goa, mining leases are spread over 43 percent of the forest area. Because the mining leases do not bind the mine owners to undertake any soil conservation measures or refill the worked over mines, a large number of abandoned mines are lying in bad shape and are under extensive gully erosion. Similar is the story in the other states.

Though EIA studies are being carried out by different mining authorities and NGO's and extensive data have been collected, but the remedial measures adopted are very few. The major minerals, which are being exploited on a large scale, are:
Coal – Fuel
Iron ore, Manganese – Metallic, and
Limestone and calcareous minerals – Non-metallic

2.1 COAL MINING

The demand for coal for the growth of industries as a main source of energy accounts for about 67 percent of the country's commercial requirements. As on January 1998, the coal resources of India have been estimated by GSI at 206240 million tonnes and lignite (low-grade coal) around 27500 million tonnes. The major coal bearing horizons are located in Bihar and other states. The main industries consuming coal are thermal power stations, steel, cement, railways, fertilizers, brick kilns and other small-scale industries for making tools.

The problem of pollution starts with the coal itself. Open cast coal mines produce large clouds of dust following blasting operations, while removing large quantities of over burden. The thermal and super thermal power plants located in the various parts of the country especially along the coal-belts are the cause of the major pollution hazards. Indian coal has a low sulphur content of 0.35 percent but ash content is around 30 to 50 percent. Electrostatic preceptors now installed on Korba thermal power house in Madhya Pradesh have reduced air pollution because of a high ash content (of the order of 50 percent). This pollution was not foreseen in the beginning.

If the pollution control measures are not taken by the thermal power plants, people should expect acid rains in next 10-15 years time. According to the studies conducted in June 1984, 31 thermal power plants out of 46 had not taken any pollution control measures and the percentage of proper functioning units was only 35 percent.

Fly ash can be put to proper use. It can be used in brick industry and ash dykes can be constructed as done by NTPC and Korba MPEB. Fly ash is usually collected in the dry form to make heaps or mounds or is carried as slurry into ponds and rivers, where it creates serious water pollution problems. It also contains toxic metals like zinc, barium, copper, arsenic, manganese etc which can leach into the groundwater. Chennai based Research Centre has rightly observed that thermal power pollutes and super thermal power pollutes super thermally. Pollution control in the super power plants is the super necessity of the hour and should be strictly enforced.

2.2 IRON ORE INDUSTRY

Iron ore mines are the largest in number, about 226. They are mainly located in Bihar, MP, Orrisa and Goa. Iron ore occurs as high grade, medium grade and low grade deposits. Bailadilla iron ore project in Bastar district in M.P has posed pollution problems. The iron ore fine from the screening plant in Kirindul have been dumped into Sankhini river turning it into a mass of red slime, and has deprived 51 villages of clean water for drinking purposes, for the tribals.

This iron ore project was undertaken to export iron ore to Japan. In this area of 48 km reach of Bailadilla range, development is another destroyer of the environment due to mining of the high-grade ore to earn foreign exchange. Methods should have been taken to restore, after reclaiming, the mine water; to revegetate the area for making it worth living for the local people. This should have been the responsibility of the exploring agency.

2.3 LIME STONE QUARRYING

It is non-metallic mineral, which finds maximum use in cement industry, which are scattered in UP, MP, AP, Bihar, Gujrat etc. The mineral resource is estimated of the order of 76500 million tonnes.

Uncontrolled quarrying for limestone is a pollution danger in areas where the cement industries are located; Katni and Maihar are the worst polluted areas. Doon valley quarries provide 3 percent of nation's limestone. The quarrying in Doon valley in UP has degraded the area. Mussoorie is being stripped bare of its verdant cover. The area under the tree cover has come down to 12% as against the officially recommended 60%. Removal of the limestone has also adversely affected the water resource in the streams and rivers of the Doon valley because the limestone belt acted as a mechanism for capturing, retaining and releasing water perennially into the streams.

It is necessary to make proper studies to minimize and control dust from the quarries and cement plants and gases from the plants. As an example, a major environmental disaster has been created for over one lakh people and thousands of animals living in limestone rich region of Krishna district of AP, which is punctured with a dozen and half cement factories, emitting dangerous cement dust into the atmosphere. The problem became more acute near the Krishna district where drinking water too is highly polluted. There has been a spurt in the no. of cases of respiratory and skin ailments in scores of villages on the Krishna-Nalgonda border. According to one estimate over 100 tonnes of cement dust is released into the atmosphere daily.

2.4 WATER USED IN MINING OPERATIONS

Water is a resource, which can be placed at a risk by the activities of the mineral industries. Water is used in many stages of mining and mineral processing. Mining pollutants in water can affect men as well as organisms and animals. Pollution control of water in mines is a major task.

Base line surveys should identify all waters, which may be risk probe from the mines. Water pollution can be reduced by appropriate methods of controlled mining. The physical control of volumes and routes of water at a mine is frequently a major task. It is required to enable the volume of water used in mining to be minimized, to prevent contamination of unpolluted water and to intercept polluted water and to divert it to the appropriate treatment facilities. Preplanning controlled mining techniques and proper design of contour mining method can help.

3.0 REMEDIAL MEASURES TO CONTROL POLLUTION IN MINING
3.1 SETTING UP OF POLLUTION ALERT CENTRES

In order to minimize the adverse effects on environment by mining operations, establishment of Pollution-Alert-Centres are necessary near major mining areas to monitor regularly pollution in air, water and land; give warnings and take remedial measures.

3.2 IMPOSITION OF CONSTRAINTS IN MINING OPERATIONS

Constraints and regulations designed to reduce adverse effects like polluted effluent, dust, noise, vibrations etc should be implemented strictly.

Stress should be laid on soil conservation measures in the mining area, landscaping, screening and restoration works like refilling of abandoned mines. Dust from blasting or water leaching of waste dumps are important sources of pollution, often-difficult to control. Methods should be evolved and implemented to tackle the problem.

3.3 CONTROLS ON THE DELETERIOUS CONSTITUENTS IN MINERAL   WASTE

It should be cautioned that the environmental problems associated with mining would grow at a faster pace than the output of minerals as population is increasing. Study of the deleterious constituents of various mineral wastes should be made by the mining industry leaders to mitigate the dangers of pollution.

Ambient air quality in respect of the fumes and also the noise should be controlled in the industrial, commercial and residential areas around mines.

3.4 INVOLVEMENT OF LOCAL PEOPLE

Involvement of the local people in the mine area in controlling the adverse damages done by the mining operations is necessary to mitigate and to some extent eliminate the possible damages.

4.0 RECAPITULATION

The mining industry is largely responsible for air, water and land pollution, where different minerals are extracted. Different minerals have different pollution problems and therefore should be tackled at their source. Strict measures should be taken against defaulting mining operations.

People's participation and public awareness will go a long way to ease the pollution problem.

Environmental Issues and Management in Coalfields of India

B.K. Samanta

General Manager (Env.),
Eastern Coalfields Limited, Sanctoria P.O. Disergash
Dist. Burdwan 713 333, India

1. INTRODUCTION

Among coalfields in India, Raniganj Coalfields is considered as birth place of coal mining in India. First Coal Company, named Summer and Heatley was started in 1774, in Sitarampur near Asansol. Next was M/s. Alexander & Co. and later it was taken over by M/s. Carr & Tagore, Calcutta. After 2 centuries of coal mining, in January'73, Govt. of India nationalised all coal mines. As a consequence 438 mines were taken over, out of which 274 were[1] operational at that time under Eastern Division of Coal Mines Authority. Renamed as Eastern Coalfields Limited, as subsidiary of Cola India Ltd. in 1975, it has now 106 underground mines and 25 opencast mines.

Environmental impacts [2]from mining are :- a) Land degradation, b) air pollution, c) landscape change, d) water pollution, e) change of natural bahitat, f) lowering of water table and g) abandon- ment of plants and buildings. In February'90, Coal India Limited, had approved creation of Environmental Management Organisation at subsidiary level, for implementation & monitoring of Environmental Management Plan. Resettlement and Rehabilitation Policy of Coal India[3] was passed by CIL Board on 4-4-94, for Project Affected People (PAP), important provisions of which are- a) compensation for cultivable land and homestead land, b) job for land acquired above 2 acres apart from monetary compensation as per prevalent legal norms, c) equivalent land/homestead @ 100 m^2 per family, c) help self-employment schemes, d) subsistence allowance for income below Rs. 1200 p.a. etc.

In March'97, CIL has passed cadre scheme of Environment Organisation, in which separate Environment Cadre has been created with office in CIL and subsidiary companies headed by CGM/ GM (Env), Area Manager(EPC&D) in each area and Environment Officers at the area level, as well as at Agent level have been authorised.

Due to mining over last two centuries, about 5112 ha. of land has subsided, but prior to 1973 there was no reclamation worth mentioning. Afforestation [4] in an organised manner started in the 80s on degraded lands, of which about 3130 ha. has been reclaimed by biological reclamation of OB dumps, filling up of
subsided area and with pisciculture/aquaculture in unused quarries. Progressive afforestation since 1985 done, by planting saplings of 108 lakhs, over area covered-3648 Ha, costing Rs.308.45 lacs. The major species covered by these afforestations are- Acacia (Akashmoni), Dalbergia Roxb (Sisso), Cassia (Bandarlathi), Leucamia Unocephala (Subabul), Faral, Polyanthia longifolia (Devdar), Babul, Palash, Eucalyptus, Teak, Alistonia scholaris (Chhatim), Terminalia arjuna (Arjun) etc. Two eco-parks have been done in Mahabir and Ratibati and one more is programmed to be made in Mohanpur, Salanpur Area.

2. INTERNATIONAL DEVELOPMENTS

The UNFCCC [2] (United Nations Framework Convention on Climate Change) was signed at the Earth

Summit in Rio de Janeiro in '94 and by '96, 164 nations signed. Periodic Conference of the Parties (COP) were envisaged and (COP-1) was held in Berlin in '95; followed by (COP-2) in Geneva, July, 1996. (COP-3) in Kyoto December 1997 adopted the Kyoto Protocol, establishing a legally binding obligation on Annex I countries, who ratify the Protocol to reduce emissions of greenhouse gases

(GHGs- CO_2, CH_4, N_2O, HFCs, PFCs, SF_6). Clean Development Mechanism (CDM) was for the certified emissions reduction.

The Buenos Aires Plan of Action was adopted at COP-4 (November 1998), with the aim of completing the programme of work on the Kyoto Mechanisms by the end of 2000. Working Group III to the Second Assessment Report of the IPCC (Intergovernmental Panel on Climate Change) reported "A global perspective is necessary, if we are to identify economically efficient strategies for achieving emission targets. The UNFCCC states that `policies and measures to deal with climate change should be cost-effective, so as to ensure global benefits at the lowest possible cost'. Emission reductions should be carried out, where it is cheapest to do so."

Opportunities for a range of efficiency gains exist across the full cycle of coal utilisation, from coal burned in open hearths for cooking and heating through to the use of coal-fired electricity, to run energy-efficient appliance. In electricity generation, technologies such as supercritical/ ultrasuper- critical steam utilisation, fluidised bed combustion (FBC), pressurised bed combustion (PFBC), and integrated coal gastification combined cycle (IGCC) offer efficiency improvements and emissions reductions.

IPCC - Panel was established by the World Meteorological Organisation (WMO) and the United Nations Environment Programme (UNEP). IPCC Working Group I (WG I) - Assesses the available information on the science of climate change. IPCC Working Group II (WG II) - Assesses the impacts of climate change on physical and ecological systems, human health and socio-economic sectors. IPCC Working Group III (WG III)- Conducts technical assessments of impacts adaptation and mitigation of climate change over both the short and long-term at regional and global levels.

Global Environment Facility (GEF)- is a Joint programme of World Bank, United Nations Development Programme (UNDP) and UNEP. GEF is a clearing house for funding programmes that support biodiversity and reductions in GHG emissions, in developing countries.

3.ENVIRONMENT MANAGEMENT PLANS

Once mining plan is decided on Geological Survey of India data, for starting a mine, at first detailed geological report is prepared with core drilling on a grid, on the basis of which Feasibility or Project Report is prepared. EMP has to be made with specific studies, assessment of environmental impacts needs to be made as EIA and management of the impacts is made as EMP (Environment Management Plan) for all industries including mining [4] approval of MoEF, by Government, so that final sanction of the Project Reports can be approved. It is prepared to guard against various environmental [9]hazards, which are likely to be faced in an industry and a classified summary is given below.

ENVIRONMENTAL HAZARDS

BIOLOGICAL | CHEMICAL | PHYSICAL | PSYCHOLOGICAL | SOCIOLOGICAL

BIOLOGICAL	CHEMICAL	PHYSICAL	PSYCHOLOGICAL	SOCIOLOGICAL
Animal	Poisons	Vibration	Stress	Overcrowding
Insect	Toxins	Radiation	Boredom	Isolation
Microbes	Irritants	Abrasion	Discomfort	Anomie
	Humidity	Depression		

It is now mandatory that before any project report is approved, EMP has to be passed by the Ministry of Environment & Forests. The standard EMP as per MoEF notification of 27-5-94, under Environment Protection Act '86, comprises of A. Summarised Data, B. Executive Summary, C.Detail EMP. Topics to be covered are:- 1) Introduction & Location, 2) Technology adopted, 3) Environment Scenario, 4)Environment Impact Assessment (EIA), 5) Environment Control Measures, 6) Environment Monitoring & Management, 7) Economics and 8) Questionnaire.

Assessment of the Environment Impacts is the basis, on which management of the environment impacts [8] can be made, combining different environment properties, in a single criteria approach. In every EMP there is proposal for creation of green [10] belts. The EMPs prepared by engaging experts are submitted to the Ministry of Environment & Forests, who examine it and approves imposing certain conditions, which are required to be monitored on half-yearly basis. As many as 22 EMPs have been made in ECL and approved by the MoEF. One of the conditions of the EMPs is quarterly monitoring of air, water, noise pollution and CMPDI, RI-1, Asansol is analysing samples for the 15 projects of ECL.

4. AIR POLLUTION CONTROL

Major steps for air pollution control, which are being taken in ECL:- a) Paving of colliery roads, coal transport roads and haul roads in quarries; b) Public roads passing through coalfields are followed up regularly with the State Govt. for maintaining, so that air borne dust in control; c) Every year ECL is programming for paving of roads for the new mines, as well as old ones; d) Regular water tanker is being used for water spraying on the unpaved haul roads and coal transport roads; e) Use of surfactant kind of chemicals for binding the dust, while spraying water on unpaved roads; f) During mining operations in OC mines, dust control is provided with the mechanical dust-traps in drills; g) Spraying of water in the coal and OB benches for dust control in dumper loading by shovels or pay loader; h) Water spraying in CHP, conveyors and in wagon loading; i) In underground mines, water spraying in coal faces, after blasting, at loading points, near tub unloading by tipplers; j) If the dust control is not effective, then dust masks are provided to the employees for use at the work site.

240 km of paved colliery roads have been made, including haul roads for opencast mines. For underground mines, 180 nos. of main ventilators and 250 nos. auxiliary fans are working to ventilate. Large CHPs have been provided with Dust-extractors as well as water spraying. Special chemical reagents are being applied on unpaved haul roads in quarries for [11] dust suppression. ECL has 9 Dust Survey Equipments, CDS Static Sampler and 700 Dust Respirators. Rows of trees on both sides of unpaved roads and around Winder, Haulage or other dust and noise sources being planted.

Emission standards for automobiles, by percent volume of vehicles in use, from Jan'87:- a) CO%- 3.5 to 5%, b) Smoke 60-70 HSU. However, Central Pollution Control Board is setting standards, specifically for the coal industry, which will be incorporated in statutes. According to Air (Pollution & Control) Act and Rules 1981, permissible limits of air pollution, SPM (Suspended Particulate Matter) in micro-gm/m^3 and gases in ppm (parts per million), as followed in Environmental monitoring are:-

Type of area	Limits of SPM	SO_2	NO_X	CO
A. Industrial	500	120	120	5000
B. Residential	200	80	80	2000
C. Sensitive	100	30	30	1000

5. WATER POLLUTION CONTROL

Chemical checks of mine water and effluents at certain sampling points are being made for projects quarterly. Bacteriological tests for drinking water are being analysed by Asansol Mines Board of Health. As per report, in summer and monsoon, water pollution is maximum and bleaching powder is used in most wells. Cemented drains are being made at suitable locations in colonies, so that environment remains clean. 3 large filtered water supply schemes, from Ajoy and Damodar rivers are in operation.

According to Water (Pollution & Control) Act & Rules 1974, limits of water pollution as per environmental monitoring [6] standards in mg/l are: pH-5.5-9.0, TSS-100, TDS-2100, BOD-30, COD-250, oil and grease-10, Chlorides-1000, Cyanides-0.2, Sulphides-2.0, Sulphates-1000, Flourides-2.0, Boron-2.0, Arsenic- 0.2, Cadmium-2.0, Lead-0.1, Copper-3.0, Chromium-2.0, Mercury- 0.01, Nickel-3.0, Selenium-0.05, Zinc-5.0 etc.

Ground water control- Although, there are 775 nos. of main pumps and 890 nos. face and auxiliary pumps, dewatering 104 underground and 21 opencast mines of ECL, the discharge is little used in domestic, stowing and more for agriculture purpose. Negligible run-off to river systems is allowed and so evaporation is less, mostly ground water is recharged through seepage in the zone. For purifying mine water for domestic use, 24 pressure filters at a cost of 38.39 Crores, have been installed in coalfield areas.

6. NOISE POLLUTION CONTROL

Persistent high noise pollution may cause hearing impairment. Prescribed limited are given below for noise below:

Type of Area	Day time	Night time
A. Industrial	76	70
B. Residential	55	45
C. Sensitive	50	40

ECL has purchased its own Noise-level Meter and continuing monitoring on regular basis in mines and CHPs. To reduce noise level, corrective actions is taken [11] by improving silencers, mufflers etc. Condition monitoring are being undertaken for machines to prevent break down and noise. Underground machineries like coal cutting machines, hauls, drills and sometimes pumps create more noise beyond permissible limit which controlled by timely maintenance. Heavy blasting in quarries calls noise vibration, which can be sequenced milli second delay detonator.

7. ECO-DEVELOPMENT
Ecological basis of reclamation is by a) restoration, b) rehabilitation and c) reuse. On the earth, there are 17 Eco-systems services for 16 biomes, including 1) marine systems (33 bill. ha), comprising of open sea, coastal waters, estuaries, continental shelve and 2) terrestrial biomes (15 bill.ha) comprising of forests, wetlands, croplands etc. Global value of services provided to the entire biosphere is roughly $33 trillion per year, which is double the global GNP of $18 trillion per year.

According to study of present ecologists, 33% of land area [9] should have forest cover and in India today, it is only around 19%. Ozone layer depletion is caused by more industrial activity, permitting greater ultraviolet radiation and so global worming, foreboding eco disaster. One tree consumes annually 3.5t of CO_2 and exudes 1t of O_2 working like many air conditioners. Total forest cover has reduced from 638,879 sq. km in 1995 to 633,397 sq. km in 1997 in India. So, tree plantation is essential for improving climate and prolong life on earth.

ECL mines are spread over 1620 sq.m. in Bengal and Bihar and is trying eco-development [13] along with continued welfare measures. Since nationalisation, 8 fires have been extinguished and fire extinguishers are provided such that as soon as fire is detected it is wiped out. ECL has also created Eco- parks at Ratibati and Mahabir and another in Mohanpur is programmed. Other achievements:-

a) *Green Belt Plan*- ECL has formulated an unique Green Belt Plan, by which unutilised company lands can be allotted to govt. agencies or co-operatives for afforestation, agriculture production and fisheries by allotting lands on yearly license agreement. There are 700 main

b) *Tree Plantation* - Tree plantation is being done by the company in a methodical way, since '85-'86 and till now about 86 lakhs of different trees have been planted over 4480 ha. of degraded lands, with progressive expenditure of about Rs. 411 lakhs.

c) *Agriculture*- In view of impending closure some of the perennially losing mines, ECL is going to take up demonstration project for development of high value agri-aquaculture in degraded land of coal mines. The results of the project will benefit the people of local area, utilising the high value seeds of cereals, pulses, food grains developed, with proper application of fertilisers and nutrients suitable for the soil.

d) *Pisciculture*- West Bengal Government had promulgated one Inland Fisheries Act '86, amended in '97, such that filling up of any water body [15] or disused quarry is punishable by imprisonment or fine. Therefore, apart from operational back filling in the quarries, water-logged disused quarries to be used by the Green Belt Co-operatives for fisheries, on yearly licence fees to ECL. Reservoir fisheries technology with proper breeds to be developed, harvesting by gill-nets with boats, in the demonstration project, will be very helpful for such pisciculture in future.

8. RECLAMATION:
At first underground subsidence control measures are being taken to minimise the need for reclamation in future. During mining operations, various underground supports are used like props, bars and cogs of timber, gradually substituted by steel supports. Hydraulic props, roof-bolts of various designs, roof-stitching by used wire-ropes are being practised. Because of desired conservation of timber and forest wealth, gradually efforts are being made to use steel, RCC etc. both in supports as well as in haulage track sleepers. During development and depillaring operations, precautions are taken to maintain recommended size of galleries, pillars, stooks, barriers etc. to obviate risk of premature collapse.

Another important aspect of ground control measures is by practising hydraulic stowing in some high grade coal mines for filling of voids, after extraction of coal. Three ropeway systems with pontoons, scrapers and many transport trucks are operating for collection of sand and transport to the mine bunkers. Till March 2000, about 32 million cu.m of sand has been stowed in mines from '90-91, now around 3 million cu.m of sand per year used.

An unique method of ground control in unapproachable critical unstable areas is by hydro- pneumatic stowing, which was indigenously developed in ECL and is continuing. About 49 such critically unstable areas, over roads, towns have been identified and programme made to stabilise them, with hydro- pneumatic stowing. About 94,273 cu.m. has been filled in 5 locations, till March, 2000 and 5 more locations are going to be started.

70

Still another important ground control measures is by limiting vibrations damage, during heavy blasting in OC mines, which is achieved by using right blast design, properly sequenced milli- second delay detonators. Similar precautions of blasting are taken for underground blasting below important surface structures or villages, under shallow cover. Safe limit of maximum strain for damage in surface structures is that vibration is kept within 3-5mm amplitude.

Reclamation being practised in ECL by two main methods :- (i) Physical- a) Filling of cracks and crevices over depillaring area; b) Back filling in operating quarries; and c) Dozing, sloping in damaged zones etc. (ii) Biological- a) Afforestation over subsided area; b) Afforestation on OB dumps; and c) Aquaculture or pisciculture in disused water logged quarries. So far, in ECL due to mining total subsided area is 5112 ha. of which 3130 ha. has been reclaimed as follows: 1) Underground mines 211 ha.

(levelling); 213 ha. (afforestation) totalling 3234 ha.   2) OC mines 500 ha. (back filling) 153 ha. (Afforestation on OB dumps) and 3) Dozing over fire area 153 ha.

9. RISK ASSESSMENT MANAGEMENT
Mainly necessitated for sustainable development of the coalfield area, ERAMP (Environment Risk Assessment Management Plan) of Asansol -Durgapur Region was undertaken by ADDA, WBPCB, through RUDO of UNO. The work has been completed [14] in May '97 and is not only the first ERAMP studied in India, but also first one including all industries.

The work was carried out in two phases:- i) For Risk Assessment by six technical groups and ii) The Management Plan by 8 technical groups. Risk Assessment through 6 technical groups were- (i) air, (ii) water, (iii) hazardous wastes, (iv) Epidemeology, (v) Household and pest problem and (vi) ecology. 8 technical groups, which worked in Phase II were based on localities- i.e. (i) & (ii) Durgapur, (iii) Asansol, (iv) Raniganj, (v) Kulti, (vi) Mining Area; (vii) Damodar river, and (viii) Jamuria.

In the report, coal reserves below townships in Ranganj Coalfields have been quantified, under 10 townships, inhabited by 0.52 million people, over 1458 sq.km, blocking 1539 million tonnes of good quality coal. In various subdivisions present land use, water flow in cu. m /sec in Damodar river in seasons of summer, monsoon and winter have been studied and risk around Damodar river, like sand accumulation/stabilisation rate im m.cm/year, with average rain fall in reservoirs of Panchet- 13.74 and Maithon- 7.38. Highest Flood Level (HFL) have been measured all along. Stabilisaton made in Maithan-Panchayet reservoirs may have to be dredged in future for reducing danger of floods, increasing power generation and irrigation. Quality analysis of the Damodar river water at different sources for TSS, TDS, DO, BOD and COD have been determined at various locations along the river.   Linking of Environmental issues in respect of % population affected in subsidence (37.4%), OC mining, land degradation (8.8%), depletion/lowering of ground water table (9%), mine fires (9.7%) have been made.

Action Plans for the 5 main townships Durgapur, Asansol, Raniganj, Kulti, and Jamuria have been made for 1) Water supply system, 2) Sanitation system, 3) Solid waste disposal system, and 4) Air quality control.   Funding mechanism will be through combination of the following sources- a) Total Government initiative, b) Government/ external/national/international funding and c) Private Sector Partnership. Proposed Environmental Projects with details have been scheduled for Ninth Plan- Short Term (1997-99), Medium Term (1999-2002) and Tenth Plan- Long Term (2002-2007).

10.CONCLUSIONS
Since nationalisation of coal mines, many other environment related works have been taken up and completed-(a) Tackling of mine fires, (b) Societal development and welfare- schools, colleges, hospitals, dispensaries, roads, colonies, water supply, banks, co-operatives, clubs, rest-shelters etc.

In fact 'Earth Day' was started to be observed in USA, from April 22, 1970, for drawing attention of people for environment problems like global worming, health risks etc. and developed Clean Energy agenda. Denis Hayes was the national student co-ordinator of the first Earth Day and currently chairs Earth Day Network, the international organisation co-ordinating Earth Day events world-wide. ECL is organising World Environment Day Competition, on 5th June, since 1997 every year, in which most collieries/ projects participate and marking is done, by teams from other areas and 3 prizes on colliery/ project level and 3 on area level are awarded each year.

ECL is gathering sufficient quantity of river bed sand for stowing in mines through a) scraper and ropeway systems, b) pontoons and dredgers, c) direct truck loading. Raniganj is the oldest coalfield of India; there are some old goaves very close to river beds and control of erosion is essential in such cases. In 1995-96, sand stowed in mines of ECL was 31.76 lac Cu.m, at a cost of Rs.35.61 Crores. There is also long term possibility of dredging of dam reservoirs [7] for stowing sand in mines, with international cooperation.

Proper pisciculture/aquaculture in the water bodies [4] of disused quarries of Nimcha, suitable plants (avifauna) had been developed with the help of officers of Geological Survey of India and Botanical Survey of India. Further, reservoir fisheries technology is proposed to be developed in a demonstration project by Government grant, for other water-logged disused quarries.

Employees of ECL, working in mines are periodically examined for occupational diseases and year-wise position:- 1996- 17269 examined, occupational disease detected-2; 1997- 11340 examined and detected -11; 1998- 18506 examined and detected -2; 1999- 25595 examined and detected–2 and so

There has been decreasing occupational diseases, as have been studied.

At the time of starting of coal mines, markets came up and sustainable development by way of growth centres began and gradually expanded. Again, after the mining of the coal reserves are completed, post-mining land-use pattern and ways of sustainable growth may be continued in some other trade, by proper selection of alternative job opportunities and commercial activities. Large coal companies have to manage environment issues, based on scientific studies, with the cooperation of local NGOs, municipalities etc. Such environment management has to be done in all cases of mining of non-renewable natural resources.

REFERENCES

1. CMPDI-RI-1, 1986- "Report on Preliminary Investigation of Reclamation of Old and Abandoned Quarries in Raniganj Coalfield"- ASANSOL- 713301.
2. "Coal Facts"- World Coal Institute, Oxford House, 182 Upper Richmond Road Putney, London-SW15 2SH, UK- Sept '98.
3. Petter D. Jenssen- " Recycling Solutions for Wastewater and Organic Waste"- International Conference in Ecological Engineering, Science City, Calcutta, Nov.23-27, '98.
4 B.B.Jana- " Biocontrol of Algal Bloom in Eutropic Lakes : Mirocystis Digestibility among Three Herbivorous Fishes"- International Conference in Ecological Engineering, Science City, Calcutta, Nov.23-27, '98.
5. `Environment Management Plan of Asansol Durgapur Industrial Corridor'- May, 1997- Ghosh, Bose & Associates with support from EHP and RUDO, USAID, Deptt. of Environment, West Bengal Pollution Control Board.
6. "Environment Protection Laws"- 1994- West Bengal Pollution Control Board, 10, Camac Street, Industry House, CALCUTTA-17.
7.Samanta, B.K.- 1996- " Environmental Management of Subsidence in Mining Areas by Dredged Sand from Dams"- First World Mining Environment Congress, New Delhi, 11-14 Dec'95- Proceedings of WOMEC'95- Oxford & IBH Publishing Co. Pvt. Ltd.- 66 Janpath, New Delhi.
8.Sarkar, S.K. et al-1996-" State of Environment and Development in Indian Coalfields"- Oxford & IBH Publishing Co. Pvt. Ltd.- 66 Janpath, New Delhi.
9.Prof.Trivedi, P.R.- 1991- "International Encyclopaedia of Ecology and Environment"- Indian Institute of Ecology and Environment, Paryavaran Complex, South of Saket, NEW DELHI-110030.
10. Khadia,K.K. and Samanta, B.K. - 1998- "Green Belt Plan for Coalfield Area" - VIIth National Symposium on Environment- Indian School of Mines, Dhanbad, Feb. 5-7.
11. Samanta, B.K.- 1997- "Practicable Environment Improvements in Coal- fields"- National Seminar on Emerging Technology in Surface Mining and Environment Challenges- IEI, Mangalore, Oct.16-18.
12. Samanta, B.K.- '98- " Legal Aspects of Environment Management in Coalfields"- National Seminar on Environment in Mining Areas- CMRI, Dhanbad, June 5-6.
13. Samanta, B.K.- '98- " Environmental and Ecological Developments in Coalfields" - 3rd International Conference on Ecological Engineering- Science City, Calcutta, Nov.23-27.
14. Samanta, B.K.- '99- " Environment Risk Assessment & Management Plan of a Coalfield"- International Symposium on Clean Coal Initiatives- New Delhi, 22-24 Jan. 15. Samanta, B.K.- '99- " Ecological Engineering for Minewater Resource in Coalfields"- 4th International Conference on Ecological Engineering- Aas, Norway- June 7-11.

Reclamation of Mining Sites : An Overview of Challenges and Applications

D.K. Ghosh, Y.G. Kale and A.S. Chachane

Indian Bureau of Mines, Nagpur, India

ABSTRACT

India is endowed with rich biological diversity and is virtually a gene bank for number of food, forest, and medicinal and aromatic plants and also for a number of animal species. The rising industrialisation including that of mining activities, endangered the very existence of natural eco-balance. Surface mining practices in our country has more potential and contribute major share in the mineral production. This trend is also likely to continue in foreseeable future. However, surface mining causes land degradation to a large extent and therefore, mining activities need to be carried out in a sustainable manner by simultaneous reclamation activities wherever feasible. There are several approaches of reclamation, which need to be considered and addressed. This paper highlights the challenges of reclamation of mining sites with particular reference to metalliferous mining sector and also discusses the policy in this matter.

INTRODUCTION

Mining activities essentially cause physical, chemical, biological and socio-economic changes in the characteristics of the area. Therefore, reclamation of mined out areas is a challenging task. Moreover, it is an intricate, complicated and site specific technique involving multidisciplinary approach. Hence, apt techniques for reclamation of mining site are very much important for planning and implementation of scientific mining methods. Of late, environmentalists also demand restoration of biodiversity by the mining operators to nearly pre-mining conditions. Proper planning coupled with sincere implementation of environmental management plans will minimise the impacts of mining on the environment and can help to preserve the diversity to large extent.

India has a total geographic area of about 3.287 million sq.km. out of which the total forest cover which includes dense forest, open forest and mangrove is estimated to be 0.633 million sq. km.[1] This constitutes 19.27% of country's geographic area which is far below the target of 33%. Although, India's mining leases are spread over less than 0.25% of geographic area, mining sector's share for degradation of land cannot be considered as insignificant because the subsequent environmental impact do not restrict within the boundaries of mining leases. Therefore, it is essential that all the mining and environmental components are taken into consideration and an integrated approach is required to combat the ill effects.

INDIAN MINING AT A GLANCE

Mining activities are carried out in India since time immemorial. In fact, Indian mineral industry plays an important role for the economic development of the nation. At present nearly 3150 mines are reporting production which includes about 550 coal mines. In non-coal sector nearly 130 mines are underground which constitutes less than 5%. Hence, the major output of mineral production comes from opencast mining. As per recent surveys[2] in non-coal mines the total area of about 440 Ha.

have been reclaimed in abandoned mines. Besides, in 350 Ha. area of working mines, reclamation and rehabilitation of mined out area have been carried out simultaneously. Reclamation by afforestation is dominated in Indian mines. More than 49 million trees have been planted in major non-coal mines covering.an area of about 22,155 Ha. with an average survival rate of 71%.

MINING AND RECLAMATION OF LAND

Land Degradation

Land degradation is the most serious environmental impact of mining. The gravity of the problem is more complicated in case of surface mining. Surface mining apart from causing air, water and other pollution problems, renders the land unsuitable for other uses unless it is restored or rehabilitated. It also reduces the land use potential. The degree of damage to the land varies with the topographic setting in which mining is carried out, the type and depth of overburden, the scale and type of mining and climatic conditions. Also these variables influence reclamation techniques and the rate at which surface mined areas can be effectively reclaimed. In short the following impacts can be summarised due to surface mining.

- Alteration in topography
- Loss of fertility of soil
- Deforestation
- Piling of large waste dump
- Alteration in ground and surface waters
- Rapid erosion of land, slope stability problems.
- Decrease of land value
- Visual impact.

Reclamation

Broadly speaking reclamation means returning a derelict land to a form and level of productivity that can sustain the prior or future land use(s) in ecologically stable state. There are several reclamation options of which agriculture, forestry, industrial or housing complex, fish farming, water works are the important ones. The immediate objective of any rehabilitation programme should be to create a landscape which is stable against the erosive forces of wind and rain in order to reduce both on-site and off-site environment insult. A second objective should be to return the mined land, where possible, to a condition that allows it to be used in a productive manner. The nature of the post mining land use will depend on a number of factors including the ecological potential of the mined environment and the dictates of society. The following principle can form the basis of successful reclamation strategy[3].

- Prepare a rehabilitation plan prior to the commencement of mining.
- Agree on the long-term post-mining landuse objective for the area with the relevant Government departments, local Government councils and private landowners. The landuse must be compatible with the climate, soils, topography of the final landform and the degree of management available after rehabilitation.
- Progressively rehabilitate the site, where possible, so that the rate of rehabilitation is similar to the rate of mining.
- Prevent the introduction of noxious weeds and pests.
- Minimise the area cleared for mining and associated facilities to that absolutely necessary for the safe operation of the mine.
- Reshape the land disturbed by mining so that it is stable, adequately drained and suitable for the desired long-term landuse.
- Minimise the long-term visual impacts by creating landforms which are compatible with the surrounding landscape.
- Reinstate natural drainage patterns disrupted by mining wherever possible. Alternatively a system of artificial irrigation has to be arranged.
- Minimise the potential for erosion by wind and water both during and following mining.

74

- Remove or control residual hazardous materials. Identify any potentially toxic overburden or exposed strata and manage them so as to prevent environmental damage.
- Characterise the top soil and retain it for use in rehabilitation. It is preferable to reuse the topsoil as early as possible rather than storing it for long. Only discard if it is physically or chemically undesirable, or if it contains high levels of weed seeds or plant pathogens.
- Consider spreading the cleared vegetation on disturbed areas.
- Deep rip compacted surfaces to encourage infiltration, allow plant root growth and key the topsoil to the subsoil, unless sub-surface conditions dictate otherwise.
- Ensure that the top one or two metres of soil is capable of supporting plant growth.
- If topsoil is unsuitable or absent, identify and test alternative substrates, eg overburden that may be a suitable substitute after addition of soil improving substances.
- Revegetate the area with plant species consistent with the post-mining landuse.
- Meet all statutory requirements.
- Make the area safe and particularly protect from grazing animals .
- Remove all facilities and equipment from the site.
- Monitor and manage rehabilitated areas until the vegetation is self-sustaining and meets the requirements of the landowners or land manager, or until their management can be integrated into the management of the surrounding area.

In surface mining, total process of reclamation of mined out areas can be completed in two phases[4].

(A) Technical Reclamation :

This includes back filling of excavation, spreading of subsoil and top soil grading on the backfill and waste dumps and depends on type of equipment used and mineral worked. Various combinations of equipment can be used for reclamation viz. backfilling, spreading, stacking scarifying, grading etc. are:

i)   Dragline
ii)   Scraper – Dragline
iii)   Scraper – Dragline – Stripping Shovel
iv)   Tractor Shovel/Frond End Loader Dumper-Dozer
v)   Shovel/Dumper-Dozer
vi)   Bucket Wheel Excavator – Spreader

(B) Biological Reclamation :

Biological reclamation of mined out pits, waste dumps and barren lands with multi species bio-diversified afforestation is an integral part of eco-friendly mining. Revegetation and reclamation of mined land offers a great challenge as more often they do not have suitable surface soil to provide bedding layer to encourage and support the biomass development while in other cases they do not support plant growth due to presence of toxic metals which ultimately affects the eco-system and ecological balance.

The main aim of biological reclamation is to restore fertility and biological productivity of the disturbed land. It takes 3 to 5 years to develop the structure of soil and its enrichment. During this phase selection of grass and other plants should be done.

There are some factors, which affect the adoption of this technique as,

i)   Climate
ii)   Depth, nature of top soil, subsoil and underlying overburden
iii)   Compaction of top soil and subsoil
iv)   Farming of local type

Reclamation of mined out areas requires a combination of engineering and biological skills. Such reclamation can be effectively implemented if the end use of reclaimed land is planned ahead and integrated into the phased mining plan right at the inception stage.

The following criteria must satisfy for considering the successful reclamation.[3]
- Physical
  Stability, resistance to erosion, proper reestablishment of drainage
- Biological
  Richness of species, plant density, conopy cover, seed production, return of fauna, weed control, productivity, establishment of nutrient cycles.
- Adherance to the water quality standards for drainage water.
- Public safety issues.

In addition to mined out areas, reclamation of filled up tailing ponds and waste dumps are also needs to be considered. The reclamation of these sites can be achieved by structural modifications, vegetation and chemical stabilization. The basic structural modifications methods that can be utilized are flattening of slope by removing the material near the crest, unloading the crest by reducing the height of tailings or mine waste, controlling surface run off, controlling phreatic surfaces, lowering the water surfaces, lowering the water surfaces of impoundment. The stabilization by vegetation is preferred as it brings improvements in appearance, imparts greater permanency to structure and helps in minimising the environmental impacts and in restoration of wildlife habitats. The land scaping may be necessary before starting vegetation. The objective of landscaping may be to
- Construct stable slopes
- Eliminate ponding unless it is planned and designed.
- To blend the topography of the disposed material with that of undisturbed landscape
- Stabilize erosion gullies
- Collect the runoff and discharge it to safe point
- Prevent concentration of run off in the area
- Conditioning of soil for supporting the vegetation

Sometimes it is necessary to neutralize the acidic material prior to vegetating. Commonly, limestone, burnt lime, hydrated lime; fly ash and municipal sewage sludge can be used. It is always preferable to allow 6 months for reaction to neutralize the soil. Many times the nutrients have to be added to stimulate plant growth. The selected species to be planted should suit the plant site and local climate and also serve the purpose to control soil erosion and support the aesthetic or self-maintaining characteristics. The resistance to disease and insect is also important consideration in selecting the species. The toxic resistant plant species based on R & D work may be useful specially over tailing dams. The chemical stabilization does not offer the aesthetic advantage of vegetative stabilization but are useful where climate or plant toxic metals prevent establishment of vegetation. Chemical stabilization is helpful as a temporary control of fugitive dust on dry part of an active site. Crushed or granulated smelter slag is useful to stabilize inactive and active tailing ponds but may not form the favourable base for vegetation. The chemical stabilization deteriorates with time and requires periodic reapplication. They are often used in conjunction with vegetative stabilization to form an abrasion and erosion resistant crest so that the vegetation grows enough to stand on its own.

Statutory Requirement

For metalliferous mines, the Mineral Conservation and Development Rules, 1988[5] spells out the detail procedure for reclamation and rehabilitation of mined out land. Some of the important provisions are abstracted below :

Rule 32 : Removal and utilisation of top soil
i)   Separate removal of top soil
ii)   Use of top soil for restoration and rehabilitation of land
iii)   If the top soil not used concurrently, it shall be stored separately for future use with adequate care to prevent loss of it.

Rule 33(4) : Wherever possible, the waste rock, overburden etc., shall be back filled into the mined excavations with a view to restoring the land to its original use as far as possible.

Rule 34 : Reclamation and rehabilitation of lands : Every holder of prospecting licence or mining lease shall undertake the phased restoration, reclamation and rehabilitation of lands affected by prospecting or mining operations and shall complete this work before the conclusion of such operations and the abandonment of prospect or mine.

Rule 41 : Restoration of flora :

1. Every holder of prospecting licence or mining lease shall carry out prospecting or mining operations as the case may be in such a manner so as to cause least damage to the flora of the area held under prospecting licence or mining lease and the nearby areas.
2. Every holder of prospecting licence or a mining lease shall –

   a) take immediate measures for planting in the same area or any other area selected by the Controller General or the authorised Officer of Indian Bureau of Mines not less than twice the number of trees destroyed by reasons of any prospecting or mining operations.
   b) restore to the extent possible other flora destroyed by prospecting or mining operations.

Based on the statutory requirement, the integrated approach for reclamation is depicted in Figure 1.

CONCLUSIONS

The land degradation posed by the surface mining could assume serious environmental threat, as the mining will continue to expand in the foreseeable future in India. The public awareness of environmental impact of mining is likely to grow with time and environmental legislations could perhaps be tightened further. Proper backfilling, waste dumps plans, effective tailing disposal, reclamation measures and rehabilitation process like retrieval of dump, afforestation, revegetation should be an integral part of the project plan rather than just for the sake of statutory compliance. Numerous small mines not in a position to take up reclamation work should be provided with some organisational set up to avoid the cumulative, environmental impact of these mines in general and land degradation in particular. The appropriate approach in this regard is to put the area under forest with native species of plants because country is short of forest resources. The services of experts in land scaping, town planning, forestry, soil conservation etc. should be utilised by the mining organisation wherever necessary. The mining industry must be cognizant at all times in future of its critical areas of social responsibilities.

ACKNOWLEDEMENT

The authors would like to thank Shri A.N.Bose, Controller General, Indian Bureau of Mines for kind permission to present this paper. The views expressed in this paper are those of authors and not necessarily of the Indian Bureau of Mines.

References

1. State of Forest Report, 1997 (1998) Forest Survey of India, Ministry of Environment and Forests pp 5,11.
2. Compilation of notes/reports. Indian Bureau of Mines (unpublished).
3. Rehabilitation and Revegetation. Best Practice Environmental Management in Mining (1995) Environment Protection Agency, Commonwealth of Australia pp 4, 30.
4. Mukhopadhyay S.K., Pal S.K. and Bhattacharya J. (1994) Environmental ramifications in surface mining with respect to land degradation under Indian context. Journal of Mines, Metals & Fuels, Aug Sept., 1994 pp 203.
5. Mineral Conservation and Development Rules, 1988 (2000) Ministry of Mines and Minerals, Government of India pp 19-20.

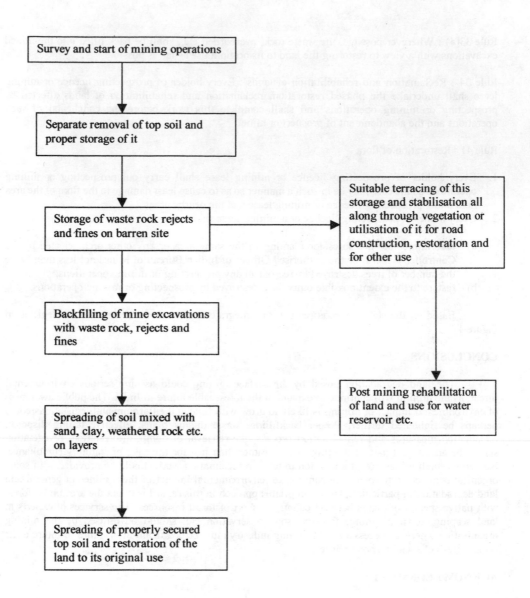

Fig. 1 : Integrated approach for reclamation of mined out land

Geoenvironmental Reclamation—Paddy Cultivation in Mercury Polluted Lands

M.J. Abdul Hameed[1], M.S. Mohamed Jaabir[2] and S. Sankaran[3]

[1]*Department of Biotechnology, Bharathidasan University, Tiruchirapalli 620 024, India*
[2]*P.G. Department of Biochemistry, Vysya College, Salem 636 103, India*
[3]*P.G. Department of Botany, Jamal Mohamed College, Tiruchirapalli 620 020, Tamil Nadu, India*

INTRODUCTION

Domestic, municipal and agricultural wastes, animal and human excreta and industrial effluents from textiles, leather, paper & pulp, fertilizer, pesticide, electrical & electro-plating, cement & asbestos, paint & dye, chemical manufacturing, pharmaceutical, coal, petroleum, mining and metal processing industries which are rich in heavy metals are routinely discharged into several of the water ways and waste lands. Being conservative and non – biodegradable, when once introduced into the ecosystem, becomes concentrated many folds, bio-transferred & bio-magnified along the successive higher trophic organisms of the food chain resulting in the contamination of the natural resources, make them unfit for human consumption and uses. In a land of ever-growing population with the demand in agricultural products, no resource can be let go waste.

Among various pollutants, heavy metals like lead, cadmium, mercury and nickel are of particular importance owing their lethal toxic effects. The problem is more acute in regions where there is scarcity of water for irrigation. In such regions, the industrial effluents are either directly used for irrigation with out any dilution or partially diluted with out any treatment. As a result, heavy metals are accumulated in the tissues of plants affecting their metabolism, growth and crop yield. A number of studies have been undertaken from time to time to investigate the effect of industrial effluents on seed germination, seedling growth, metabolic activities and crop yield, in many crop plants like paddy, wheat, sorghum and others.

The ions of heavy metals are taken up by the plants through their root system or by foliar absorption, when their concentration are elevated (Larger et al, 1970). These ions are translocated to different plant parts and interfere in normal metabolic processes and thus lead to reduction in growth (Allison et al, 1981 & Shaeran et al, 1990). Such interference of lead and mercury has been reported (Valle and Ulmer, 1972) to have adverse effects on the physiological and biochemical activities of plants. Pb^{2+} and Hg^{2+} has been reported to increase the permeability of tissues due to membrane damage in some aquatic plants (Jeana and Choudhury, 1982). The membrane damage is often related to membrane lipid peroxidation caused by free radicals and hydroperoxidases (Thompson, 1988; Hendry et al, 1992). Malonedialdehyde (MDA) which is one of the decomposition products of polyunsaturated fatty acids of biomembranes has been shown greater accumulation in heavy metal stressed seedlings of (rice) paddy.

Toxic effect of mercury in plant growth is also reported due to its inhibitory action on nitrate reductase activity by interacting with the enzyme protein which is counteracted by the presence of Hg^{2+} (Madhu Pandey and H.S. Srivastave, 1994). Lower concentrations (1mM) of cobalt has been shown (Agarwala et al, 1961) to stimulate germination in barley where as Cd, Ni, Cr, reduce it at higher concentrations (5mM & 25mM). Attempts in paddy for germination at low concentration (0.1mM & 2mM) and at concentration greater than 4mM had comparatively similar effect (Mukerji and Mukerji, 1979). Cadmium and mercury reduced the germination rate, total performance and long term viability at various concentrations in Spartina alternifolia seeds (Mrozek, 1980). In most of the metal toxicity studies, toxicity was first experienced in root tips and lateral root development. Studies on Phaseolus plant has shown the toxic effects of mercury on growth, chlorophyll, saccharides and soluble nitrogen (M.A.A. Gudallah, 1994). Similar inhibitory effects of Hg has also been reported in a number of experimental works (Sundaravelu et al, 1997) in Cucumis sativus, Vigna mungo (Subramani et al, 1997) etc.

Though pollution can be controlled with stringent rules and regulations, it is difficult to remediate the damages done by the effluents discharged in the past. Our present investigation aims in adding to the concerted effort targeted towards reclamation of such heavy metal polluted lands by cultivating suitable varieties of the agricultural crops.

MATERIALS AND METHODS
Healthy seeds of Oryza sativa, Linn. Variety IR-36, IR-50, IR-72 and IR-80 were procured. Uniform sized, good looking seeds were selected in each variety and soaked in different concentrations viz. 20, 50, 100 and 200µg/ml of mercuric chloride solution for 12 hrs. Control was maintained simultaneously in distilled water. Germination was carried on moistened cotton pads in Petridishes for a period of 48 hrs. The germination percentage was calculated considering the seeds having 2mm or more plumule length as germinated. Each treatment was replicated thrice.

Germinated seeds from each concentration were transferred to Petridishes containing moistened cotton pads. Seeds were supplied with Hoagland nutrient solution and grown under diffused light at room temperature. The average root and shoot lengths were recorded after 72 hrs from the date of sowing unto the eleventh day.

RESULTS AND DISCUSSION
Control of IR-36 recorded 98% seed germination while the experiments recorded gradual decrease as 95% in 20µg/ml, 93% in 50µg/ml, 89% in 100µg/ml and 76% in 200µg/ml concentration of mercuric chloride. Control of IR-50 recorded 99% seed germination while the experiments recorded gradual decrease as 93% in 20µg/ml, 84% in 50µg/ml, 81% in 100µg/ml and 76% in 200µg/ml concentration of mercuric chloride.

Control of IR-72 recorded 96% seed germination while the experiments recorded gradual decrease as 80% in 20µg/ml, 76% in 50µg/ml, 70% in 100µg/ml and 64% in 200µg/ml concentration of mercuric chloride. Control of IR-80 recorded 98% seed germination while the experiments recorded gradual decrease as 94% in 20µg/ml, 85% in 50µg/ml, 73% in 100µg/ml and 65% in 200µg/ml concentration of mercuric chloride. Thus there is an inverse proportion between the concentrations of mercuric chloride and percentage of germination (Table I. & Fig. I).

TABLE – I

EFFECT OF MERCURIC CHLORIDE ON THE SEED GERMINATION
OF FOUR VARIETIES OF PADDY

No.	PADDY VARIETY	CONTROL	20µg/ml	50µg/ml	100µg/ml	200µg/ml
1.	IR-36	98%	95%	93%	89%	76%
2.	IR-50	99%	93%	84%	81%	76%
3.	IR-72	96%	80%	76%	70%	64%
4.	IR-80	98%	94%	85%	73%	65%

FIG. I EFFECT OF MERCURIC CHLORIDE ON SEED GERMINATION

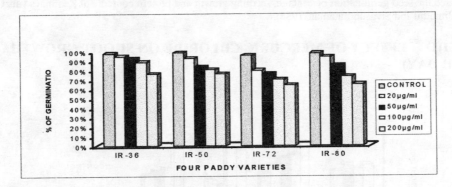

Seedling growth in IR-36 and IR-50 (Table.II & III) under mercuric chloride treatment exhibited gradual increase in root and shoot length at concentrations of 20, 50 and 100µg/ml while 200µg/ml of mercuric chloride treatment affected the growth drastically. However, these seedling growths were less when compared to the normal healthy growth of the control plants. IR-72 and IR-80 seemed much affected by the mercuric chloride treatment that its growth was seriously inhibited or retarded even at lower concentrations such as 20 and 50µg/ml (FIG.II).

TABLE – II

EFFECT OF MERCURIC CHLORIDE ON THE SEEDLING GROWTH IN PADDY VARIETY IR-36

N O	NO. OF DAY	CONTROL ROOT, SHOOT Length (cm)		20µg/ml ROOT, SHOOT Length (cm)		50µg/ml ROOT, SHOOT Length (cm)		100µg/ml ROOT, SHOOT Length (cm)		200µg/ml ROOT, SHOOT Length(cm)	
1	3	2.26	1.09	1.85	1.15	2.43	1.47	2.75	1.33	0.47	0.49
2	5	4.21	2.76	5.82	4.68	4.15	2.47	4.04	2.00	1.73	1.27
3	7	5.99	5.06	6.95	6.38	5.92	5.14	5.10	4.29	2.79	2.01
4	9	3.69	7.09	4.14	7.15	3.03	7.05	3.87	6.14	2.66	3.79
5	11	4.59	7.71	4.27	7.26	4.00	7.78	4.16	6.99	2.95	3.86

FIG.II EFFECT OF MERCURIC CHLORIDE ON SHOOT GROWTH (3rd DAY)

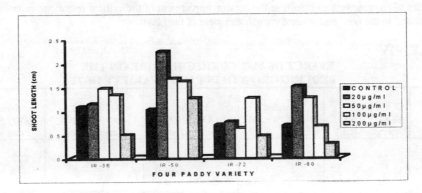

On the whole, higher concentrations of mercuric chloride, particularly 100 and 200µg/ml had adversely affected the seed germination in all the varieties of paddy tested. This is in tune with the results of earlier studies of Mukherji and Mukherji (1979), Mrozek, E. Jr, (1980) and K.Kalimuthu and Sivasubramanian (1988), stating that mercuric chloride and lead acetate at 20, 50, 100 and 200 µg/ml decreased the seed germination percentage, seedling growth and protein contents of Zea mays var. Co.1. (K.Kalimuthu and Sivasubramanian, 1988).

FIG. III   EFFECT OF MERCURIC CHLORIDE ON SHOOT GROWTH (11TH DAY)

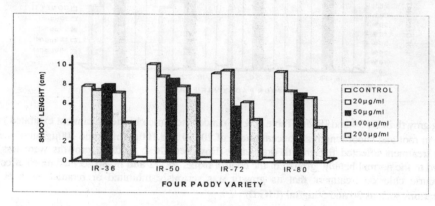

TABLE – III

EFFECT OF MERCURIC CHLORIDE ON THE SEEDLING GROWTH IN PADDY VARIETY IR-50

N O	NO. OF DAY	CONTROL ROOT, SHOOT Length (cm)		20µg/ml ROOT, SHOOT Length (cm)		50µg/ml ROOT, SHOOT Length (cm)		100µg/ml ROOT, SHOOT Length (cm)		200µg/ml ROOT, SHOOT Length (cm)	
1	3	2.11	1.03	3.88	2.24	2.84	1.67	2.64	1.67	2.21	1.26
2	5	5.79	4.62	6.45	5.43	6.81	5.76	6.40	5.27	6.40	2.51
3	7	6.08	5.32	6.16	5.86	6.60	6.02	6.61	6.05	6.05	5.40
4	9	4.27	8.07	3.91	6.82	3.34	6.65	4.43	7.04	3.51	6.23
5	11	5.14	10.02	5.84	8.73	5.77	8.43	4.74	7.68	4.19	6.75

Despite a gradual increase in the root and shoot length from the third to the eleventh day, the growth which was very much reduced when compared to the control reveal the inhibitory effect of mercuric chloride on plants. This is probably due to the toxicological properties of mercury (Valle and Ulmer, 1972) that has a strong affinity for ligands such as phophates, cysteinyl and histidyl side chains of proteins, purines, pteridinies and porphyrins. Such mercurial – SH/S-S interactions occurring in multitudes of enzyme system results in the serious impairment of the critical metabolic pathways which is manifested in the poor and stunted growth and yield of the plants.

TABLE – IV

EFFECT OF MERCURIC CHLORIDE ON THE SEEDLING GROWTH IN PADDY VARIETY IR-72

N O	NO. OF DAY	CONTROL ROOT, SHOOT Length (cm)		20µg/ml ROOT, SHOOT Length (cm)		50µg/ml ROOT, SHOOT Length (cm)		100µg/ml ROOT, SHOOT Length (cm)		200µg/ml ROOT SHOOT Length (cm)	
1	3	1.69	0.70	1.47	0.76	1.40	0.62	2.27	1.25	1.22	0.46
2	5	3.28	1.94	6.12	4.79	5.30	4.07	4.94	3.82	1.28	1.03
3	7	6.03	5.03	4.82	5.14	5.22	4.53	5.23	4.48	4.54	3.59
4	9	4.04	6.21	4.16	6.69	3.55	4.72	3.92	5.41	2.86	4.16
5	11	4.59	9.07	4.87	9.40	3.99	5.50	3.18	6.09	2.42	4.24

TABLE – V

EFFECT OF MERCURIC CHLORIDE ON THE SEEDLING GROWTH IN PADDY VARIETY IR-80

NO	NO. OF DAY	CONTROL ROOT, SHOOT Length (cm)		20µg/ml ROOT, SHOOT Length (cm)		50µg/ml ROOT, SHOOT Length (cm)		100µg/ml ROOT, SHOOT Length (cm)		200µg/ml ROOT, SHOOT Length (cm)	
1	3	1.43	0.68	3.08	1.50	2.46	1.25	1.65	0.66	1.0	0.29
2	5	4.26	2.98	4.95	3.82	3.89	2.42	3.28	1.97	1.45	0.66
3	7	4.58	3.96	4.69	4.06	4.79	4.15	4.52	3.81	2.76	2.14
4	9	3.97	6.04	4.24	6.40	4.34	6.30	3.21	4.62	2.42	3.10
5	11	4.76	9.30	4.28	7.23	3.65	6.97	4.06	6.59	2.12	3.46

However, our study reveals that IR-36 and IR-50 are resistant to the toxic effects of mercuric chloride than the other varieties of paddy IR-72 and IR-80 (Table.IV, V & FIG. III). Moderately polluted lands can therefore be reclamated by cultivating the lesser susceptible IR-36 and IR-50 paddy varieties for optimum utilization of resources. Else, the concentration of mercury present in the land or irrigating (effluent) water has to be diluted heavily before such selective cultivation.

ACKNOWLEDGEMENT

The authors are profoundly thankful to Prof. M.Vivekanandan, Head, Dept. of Biotechnology, Bharathidasan University, Tiruchirapalli, Tamilnadu, for his constant encouragement and valuable guidance.

REFERENCES

1. Abdel-basset, A.A., ISSA and Adam, M.S (1995). Chlorophyllase activity: effects of heavy metals and calcium. Photosynthetica. 31(3): 421-425.

2. Agarwala, S.C., et al (1961). Effects of excess supply of heavy metals on barley during germination with special reference to catalase and peroxidase. Nature. 191:726

3. Anil, K.DE, Asit, K.Sen, and Deb, P. Modak (1985). Studies of toxic effects on Hg(II) on Pistia stratiotes. Water Air and Soil Pollution. 24: 351-360.

4. Anitha Mishra and Choudhuri, M.A. (1996). Possible implications of heavy metals (Pb^{2+} and Hg^{2+}) in the free radical – mediated membrane damage in two rice cultivars. Indian J. Plant Physiol. 1(1): 40-43.

5. Clijsters, H., Van Assche, F. (1985). Inhibition of photosynthesis by heavy metals. Photosynth. Res. 7: 31-40.

6. De Filippis, L.F. (1979). The effect of heavy metal compounds on the permeability of Chlortella cells. Z. Pflanzenphysiol. 92: 39-49.

7. Gudallah, M.A.A. (1994). Interactive effect of heavy metals and temperature on the growth, chlorophyll, saccharides and soluble nitrogen content in Phaseouls plants. Biologia Plantarum. 36(3): 373-382.

8. Kalimuthu, K. and Sivasubramanian. (1990). Mercuric chloride and lead acetate at 20, 50, 100 and 200µg/ml decreased seed germination percentage, seedling growth and protein contents of a Zea mays. Var. Co.1. Indian J. Plant Physiol. 33: 242-244.

9. Madhu Pandey and Srivastava, H.S. (1994). Hg inhibits nitrate reductase activity by Hg interaction by the presencese of Mg^{2+} . Indian J. Envi. Hlth. 36(1): 13-18.

10. Mohapatra, P.K., Liza Mohanty and Mohanty, R.C. (1993). Reduction in mercury toxicity to Scenedesmus Bijuga with addition of glucose, glutamase and 2-oxoglutarate. J. Ecotoxicol. Environ. Monit. 3(1): 019-026.

11. Mrozex, E.Jr., (1980). Effects of mercury and cadmium on germination of Spartina alternifolia loisel seed at various salinities. Env. And Exp. Bot. 20: 367-377.

12. Mukherji, S. and maitra, P. (1977). Growth and metabolism of germinating rice (Oryza sative.L) seeds as influenced by toxic concentrations of lead. Z. Pflanzenphysiol. 81: 26-33.

13. Mukherji, S. and Mukherji, C. (1979). Characterization of cadmium effects in different plant materials. Indian J. Exp. Biol. 17: 265-269.

14. Murty, S.D.S., Bukhov, N.G., Mohandy, P. (1990). Mercury induced alterations of chlorophyll a fluorescence kinetics in cyanobacteria - Multiple effects of mercury on electron transport. J. Photochem. Photobiol. 6: 373 -380.

15. Vallee and Ulmer. (1972). Biochemical effects of mercury, cadmium and lead. Annual review of Biochemistry. 41: 91-128.

16. Van Ass he, F. and Clijsters, H. (1990). Effect of metals on enzyme activity of plants. Plant Cell Environ. 13: 195-206.

17. Anand, S.J.S., Khandekar, R.N. and Krishnamoorthy, T.M. (1995). Estimation of mercury in environmental samples using Radioanalytical Neutron Activation Analysis (RNAA). Proceedings of 4th National Symposium on Environment, Madras, Tamilnadu. 117-120.

18. Bharani, A. and Ramanathan, G. (1997). A short term study on heavy metal toxicity to minor millets.· Proceedings on National Symposium on Recent trernds in Biotechnology, Tiruchirapalli, Tamilnadu.

19. Rao, V.N.R. (1997). Detoxification of heavy metals by microalgae. Proceedings on National Symposium on Recent trends in Biotehnology, Tiruchirapalli, Tamilnadu.

20. Shaik Nagoor, Alex Joseph and Vyas, A.V. (1995). Toxic effects of cadmium and mercury in Maize seedlings. Proceedings of 4th national symposium on Environment, Madras, Tamilnadu. 72-74.

21. Subramani, A., Saravana, A. and Lakshmanachary, A.S. (1997). Influence of heavy metals on germination and early seedling growth of Vigna mungo (l). Hepper. Proceedings on National Symposium on Recent Trends in Biotechnology, Tiruchirapalli, Tamilnadu.

22. Sundaravelu, S., Selvaraju, M. and Muthukrishnan, T. (1997). Response of mercuric chloride on germination early growth and pigmentation changes of Cucumis sativus and luffa pentrandra Roxb. Proceedings on National Symposium on Recent Trends in Biotechnology, Tiruchirapalli, Tamilnadu.

Geoenvironmental Pollution due to Fertilisers

K. Jeevan Rao

Department of Soil Science & Agricultural Chemistry, Agricultural College, Acharya N.G. Ranga Agricultural University, Mahanandi, Nandyal (R.S.) 518 502 (A.P.), India

INTRODUCTION

India's population is variously projected at 1330 million to 1620 million by 2020 (16-17 million population added each year). Foodgrain demands by 2020 is estimated at 260-300 million tonnes. The challenge during the next millenium is to achieve and sustain growth rates high enough to feed the growing population without degrading the environment[1]. Irrigation, fertilizer and seed are the contributing factors for growth in agricultural production and productivity. Of the 142 million has (Mha) area under cultivation, 50Mha is irrigated and about 125 Mha is under cereals and pulses (i.e. 87% of area). NAAS (1996) in its document entitled 'Agricultural Scientists" Preception and Plant Nutrient Needs, Supply, Efficiency and Policy issues: 2000-2025" has projected that 30-35 million tonnes of fertilizer nutrients would be required to meet the food grain demand by 2020. This demand will strech by almost 15 million tonnes more, if requirements of other crops are included[2]. Foodgrain production in 1998-99 was 200 million tonnes and in 1999-2000, it is expected to be around 204 million tonnes.

FERTILISER CONSUMPTION

India has come a long way, Since independence, in respect of production and consumption of fertilizers. In the year 1951-52, the country produced a mere 27 thousand tonnes of fertilizers and consumed 59,000 tonnes of nitrogenous and 7,000 tonnes of phosphatic fertilizers totaling 66,000 tonnes. Capacity expansion in fertilizer production, through the establishment of additional fertilizer manufacturing units, started during the years 1967-68 with the production of 0.6 million tonnes of fertilizers. The country has produced 10.44 million tonnes and consumed 13.83 million tonnes of NPK fertilizers in the year 1994-95.

The per hectare consumption of fertilizers, which was as low as 0.5 kg during 1951-52 has now increased to 89.9 kg. in 1998-99. The increased use of fertilizers beginning with the year 1966-67 coupled with the use of high- yielding varieties of field crops and expansion of area under irrigation was responsible for the " Green Revolution '' leading to appreciable shifts in the productivity and production of food grains such as wheat, rice , maize and sorghum. The adoption of fertilizer responsive high-yielding varieties and hybrids of field crops for cultivation on large scale in irrigated areas is one of the principal reasons for increased fertilizer use[4].

Fertilizer has certainly played a very important role in India's green revolution, but yet pehectare consumption of fertilizer is still much less as compared to neighbouring countries in Asia[5] (Table 1) . While the all – India average consumption is at 90kg/ ha, Punjab used 177 kg/ha , Delhi 229 , Pondicherry 481, Andhra Pradesh and Tamil Nadu both around 150 kg / ha during 1998-99 . However a disturbing feature in fertilizer consumption is an apparent imbalance in the use of N, P_2O_5 and K_2O (nutrient consumption ratio) . The N : P_2O_5 : K_2O consumption ratio figures are 5.9:2.4:1.9., 7:2.3:1; 7.9:2.9:1 and 8.5:3.1:1 for 1991-92, 1993-94, 1997-98 and

Table 1: Fertilizer consumption in India and some Asian countries [5]

Country	Consumption (kg ha^{-1})
India	84.3
Bangladesh	142.8
China	266.4
Japan	360.5
South Korea	479.4
Pakistan	115.5
Sri Lanka	108.3

1998-99, respectively for the country as a whole[6]. Consumption of fertilizer per unit cropped area in some Indian states has been shown in Table 2.

Table 2: Fertilizer consumption (kg/ha) in some states of India (1997-98)[7]

Sate	N consumption (kg/ha)		State	N consumption (kg/ha)	
Delhi	347.5	(3.4)	Gujrat	62.8	(1.5)
Pondicherry	260	(1.1)	West Bengal	62.7	(1.3)
Punjab	130.6	(1.1)	Karnataka	50.4	(1.5)
Haryana	109.0	(1.1)	Maharastra	45.7	(1.3)
Chandigarh	105.0	(-3.0)	Jammu and Kashmir		
Uttar Pradesh	91.9	(1.4)	Manipur	40.7	(1.3)
Andhra Pradesh	84.1	(1.0)	Madhya Pradesh	30.3	(1.8)
Tamil Naduu	72.2	(1.2)	Rajasthan	29.0	(1.9)
Bihar	65.4	(1.6)	Kerala	28.5	(1.0)

(   ) Indicate the increase (in times) in nitrogen consumption in 1997-98 (kg/ha) as compared to 1989-90.

ENVIRONMENTAL CONCERNS

Fertilizer in general and N fertilizer in particular are at the center of a controversy. Well managed fertilizer use can contribute to increased food production and reduced degradation of natural resources. But excessive use of fertilizer can also contribute to nitrate leaching, eutrophication, cadmium uptake by plants, and greenhouse gas emissions. Hence balanced perspective on fertilizer is required so that food production is maximized, soils are maintained in a productive state, and environmental pollution is minimized.

Environmental impacts associated with fertilizers can be divided into the effects associated with fertilizers use and those associated with fertilizer production. In each category, impacts can be further distinguished by the source of the pollution where it occurs. Most of the pollution problems associated with fertilizer production occur at the point of production. They include disposal of phosphogypsum; emissions of fluorine, sulfurdioxide, and nitrous oxides; and waste water disposal from production facilities. Because these pollutants occur near the factory, they can easily be controlled. On the other hand, nitrate leaching and entrophication are associated with fertilizer use; they fall in the category of nonpoint pollution. In these cases, the polluter is separated from the polluted objects, and therefore the pollution problems are difficult to control. For example, when fertilizer is carried away by water runoff or soil erosion, it may contaminate lakes or rivers and result in eutrophication, leading to algae growth. It is difficult to fix the responsibility for that pollution because the damage occurs far from the site of pollution: the "polluter pays" principle cannot be easily applied. Similarly, ground water can be contaminated by excess N from organic matter such as animal manures, from inorganic fertilizer, or from natural sources such as mineralization of soil as well as from sewage sludge, septic tank drainage, or industrial waste. Hence, it is difficult to implicate a single source, say fertilizer, without adequate monitoring, measurement, and analysis [8].

Of the three main fertilizer types, only K_2O fertilizer has no known adverse effects on the environment. N and P_2O_5 fertilizer, when not managed properly, can affect the environment adversely.

Nitrate Leaching

High levels of nitrates in drinking water are considered harmful to human health, especially to infants, because they can cause a rate condition called methemoglobemia, or what is known as blue-baby syndrome. The World Health Organization (WHO) has recommended that nitrate levels in drinking water should not exceed 50 miligrams of nitrate per liter of water. Both the European Union (EU) and the Environmental Protection Agency of the United States (USEPA) have recommended monitoring nitrate levels in water and taking corrective measures in those areas where nitrate levels exceed the recommended limits [9,10]

Excessive use of fertilizers in the cultivation of crops like rice, maize, sugarcane, cotton, grapes and other commercial crops like chillies has grave consequences. Concerns are voiced, both in respect of health and environmental quality. The major focus has been on issues related to pollution due to injudicious use of the very mobile and highly water soluble fertilizer nutrient nitrogen [4].

Urea is the major nitrogenous fertilizer used in the world because of its low unit cost, high solubility and non-polarity. However, it is susceptible to various N loss mechanisms, namely, ammonia volatilization, leading, and denitrification. Ammonia that emanates from the urea applied to agricultural fields contributes to acid rain, while nitrates produced in sol contribute to ground water pollution due to leaching of nitrates, and ozone depletion because of the release of nitrous oxide by denitrification. Consumption of urea may increase several fold by the end of the 21st century. Available data on nitrogen use efficiency indicate that 30 to 50 percent of the applied urea-N is lost, contributing to health and environmental hazards [11].

In a study of 236 ground water samples, 21-38m deep, collected from different blocks (Table 3), 78.4 percent samples contained less than 5mg NL^{-1} and 21.6 percent showed nitrate levels above 5 mg NL^{-1} [12]. In 367 ground water samples, collected from 9 to 18 m deep had pumps located in village inhabitations showed appreciable nitrate concentrations. Sixty-four percent samples showed nitrate between 5-10mg NL^{-1} and 2 percent showed nitrate levels above 10mg N/1 against safe limit of 10mg N.

It may be interest that nitrate in water supplies in concentrations above 10mg/1 have led to numerous cases of methaemoglobemia in infants. Some nitrosamines, formed by the reaction of nitrate with secondary amines, are suspected to be carcinogenic. However, research in this field is inconclusive. Very high nitrate concentrations, coupled with increasing trend have been reported in some parts of the world. In India, nitrate concentration, as high as 530 mg/1, has been reported in Churu district of Rajasthan (Table 4) [13].

Table 3: Nitrate concentration in ground water samples from tube wells located in cultivated areas of Punjab [12].

Block	N-Fert. Application (kg $ha^{-1}yr^{-1}$)	No.of samples (Wells)	Range (NO_3 mgL^{-1})
Dehlon	249	84	1.14-6.72
Ludhiana	258	33	0.50-5.9
Sudhar	242	43	1.2-6.48
Kartarpur	193	34	1.14-6.72
Jandialaguru	172	24	2.20-6.42
Lalerkotla	151	18	1.40-5.8

Table 4: Nitrate concentrations in water supplies of few cities in India[13]

Place	Highest nitrate concentration observed (mg L^{-1})	Possible nitrate source
Churu	530	Geological deposits
Meerut	156	Fertilizers, insanitary Waste disposal
Jaipur	180	Leaching from septic tanks
Nagpur	77	Leaching from septic tanks
Hyderabad	78	Insanitary waste disposal

The nutrient imbalances arising due to excessive use of single nutrient (N-fertilizers) have serious consequences. Prior to 1965, in India, the native soil fertility supported sub-optimal yield levels. But, with the advent of high yielding fertilizer responsive varieties of crops like wheat, maize, rice, sugarcane, sorghum, cotton, chillies and others it was possible to obtain high biomass and grain yields of the order of 10 to 12 tha^{-1} with the application of nitrogen alone because of its priming effect on other nutrients which enabled crops to absorb macro and micronutrients from the native nutrient pool of soil. But, with continuous cropping the native nutrient pools got depleted fast leading to deficiencies of phosphorus and many secondary and micronutrients such as sulphur, magnesium, zinc, iron and manganese[4].

Eutrophication

When P_2O_5 and N fertilizer is carried away by water runoff and soil erosion to lakes, rivers, and other water bodies, it contributes to excessive growth of algae, which can result in oxygen depletion and fish mortality. The aesthetic and recreational value of water bodies is also reduced. Eutrophication is basically a fertilizer management problem and can be prevented by reducing erosion and runoff[8].

Cadmium

When taken in large quantities, cadmium is hazardous to human health. The World Health Organization guidelines [14] suggest that cadmium intake of up to 1 microgram per kilogram of body weight per day is not harmful to humans. Just as nitrate leaching can occur from several sources, cadmium levels in the soil can also increase from P_2O_5 fertilizer, animal waste, sewage sludge, and industrial and atmospheric deposits. In P_2O_5 fertilizer cadmium comes from phosphate rock. Many sedimentary phosphate rocks found in Morocco, Togo, Tunisia, and the United States have high levels of cadmium, ranging between 35 and 55 milligrams per kilogram of rock[15]. However, the mechanisms by which cadmium is transferred from direct applications of phosphate rock or finished P_2O_5 fertilizer to soil, plants, and humans are complex and require further research. At the Rothamsted Station (in Hertfordshire, U.K.), even after 100 years of use of P_2O_5 fertilizer, an insignificant amount of cadmium was found in grains, whereas leafy crops, such as tobacco and spinach, had picked up a considerable amount of cadmium from the soil [16]. Additional research is needed to develop proper regulatory measures. In any case, because P_2O_5 fertilizer use is still limited, it is unlikely to pose a major health risk in the near future. Even in Western Europe, where P_2O_5 fertilizer has been applied in higher rates over longer periods, cadmium intakes have, on average, been low [17]

The Greenhouse Gases

Some scientists have predicted that increasing concentrations of gases such as carbon dioxide, nitrous oxide, and methane in the atmosphere will cause rising temperatures and global warming. Fertilizer use and production have the potential to contribute to several of these gases. Oxides of N can emanate from N fertilizer use, especially in paddy fields, and CO_2 from fertilizer production facilities, However, the contribution made by fertilizer to greenhouse gases is likely to be negligible[18,19]. Better understanding through more research is needed rather than regulatory measures. Well managed fertilizer use can actually reduce global warming by sequestering carbon in the soil organic matter[8].

Environmental Concerns Associated with Fertilizer Production

The environmental concerns associated with fertilizer production are few and well understood; technologies to minimize their adverse effects in most cases are also well known. The major issue remaining pertains to the cost of installing technologies and who should pay those costs.

Pollutants associated with nitrogen production

The pollutants associated with ammonia and urea production come in various forms gases, liquids, and solids; discharge of these pollutants can adversely affect the community and the atmosphere. For example, the discharge of waste water from ammonia plants can add to nitrate levels, and the emission of nitrous oxide, sulfur dioxide, and CO_2 can contribute to greenhouse gases and acid rain. With proper technologies and regulation, these emissions and pollutants can be minimized [20]

Pollutants associated with phosphate production

Phosphogypsum is a by product of production of phosphoric acid. For every ton on phosphoric acid produced, 4 to 5 tons of phosphogypsum are produced. Because phosphogypsum contains radium, it can emit radon, a radioactive gas, which is hazardous to both humans and animals. To safeguard against its harmful effects, phosphogypsum should be deposited in covered stacks or ponds. Although it can be used for agricultural, industrial, and road building purposes, economic considerations do not justify such uses on a large scale[17,21]. Disposing of phosphogypsum in an environmentally friendly way would have cost \$5 to \$80 per ton of P_2O_5 in the United States in 1988/89 [21]. In addition, land reclamation would have added \$1 to \$5 per ton and process water treatment another \$ 20 to \$70 per ton. Treating these pollutants in the phosphate industry in 1988/89 would have increased the cost of production of DAP by \$175 per ton of P_2O_5 in the United States alone. Although these cost estimates pertain to the United States, they are indicative of the cost implications of environmental measures in other countries.

Policy Measures for Environmental Protection

Environmental problems in general and those related to the fertilizer industry in particular can be attributed to three factors; market failure, policy failure, and the knowledge gap. The market failure argument suggests that environmental problems are caused by the nonexistence of markets for environmental goods. For example, a fertilizer factory dumps waste products say, phosphogypsum in the river because it is a free good and no one owns it. Subsequently, if the factory is required to pay the cost of treating the river, then it will find ways to prevent the damage caused by the pollutants. In economics, this is known as "internalizing the externality". The policy failure argument suggests that the pursuance of wrong policies can lead to environmental damage. For example, excessive crop price support programs and input subsidies contribute to excessive use of agrochemicals such as pesticides, causing harm to both humans and the environment. The knowledge gap argument implies that lack of proper knowledge about technologies, products, and practices leads to environmental damage. Eutrophication resulting from fertilizer runoff is an example of knowledge failure. Based on these and other factors, the following policy measures should be implemented.

➢ To reduce excessive input usage, it is necessary to rationalize the application of these inputs based on scientific knowledge already generated. Alternative strategies like integrated plant nutrient management, biological control, cropping systems, sustainable farming systems, sustainable farming systems, multiple cropping and judicious water management should be adopted without delay. Considerable amount of information is available on these aspects for easy access.
➢ Organic manuring, green manuring and compost utilization should be encouraged not only to improve soil productivity but also to improve and increase the efficiency of microbial activity and fertilizer use efficiency. Composting of organic residues and their application to soil reduces the problem of methane emission from submerged soils as compared to direct application of organic residues.
➢ Land, water bodies and atmosphere are getting polluted irreversibly damaged by industrial activity. The Environmental Impact Assessment guidelines already available should be made mandatory for the industry to comply.
➢ Development of a sound database on natural resources and pollution problems has not received the attention it deserves by many member governments and the decision making process have been evolved without the benefit of any access to detailed information. In order to arrive at correct judgement based on data analysis, a sound database is essential.
➢ Sustainability issues and resource degradation should not be treated in isolation.
➢ National land use policies have a direct impact on environment sustainability. The need for a national land use plan should be urgently addressed to by the national governments.
➢ Environment Impact Assessment (EIA) should not be confined to project level. It should address policies and programmes as well.
➢ There should be a greater interdependence between agricultural development and environment sustainability to ensure that limits are observed and respected.
➢ Efficiency of fertilizer (nutrient) use has to be enhanced. All possible technologies and methods including modification of fertilizer materials, time, method and dose have to be employed.
➢ Appropriate mathematical models should be developed to study the direct, residual and cumulative effect of fertilizer, so that the fertilizer use efficiency can be evaluated in the proper perspective.

> Research that leads to development of technologies and methods for reduction of losses of fertilizer N through denitrification, leaching and volatization shall be one definite step to economize on the use of fertilizer.
> Technologies developed to deal with fertilizer pollutants in developed countries to be tested under our local conditions before their adoption.

> Policies for environmental monitoring should be introduced, especially in those developing areas where fertilizer use levels are excessive, and proper fertilizer management practices should be encouraged through research, extension, and education of farmers. Further research should be conducted to increase understanding of the environmental interactions of fertilizer production and use.

Many countries are introducing policy reforms to restructure their economies in general and fertilizer sector operations in particular. Unless these policy changes are phased and sequenced properly, they may cause steep reductions in many countries[22,23]

Although several policies affect fertilizer sector operations, policies dealing with devaluation, subsidy removal, and privatization seem to have a profound impact[23]. The depreciation of domestic currency (devaluation) generally results in increased fertilizer prices for farmers and higher raw material and equipment prices for fertilizer producers. Consequently, fertilizer use decreases because crop prices generally do not keep with the resulting inflation. The removal of fertilizer subsides at the same time only adds to the declining trend. Thus, during rapid devaluation, some safety nets should be put in place to prevent too sharp a decrease in fertilizer use. Further, when the fertilizer market is shrinking due to devaluation and subsidy removal, sudden withdrawal of the government from production, import, marketing, and distribution to promote privatization is not desirable. Successful privatization is a slow and time-consuming process, requiring investment in institutional and physical infrastructure and management skills.

On the other hand, when devaluation, subsidy removal, and privatization are introduced gradually and are supported by the development of adequate human and institutional infrastructure and regulatory mechanisms, they promote growth in fertilizer use. Thus, policy reforms should be introduced in such a way that they promote growth in fertilizer use[8].

CONCLUSIONS
Because fertilizer use levels were low in the past, there was little need to be concerned about the environmental impacts associated with fertilizer use. In the future, however, environmental monitoring should receive higher priority because fertilizer use levels are climbing rapidly.

Promoting future growth in fertilizer use without causing harm to the environment will pose several challenges. First, to reduce the knowledge gap, farmers will have to learn how to apply fertilizer properly and efficiently. Because many farmers are illiterate, this will require a large effort. Second, additional research will be needed to understand the dynamic interaction of fertilizer use with environment. Developing programs to reduce runoff losses, cadmium uptake by plants, and nitrate leaching from fertilizer and non fertilizer sources will require further research and technology development work. New management practices will have to be developed. Third, policies will have to be redesigned: Although the removal of fertilizer and crop subsidies that lead to excessive fertilizer use is a high priority, new incentives to promote environmentally friendly agronomic practices, such as cereal – legume rotations, should also receive adequate attention. Generally, pricing policy is a better tool than regulatory policy (such as quantity restrictions on fertilizer use), although Pricing policy alone (including taxes on fertilizer use) may not be sufficient to reduce or eliminate environmental damage; other policies and programs leading to better knowledge and practices should be encouraged[8].

90

REFERENCES

1. NAAS. National Academy of Agricultural Sciences. (1999). Agricultural Scientists perception on Indian Agriculture Scene, Scenario and vision, New Delhi.

2. NAAS. National Academy of Agricultural Sciences. (1996). Agricultural Scientists perceptions – plant nutrient needs, supply, efficiency and policy, issues 2000-2005 New Delhi.

3. FAI. Fertilizer Association of India. (1999). Fertilizer News 44:105.

4. Subba Rao, I.V. (1999). Soil and environmental pollution – A threat to sustainable agriculture. J. Indian Soc. Soil Sci. 47:611-633.

5. FAI. Fertilizer Association of India. (1998). Fertilizer statistics. 1997-98, New Delhi.

6. Goswami, N.N. (1999). Priorities of soil fertility and fertilizer use research in India. J.Indian Soc. Soil. Sci. 47: 649-660.

7. Deepanjan Majumdar and Navindu Gupta. (2000). Nitrate pollution of ground water and associated human health disorders. Indian J. Environ Hlth. 42: 28-39. .

8. IFPRI. International Food Policy Research Institute. (1996). The role of Fertilizer in sustaining food security and protecting the environment to 2020. Food, Agriculture and the environment discussion paper 17:38-43.

9 USEPA. Environmental protection agency of the United States. (1990). National Pesticide Survey Project – Nitrate, Washington D.C.

10. CEC. Commission of the European Committees. (1991). Council directive of 12 December 1991 concerning the protection of waters against pollution caused by nitrates from agricultural sources. Official Journal of European Communities. 34:1-8.

11. Rajendraprasad (1998) Current Science 75:677.

12. Bajwa, M.S., Singh, B. and Singh. P. (1992). Nitrate Pollution of Groundwater under different system of land management in the Punjab. Contributed papers First Agril. Sci. Congress. 223-230.

13. Mathur, Y.P and Pradeepkumar. (1990). Indian J. Environ. Health. 32:97.

14. WHO.World Health Organization. (1972). Sixteenth report of the joint FAO /WHO expert committee on food additives. Technical report service no. 505. Geneva.

15. Bockman, O.C., Kaarstad, O., Lie, O.H. and Richards, I. (1990). Agriculture and fertilizers: Fertilizers in perspective. Oslo , Norway, Norsk Hydro.

16. Johnston, A.E., and Jones, K.C. (1992). The cadmium issue – long term changes in the cadmium content of soils and the crops grown an them. In phasphate fertilizers and the environment – proceedings of a international workshop, (ed). J.J. Schultz.SP-18. Muscle shoals, Ala. U.S.A. International Fertilizer Development Center.

17. Isherwood, K.F. (1992). Phasphate Industry and the environment. In phasphate fertilizers and the environment proceedings of an international workshop, ed. J.J. Schultz. SP-18. Muscle shoals, Ala. U.S.A. International Fertilizer Development Center.

18. Byrnes, B.H. (1990). Environmental effects of N fertilizer use – an overview. Fertilizer Research. 26:209-215.

19. IFDC. International Fertilizer Development Center. (1993). IFDC Annual Report, 1992, Muscle shoals . Ala. U.S.A.

20. Frederick, M.T., and Lazo delaveg, J.R. (1992). World ammonia / urea production outlook – The magnitude of the environmental impact problem . In environmental impact of ammonia and urea production units. Ed. D.R. Waggoner and G. Hoffmeister. Proceedings of an International Workshop – SP-17. Muscle shoals. Ala., U.S.A. International Fertilizer Development Center and The Fertilizer Association of India.

21. Schultz, J.J., Gregory, D.I and Engelstad, O.P .(1992). Phesphate fertilizers and the environment – A discussion paper. P-16. Muscle shoals, Ala., U.S.A International Fertilizer Development Center.

22. Bumb, B.L. (1989). Global Fertilizer perspective, 1960-95: The dynamics of growth and structural change. T-34 and T-35. Muscle shoals, Ala., U.S.A. International Fertilizer Development Center.

23. Bumb, B.L. (1995). Global Fertilizer perspective, 1980-2000: The challenges in structural transformation. T-42. Muscle shoals, Ala., U.S.A International and fertilizer development center.

Land Corridor for Environmental Sustenance of—to be Reclaimed/Reclaimed, Decoaled Areas in the North—Eastern Part of Wardha Valley Coalfield—A Concept

M.K. Shukla
Western Coalfields Ltd., Nagpur, India

INTRODUCTION

Wardha Valley Coalfield constitutes the most productive of the four geographically identifiable Coalfields within the administrative domain of Western Coalfields Ltd. While the readily accessible parts of the coalfield have been scientifically explored by drilling, the environmentally sensitive i.e. those areas which lie adjacent to or within the forest areas, are being targeted to meet the future demands.

Amongst such areas, the North-Eastern part of the coalfield, sections of which are already under coal extraction and where additional strike extension of coal has been proved, are slated to go into production to meet and sustain coal demand from this coalfield. This has become necessary, more so, on account of the geo-techno-economic attributes of the coal inventory available there as compared to those available elsewhere.

In order to extend the already available sites and to open up additional areas, it has been the endeavour of the Western Coalfields Ltd., to proceed cautiously and systematically allaying fears of deterioration and destruction of environment which gets assigned to the coal mining operations/transportation/utilization etc. It is in this context, a constant appraisal and assessment of the inputs required to provide for environmentally friendly approach is undertaken.

Keeping the above in view, preliminarily, the Western Coalfields Ltd., has been pursuing a programme * (in collaboration with National Environmental Engineering Research Institute Nagpur) Wherein evaluation of approaches to reclaim and develop mine spoil dumps has been undertaken since late 1993. This has resulted in developing an Integrated Biotechnological Approach (I.B.A.) wherein locally available ameliorative materials have been successfully utilized in enhancing plant growth and their proliferation supplemented by inoculation of plants with specialized cultures of endomycorrihizal fungi.

- Reclamation and Development of Mine Spoil Dumps at W.C.L. Mines – Phase I & II.

The added dimension to the study above, is the latest project ** in progress wherein it has been proposed to develop the biodiversity of Tadoba National Park and Tiger Reserve on overburden dumps and backfilled areas of adjacent Padmapur Open Cast Mine.

Nevertheless, while ways and means of extracting coal without disturbing the biota are under consideration, rapid reclamation and restoration techniques for areas bordering forest where coal extraction is in progress or likely to take place in near future, are being developed. The necessity of these have been felt in view of the location of these and the requirements to put in place environmentally sensitive/friendly systems.

This paper deals with an aspect of reclamation and restoration of these coal bearing lands preferably to

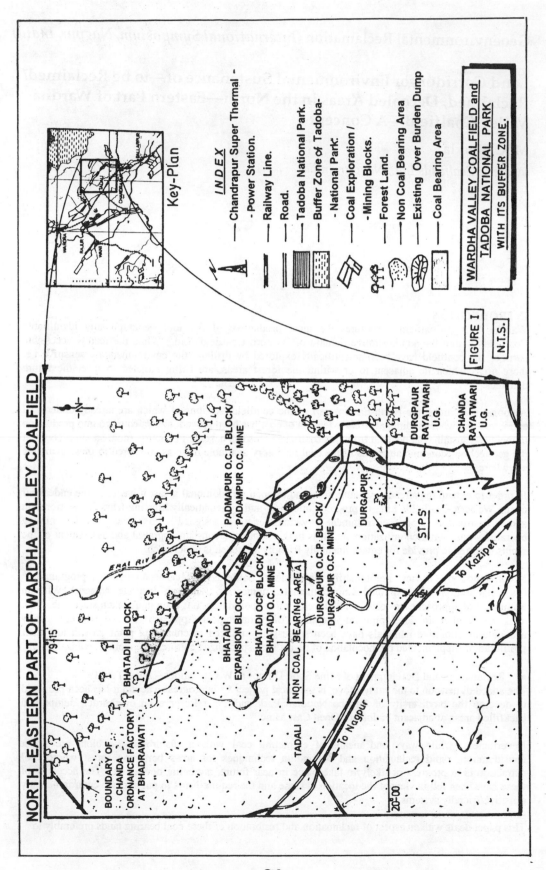

NORTH-EASTERN PART OF WARDHA-VALLEY COALFIELD

Key-Plan

INDEX

- — Chandrapur Super Thermal - Power Station.
- — Railway Line.
- — Road.
- — Tadoba National Park.
- — Buffer Zone of Tadoba-National Park.
- — Coal Exploration /-Mining Blocks.
- — Forest Land.
- — Non Coal Bearing Area
- — Existing Over Burden Dump
- — Coal Bearing Area

WARDHA VALLEY COALFIELD and TADOBA NATIONAL PARK WITH ITS BUFFER ZONE.

FIGURE I

N.T.S.

BOUNDARY OF CHANDA ORDNANCE FACTORY AT BHADRAWATI

ERAI RIVER

BHATADI II BLOCK

BHATADI EXPANSION BLOCK

BHATADI OCP BLOCK / BHATADI O.C. MINE

NON COAL BEARING AREA

TADALI

To Nagpur

PADMAPUR O.C.P. BLOCK / PADAMPUR O.C. MINE

DURGAPUR O.C.P. BLOCK / DURGAPUR O.C. MINE

DURGAPUR.

STPS

To Kazipet

DURGPAUR RAYATWARI U.G.

CHANDA RAYATWARI U.G.

79·15

20·00

94

put them back to forest use after coal extraction, utilising the data already generated in the three projects undertaken by Western Coalfields Ltd., into a package providing environmentally sustained solutions.

SURFACIAL COAL EXTRACTION Vs FOREST LAND AND LAND CORRIDOR
The areas which have been covered in the present paper (Figure 1) consist of the contiguous geological blocks namely;

(i)   Northern part of Durgapur Open Cast Project (O.C.P.) Block/Durgapur Open Cast Mine (towards South)
(ii)   Padmapur OCP Block/Padmapur Open Cast Mine.
(iii)   Bhatadi OCP Block/Bhatadi Open Cast Mine.

** Restoration of Biodiversity on overburden Dumps and Backfilled Areas using Integrated Biotechnological Approach. (in collaboration with National Environmental Engineering Research Institute, Nagpur)

(iv)   Bhatadi Expansion Block
(v)   Bhatadi II Block (towards North)

Further towards North of Bhatadi II Block, the coal seam appears to have been preserved at a depth of more than 150 meters, apparently within a faulted trough upto the southern boundary of Ordnance factory at Bhadravati.

In the four identified blocks out of five above, the major thick coal seam (15 m to 25 m) is quarriable except in part of Bhatadi Expansion Block where it does not incorporate at shallow depth but is available at a depth of >60m. and beyond.

The coal seam which strikes almost North-South (Northern part of Durgapur Block) becomes NNW – SSE to NW –SE direction in the contiguous blocks upto Bhatadi II Block. The coal deposit dips towards East (in Southern Blocks) and NE (in Northern Blocks). The definition of the individual blocks has been on the basis of major structural dislocation except between Padmapur OCP Block and Bhatadi OCP Block where the Erai river is present in between.

The dip side of all these block consist of forest areas, parts of which are extremely dense. These forest areas lie incontinuity with the Tadoba National Park and Tiger Reserve and as per available records provide for sighting of wild life too.

The time project coal production schedule from the blocks identified above is as provided in Table – I.

Table-I : Showing Blocks/Mines in North-Eastern part of Wardha Valley Coalfield and their time projected production schedule.

Sr. No.	Name Of Block/ Mine	TIME PROJECTED PRODUCTION SCHEDULE								
		1999 2000	2000- 2001	2001- 2002	2002- 2003	2003- 2004	2004- 2005	2005- 2006	2006- 2007	2011- 2012
1.	Durgapur OCM	(already in operation)								
		(1.74)*	(1.80)	(1.80)	(1.80)	(1.80)	(1.00)	(0.70)	(0.50)	
2.	Padmapur OCM	(already in operation)								
		(1.27)*	(1.20)	(1.11)	(1.11)	(1.11)	(1.11)	(1.11)	(1.11)	
3.	Bhatadi OCM	(already in operation)								
		(0.30)*	(0.30)	(0.30)	(0.30)	(0.20)	(0.05)			
4.	Bhatadi Expn.	(to come into operation)								
				(0.10)	(0.25)	(0.25)	(0.55)	(0.55)	(0.55)	(0.55)
5.	Dhatadi-II			(to come into operation)						
										(0.25)

Figures within bracket ( ) are projected production in million tonnes.( )*-Actual production figures in million tonnes.

The entire overburden of these quarriable blocks is presently being stacked on the rise side, as it is a

95

non-coal bearing area. Over a period of time, the ultimate scenario would consist of overburden dumps towards East/North-East of the quarries. Of course, depending upon planned approaches and it's implementation, backfilled/filled-in quarried area would also be available providing for levelled land upto forest area in between or adjacent to working quarries.

In this most likely scenario or even before the consummation of the picture as above, it is suggested that such levelled off areas be developed as (protected) "LAND CORRIDOR" linking the forest area with the reclaimed/to be reclaimed areas allowing for unrestricted and protected propagation of both plant and animal species having provided a catalytical growth of plants at the same locations as has been done at experimental overburden dumps sites etc.

LAND = CORRIDOR = CONCEPTUALISATION – PLANNING AND IMPLEMENTATION.

"Corridor " as the name suggests, consists of a passageway with an impinging connotation of security attached to it. In the present context, corridor on land implies secured land piece with/without positive environmental attributes – which allows for free and unfettered movement, exchange, etc. Of biological element alongwith attending upgradation of environmental parameters of the passageway between forest land and the reclaimed/to be reclaimed lands.

It has now become a part and parcel of the planning/exploitation/closure procedure of any coal mining project to possess an inbuilt mechanism of simultaneous/later-backfilling/filling of an adjacent quarry leading as far as possible to release of land to it's original/other productive uses. Normally on economic consideration, it is preferred to simultaneously backfill/fillup an adjacent already excavated/decoaled area rather than to generate external overburden dumps. It is also evident from the schedule of coal production programme that areally time staggered adjacent excavated and filled in areas can/would exist in the area under reference. Under these circumstances, such areas can be identified and developed as "corridors" between forest area and filled in areas/overburden dumps for environmental regeneration of land within and beyond it.

It is nevertheless important that such identified areas should primarily possess the following characteristics.
I.   The area should be wholly and totally free of any type of activity except those which are primarily necessary to catalyse biota related generation schemes. These will serve as add-on to the process of biological transgression from the forested area into the passage way and beyond.
II. In order to put the aforementioned activities into place, the are would have to be sufficiently insulated, preferably with a bufferzone since development-driven procedures would be active in adjacent areas. The elongated core zone should possess sufficient width to sustain growth and allow for endemic plant species to transgress in space over time even at the expense of exotic species, which would have been planted to act as a precursors to the transgression.
III. The plant related growth would invariably attract movement of animals and if water bearing quarries beside the corridor are available as water holes, a more active animal transgression cannot be ruled out. Adequate protection as well as space would have to be provided/created for proliferation and propagation of biota depending upon guestimated nature and type of biotic population. The appropriate environmental enhancement of the peripheral zone around the main forested area would add to the sustenance of environmental growth of the are.

1)   In addition to the above factors uniqueness of individual sites would determine additional prerequisites required by the corridor.

PRESENT STATUS AND FUTURE PROJECTIONS
The Integrated Biotechnological studies on land reclamation and development of mine spoil dumps. Phase II & I along with the project on restoration of biodiversity on overburden dumps and backfilled areas etc. Have provided for-
a)   Cost of plantation with consequent reclamation of land @2500 trees/Ha comes to around Rs.1, 25,000/Ha.

b)   Availability of information that even ecologically desirable and economically important trees could also be planted for reclamation.

c)   Information on gainful utilisation of discharged/wasted material from nearby industries like paper and sugar mills etc. for amelioration of overburden material aiding plant growth.

d)   Identification of specialised cultures of endomycorrihizal fungi for bio-remediation of degraded lands. This in conjunction with (c) above constitutes the Integrated Biotechnological Approach (I.B.A.) which appears to have helped achieve around 90% and above survival rate of the plants and 4-6 times higher growth rate.

e)   Data suggesting that within a period of around 4 – 5 years, an area wherein Integrated Biotechnological Approach aided reclamation is undertaken, the inter-plantation area together with its surroundings regenerates and catalyses to accept increased biotic growth elements.

f)   Inventory preparation of ground flora and microbial population of Tadoba National Park and Tiger Reserve and data on the interaction between different components of the forest eco-system, and

g)   In addition to the other details, a well defined scheme for establishment of biodiversity of Tadoba National Park and Tiger Reserve on overburden with the conclusion of the presently running programme by July, 2001.

It is noteworthy, therefore, that a lot of data has been generated and that with additional linking inputs an integrated approach utilising land corridors could be evolved to revert such lands, as are under reference, back to the forest. What is more important is that such reversion would take place at a much faster pace and with a qualitative difference.

CONCLUSION

While the specifics of advantages accurring to the company out of such reversion could be determined over time, Western Coalfields Ltd. would have achieved another landmark in "ENVIRONMENTAL STRATEGY IN COAL MINING INDUSTRY' using the suggested concept of "Land Corridor".

ACKNOWLEDGEMENT

Thanks are due to Shri C.M. Singh, Sr.Survey Officer, and his team and Shri B .R. Vijay Kumar Naidu, Dark Room Assistant/Ferro Printer for preparation of figure

Biogas Generation Potential in India Through Anaerobic Digestion of Agrobased Industrial Wastewater—A Step Towards Environmental Reclamation

Narendra M. Kanhe[1], Arun Kansal[2], Rajesh Gupta[3] and Anand G. Bhole[4]

[1]*Ph.D. Scholar, V.R.C.E., Nagpur and Lecturer in Civil Engg., A.S.T.S., Wardha, India*
[2]*Lecturer in Env. Engg., School of Env. Management Indraprastha Univ., New Delhi, India*
[3]*Asstt. Prof., Department of Civil Engg., V.R.C.E., Nagpur, India*
[4]*Emeritus Fellow, L.I.T., Nagpur and Formerly, Prof. of Env. Engg., V.R.C.E., Nagpur, India*

INTRODUCTION

We are in the age of energy crisis and environmental pollution. Fossil fuels (coal and petroleum), our conventional energy sources are getting exhausted due to over-mining. These resources are not reproduced quickly. Millions of years are required in conversion of plant and animal remains to coal and petroleum. The rate at which these are utilised today, these may last a few decades more. Hence there is an urgent need to find alternative source of energy and preserve the conventional energy resources. Another major problem of today's concern is of environmental pollution due to the discharge of untreated industrial wastewater.

Biogas generation through anaerobic digestion of industrial wastewater is one of the cost effective and eco-friendly solution for above problems. Any wastewater, which is rich in biodegradable organic matter with adequate nutrients and not containing toxic compounds, is suitable for anaerobic treatment. India has a significant potential for the implementation of anaerobic digestion technology for biogas generation as explained later in this paper.

A survey was conducted to know the potential of biogas generation in India through anaerobic digestion of different agrobased industrial wastewater. Likely production of biogas generated through anaerobic digestion of wastewater from different industries such as distillery, sugar, dairy, edible oil and coffee was estimated. It was found that out of the total potential of biogas generation through anaerobic decomposition of agrobased industrial wastewater, 80% potential lies in the distillery, sugar and dairy wastewaters. This paper provides details of distillery, sugar, and dairy industries regarding production of biogas.

DISTILLERY INDUSTRY

There are about 218 large and medium scale distilleries in India (CIER's Industrial Data Book, 1998). Alcohol manufacture in Indian distilleries is based either on molasses or malt-barley. Rectified spirit or neutral alcohol is obtained as the final product of distillation process. The residue of the distillation process is the spent wash which is a strong organic effluent. The other wastes from the process include yeast sludge (which is usually mixed with spentwash), floor washes, waste cooling water and waste from the operations of yeast recovery or by-products recoveries processes. The characteristics of the composite distillery effluent are depicted in Table 1 :

The very high content of organic matter in the wastewater makes it unsuitable for direct aerobic treatment and therefore a two stage biological treatment consisting of anaerobic followed by aerobic treatment is usually adopted. However, to make the anaerobic treatment process viable, the effluent needs pretreatment for pH and temperature correction and lime scrubbing of biogas for H_2S removal (if the biogas is to be used for power generation.)

A 16 Mld diphasic anaerobic plant based on hybrid reactor system has been installed at Daurala Sugar Works, Daurala (U.P.) . The acid phase digestor is designed for 1.5 days HRT with COD loading rate of 50 Kg/m$\geq$ day while the methane phase reactor has an HRT of 9-10 days with a COD loading rate of 9-11 Kg/m$\geq$ day. A COD reduction of about 70%, with a biogas production rate of 33-35 m$\geq$/m$\geq$ of effluent with a methane content of 63-65% has been reported.

TABLE 1 Characteristics of composite distillery effluent

Parameter	Value
Wastewater generation	12-16 liters per liter of alcohol production
Temperature	70 – 95 ° C
pH	4.0 to 5.5
Colour	Dark Brown
COD	1,00,000 – 1,30,000 mg/l
BOD	40,000 – 60,000 mg/l
Total Solids	80,000 – 1,20,000 mg/l
Volatile Solids	4000 – 9000 mg/l
Nitrogen	900 –1000 mg/l
Phosphorous	40 – 50 mg/l

Manihar (1995) conducted pilot studies on a 100 liter fixed film reactor with varying COD loading in order to optimise the treatment efficiency and methane production. The unit consisted of separate reactors for acidification followed by biomethanation. The 20-liter acidification reactor was constantly agitated and operated at an HRT of 2.5 days to obtain a COD reduction of 10-15%. The fixed film methane reactor consisted of a 100 liter fixed bed filled with random PVC rings which has a void volume of 97% and a surface area of 320 m''/ m$\geq$. The reactor was operated in upflow mode with a varying recirculation rate of 15-25 times the inlet effluent rate. The maximum COD loading rate achieved was 25 kg /m$\geq$ day with an overall COD reduction of 60-70%. The gas yield was 0.4 m$\geq$/kg of COD removed with methane content of 60-70%.

Libin et al (1997) conducted experiments on UASB reactor system. The unit was operated at 35° C with an HRT of 2.5 days and a COD loading rate of 8.9 kg/m$\geq$ days. Under these conditions, the COD reduction obtained was 92% and the volumetric gas production rate was 3.2 m^3/m$\geq$ of effluent with a methane content of 63%.

TABLE 2 : Energy generation potential of distillery wastewater

STATE	Production (Million Liter)	Energy Potential (Ton LPG)
Andhra Pradesh	86.88	23457
Bihar	31.50	8505
Goa,Daman & Diu	8.23	2222
Gujarat	67.50	18225
Haryana	34.00	9180
Himachal Pradesh	1.11	297
Karnataka	88.00	23760
Maharashtra	290.00	78300
Nagaland	0.80	216
Orissa	3.62	977
Pondichery	7.60	2052
Punjab	38.34	10351.8
Rajasthan	8.80	2376
Tamilnadu	80.00	21600
Uttar Pradesh	431.55	116518.5
ALL INDIA (Total)	1177.92	318038.4

Source : Production figures : Sugar Statistics, February 98, Vol. 29, No.6

Based on the experimental results as above it is estimated that COD reduction is about 70% and gas yield is about 0.5 m^3/kg of COD removed. Using the formula given in Appendix I, it is estimated that distilleries have the potential to generate 0.27 kg Liquefied Petroleum Gas (LPG) equivalent energy from wastewater per liter of alcohol production. Statewise biogas generation potential in India from the anaerobic digestion of distillery wastewater is obtained and shown in Table 2. The energy potential is calculated on the basis of LPG equivalent, with the calorific value of LPG being taken as 10,800 kcal/kg.

At all India level, about 0.32 million ton LPG equivalent energy could be generated if all distilleries opt for anaerobic treatment of wastewater.

SUGAR INDUSTRY

There are about 393 large and medium scale sugar mills in India (Sugar Statistics, February 98, Vol. 29, No.6). The sugar manufacturing process consists of the following steps: (a) extraction of sugarcane juice; (b) clarification of the juice by the addition of lime and sulphur dioxide to heated juice at 70° C.; (c) concentration of clear juice after clarification to 60% solids; (d) syrup sulphitation and crystallisation; and (e) centrifugation, drying and bagging the sugar crystals. The manufacturing process produces wastes such as baggase and press mud. The former is normally used as fuel in boilers while the later is employed either for soil enrichment or for biomethanation. In addition the process generates wastewater with the characteristics as shown in Table 3:

Table 3 : Characteristics of sugar industry wastewater

Parameter	Values
Wastewater generation	0.2 to 1.8 m³/ton of sugar manufacture
pH	4-7
COD	1800-3200 mg/l
BOD	700-1500 mg/l
Total Suspended Solids	500 -700 mg/l
Total Dissolved Solids	2500-3500 mg/l

Apart from the effluent, the sugarcane press mud is also treated by anaerobic digestion. About 4 tons of press mud becomes available for every 100 tons of sugarcane crushed. Sugar mill wastewater is acidic in nature and thus requires pH correction prior to anaerobic treatment. Since sugar manufacturing is a seasonal activity, the process needs to be restarted at the beginning of every season. A two stage process (anaerobic followed by aerobic) for wastewater treatment is applied in almost all the sugar Industries in India. The COD reduction of 70-75% is reported in the anaerobic process (Sugar statistics, February 98). Radwan and Ramanujam (1995), have attempted the use of rotating biological contactor (RBC) for the treatment of synthetic sugar effluent. COD reduction was reported in the range of 90-97%.

Table 4 : Energy generation potential of sugar industry :

STATE	Production (Million Litre)	Energy Potential (ton LPG)
Andhra Pradesh	6.35	10661.65
Bihar	4.12	6917.48
Gujarat	8.11	13616.69
Karnataka	8.37	14053.23
Maharashtra	38.34	64372.86
Madhya Pradesh	0.99	1662.21
Punjab	6.25	10493.75
Rajasthan	0.23	386.17
Tamilnadu	11.84	19879.36
Uttar Pradesh	32.06	53828.74
ALL INDIA (Total)	122.20	205173.8

Source : production figures : CIER's Industrial Data Book, 1998

Khursheed et al, (1997), conducted a pilot scale study on a 2.83 m³ capacity UASB reactor at a sugar mill at Satha, Aligarh in northen India. The unit was operated at 34° C and 730 mm mercury pressure with an HRT of 0.25 day and an average COD loading rate of 13 kg/m³ d. The COD reduction of about 80% was observed. The average methane gas recovery was 0.22 m³ /kg COD removed. The methanogenic activity of the sludge was 0.56 kg COD CH₄/kg VSS d, sludge yield coefficient was 0.34 kg VSS/kg COD and SRT was 34 days respectively.

A ton of sugar produced has the ability to generate energy equivalent to 0.19 kg LPG from wastewater and 16.6 kg LPG from pressmud if treated anaerobically. The potential of energy generation from sugar industry waste is given in Table 4. It is possible to produce energy equivalent to 0.2 million ton LPG per year from sugar industry waste alone.

DAIRY INDUSTRY

The liquid waste from dairy originates from several sources such as the receiving station, bottling plant, cheese plant, butter plant, condensed milk plant, dried milk plant and ice-cream plant. The characteristics of the composite wastewater of an integrated dairy are depicted in Table 5.

Table 5 : Characteristics of dairy wastewater

Parameter	Values
Wastewater generation	2 liters per liter of milk processed in Case of chilling plant and 4.5 liters per liter milk processed.
pH	5.6 – 8.0
Oil and Grease	68 – 240 mg/l
Suspended Solids	28 – 1900 mg/l
COD	1100 –3500 mg/l
BOD	300 – 1700 mg/l

Roy and Chaudhary (1996) conducted pilot scale experiments on a 20m³/d fixed film reactor for biomethanation of dairy wastewater. The reactor was packed with plastic media up to a height of 2.23 m leaving a gap of 0.4 m at the bottom. COD loading rate was 2 Kg/m$\geq$ d (40 kg/day) and the designed mean cell residence time was 25 days. About 77% COD reduction was achieved and the methane gas production at STP was 0.33 m³/m³ of wastewater.

Mehrotra and Jain (1997) carried out feasibility studies on the treatment of simulated dairy wastewater in a 2.8 L capacity upflow anaerobic expanded sludge bed reactor. The reactor was found to remove COD load of 90% at a loading rate of 8 kg /m³ d and optimum HRT of 0.3 days. Specific gas yield was observed to be 0.41 m³/kg COD with a sludge loading rate of 0.6 g COD/g VSS d.

The potential of energy generation from dairy industry effluent (assuming 3L wastewater produced per liter of milk processed) is 0.84 kg LPG equivalent per m³ of milk processed. The effluent needs simple post – treatment to meet the discharge standards. The Statewise energy generation potential from dairy industry wastewater is depicted in Table 6. From these figures it is estimated that 0.05 million ton LPG equivalent energy can be generated on a countrywide basis by the anaerobic digestion of dairy industry wastewater.

EDIBLE OIL AND COFFEE INDUSTRIES

Collection of sufficient information pertaining to the anaerobic treatment of these industries is in progress.

However, the effluent characteristics of these industries reflect that the anaerobic treatment would be feasible and hence needs to be further investigated.

Assuming that a suitable reactor could be developed for 70% reduction in COD, with a biogas yield of 0.2m³ /kg COD removed, the energy generation potential from wastewater of an edible oil manufacturing plant would be 0.34 kg LPG equivalent per ton of oil produced. Production of edible oil during 1994 – 95 had touched 7 million tons (CIER's Industrial Data Book, 1998). From this the potential of energy generation from edible oil manufacturing industries would be 0.0024 million tons of energy equivalent.

Similarly, for the coffee industry, if we assume a 70 % COD reduction by anaerobic digestion alongwith a gas yield of 0.3 m³ /kg COD removed and the methane content of 65%, then coffee effluent has the potential to generate 1.45 kg LPG equivalent energy per ton of coffee produced. In the year 1994 – 95, India had manufactured about 43 thousand tons coffee (CIER's Industrial Data Book, 1998) which reflects that 0.062 million tons of LPG equivalent per annum from the coffee industry effluent.

TABLE 6 : Energy generation potential of dairy industry wastewater

State	Production (Million Liter)	Energy Potential (ton LPG)
Andhra Pradesh	3223.00	2707.30
Assam	713.00	598.90
Bihar	3582 .00	3008.90
Gujarat	3995.00	3355.80
Haryana	4140 .00	3477.60
Himachal Pradesh	707.00	593.90
Jammu and Kashmir	701.00	588.80
Karnataka	2938 .00	2467.90
Kerala	2301.00	1932.80
Madhya Pradesh	5662.00	4756.00
Maharashtra	4810 .00	4040.40
Manipur	91.00	76.44
Meghalaya	51.50	43.26
Nagaland	67 .00	56.28
Orissa	591.00	496.40
Punjab	6249.00	5249.10
Rajasthan	4951 .00	4158.80
Tamilnadu	4184.00	3514.60
Tripura	38 .00	31.90
Uttar Pradesh	11288 .00	9481.90
Sikkim	34 .00	28.60
West Bengal	3733.00	3135.70
ALL INDIA (Total)	64049.50	53801.60

Source : Production Figures : CIER's Industrial data book. (1998)

SUMMARY AND CONCLUSIONS

This survey investigated the potential of bioenergy generation from the anaerobic digestion of major agrobased industrial wastewater in India. This survey provides only a preliminary indication of the biogas generation potential in India from various industrial effluents.

Anaerobic decomposition of wastewater of these industries solves the dual purpose of reducing the pollution potential which was confirmed from the data of percentage removal of COD (70%) after the treatment in anaerobic digesters and the production of energy in considerable LPG equivalent. Thus simultaneously environmental reclamation and alternative source of energy is obtained through anaerobic digestion of agrobased industrial wastewater treatment.

The conclusions of this survey are:

1. The total energy generation potential from the anaerobic digestion of agrobased industrial wastewater is estimated to be 0.635 million tons LPG equivalent per annum. If the biogas is further used for power generation in gas turbines, The total electricity production (assuming 38 % efficiency) works out to be 360 MW.
2. The maximum bioenergy generation potential exists in distilleries and sugar mills, which together can generate 0.52 million tons LPG equivalent per annum. However, the technology for using press mud for biomethanation in the sugar industry requires further investigation. The bioenergy generation for the dairy industry works out to be 0.05 million tons.
3. Effluents from edible oil manufacture and coffee pulping units can be treated anaerobically. They together can generate about 0.062 million tons LPG equivalent per annum. However, little work has been done in this area and therefore pilot scale testing is required.

APPENDIX I

LPG Equivalent is calculated as

Total wastewater generation x COD conc. = Total COD produced x Efficiency of the reactor

= Total COD removed x Biogas generation factor x % methane

= Total m^3 of methan produced x density

= Kg of methane x calorific value

= Total energy per calorific value of LPG

= Kg of LPG x (55/60)

= Correct Kg of LPG

REFERENCES

1. Amritkar, S. R., (1995), "Introduction of anaerobic pretreatment in treating dairy effluents - A positive step towards conservation and cogeneration of energy", *Proceeding of the Third International Conference on Appropriate Waste Management Technology for Developing Countries, Nagpur, 127 – 132.*

2. C I E R'S. 1998, (Centre for Industrial and Economic Research), *Industrial Databook*, 169–172

3. Kaul, S. N., Nandy, T., and Trivedy, R. K., (1997), "Pollution control in distilleries", *India: Enviro Media, 233.*

4. Kaul S. N., and Nandy, T., (1998), "Biogas recovery from industrial wastewater", *Journal of Indian Association for Environmental Management, 24(3), 113–189*

5. Khursheed, A., Farooqi, I. H., and Siddiqui, R. H., (1997), "Development of granular sludge on cane sugar mill waste treatment using a pilot scale UASB reactor", *Indian Journal of Environmental Health, vol. 39 (4), 315–325*

6. Libin Wu, Xiangqiang Wu, Yibo Qian, Xluqin Yan, Yingming Fan, and Hul Zu, (1997), "A study on anaerobic treatment of distillery wastewater with UASB digester", *Proceeding of the 8th International Conference on Anaerobic Digestion, Sendai, Japan, 25 – 29 May, vol. 3, 133 – 135.*

7. Manihar, R., (1995), "Two phase Anaerobic digestion of distillery effluent – a case Study", *Proceeding of 3rd International Conference on Appropriate Waste Management Technologies for Developing Countries, NEERI, Nagpur, 25–26 February, 36-40*

8. Manihar, R. (1995), "Fixed Film anaerobic digestion of distillery effluent", *Proceeding of the 3rd International Conference on Appropriate Waste Management Technology for Developing Countries, NEERI, Nagpur, 25 – 26 February, 155 - 159*

9. Mehrotra, R., and Jain, P. K., (1997), "Treatment of dairy waste by upflow anaerobic expanded sludge bed reactor", *Proceedings of the 8th International Conference on Anaerobic Digestion, Sendai, Japan, 25 – 29 May*

10. Roy, B., and Chaudhuri, N., (1996), "Appropriate treatment option for dairy wastewater in India", *Journal of Indian Association for Environmental Management, vol.23 (1),, 1–8*

11. Sugar Statistics, February 1998, Vol. 29, No.6

Land Reclamation with Industrial Wastes

L. Behera[1], S.K. Chand[2], S. Mukherjee[3] and C. Subbarao[4]

[1]*Ex-Post Graduate student*, [2]*QIP Research Scholar*, [3]*Design Engineer and* [4]*Professor Civil Engg. Dept., Indian Inst. of Technology, Kharagpur 721 302, India*

INTRODUCTION

Steel manufacturing and thermal power generation are two of the most important activities in the industry sector. But both these processes bring, in their wake, a lot of adverse effects. Smelting process in steel industry produces waste by-products known as slag, while thermal power plants use coal as raw material and produce waste by-product known as coal ash. These waste materials are usually disposed off in the vicinity of the plants concerned. These waste materials, produced in enormous quantities, occupy huge areas, which could otherwise be put to much more profitable use. Again, these huge dumps of waste by-products cause pollution of soil water and air and hence pose severe threat to the environment.

So, alternative use of these two types of waste materials has become a crying necessity. In the present investigation, an attempt is made to intermix these two waste materials by suitable proportion so as to use it as a fill material in land reclamation projects and also to store more material in disposal areas in a way that will produce usable land when completed.

MATERIALS

Blast Furnace Slag

During the smelting process in the furnace, lime and magnesia combine with silica and alumina along with ash produced from coke and this forms molten oxide known as slag. Cooling of molten slag is required before its disposal. According to the method of cooling, slag may be either air-cooled slag or granulated slag. In the case of granulated slag, molten slag is suddenly chilled by immersing in water by either pit process, jet process or dry granulation process. In dry granulation process, the stream of molten slag is broken into relatively small particles by impact using a mechanical granulating device and sufficient water is fed to the molten slag to quench it.

Blast furnace slag is used in manufacture of cement, as precast paving blocks and as railway ballast. It is suitable as sub-base material for roads, either in its entirety or in combination with other road materials (Srinivasan, 1993). This can also be used in combination with lime as a stabilizer for soils (Kamon & Nontananandh, 1991).

Pond Ash

In the wet disposal system adopted by most thermal plants, the fly ash and bottom ash are mixed together with water and disposed off by wet slurry system in ponds / lagoons. This mixture is known as pond ash. In India, pond ash is available in large quantities at thermal power plants.
The use of fly ash for stabilization of soil is well documented (Chu et al, 1955; Mateos, 1963; Ghosh et al, 1973; Joshi & Nagaraj, 1987; Sastry, 1989). This is also very suitable as structural and backfill material (DiGioia & Nuzzo, 1972; Joshi et al, 1975; Leonards & Bailey, 1982).

For the present work, blast furnace slag (BFS) was procured from Tata Metaliks Ltd., Kharagpur. Pond ash was brought from Kolaghat Thermal Power Station, Kolaghat. The pond ash was found to have been mixed with some amount of the local soil at the disposal site. So in this paper, the material has been designated as recycled pond ash (RPA). The physical and engineering characteristics of these two materials are given in Table 1 and the chemical composition in Table 2.

Experimental Program and Results

First of all, a detailed characterisation, with respect to the engineering properties and chemical composition of the two materials, i.e., BFS and RPA were done and these are presented in Tables 1 & 2. Then both BFS and RPA were blended together in different proportions and for each of the mix, tests were conducted for various geotechnical parameters such as grain size analysis, specific gravity, consistency limits, compaction characteristics, unconfined compressive strength, constrained compression modulus, free swell index and permeability.

The chemical analysis was done in accordance with BS 4550 : 1970, while the geotechnical tests were carried out as per the relevant provisions of IS 2720. The constrained compression modulus test was conducted according to the method given by Lambe and Whitman (1983) Because of limitation of space, only a part of the results are presented herein.

Table 1 Properties of BFS and RPA

No	Parameter	BFS	RPA
1	**Grain Size Analysis**		
	Percentages of sand	**88.0**	**64.0**
	Percentages of silt	**11.8**	**34.6**
	Percentages of clay	**0.2**	**1.4**
	Coefficient of uniformity	**33.3**	**18.8**
	Coefficient of curvature	**6.3**	**0.4**
2	**Specific Gravity**	**1.69**	**2.21**
3	**Consistency Limits**		
	Liquid limit (%)	**69.0**	**36.0**
	Plastic limit (%)	**9.0**	**21.0**
	Plasticity index (%)	**60.0**	**15.0**
4	**Compaction Characteristics (IS light weight)**		
	Optimum moisture content (%)	**32.6**	**18.8**
	Maximum dry density (kN / m^3)	**11.8**	**14.5**
5	**Unconfined Compressive Strength (kPa)**	**30.9**	**23.5**
6	**Coeff. of Permeability (m / s)**	**5.8×10^{-6}**	**3×10^{-7}**
7	**Free Swell Index (%)**	**-28.6**	**-30.6**

Table 2 Chemical Composition of BFS and RPA

Element	Element Content (% by dry weight)	
	BFS	RPA
SiO_2	32.7	51.2
CaO	29.5	18.8
Al_2O_3	23.3	24.4
MgO	11.4	2.6
Fe_2O_3	0.6	0.1
Others	2.5	2.9

Table 3 Particle Gradation and Specific Gravity of Mixes

No	Mix BFS (%)	Mix RPA (%)	Gravel Size (%)	Sand Size (%)	Silt Size (%)	Clay Size (%)	Specific Gravity
1	100	0	7.7	80.1	12.0	0.2	1.69
2	83	17	6.0	77.5	16.2	0.3	1.76
3	75	25	5.9	76.1	17.7	0.3	1.87
4	67	33	5.5	73.9	20.3	0.3	1.93
5	50	50	3.5	71.2	24.9	0.4	2.02
6	25	75	2.6	66.9	30.1	0.4	2.11
7	0	100	1.0	63.0	34.6	1.4	2.21

DISCUSSION AND CONCLUSION

The major constituents in both BFS and RPA are silica, calcium oxide and alumina. The percentage of CaO in BFS is more than that in RPA. Depending upon the availability of free lime, this may help form cementitious compounds.

In all proportions of mix, major fractions are in the range of sand and silt, but the percentage of finer fraction increases on increasing the percentage of RPA in the mix. There is not much variation in maximum dry density values for different types of mixes. The unconfined compressive strength of only BFS is slightly higher than that of only RPA. But the strength increases when the two are blended in different proportions and the maximum UCS value of 103.6 kPa at 28 days is obtained with a mix ratio of 67% BFS + 33% RPA. Also for all the mixes, the UCS increases with curing time indicating possible pozzolanic reaction between lime and silica present in both components. Again, the free swell index is negative and swelling pressure is zero for all the mixes showing that these are non-swelling in nature.

On the whole, it is seen that by properly blending these two industrial wastes, a non-swelling material of sufficient strength is obtained which may be used as structural fill for land reclamation.

REFERENCES

Behera, L. (1997). " Geotechnical engineering properties and behaviour of blast furnace slag - pond ash mixes. " M. Tech. Thesis. IIT, Kharagpur, India.

Ghosh, R. K., Chadda, L. R., Pant, C. S., and Sharma, R. K. (1973). " Stabilization of alluvial soil with lime and fly ash." Jl. of Indian Road Congress. 35 (2), 489-511.

Joshi, R. C., Duncan, D. M., and McMaster, H.M. (1975). " New and conventional engineering uses of fly ash ." Jl. Of Trans. Engg. Div., ASCE, 101 (4), 791-806.

Lambe, T. W. and Whitman, R. C. (1983). " Soil Mechanics ", John Wiley & Sons.

Mateos, M.. (1963). " Stabilization of soils with fly ash alone. " Highway Research Record, vol.29.

Status of Environmental Degradation—A Case Study of Mining Sector

P.K. Singh[1], T.B. Singh[1], B.K. Tewary[1] and R. Singh[2]

[1]Environmental Management Group, Central Mining Research Institute, Barwa Road, Dhanbad 826 001, India
[2]Head of the Department, Department of Geology, V.B. University, Hazaribagh, India

1.0 INTRODUCTION:

The unprecedented increase in the number and the activities of human beings, since the industrial revolution, and particularly in this century, have given rise to a deterioration of the environment and the depletion of the natural resources which ultimately threatens the future of the planet. The gross imbalance that has been created by the rapid industrialization and the population growth - these issues are at the center of the current dilemma in the developing countries like our India. And redressing this imbalance will be the key to the future security of our planet.

Thus, environmental protection is a global issue today. For the last three decades there has been gradual increase in concern for the quality of environment.

2.0 STATUS OF ENVIRONMENTAL DEGRADATION IN INDIA:

Environmental problems have attracted the attention of a wide cross-section of the people all over the world during the last three decades. People are becoming increasingly conscious of a variety of problems like global warming or ozone layer depletion, acid rain, droughts, floods, pollution of air and water and problems from hazardous chemicals, which have adverse effects on the environment.

It was in early seventies that India along with the rest of the countries in the world realized the magnitude of this problem. India, which is already engaged in bringing people above the poverty line, is put to still greater pressure towards environmental protection.
The major challenge to environmental crisis in India is the rapid growth of population. By the census of 1991, the population of India was 850 million and it is constantly increasing and has exceeded one billion at the present time.

The industrial revolution that was ushered in the last quarter of 18th century and progressed at an ever-increasing pace through 19th and 20th centuries proved to be a vital factor in the degradation of environment. Today, India has become the 10th largest industrial nation of the world, but at the same time, it is one of the most polluted countries of the world as far as industrial pollution and hazardous wastes are concerned.

Thus, the cities in India are facing acute environmental problems like lack of sanitation, polluted water and air, lack of recreational areas, traffic congestion etc. Moreover, the domestic and industrial waste disposal in the urban areas is very serious. The level of degradation can be understood clearly by the following facts in brief:

Air-Quality degradation: The congestion and deteriorating air-quality in the urban areas is a result of the increase of industrial and construction activities, and increase in vehicular traffic density.

India is facing a major crisis-rise in health problems due to increasing vehicular air pollution has become a cause for serious concern. Most of the 23 Indian cities with a population of over 1 million have air pollution levels which are dangerously higher than the standards recommended by the World Health Organization (WHO). Infact, in Delhi, Calcutta, Mumbai, Ahmedabad, Kanpur and Nagpur, levels of suspended particulate matter (SPM) are three times higher than the WHO standards. Vehicular pollution is the main culprit in most cities - in Delhi alone, it accounts for 65% of the total pollution. Industry and thermal power plants contribute only 25%, while the remaining 10 % is due to domestic activity. In Calcutta and Mumbai, vehicles are responsible for 52% and 30% respectively of the total population.

Water-quality degradation: Many rivers in India have become highly polluted due to dumping of domestic, municipal and industrial wastes in river directly. The primary source of water pollution in India, especially in and around urban areas, is pollution from municipal and industrial sewage and is largely a consequence of inadequate sewage collection and treatment facilities, as well as the lack of sanitation.

A study on water pollution in the Ganges basin found that about 75% of the wastewater are generated from municipal sources, with 88% of the municipal sewage coming from different cities. Few rivers meet the standards for safe drinking water (CPCB, 1990). This is an important issue because rivers and lakes are the primary source of drinking water, most of, which is untreated.

Groundwater, particularly in regions with a high concentration of irrigated agriculture, has turned brackish and is chemically contaminated with excess of fluoride, iron, arsenic, and, nitrates from fertilizers and pesticides.

Soil - quality degradation: Increased use of chemical fertilizers, pesticides, insecticides have change the soil ecology and have degraded it so much that it has lost its basic elements which maintain the soil fertility

Deforestation: Forests are also vanishing regularly due to number of causes. And the effects of deforestation may be observed as loss of bio-diversity, adverse effects on natural phenomena like rainfall, atmospheric quality, floods etc.

Noise pollution: Noise-levels in Indian cities are extremely high due to loudspeakers, construction activities, use of generators for commercial activities and several fold of increase in the automobiles.

Thus, in the sum and substance, we can say that already the status of environmental degradation in India has become acute and there is a serious need to think over by all level of thinkers, so that, the level of environmental degradation can be reduce in the next millennium.

3.0 CHALLENGES OF THE NEXT MILLENNIUM:

Before two and a half million years, when man started his life, he simply utilized a small portion of God-gifted natural resources. As his number increased and his culture and technology became advanced, he modified the natural system into an artificial and highly productive system to get more energy and more nutrient sources and it resulted in the production of more by-products and wastes which naturally mixed up in the nature gradually. This enormous exploitation of natural resources and gigantic and unmanageable amount of by-products and wastes has resulted in today's environmental crisis, which has endangered not only the human existence but the mother earth as well.

To all Indians, 21st century looks like the promised future, the era where their aspirations are going to be achieved. But, there are certain challenges for creating pollution-less environment in the next century, which are as follows:

i)   Need of controlling population explosion.
ii)   Checking of deforestation & loss of bio-diversity.
iii)   Reducing the poverty to a certain extent.
iv)   Need of planned urbanization & industrialization or its modification.
v)   Use of Eco-friendly technology.
vi)   To control the load on transportation.
vii)   Countrywide awareness of environmental education.
viii)   Strict imposition of environmental legislation.
ix)   Routine-wise monitoring of environmental pollution.

4.0 RAPID GROWTH OF INDUSTRIALIZATION IN DHANBAD DISTRICT:

Dhanbad district in Bihar being pioneer in coal mining and allied industrial activities. Dhanbad, being in the heart of Damodar River Basin, is one of the most industrialized district of South Chotanagpur region. The world famous prime-coking coal centre i.e., Jharia coalfield falls within this district. The rapid growth of industrialization in Dhanbad coalfield area has made an impact in the local environment.

Within the last few years, there has been a steady growth in the small industries sector based on coal as a basic raw material. In Dhanbad district, there are several industries besides Jharia coalfield, such as Beehive coke ovens, coal washeries, coal briquetting plant, power plants, fertilizer industry, explosive and cement industries etc.

The main reason for the rapid growth of industrialization is the high density of population in the adjoining plains of Bihar, U.P., and West Bengal are sources of cheap labour and provide markets for agricultural, fertilizers and consumer goods. Another factor is that a good network of broad-gauge double track electrified or dieselized railways with main lines providing direct lines to Calcutta, Delhi and Bombay and the branches extending directly to mineral sources, are the most advantageous to the industrial plants. Road network is also providing easy accessibility to different urban centres in the region.

5.0 ENVIRONMENTAL IMPACT DUE TO MINING AND OTHER INDUSTRIAL ACTIVITIES:

The industrial and technological developments take a nation forward but at the same time give rise to many environmental problems. The exploitation of mineral resources through surface or underground mining is invariably associated with a number of wide ranging environmental problems. The **main environmental problems in mining industry** may be classified as:

- Air pollution due to increase in opencast mining projects.
- Noise pollution in mining and industrial areas.
- Change in topography.
- Lowering of groundwater table.
- Mine fire.
- Land degradation due to dumping of overburden dumps.
- Land subsidence.
- Deterioration of surrounding water quality.
- Acid mine drainage problem in metal mining and some of the coalfields.
- Leaching of pollutants from tailing and disposal areas.
- Occupational health problems in mining areas.

After the nationalization of the coal industry (coking coal in 1971 and non-coking coal in 1973), a huge amount of money has been invested by the private entrepreneurs in installing briquetting industries, beehive coke ovens and fire-bricks industries **in Dhanbad district** without taking care of the locations and the discharge of the effluent from coal washeries, fertilizers plants and other industries have further polluted the environment. The bad conditions of the roads and plying of large number of vehicles are adding the additional pollutants in the environment.

6.0 POLLUTION STATUS OF DHANBAD - JHARIA MINING AREAS :

Air pollution : The air pollutants encountered in the coalfield region are both particulate and gaseous in nature, as shown in Table 1, which is coming out as emission from different industrial and mining processes.

Water pollution : Inadequate environmental protection measures in coal mining and related industries have created an alarming water pollution problems in the region. Major source of water pollution is the industrial effluent discharge. The river Damodar has now become so much polluted that it is termed by many as the **industrial sewer of the area. Table 2 describes the water quality of Damodar river.**

Noise pollution : Deployment of large number of vehicles for transportation of men and materials, various commercial activities, use of loudspeakers and diesel generators etc., all put together creates its acoustic environment vibrant. The noise level in Dhanbad – Jharia coalfield region is presented in **Table 3.Land degradation** : One of the major environmental impacts of surface mining operation is that it causes damage to the land, forest, agriculture and disturbance of the drainage system and hydrology of the area. By creating ugly dumps at any site, the soil stability and productivity are adversely affected and erosion takes place. The land is mainly degraded by fire, subsidence, dumping of municipal solid waste,

111

overburden dumps as well as industrial solid waste from coke oven plants, coal washeries and thermal power plants. The area affected in the region has been shown by Table – 4.

The change in landuse pattern of Jharia coalfield of Dhanbad district can be clearly seen with the help of toposheets of SOI, and aerial photographs of three different periods (1925, 1974, and 1993). From study, it has been found out, **that till 1974, the area under forest and agriculture in the coalfield region was more than half of the total land areas, thereafter, it decreases due to rapid industrialization.**

7.0 STRATEGIES TO BE ADOPTED FOR MITIGATIVE MEASURES :
The important strategies to be adopted for checking pollution in coalfield areas area as follows :

a) There is a need of planned efforts in respect of pollution control measures over the years. This will not only check the pollution level due to increased industrial activities , but will reduce it marginally.

b) Tree plantation in the various parts of the coalfield area, particularly on degraded land, is needed. These trees provide as sinks for particulate matter and also control noise pollution to a certain extent.

c) Mitigative action is required at micro-levels especially for fire areas, chemical based industries, coke ovens and increasing vehicular movement which emit toxic gases.

d) Serious attention is needed for repairing/ constructing of broken/ damaged roads within the area by concerned agency.

So, if these minor remedial action is not undertaken in time, serious environmental degradation will remain in the next century and all progress and development will be of no value.

8.0 THE MAIN R&D EMPHASIS SHOULD BE LAID IN THE NEW MILLENNIUM IN THE FOLLOWING AREAS :
■ Control of dust pollution from haul road of opencast mines.
■ Reclamation of abandoned mines.
■ Mine fire control especially in Jharia coalfield & Northern coalfield.
■ Computer modelling for locating mining waste dumping site to avoid groundwater contamination.
■ Control of acid mine drainage especially in metal mining as well as in some of the coalfields like Northern coalfield, North-eastern coalfields and Western coalfields where coal contain greater amount of sulphur.
■ Biological reclamation of over-burden dumps.
■ Development of a green belt area in active mining sites for attenuation of dust and noise pollution.

9.0 CONCLUSION :
Development of a country depends very much on the advancement of industrialization and is essential for better standard of living of the mankind. But as a result of mass scale industrialization, world's environment is being polluted day by day and is endangering the existence of living beings of the earth. This has attracted the attention of environmental engineers, medical practitioners, researchers and planners of throughout the world and therefore attempts should be made to make air, water and living environment clean and prevent diseases arising out of various industrial activities in the next century.

India is suffering from same problem of environmental degradation. It is essential at this stage, because we have already switched in the 21st century, that there should be clear profile of causes of degradation so that mitigative strategies can be suggested at the micro-level. And after getting the mitigative strategies, there should be strict rules and legislations for applying it. Then only we can get - *"Pollution Free Country of 21st Century"* .

ACKNOWLEDGEMENT :
The authors thank to Director, CMRI, Dhanbad for giving his kind permission to present this paper in the International symposium The opinion expressed in this paper are of the authors only and not necessarily of the CMRI.

REFERENCES :

1. Brandon C. & Hommann K. (1995). The cost of inaction : valuing the economy-wide cost of environmental degradation in India; Asia Environment Division, World Bank.
2. CMRS Findings; Status of research on Environmental pollution in mines and mining areas & miners health; May 1996.
3. Khitoliya R.K. & Khitoliya J.(1996). Environmental Protection and Legislations ; The E.N., dated : 22-28 June .
4. Reddy S.J. ; Encyclopedia of environmental pollution and control ; Vol. 2 ; R.K.Trivedy ; Enviro Media ; Karad (India)
5. Report (1998) ; CCDRB - Existing Scenario, Vol.II ; Prepared by CMRI, Dhanbad.
6. Sharma A. & Roychowdhury A. ; Slow murder : the deadly story of vehicular pollution in India ; State of the Environment Series ; Center for Science & Environment, New Delhi, November.
7. Singh P.K., Prasad S.S. and Singh T.N. - ' The status of noise pollution in   Dhanbad Municipal Area, ; Indian Journal of Environmental Protection, BHU,   Varanasi, Vol. 20, No. 1, January 2000 .
8. Singh Y.P. & Aggarwal A.(1998). Environmental Crisis : a Challenge Ahead ; The E.N., dated 30 May- 05 June .
9. Tewary B.K. ,Tiwary R.K., Nityanand B. and Singh P.K.,   - ' Status of environmental degradation in a coalfield area and suggestions of mitigative measures' ; presented in the Seminar on Management of Environmental Pollution organised by B.I.T., Sindri on 15-16 th November '1999.

TABLE - 1

STATUS OF AIR POLLUTION IN COALFIELD REGION

Parameters (ug/m^3)	Locations		
	CMRI (#)	Dhansar (*)	Lodna ($)
SPM	139.8	561.9	778.7
SO$_x$	561.9	88.2	90.2
NO$_x$	39.5	89.4	102.8
CO	850	1687	1415

Where, # - Protected place.

*- Mining and transportation activities.

$ - Mining, fire and transportation activities.

TABLE - 2

STATUS OF WATER QUALITY OF DAMODAR RIVER OF COALFIELD REGION

Sl. No.	Parameters	D/s Jamadoba	U/s Patherdih Mahulbani Ghat	D/s Patherdih Domgarh Ghat
1.	PH	7.90	7.50	7.60
2.	Temperature (°C)	21.0	21.3	21.6
3.	TSS (mg/L)	523.0	290.0	280.0
4.	TDS (mg/L)	460.0	375.0	340.0
5.	DO (mg/L)	5.20	5.60	7.10
6.	BOD at 20°C (mg/L)	3.10	2.10	1.80
7.	COD (mg/L)	213.0	102.0	119.0
8.	Hardness (mg/L)	228.0	156.0	172.0
9.	Alkalinity (mg/L)	44.0	41.0	41.8
10.	Nitrate (mg/L)	10.94	10.54	9.80

TABLE - 3

STATUS OF AMBIENT NOISE LEVEL IN COALFIELD REGION

Sl.No.	Location	Noise Level (in dBA)			
		Day Time		Night Time	
		Leq	Std.*	Leq	Std.*
INDUSTRIAL AREAS		-	**75**	-	**65**
1.	Bastacolla industrial area	79		72	
2.	Block II O/C Mining area	84		77	
3.	Near Lodna washery	81		75	
COMMERCIAL AREAS		-	65	-	55
4.	Bartand market	71		61	
5.	Bank more, Rajendra Market	73		57	
6.	Purana Bazaar	76		71	
RESIDENTIAL AREAS		-	55	-	45
7.	CMRI Colony	53		45	
8.	DGMS Colony	65		55	
9.	Housing Board Colony	68		58	
SILENCE ZONE		-	50	-	40
10. Central Hospital		59		55	
11. Patliputa Medical College		65		54	

TABLE - 4

STATUS OF AFFECTED AREA OF LAND IN COALFIELD REGION

Sl.No.	Causes	Affected Area (in sq.km.)	Total area (in Sq.Km.)
1.	By fire	17.32	
2.	By subsidence	39.47	71.27
3.	By excavation	12.68	
4.	By dumps	6.30	

114

Mine Land Reclamation—A Case Study of Sesa Goa Ltd.

*****A.K. Rai and ****Mahesh Patil

*Director Production & Logistics, **Manager-Environment Management
Sesa Goa Ltd., Sesa Ghor, 20 EDC-Complex , Patto Panaji, Goa, India

INTRODUCTION

As we move on to the 21st century, the global challenge of finding paths of development that are economically and environmentally sound continues. Living within our ecological and economical means demands a far-reaching policy, institutional and technological reforms complemented by shift in individual values and behavior. For sustainable development, we must implement policies that reflect the links between prosperity, a healthy environment and social equity.

Sesa Goa is a public limited company and a subsidiary of Mitsui & Co. Of Japan. It has been in the business of mining, processing and export of iron ore for the past 45 years and is the highest exporter of iron ore in the Indian private sector for the last four years. It has been certified to ISO-9002 for Quality Management and ISO-14001 for Environment Management System. Sesa Goa has a special concern for environment since its inception and environment management was an integral part of mining operations much before such a concept was known in this part of the world. In fact, the afforestation and reclamation efforts that were started in the late sixties at Orasso Dongor mine at Advalpale and Sanquelim mine have converted these mines into orchards and bears the testimony to initiatives that were taken thirty years ago.

Sesa Goa is aware of its special responsibility towards environment and recognizes the fact that to sustain our future generations, there has to be a balance between environment and development. The traditional approach for mine environment management was very narrow and centered with two things in mind i.e. erosion control and mine land reclamation. However, at Sesa Goa a holistic approach is adopted wherein combinations of several of various techno-bio methods are utilised so that desired results are achieved in the shortest period of time.

MINING:

Mineral exploitation (Mining) is second only to agriculture as the World's oldest and most important industry. Mineral and mineral-based products are integral part of the economic and social fabric of modern society. Mining is a unique industry wherein project site is determined by location of deposit. It is unchangeable; therefore mining has to be planned around where the mineral deposit exists.

Mining cause damage to land, soil, water, flora, fauna as for mining land has to be broken open. However mining does not mean permanent loss of land for other use. On the other hand it holds potential for altered and improved use apart from restoring for agriculture, forestry and irrigation.

ECOSYSTEM AND PLANT SUCCESSION

Vegetation together with the soil in which it has its roots, the associated fauna and the environment that surrounds them are closely interrelated and interdependent. They interact and support each other to constitute an ecosystem. Although ecosystems are sensitive to the outside influences they are self sustaining. Once properly established they need no further support. This is because of natural cycling of accumulated materials, which maintains the vegetation and the other organisms within it. Ecosystem

115

has a capacity to develop. In nature after major disturbance vegetation will slowly and gradually develop over a period of time. This process is termed as plant succession

ENVIRONMENT MANAGEMENT AT SESA GOA LTD.

At Sesa Goa Planning for Environment Management and mine operation go hand in hand. Initially mining area is fully explored and mine design/ plans are prepared using computer model. At Codli mine by computer design the mine is divided into fourteen basins (Synformal Synclines) and these basins are exploited in sequence. This ensures that entire area is not broken open at the time. The basin is mined as a mining pit and after exhaustion, it is initially used as a Tailing pond and then for reject dumping. Thus reclamation is undertaken concurrently with mining.

Tailing Management:

Most of the iron ore requires beneficiation. After wet beneficiation, plant tailings are disposed into worked out pits, and are treated with lime and flocculent to recover 70% of water. Recovered water along with rain water accumulated in working pit is used for ore beneficiation. This ensures 100% reuse of water and no discharge of effluents/ tailings outside mine. Once the pits are filled with tailings rejection dumping is undertaken and later afforested.

Geo-Textiles For Erosion Control:

Geotextiles are thin, permeable material used in civil engineering to improve the structural performance of soil and of works such as road pavements. These fabrics are also applied increasingly for specific functions in agricultural sector. At present the geotextiles market is dominated by synthetic material which are produced from polypropylene. Natural fibers that are used in geotextiles include jute, coir, cotton etc. and have wider use in mine land reclamation. At Codli mine, experiments are conducted to study use of coir geotextiles for soil erosion and also for establishing grass cover. At the dump surface geotextiles are laid before onset of monsoon. Top end of geotextiles is enclosed to dump either by pegs or using boulders. Grass seeds and saplings are sown in between two geotextiles. Geotextiles reduces impacts of rain on dump surface thus preventing erosion and also conserving moisture by acting as mulch. Geotextiles decomposes within three years by which time the dump is stabilised with vegetation. Plate 1depicts dump with Geotextile.

"Root Trainer" – New Technology In Forest Nursery:

Large-scale production of seedlings in polybags is in vogue in all nurseries at present but inherent problems like improper development of root structure has not been realised so far. The polybags as nursery containers have technical and logistical limitation. Polybags require large amount of soil and are difficult to handle due to their size and weight, are poorly aerated, discourage lateral root development, encourages coiling of roots, etc. Commonly, root pierces through the polybags and grows into the soil below. Therefore these roots break off when seedlings are disturbed resulting in severe shock to plants and postponing growth. To circumvent the shortcomings of polybag containers, the root trainer (RT) concept is introduced at Codli mine. Basically, a root trainer is a container of more or less conical shape made up of high-density polyethylene or expanded polystyrene material with openings at both, top and bottom. The bottom opening is called drainage hole. The RT is provided with 6 vertical ridges inside which are meant for guiding the root growth to the drainage hole at the base. The root trainers are placed in a RT stand or tray in such a way that a drainage hole is exposed above the ground that facilitates free flow of air. The free flow of air constantly sloughs up the primary root when it comes out of drainage hole resulting in proliferation of lateral roots. The more the lateral roots developments, the more the tertiary roots and the root hairs resulting in the developments of vigorous fibrous root system. In this process, the overall surface area of absorption zone of roots increases and when planted in the field, the seedlings get established within 48 hr. and starts showing growth. It survives the prolonged drought period after planting because the root system development is complete in the nursery itself.

Case Studies Of Mined Out Areas:

Sesa Goa has taken special care to rehabilitate mined out areas without abandoning them in "as is where" condition, a rare approach in mining industry.

➤ **Orosso Dongor Mine:**

Orosso Dongor mine is situated in Advalpale village, North Goa having an area of 100 Ha. It was extensively mined for iron ore in the 60's and abandoned in 1974. Dumping was done in non – mineralised area in step manner, with broad facced terraces provided to minimize erosion. Laterite was stored separately during mining which was utilised for covering the clay portion of the waste dump before onset of monsoon. At the toe of the dump laterite walls of suitable height and width are constructed. Cross walls are also constructed at a distance of 15 meters from each other. The area between the laterite walls acts as a settling pond thus retaining silt and allowing clear water to flow out.

116

Mining waste are sterile with nil organic matter and biological activity. The waste dump and garden soil analysis are shown in in table I, which indicates that soil is deficient in basic nutrients like Nitrogen, Phosphorus and Potassium. The low organic matter content and water holding capacity of mine rejects makes the environment hostile for plant growth. To overcome this problem initially fast growing trees adopted to lateritic conditions and not browsed by cattle were selected.

Over a period of time our experience showed that cashew plantation helped not only for stabilisation of soil but also gave economic returns to local villagers . Now entire area is under cashew plants which could be seen in Plate No2.

Table I

Sr. No	Characteristic	Garden soil	Dump Reject
1	PH	5.5	5.5
2	Loss on ignition	15.50%	6.23%
3	Organic carbon	2.40%	0.80%
4	Iron	20.80%	46.00%
5	Alumina	15.52%	7.13%
6	Manganese	0.20%	.73%
7	Copper in ppm	6.00	4.00
8	Zinc in ppm	7.60	18.60
9	Nickel in ppm	14.00	10.0%
10	Lead	Traces	Traces
11	Calcium	0.15%	0.8%
12	Magnesium	0.07 %	0.01%
13	Available nitrogen in Kg/Ha	51.52	15.66
14	Available Phosphorous in Kg/Ha	>448	33.80
15	Available Potassium in Kg/Ha	53.76	2.24
16	Silica	38.17%	34.17%

> **Sanquelim Mine :**
The Sanquelim mine is situated at Sanquelim, North Goa and is spread over an area of 200 Ha. It was started in 1960 and abandoned in 1988 after exhausting mineable ore reserves.
The experience of rehabilitation of Orosso Dongor mine was utilised at Sanquelim for covering clay with laterite and planting Acacia and cashew. Apart from this, "POT CULTURE" experiment was conducted with the objective to identify plant species that can tolerate inhospitable condition of mine sites and grow without artificial aid such as fertilizer and irrigation. The results revealed that local flora can grow on mine rejects if the soil is supplemented with organic matter in the form of Neem cake, Organic manure –"Myceameal" etc..These findings were used to make plantation programme more scientific and to introduce local flora thus improving the biodiversity of reclaimed area. Success of programme is revealed in the survey conducted by Botony Department, Goa University in year 1994, which recorded that in all total of 164 species belonging to 138 genera distributed among 55 families exists, consisting of grasses, legumes, climbers, shrubs and trees.
At present more than 4 lakhs saplings have been grown over an area of 200 Ha.

Agri-Horticultural Approach:
To utilise the mine rejects for more productive purpose, the Agri - horticultural approach was adopted on experimental basis at our Sanquelim mine. Initially dump is stabilised by planting acacia at a very close spacing. Acacia being hardy is not browsed by cattle and fixes atmospheric nitrogen. The dump soil becomes rich in humus due to foliage of acacia thus resulting in initiation of soil microbial activity. A leguminious creepers which is used as a cover crop in rubber plantation was also introduced . The creeper apart from fixing Nitrogen, also acts as a mulch for water conservation, prevents soil erosion and encourages microbial activity. Once the entire area is a stabilised (after 4 years), economic, horticultural crop like Cashew, Jackfruit, Coconut, Banana, Manago etc. are introduced by thinning out Acacia tree to the extent needed. Irrigation is provided from exhausted pits. The results are very successful. We have around 5 Ha area covered under AgriHorticultural approach, which can be seen as depicted in plate 3. Our experience in iron ore mine reclamation has shown that following species are suitable for mine land reclamation viz.

Grass : Congo grass, Stylo grass, Kudzu, & Vettiver szenoides

Plant Species

Common names	Botoanical Names
Cashew	*Anacardium occidentials*
Accacia Spp	*Accacia mangium, Accacia auriculoformis*
Saton	*Alistonia scholaris*
Shisum	*Delbergia latifolia*
Awla	*Emblica officialis*
Khair	*Accacia catechu*
Shivan	*Gmelina arborea*
Karanj	*Pongamia pinnata*
Gulmohar	*Delonix regia*
Rain tree	*Samania saman*
Amaltus	*Cassia fistula*

At present more than 15 lakhs saplings have been grown over an area of 600 Ha.

Pisciculture:

Pisciculture is one of the best way of pit reclamation. An experiment was launched at Sanquelim mine to introduce Pisciculture in collaboration with National Institute of Oceanography (NIO) Dona Paula, Goa in year 1990. One of the worked out pits (Lisboa) was terraced with loose soil to facilitate afforestation, and the pit is used for Pisciculture. The pit is 150m*30m with an average 6 m depth of water. The pit receives fresh water from adjacent rivulet and good volume of water is always maintained throughout the year. The pit water which was analysed by NIO as shown in table 2 was found not suitable for fish culture .

Table No:II – Environmental parameters and chemical analysis of water.

Parameters	Sample 1		Sample 2		Sample 3	
	Surface	Bottom	Surface	Bottom	Surface	Bottom
Water Temp 0c	28.5	28.5	28.6	28.0	29.0	29.0
PH	6.3	6.3	6.3	6.2	5.4	5.2
Dissolve oxygen (ml/l)	5.3	4.4	4.4	4.4	4.8	4.5
Oxygen demand (mg/l)	1.72	0.54	0.62	0.46	1.27	0.59
POC(ug/l)	223	178	300	230	90	380
Suspended load 9mg/l)	5.8	14.3	5.2	7.6	2.6	1.2
Sediment pH	4.9		-		4.6	

To make it suitable following amendments were made.

➢ By reducing acidity levels by application of lime.
➢ By improving nutrient contents in the water by application of organic/inorganic fertilizers.

The fingerlings of Rahu, Mrugal and Carp (around 15000) were released in the open pit and were fed daily at the rate of 5% of the body weight with soaked groundnut cakes. The temperature was monitored daily at 1m depth for the adjustment of feed as per feed chart. The results are very encouraging, and now the pond is full of fish. Also ducks are released into the pond and thus a self sustaining ecosystem is established. Plte No.4 depicts pisciculture pond established at Sanquelime mine.

Integrated Bio-Technological Approach For Mine Land Reclamation:

The objective of this approach is two front i.e. prevention of heavy metal leachates from dump and rejuvenation of productivity by vegetation of dump with organic waste and biofertilizer. In this approach mine dumps are evaluated for physical, chemical and biological characteristics. Based on field studies, heavy metal protocol is developed to achieve separation of different metals from mixed pollutants/. Various blends of organic waste, mine spoil and biofertilizer are evaluated for build up of nutritive and supportive capacities. All these studies are carried out in laboratory and green house. Based on above studies 8000 saplings of 21 varieties are planted on five hectares of rejection dump. Grass species " Congo" is introduced for ground cover. Saplings before planting are treated with Nitrogen fixing bacteria, Rhizobium and Azaetobacter. Later," Mycorrhizae" is introduced in rhizosphere. The performance of plants is being monitored and will continue for 3 years.

This project is a collaborative research work in association with National Environmental Engineering Research institute (NEERI) Nagpur, Department of Biotechnology and Lund University, Sweden. The experimental plantation after one year is shown in plate No. 5.

CONCLUSION:

The disturbance of land surface due to mining alters the potential for revegetation growth, such that it is not possible for pre-existing plant community to be exactly recreated.

Although the degraded mine wasteland can be stabilised, the use of expensive high input methods is inappropriate for developing countries where there is a general reluctance to increase mining cost because of limited financial resources. Low input approach include proper planning of waste dump using locally available boulders and jute or coir for erosion control, use of hardy, climatically adopted species having low nutrient requirements and use of Nitrogen -fixing plants, especially Legumes to provide sustainable source of Nitrogen. Rehabilitating mining sites is a way to provide speedy vegetation. It aims to achieve vegetative cover within few years so that later succession will be quick. The ecosystem should be sustaining with increase in Bio-diversity and capable of maintaining itself even if unattended.

Achieving a self-sustaining ecosystem in degraded areas requires careful planning prior to the start of mining activity. Thus mining industry will not lead to degradation of environment if a combination of imagination, care and scientific skill is employed. Infact the sites could become more productive than pre- mining.

Stabilisation of "Abandoned Tailing Dam" for Environmental Protection at Zawar Mines

V.P. Kohad

Hindustan Zinc Limited, P.O. Zawar Mines, Udaipur, Rajasthan 313 901, India

1.0   INTRODUCTION

Minerals and metals are part of our life. Mineral products are components of our homes, appliances, food containers, workplaces, agriculture, energy, transportation, communication, safety, defence, etc. Are made of some metal or other. Globally, mineral resources have generated substantial wealth and become a powerful catalyst for economic development. Ore beneficiation plays vital role in overall economics of ore to metal processing. Continuing R & D, instrumentation, computerised process controls new regents and concentrating equipment have made complex low grade ores amenable for beneficiation to produce specified quality concentrates for smelting extraction of metals, such deposits were previously considered economically unviable for exploration.

In beneficiation process of non-ferrous metallic minerals, irrespective of separation technology adopted – gravity, flotation, magnetic or electrostatic, waste generated may be as high as 98% of the ore mined, this waste material of no commercial value is generally called 'tailing'. Disposal and safe storage of tailing had been problematic in view of safety & long term environmental impact. In general, there are three modes of tailing disposal viz. Under water disposal, back filling of excavated pits/stopes in mines and storage dams. Due to global awareness, access to information and concern about environment problem in mining industry. Incidents of tailing dam failures in Guyana, South Africa, Phillippines and recent one (1997) in Spain, affected human, wild and acquatic life severely beside loss of property & restoration cost. Management of tailing dam, post closure phase in particular, has been under review throughout the world. Rehabilitation of abandoned tailing dam has become essential in view of risks related to its stability, air and water pollution. Besides this, waste land can be restored for land use like agriculture, forestry, public place for recreation and other developments. In this paper successful project of old tailing dam stabilization at Zawar Mines has been discussed.

2   TAILING DISPOSAL & STORAGE SYSTEM AT ZAWAR MINES

First Indian lead-zinc ore beneficiation plant of 175 TPD capacity was installed at Zawar Mines, Rajasthan (India) in 1950 adopting froth flotation technique. Annually about 0.8 million tons tailing is generated in ore beneficiation process, which is disposed off in a slurry (28 to 30 Wt.% consistency) & stored in a safe impoundment so that there is no risk and impact on environment.

Earlier, when first beneficiation plant was installed, not much engineering thought was given or was invested for tailing impoundment. Tailing was pumped to a valley about 500 mtrs. Away from the plant. Initial dykes were raised with mine waste rock and tailing was deposited near the dyke, coarser tailings got deposited near to form dyke and finer particles moved away. Same method was continued for subsequent raising of the dam. Later on, water recovery plant was installed near mill & thickened tailings were pumped to tailing dam. In the dam, cement masonry decant wells were constructed at intervals of 50 to 100 mtrs. These wells connected at the bottom with rectangular channels, and the central channel was taken out at down stream toe of the dyke. Ref. Fig.1 Water percolated in these wells

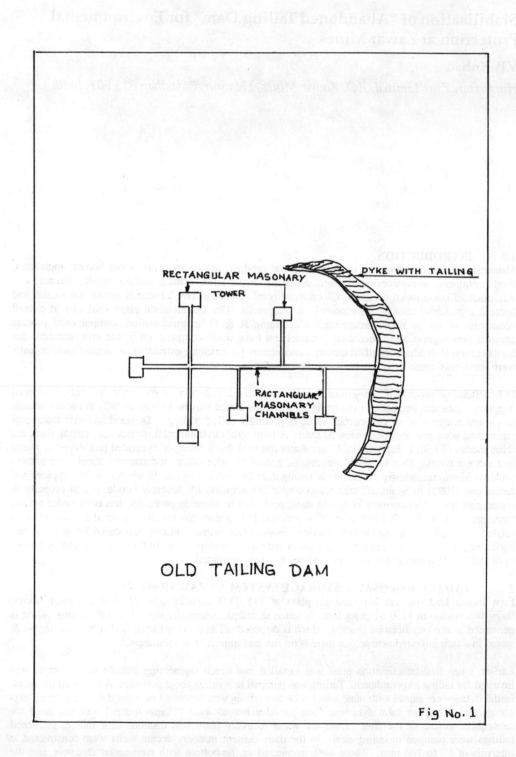

RECTANGULAR MASONARY

TOWER

DYKE WITH TAILING

RACTANGULAR
MASONARY
CHANNELS

OLD TAILING DAM

Fig No. 1

was drained out through the network of channels and discharged to nearby Tidi river flowing through mining complex.

As second 2000 TPD ore beneficiation plant installed in 1977, quantity of tailing generated increased & it was not possible to build up additional storage capacity in the old tailing dam site. It was essential to develop a new scientifically designed tailing storage facility in view of long term environmental impact. Therefore new dam was constructed in a valley at a distance of 2 km from the plant covering 96 hectares area. A system of recycling of recovered water from new dam was introduced for water reuse in the beneficiation plant.

3.0   ENVIRONMENTAL PROBLEMS OF TAILING DAM

3.1   Environmental problems associated with the tailing dam are mainly air & water pollution. Tailings contain heavy metals like Pb,Zn,Cu,Cd,Fe which lead to oxidize & may get leached during long storage due to presence of pyrite/pyrrotite & bacteria, may pollute underground water on formation of soluble salts of these metals. Similarly tailing water may contain residual hazardous chemicals like xanthate, cyaride, copper sulphate, cresylic acid, MIBC used in the beneficiation process. Dust caused by drying out of the surface may be sweft into neighbouring habitation & eco-system. When the tailing dam is active, it is monitored regularly to minimise safety hazards & environmental impact. After cessation of its use, not much attention is given when it is abandoned. Now, due to global awareness, this area is getting importance in view of long term impact on ecology.

3.2   HAZARD OF ABANDONED DAM

If the dam is not properly stabilized, due to erosion of dam sloped during mansoon /rain water area may carry solid to nearby water resource. If water discharge on the dam surface is not properly levelled and quick drainage system not provided, there are chances of trapping of cattles, wild animals in the dam or percolation towers. Due to weathering, some of the rocks may get decomposed and gullies may form below surface & result in sudden collapse of dam edges. All above problems were experienced with old tailing dam at Zawar, covering 40 hectare of land near to residential colony, dust storms was major issue. Besides, due to improper design at the time of construction, dam slope failure/erosion was very common. During mansoon, tailing used to wash away to nearby Tidi River, Baron tailing dam, looking like a desert, spoils scenic beauty and was like a scar in this green hilly area. In view of above environmental issues, it was essential to take action for stabilization and rehabilitation of old tailing dam for its after-use.

4.0 ATTERNATIVE METHODS CONSIDERED FOR STABILISATION

For maintaining safety and environmental integrity of decommissioned tailing dam at Zawar following alternatives were considered:

- Covering of dam surface with mine waste rock.
- Coating/cladding with cheaper binder material like used oil, tar, smelter/refinery residues, sludges molasses etc. To form a hard layer which can not be blown off by wind or water.
- Establish vegetation cover

Considering; possible future end use of the land, cladding with binder material or dumping waste rock was not chosen, besides, it was uneconomical, Possibilities were also explored, if the tailing could be used as substitute to building material. Central building research institute conducted studies, but it was found uneconomical in view of cheaper alternate conventional materials available in the area. Therefore, establishing vegetation coverage was considered most cost effective, long lasting, Eco-friendly and beneficial way for stabilization of the dam.

5.0 POST CLOSURE PREPARATORY WORK FOR STABILISATION

Before taking action for growing vegetation on the dam, important prerequisite was to maintain integrity of the tailing deposit. First of all, old deceant wells were plugged with mixture of graded waste rock, lime and tailings in order to check internal erosion or piping may be caused by flow of water through weathered tailing material.

Slope stabilization was another important area of action as it was most vulnerable, most of the erosion occurs from the slopes of tailing impoundment and relatively little from the top surface. Dolomite i.e. waste rock was dumped at toe and at critical places compaction was done to strengthen. Slopes were dressed to angle 40 to 45 deg.. If slopes are protected from water cascading down from the top of the dam wall, gullies erosion is reduced. Old gullies were repaired and filled with rock and lime. Used gunny / HDPE bags filled with tailing material were put along the slopes and benches were prepared at specified intervals.

Gradient of top surface of tailing dam was dressed for quick flow of rainwater wherever possible and the area was divided in segments because of high level difference at many places. A grid of storm water drains was created for quick flow of rainwater from the surrounding hills. This was also essential for quick evaporation of surface water and to protect vegetation from submergence during initial stage when much care is required.

6.0 LAB SCALE & FIELD PLANTATION EXPERIMENTS

Growing vegetation in the tailing is difficult, because of its physico-chemical characteristics. Unlike earth or soil which has undergone millions of years natural changes and regeneration process, tailing being man-made/artificial mix of finely ground mineral and rocks, does not have natural properties of soil. As given in table No.1, the material is totally devoid of nitrogen, potash, phosphorus, micro bacteria and organic matters. It has very poor water retention capacity & contains heavy metals & salinity. Presence of toxic metals, residual chemicals, extreme temperature upto 48 deg. C during summer and low temperature upto 4 deg. C in winter, low rain fall in this semi-arid zone are negative points for survival and growth of vegetation on the tailing dam. While selecting species, other consideration were dust arresting capacity, fast growth, atmospheric nitrogen fixing efficiency and low water requirement.

Initial field plantation, experiments carried out departmentally n 1980. Local varieties of plants and shrubs found growing in the vicinity of tailing dam, effluent drain hear beneficiation plant and Tidi river bed where accumulation of tailing is generally observed, were selected for this purpose. Mixture of tailing material soils and manure was used in pits and pots at tailing dam. Mostly, local Julli flora species were tried for field experiments, the results were encouraging, thereafter group plantation was taken up on the top near edges of tailing dam in pits 60 cm x 60cm x 75cm depth using various proportions of soil, tailing and bio-manure. On slopes/edges of the dam embankment, benches were made and rainwater drains were provided. Slopes were protected by using gunny bags filled with tailings in grids, leaving some pockets for plantation of saplings and grass on shown in fig.2. First Scientific study was made through department of Botany, Institute of Science, Bombay University, in the year 1982 - 83, extensive work was carried out on various species of plant & grass using different soil / manure mix, growth of species, etc. This work provided valuable guidance for field plantation in the later years.

7. LARGE SCALE PLANTATION

After 1983, every year in mansoon, group plantation was undertaken. Post plantation maintenance & protection of the area on contract basis was introduced. Table No.2 gives various species planted on the tailing dam. Total 1.5 lacs trees are planted till last year. Saplings of 1981-82 have grown to height upto 15 meters. Survival is above 90%. For irrigation purpose, beneficiation plant effluents and tailing dam recovered water is used.

Since tailing dam area was large, it was not possible to undertake plantation to cover it at once besides cost involved. It was also essential to study growth & survival of the species. Therefore a decision was taken to grow local variety of grass (Cyanadon Dactylon) in early stages of establishment on experimental basis. Initially 1.5-hectare area was selected, topsoil was quite hard and salinity was there. Therefore ploughing to the depth of about 30-cm was done, soil from local area was mixed alongwith farmyard manure. After levelling, bunches of grass from nearby riverbank were transplanted. It survived and growth was satisfactory, thereafter regular irrigation and doses of urea and super phosphate was applied which further boosted growth. Gradually total 25 hectares area was covered with grass. Along with tree plantation, grass plantation was also taken up between tree plant rows & seeding in open area.

8. RECREATIONAL FACILITIES DEVELOPED

Efforts have been made to attract residents of Zawar Mines, basically to educate and create awareness about importance of environment & ecology. A cricket pitch and ground covered with grass was developed on the tailing dam. Picnic spot was created with a shed; drinking water and telephone facilities were made available at pump house about 500 mtrs. away from the picnic spot. A sunrise point was developed to view picturesque valley, green hills, residential colonies and mining complex. Access road and pathways have been developed in plantation area.

On local festivals like Makar Sankranti, Vijaya Dashami, residents flock to the plantation site. Every year, world environment day 5th June is celebrated on stabilised tailing dam plantation site. Plantation by family members of residents is done, programmes on environmental awareness are organised. This project has been appreciated by many National & International institutions & awarded by Rotary International.

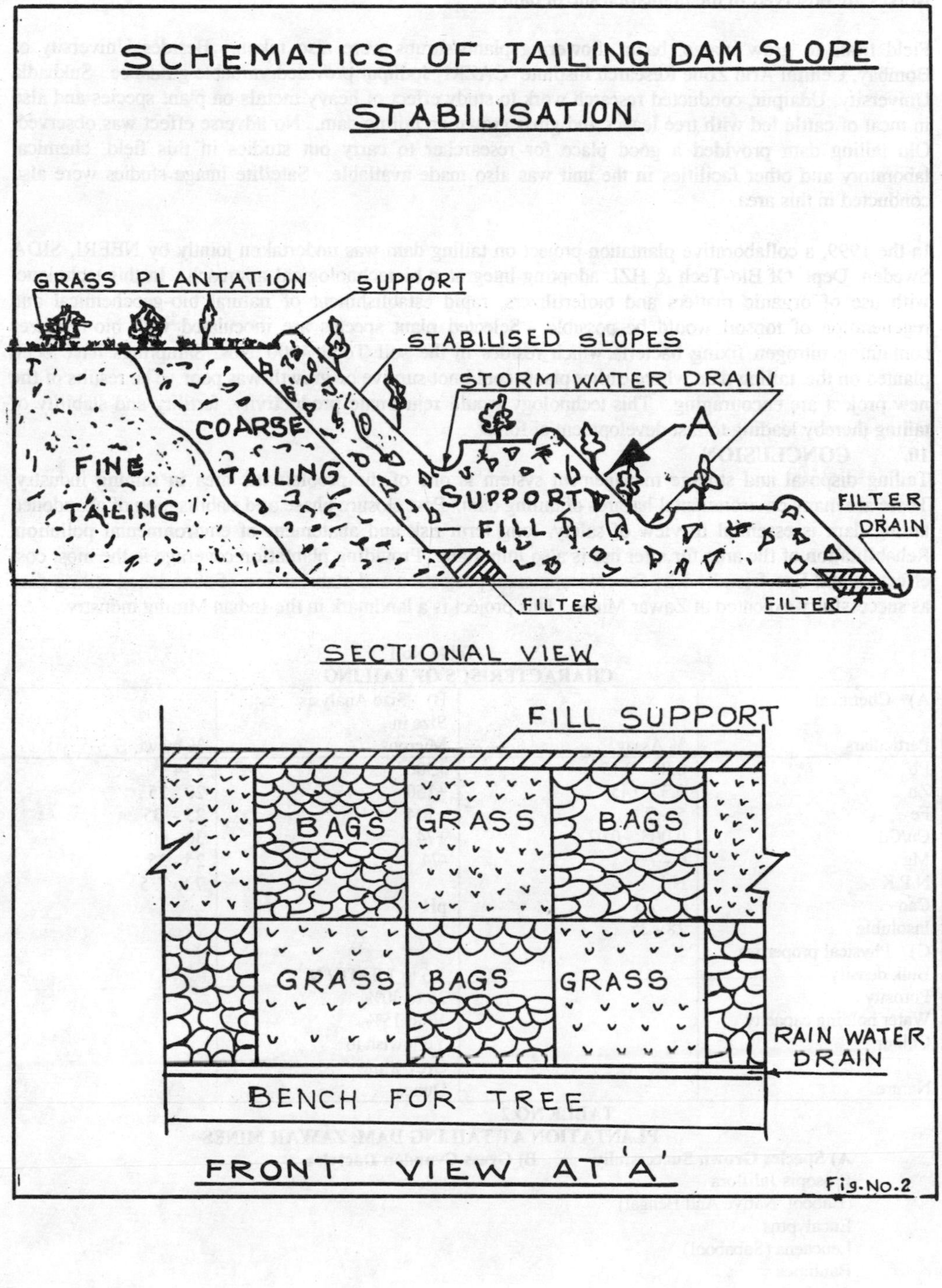

SCHEMATICS OF TAILING DAM SLOPE STABILISATION

GRASS PLANTATION   SUPPORT

STABILISED SLOPES

STORM WATER DRAIN

A

FINE TAILING

COARSE TAILING

SUPPORT — FILL

FILTER DRAIN

FILTER

FILTER

SECTIONAL VIEW

FILL SUPPORT

| BAGS | GRASS | BAGS |
| GRASS | BAGS | GRASS |

RAIN WATER DRAIN

BENCH FOR TREE

FRONT VIEW AT 'A'

Fig. No. 2

9. RESULTS, SCIENTIFIC STUDIES & NEW PROJECT

About 80% area of abandoned tailing dam has been covered with trees & grass. Regular air quality surveys are conducted, the change in the dam area is clearly visible and experienced. Dust problem in near residential area is totally solved. It has been observed that there is increase in humus, grass and saplings of tress planted earlier are naturally regenerated in tailing. Varieties of birds, rabbits, foxes and wolves are observed in the plantation site of dam.

Field trials to grow maize, bazra, flowering plants/shrubs were also taken. Besides University of Bombay, Central Arid Zone Research Institute 'CAZRI' Jodhpur provided valuable guidance. Sukhadia University, Udaipur, conducted research work to study effect of heavy metals on plant species and also in meat of cattle fed with tree leaves and grass grown on tailing dam. No adverse effect was observed. Old tailing dam provided a good place for researcher to carry out studies in this field, chemical laboratory and other facilities in the unit was also made available. Satellite image studies were also conducted in this area.

In the 1999, a collaborative plantation project on tailing dam was undertaken jointly by NEERI, SIDA Sweden, Dept. Of Bio-Tech & HZL adopting integrated biotechnological approach. In this technique, with use of organic matters and biofertilizers, rapid establishment of natural bio-geochemical and regeneration of topsoil would be possible. Selected plant species are inoculated with bio-fertilizer containing nitrogen fixing bacteria which resides in the soil Total 4000 Nos. Samplings have been planted on the, tailing dam where earlier plants could not survive or growth was poor. The results of the new project are encouraging. This technology would rejuvenate productivity, fertility and stability of tailing thereby leading to fast development of forest.

10. CONCLUSION

Tailing disposal and storage management system is one of the problematic area in mining industry. There are many environmental hazards of tailing dam. Post closure phase and stabilisation of abandoned tailing dam is essential in view of safety, long term risk and abatement of environmental pollution. Rehabilitation of the area for after use is also important. Providing plantation coverage is the most cost effective and Eco-friendly way for environmental protection and stabilisation of abandoned tailing dam as successfully executed at Zawar Mines. This project is a landmark in the Indian Mining industry.

Table No.1
CHARACTERISCS OF TAILING

A) Chemical Particulars	% Assay	B) Size Analysis Size in Microns	% by wt.
Pb	0.05 - 0.15	+300	2 -4
Zn	0.3 - 0.45	+150	20 - 25
Fe	5 – 7	+104	25 - 35
Cu/Cd	0.005 - 0.02	+74	30 - 40
Mg	6 - 7	-74	25 - 35
N.P.K	Nil		7 to 7.5
Cao	7 - 7.5	pH	
Insoluble	28 - 35		
C) Physical properties			
Bulk density		1.6 to 1.8 MT / M^3	
Porosity		25 to 30%	
Water holding capacity		10 to 15%	
Colour		Yellowish to Grey white	
Nature		Dunes	

TABLE NO.2
PLANTATION AT TAILING DAM, ZAWAR MINES

A) Species Grown Successfully **B) Grass Cynadon Dactylon**

Prosopis Juliflora
(Babool Native And Bengal)
Eucalyptus
Leucaena (Sababool)
Bauhinea
Acacia Auriculiformis
Banyan
Ingadulee
(Kinker / Jungle Jalibi)
Cassia Siamea
Ipomoca Garnea
Bauhinea Purpurea
Marisum Odorium (Kaner)

126

Geoenvironmental Reclamation (*International Symposium, Nagpur, India*)

Land Reclamation in Coal Mines—Case Study in WCL

K.C. Vijh

CMD, W.C.L., Nagpur, India

1.   INTRODUCTION

All developmental activities have their impacts on the earth's elemental resources like land, air water etc.   While development is an ongoing process, of late its impact on these resources forced mankind to give a serious thought on the very process of development.   People to-day question how much development is desirable? Is the cost of development out weighing its benefits? This is all because the processes lack in one vital aspect in the whole planning & implementation process – the aspect of adequate mitigation of the environmental impacts. And the results are showing. The danger signals are to be seen all over.

Out of all the environmental attributes & Nature's resources, Land is of primary importance. This paper deals with a few aspects of degradation, reclamation & a case study.

It is imperative that the planning of all types of projects start with the approach of "working on other's property" – meaning that the project in question is being set up on some one else's property, and hence we have to make good the damages being caused – or else the 'other person' will disallow the workings.

COAL SECTOR

Coal Mining Industry in our country has witnessed phenomenal growth since nationalisation. Coal production has spurted from 72.72 million tonnes in 1971-72 to about 300 million tonnes in 1999-2000. At the time of nationalisation production from open cast mines was only 25% and today, open cast mines contribute 70% of production. We are handling about 600 million cubic metres of overburden annually and this has brought out coal mining related environmental problems.

Open Cast mining results in serious disruption of surface topography and water courses, despoliation of land by dug open cast pits and massive overburden piles, land erosion & degradation, rehabilitation & resettlement of project affected people, loss of flora and fauna, dust pollution noise pollution and blasting vibrations. Derelict land due to mining activities not only affects visual aesthetics but leaves permanent scars on the habitats.

Underground mining is associated with subsidence of mined land, lowering of ground water table, often adversely affecting adjoining villages, pollution of water and air.

For appreciating complexity and magnitude, coal mining industry's environmental problem, you have to take into account the following situations !

1. Pollution due to pit-head power stations – Due to obvious advantages of pit-head thermal power stations and other industries, environmental scenario in the coalfields would be getting more and more complex in the years to come.

2. Thrust on large scale open cast mines with capacity of 10-14 million tonnes per annum to meet ambitious coal production targets poses problems of large scale degradation of land and waste management problems specially due to worsening of the stripping ration.

Environmental degradation due to past coal mining by underground and often manual of semi-mechanised open cast prior to nationalisation of the coal industry has left legacy of massive environmental backlogs.

The degradation of land takes place due to following activities :

- Excavating the earth to reach coal benches (opencast mining)

- Dumping the excavated earth (overburden dumps) on another tract where no coal is available below

- Making coal stock yards, sidings etc.

- Making roads for movement of heavy vehicles etc.

- Construction of industrial structures like Coal Handling Plants, Workshops, Stores etc.

Out of the above, overburden dumps present most conspicuous signs of land degradation since

a) they comprise of loose soil, stones, boulders etc. with steep slopes and with practically no soil nutrients.

b) flow of eroded soil of no nutritional value downstream alongwith natural water courses with rains, thus degrading land further.

The degraded land is also a source of dust and hence requires to be tackled at the first place.

As you know land is a finite and non-renewable resource which sustains the life forms on the earth. Western Coalfields Ltd.'s focus is to restore this precious commodity at the end of the mining operations so that the final land form at the end of mining activity is as much productive as possible for use of the community at large. Development, unless it aims at sustainability, is more likely to lead to destruction.

MINING & SUSTAINABLE DEVELOPMENT

The concept of Sustainable Development came to prominence 10 years ago with the release of the World Commission on Environment and Development's Report – "Our Common Future" (commonly known as the Brundtland Report). Since then, there has been much debate, but little agreement, on whether extractive industries, such as the minerals, coal and petroleum industries, can be a part of sustainable development.

The initial response was that, because mining depletes a finite resource (an ore body, coal deposit or petroleum reservoir), it cannot be sustainable. Once the resource is extracted and used, it cannot be renewed. Therefore, not only does mining not contribute to sustainable development, it is counter to the very principles of sustainability. But is sustainable development only one-dimensional or does this oversimplify an important, and complex, concept. To decide whether there is a role for mining in sustainable development, we need first to answer two critical questions: what exactly is sustainable development, and can mining be part of it ?

Sustainable development has not been an easy concept to define and has tended to mean all things to all people. This fluidity of definition may have contributed to its resilience as a concept and its broad acceptance as a benchmark against which human activity should be measured. Whether the Mineral Industry, let alone particular activities within an Industry, are contributing to sustainable development.

The most widely used definition comes from the Brundtland Report, where sustainable development was defined as **"development that meets the needs of the present without compromising the ability of future generations to meet their own needs"**. An alternative way of presenting the concept was put forward by the World Conservation Program. It described the sustainable development as **"improving the quality of human life while living within the carrying capacity of supporting ecosystems**

A number of attempts have also been made to define the concept specifically in terms of the mining industry. The Final Report of the Ecologically Sustainable Development Working Group on Mining, published in 1991, described sustainable development as **"ensuring that the mineral raw materials needs of society are met, without compromising the ability either of future societies to meet their needs, or of the natural environment to sustain indefinitely the quality of environmental services such as climate systems, biological diversity and ecological integrity"**. More recently, the Governments of Developing countries in their Environment Policy commitment to the principle of sustainable development, state **"that our economic decisions will not take priority over considerations of health, safety and the environment"**. In Australia, Canada, and many other countries sustainable development is seen as a system in which the Mineral Industry operates in a manner that respects and responds to the social, environmental and economic needs of present generations and anticipates those of future generations.

WCL AND LAND RECLAMATION

Having realised the situation as described above, we in WCL are taking care of this vital aspect in the following manner -

(1)   Planning for land – Today increased thrust have been put to minimise the requirement & acquisition of land for new projects. The concept of concurrent back filling and creation of O.B. dumps to a minimum are inbuilt at planning stage.

(2)   Separate stacking of topsoil – Today, all EMP's provide for stacking the topsoil separately as a precious material to be used during reclamation.

(3)   Reclamation – Post mining land use plan and mine-closure plans are gradually becoming more pronounced and they now help to visualise creation of assets at the end of mining operations in the shape of reclaimed land forms – more productive than at pre mining stage.

Recently we have carried out study to find out the nutrition contents and plant sustainability criteria of our O.B. dumps by analysing the extent of nutritional elements like Nitrogen, Phosphorus, Potassium etc. that are required for their plant productivity. This study has been carried out in the OB dumps of Durgapur, Padmapur, Sasti, Niljai & Umrer.

The above steps are taken since the developmental activities need to be sustainable so that development does not lead to destruction.

We have created massive Topsoil dumps in Niljai, Sasti, Umrer etc. which can be used at the time of final reclamation.

However, it is the old barren dumps & the projects where top soil has not been stacked separately that need to be tackled as thrust areas. In this direction a significant step has been taken by WCL by making an R&D Study through the Institute of Science, Nagpur University.

A brief description & results of the study are as below.

RECLAMATION OF MINED OUT LAND THROUGH RESEARCH & DEVELOPMENT – CASE STUDY

Development of Technology for conversion of Backfilled Area in Ballarpur Open Cast Mine for Agriculture: Its Demonstration

R&D study was conducted through Institute of Science, Nagpur to find out whether backfilled area can be utilised for agriculture purpose or not. Keeping this in view, R&D Project was started at Ballarpur OC Project, in the year 1993. About three-hectares of land was backfilled in decoaled voids and subsequently levelled for utilisation of R&D experiment.

OBJECTIVE

Main objective of the R&D was to verify whether or not backfilled coalmine area could be utilised for sustained agriculture under various conditions as described below:
R&D studies and demonstration project involved:

1) Carpeting the reclaimed area with top soil with different thickness with a view to ascertain the optimal top soil requirement in the area to convert the mined out area for agricultural purposes.

2) Arriving at optimum conditions of soil amendments in the absence of availability of topsoil for making it suitable for agricultural purposes.

3) Intermixing judiciously the black cotton soil with overburden in the sub layers during the final stages of reclamation to avoid top soil (in view of practical situation obtaining in the Project areas) and arriving at the optimal conditions of soil amendments with a view to make suitable for agricultural purposes.

METHODOLOGY

The methodology adopted for reclamation and reuse of land and the experimental design for studies at pot level, micro pot level and field level is carried out by carrying out the experiments in the pots, in the small pots of size 1m X 1m X 1m and at the field level at Ballarpur open cast site. Mainly following four conditions were studied:

1. OB (Overburden Material) alone
2. OB+Black cotton soil
3. OB+Lime Sludge
4. OB+Farm Yard Manure

Plants with monocotyledon and dicotyledon roots were tried repeatedly. In all, 15 crops harvested such as sunhemp, moong, maize, wheat, soybean etc. were tried during the period 20th August 93 to 30 th November, 1997.

It is observed that the changes in spoil characteristics were rapid in second and third year.

FINDINGS

The environmental pollution problem associated with disposal of solid wastes i.e. lime sludge from pulp and paper mill can be successfully solved by using amendments such as farm yard manure and lime sludge.

The best treatment was found to be the amendment of OB material with farmyard manure i.e. condition 4. Healthy growth of crops was observed and at the end of third year the results were compared with normal agricultural soil. The closest values were registered with this treatment.

The only precaution to be taken is proper irrigation facility as the water holding capacity is low. It increases with time but the increase is low in controlled conditions while in fields due to mixing of soils from adjacent plots it was quiet high.

Second best method was found to be OB material with lime sludge i.e. condition 3. Nearly parallel results were observed with soil. This method has the added advance of using waste/ effluent material thus partially solving waste disposal problem of paper & pulp mills. Besides It helps in reclamation of opencast mines by neutralising its acidity and increasing its nutrient level.

Black cotton soil was poor in nutrient status and was similar to OB as it was dumped from several years and no plantation was done on it. But the water holding capacity was high. Interestingly, it was found that the crops grew well in OB alone with other amendments where the porosity was more. Retarded growth of crops was observed in first two years. Improvement was rapid in this treatment in the third year.

Thus it can be concluded that this method can also be applied but to enhance the rate of improvement some fertilisers shall be added so that the biological activity gets enhanced and ultimately the nutrient status.

The fourth combination i.e. the overburden alone (condition 1) registered improvement but it was very slow, as the condition of the nutrient status was critical. But the proper selection of crop species, irrigation facilities and rotation of crops can improve the structure and regain its fertility back but at a slow rate.

The results have shown that the lime sludge a waste from pulp and paper mill, and farmyard manure can successfully bring back the fertility of overburden material if proper doses are added. It not only minimises the dust pollution problem but also brings back its bio-diversity. Place where no grass was seen and places totally barren full of sandy material can be altered to a green patch with healthy crops and birds chirping, butterflies flying around.

SCOPE OF APPLICATION OF THE STUDY IN COAL INDUSTRY

The technology developed can be successfully applied to other abandoned patches with overburden material and mine areas can be reclaimed by adoption of environmental friendly technology. This technology has twin advantages it solves solid waste disposal problem to some extent and overburden reclamation, ultimately helping and encouraging sustainable development.

FUTURE SCENARIO

The land scenario is presently quite discouraging as of now. The increasing pressure by environmentalists to bring more land under forest cover and increasing bargaining position of revenue land owners present a grim situation in respect of new projects.

The priority there fore should be to convince the forest departments & people at large that it is not only acquisition which the coal companies are doing but are returning land to the society. WCL is actively pursuing the approach of reclaiming lands which can be converted and used for income generation, forestry and even rehabilitation & resettlement. This will require scrutiny from legal angle as well as involvement of state authorities. But this appears to be the only approach left if future land acquisition are to be hassle tree & expeditious.

CONCLUSION

It is now established that the land degradation by mining activities is no cause for worry provided a proper post mining land use plan is formulated and followed up. At the end of the mining activity a lush green productive land being returned to the community is a reality.

REFERENCE: 1. Centre for Environmental Studies, Indians School of Mines, Environmental Training Study Materials.

2. Final Report of Institute of Science, Nagpur University on "Development of Technology for conversion of Backfilled Area in Ballarpur Open Cast Mine for Agriculture: Its Demonstration".

Reclamation and Remediation—A Broad based, Long-term, Mining Approach

Keith Atkinson

Director, Camborne School of Mines, University of Exeter Pool, Redruth, Cornwall, TR15 3SE, UK

INTRODUCTION

Agricola's *De Re Metallica* (1556) probably gives the first major insight into contamination due to mining when reflecting the generic view of mining over four hundred years ago. Detailed study of Agricola, particularly some of the illustrations, will demonstrate toxicity in soils and windborne pollution resulting from mining and processing activity in the drawings of the diseased and the stressed nature of some of trees. Despite this, for over four hundred years after its publication if the term "environment" was used, in mining areas it was usually linked to the *mining environment*, i.e. the underground environment in which the miners worked. Similarly the concept of contamination was thought of in terms of contamination of samples or impurities within a product rather than contamination of land. Obviously this has all changed and increasingly during the latter half of the Twentieth Century the environmental and social aspects of mining became paramount, certainly in the industrially developed nations of the world.

As much of the environmental contamination just described results from the activity of mining, the professionals employed in the mining industry - geologists, mining engineers and mineral processing engineers - will be best placed to define the causative processes. However, even in this very specialised case of contamination, when the contaminating industry is closely defined, reclamation must be seen as a multi-disciplinary, or multiprofessional, task involving *inter alia* scientists, engineers, economists, planners and, occasionally, politicians. Within these general descriptions however there may be a large range of disciplines represented, geologists, biologists, botanists, chemists, hydrologists, surveyors, geotechnical engineers, civil engineers, chemical engineers, mining engineers and minerals processing engineers, to name but a few. Therefore, such an interdisciplinary team will be needed to solve the problems of mining remediation well the twenty-first Century.

Land contaminated by previous metal mining presents numerous challenges to those wishing to remediate it, challenges that have more in common with industrial contamination than coal mining contamination. In common with all reclamation, of prime importance is establishing the history of the site - "know your enemy" is a very good maxim when undertaking remediation. Unfortunately determining accurately the history of a site contaminated by past mine working is not as easy as might be thought. Industrial contamination, for instance, might be considered to have originated with the Industrial revolution whereas mining, even in England, extends back to Roman, if not Phoenician, times; if one considers a country such as China, for example, then contamination could extend back four millennia! The same site may have been reworked on numerous occasions and at each working the extraction and processing technologies will have changed. This tends to leave a very inhomogeneous material with a complex stratigraphy and chronography. Establishing what may have been worked at one particular time may be difficult enough while the presence of trace metals and gangue minerals may never have been recorded information.

Therefore a thorough chemical analysis is a primary requirement. Once this has established the likely suite of minerals, elements and contaminants present it is important to establish how they are distributed and whether any are in forms capable of entering the food chain or otherwise giving rise to contamination. These simple statements conceal significant pitfalls and major research areas in themselves and demonstrate the need for the multi-disciplinary approach. For instance, how should sampling be carried out , how representative will it be, how statistically viable will it be, would the services of a geostatistician help? Also how is 'available metal' established, what protocol should be adopted that is internationally acceptable and meaningful in its results? Here a whole range of disciplines can interact

Fundamental to any reclamation strategy is whether the contamination is to be dealt with off site - by removal for processing or to landfill, for example, - or on site - by cover down, soils washing or modifying the local environment. A review of some of the previous reclamation techniques used by various authorities in Cornwall has been given by Mitchell(1990). Details of novel methods of treating contamination in situ, which were devised at the Camborne School of Mines (CSM), are given in Atkinson et al (1990), Mitchell and Atkinson (1991), Atkinson and Mitchell (1994). The technical and economic background to soils washing has been summarised by Atkinson et al (2000). To illustrate the diversity of mining-oriented reclamation, some of the research currently underway at CSM, and its future potential, will be reviewed.

THE USE OF MINERAL AMENDMENTS IN RECLAMATION

The concept of using one naturally occurring material (such as an industrial mineral) to counteract contamination arising from the working of another natural material (metal ores) has an element of sustainability to which the author was particularly attracted when establishing the reclamation research at CSM. This original concept has been considerably diversified over the intervening years but is still a tenet to which the author would subscribe. The technology developed has been exported, most recently in the use of indigenous diatomite to ameliorate arsenic contamination from past gold mining in Thailand (Whitbread - Abrutat et al 1997).

Considerable research related to reclamation using industrial minerals has been undertaken at CSM some of which is quoted above. As a result of extensive laboratory tests on a variety of naturally occurring, and synthetic, materials, selective groups of amendments have been used in the field. Having assessed the suitability of the amendments to control acidity and toxic cations in solution, diatomaceous earth has been shown to be one of the most successful in the laboratory but in the field it does have a somewhat negative effect on permeability (Atkinson & Mitchell, 1994).

Field trials of mineral amendments at Geevor Mine, west Cornwall, have been running since 1995. The amendments were hand ploughed into the trial plots without the addition of topsoil and without artificial seeding. Natural growth of vegetation has been regularly monitored since initial amendment. At this site the amendments were sepiolite ("cat litter"), montmorillonite and vermiculite at two different concentrations. The field site at Geevor is now part of a major heritage site, attracting large numbers of tourists. It is coastal, quite exposed and prior to amendment had been contaminated, severely in places, by wind-blown processed material. Some organic matter was present on the site from the original scrub vegetation. The amendments have succeeded in changing the pH at the treated sites and in reducing the amount of bio-available metals entrained in the pore waters. As a result, the amendments have significantly enhanced the establishment of vegetative cover when compared to the non-amended, control plots and the surrounding areas. Sepiolite and montmorillonite have resulted in very significant ground cover whereas the vermiculite plots, which appear to have been more susceptible to erosion, demonstrate less success (Jenkin, 1998).

One aspect of the use of industrial mineral amendments that the author has long held would greatly enhance the future reclamation potential of mine tailings is the use of the amendments as end of pipe additions. This would entail co-deposition from the mineral processing plant of tailings and amendments at predetermined ratios. This should not only lock-in the metal contaminants, preventing them becoming bio-available when reclamation is attempted, but also could prevent the formation of diagenetic minerals within the tailings pond. Recently Lillepage et al (2000) have described the use of phosphate based additives to stabilise the metals in contaminated sludges. Their Environbond metals treatment process utilises a patented phosphate chain that sequesters a wide range of metal ions and results in metal phosphate complexes that have extremely low solubility. Undoubtedly there are fruitful avenues still to be followed in the development of environmentally sustainable pipe-end technology.

IMPACT OF MINING ON SEDIMENTATION IN RIVER ESTUARIES AND THE POTENTIAL DANGERS DUE TO GLOBAL CLIMATE CHANGE.

While mining contamination at inland sites is easily visible, if not always easily remediated, the impact on local estuaries is less readily discernible. Only dramatic illustrations such as the Wheal Jane incident in Cornwall in 1991, when a very large plume of ochreous water was seen in the Carrick Roads, and the estuary of the river Fal, focus attention on the sediment contamination in rivers caused by mining. The Wheal Jane incident amply demonstrated the impact of a sudden influx of mine waters directly into a major river; what the sediments in the estuaries conceal is the degree of impact caused in historical times by mining. In the way that contaminated sites are being located, reclaimed and remediated on land, it is becoming increasingly evident that a thorough study of historical fluvial contamination by mining is required, not only to anticipate any potential pollution from reworking but also to provide valuable geochemical evidence for the long term stability of subaqueous disposal of tailings.

The importance of mining as a source of sediments in certain estuaries in Cornwall has been described by Pirrie et al (2000). This work demonstrates that at specific periods during the last two centuries the release of particulate mine waste, probably representing periods of increased mining and milling as well as following mine closure had a significant impact on downstream estuaries. Citing Reid and Scrivenor, Pirrie et al (2000) record as evidence of the environmental effects of these pulses of mining related sediments the loss of quays and shipping trade, due to enhanced siltation. Further they recognise that those sediments with abundant cassiterite, such as in the Camel estuary from the tin mines around the Lanivet area, are less likely to cause environmental damage than sediment derived from Pb, Zn operations such as those found in the Gannel estuary. They also recognise diagenetic mobility of Pb, Zn and Cu in the Gannel estuary indicating that these metals may be still bioavailable.

With the possibility of rising sea levels consequent upon global warming it is necessary to examine the distribution of such metal "fluxes" in the estuarine sediments to see which, if any, are bioavailable and which may be in danger of being reworked during a future sea level rise. The depths of the main metal-enriched horizons need to be determined for all Cornish estuaries to minimise the problems that would be caused by reworking. Also the potential for future environmental impact by re-release of those historic mine wastes should be assessed in the light of increased data on the affects of climatic change and associated rising sea levels. Such an approach is also necessary for river systems draining other major metal mining regions of the world.

WETLANDS AND ACID MINE DRAINAGE

The treatment processes established at Wheal Jane in the aftermath of its closure, and the ochreous plume mentioned above, have been documented by Hallet et al (1999). While the reed beds and associated infrastructure established at that site follow accepted procedures, it cannot be over emphasised that reclamationists need to understand the processes in wetlands, rather than just treat constructed wetlands as "physico-chemical cells" or "black box" solutions for mine water treatment. The author believes that to understand such processes requires the detailed study of natural wetlands which are complex chemical, physical, biological ecosystems. Without such detailed knowledge there is a danger of over-simplification and loss of treatment efficiency and robustness (Wetzel, 1993). Brown (1996) attempted to identify and quantify various amelioration processes occurring in a semi-natural wetland receiving contaminated minewater. This is an extremely complicated task since even such rudimentary factors such as inflow and outflow volumes are more difficult to quantify without the control mechanism present in a designed cell.

Brown (1996) showed that semi-natural wetlands associated with old mineral workings acted as sinks for tin, copper and zinc. She established that the plant debris within the system played a major role in its proper and continued functioning. One major problem associated with older wetlands is their potential to export iron and manganese and conditions when this may occur have begun to be addressed by Stoddern and her fellow workers (Stoddern et al 1999). These workers describe the development of a mesocosm experimental system and its functioning under different conditions. This work is necessary as the mobility of these two metals not only causes a colour and taste nuisance when water is used as a domestic supply but also can result it plumes of discoloured water in nearshore environments. Such discolouration used to be the case for the Red River in Cornwall, the Rio Tinto in Spain, and was the case during the Wheal Jane discharge mentioned above. Furthermore the armouring of anoxic limestone drains associated with engineered wetlands is a serious long-term problem which this research could help to overcome or at least ameliorate. The work by Brown (1996) and by Stoddern et al (1999) demonstrates the possibility of developing sustainable natural wetland treatment systems for successful metal containment.

An assessment of existing wetlands in Cornwall, their vegetation, water and sediment chemistry was undertaken by Jenkin et al (2000) as background to their detailed work on the Red River valley. The object of this work was to establish sustainable wetlands that would overcome problems associated with periodic flushing of metal contaminants from the banks and valley floor while overcoming the periodic wet/dry cycles that allowed the mobilisation of metals. Therefore the design, which involved extension of the existing wetland areas by 'soft engineering' more wetlands, would result not only in metal containment and removal but also the development of rush pasture, reed beds, marginal plants and willows. These soft engineered wetlands would improve water quality as well as enhance biodiversity and perform a flood buffering role. This approach reflects the botanical approach that this author believes is crucial to the long-term stability of wetlands and must be given equal, if not greater, weighting than the apparently empirically derived, phyisco-chemical, engineering stance that seems to have dominated constructed wetlands to date.

USE OF WASTE MATERIALS IN NATURAL RECOLONISATION

Until relatively recently reclamation in Cornwall has involved some attempt at revegetating areas of mine waste in order to create "amenity areas". Most revegetation has included the importation of topsoil, seeding with grass and planting trees. Incredibly, sometimes naturally recolonised heathland has been removed during site works. In many cases the new grass and trees grew poorly and frequently died. This is not surprising and indeed was forecast (Atkinson et 1990) since the toxic metal content had not been ameliorated, merely covered down. Once the root systems penetrated the contaminated subsoil, or capillary action caused the metals in solution to rise to the rooting zone, then failure was inevitable. Very little work appears to have been done on the production of naturally based barrier layers to prevent such upward migration. It was for this reason that Atkinson and Mitchell (1991) pursued the approach, quoted above, of high CEC mineral amendments ploughed to rooting depth into the soil. There are, however, probably still opportunities to develop other approaches to inorganic buffer, or barrier, layers. Selected soil amendments have been shown to lock up metals, at least in the short term, making these metals less plant-available. The effects of these amendments on the soil pH have also been documented by Mitchell and Atkinson. Further, as might be anticipated, the addition of organic materials can enhance vegetation growth on these amended sites (Jenkin, 1998). Jenkin describes the addition of a variety of cheap, locally available, organic materials to a soil substitute made up of a mixture of river dredgings and inert, quartz - sand derived from China Clay waste. Chopped green waste, or spent mushroom compost, improved soil texture, water holding capacity and germination of trial seeds. These additions also diluted metal concentrations and reduced metal mobility in the short term. Given the work done by Pirrie et al (2000) quoted above, one further aspect of using river dredgings as an organic treatment during reclamation that must be considered is the likely content of contaminant metals. The same would apply to the use of sewage sludge in this capacity. If the river dredgings, or sewage sludges, are mixed with the industrial mineral amendments such as those quoted above, the metals are likely to be made non bioavailable.

Unfortunately, with at least 40% of Cornwall's 700 prime wildlife sites containing, or adjacent to, land classified as 'derelict' the replacement of areas of heather and scrub by such amenity areas and car parks has not always found favour with the local population. Furthermore, local ecologists tend to disapprove of this procedure as discrete faunal and floral habitats are destroyed irreparably. Consequently a move has begun to revegetate more sympathetically with the re-establishment of heathland and to appreciate the conservation value of these sites (Spalding et al 1996). As evidence from sparsely populated long existing mine waste dumps testify, however, such re-colonisation is slow and therefore the use of mineral amendments coupled with selected waste addition should greatly increase the rate and success of such re-colonisation.

CONCLUSIONS

The diversity of disciplines encompassed in a mining activity allows not only a broad approach to the reclamation of mining contaminated land but also is capable of identifying and researching many other potentially contaminating situations in the environment. Within the area of mining contamination reclamation s.l., there are numerous areas for continued research, these include more statistical approaches to sampling protocols, establishing procedures for the recognition of bioavailabiity of contained toxic metals, further investigation of the historic impact of mining on estuaries and the potential for remobilisation by reworking. The likely effects of global climatic change, not only on estuary sediments but also on contaminated land sites due to changing water table levels needs serious research attention. As a long term benefit in reducing the potential environmental impact of mining, further work should be undertaken on the use of industrial minerals as amendments, their use in pipe-end tailings disposal and in ameliorating river, or harbour, dredgings, as well as sewage sludge, to upgrade waste materials for improved soil construction on reclaimed sites. If procedures involving importation of topsoil placed directly over unmodified contaminated material are to be used with long term success then more research may be required on the technology and use of buffer or barrier layers.

136

REFERENCES

Agricola, G. 1556. De Re Metallica. 1950 translation by Hoover, H.C. and Hoover, L.H., Dover pubs.inc., New York.

Atkinson, K., Edwards, R.P., Mitchell, P.B. and Waller, C.P. 1990. Roles of industrial minerals in reducing the impact of metalliferous waste in Cornwall. Trans. IMM 99: A158 -172.

Atkinson, K. and Mitchell, P.B. 1994. The environmental impact of mining and some novel methods of reclamation. In: The impact of Mining on the Environment - Problems and solutions. A.A.Balkema, Rotterdam. pp. 305 - 310

Atkinson, K., Barley, R.W., and Pascoe, R.D. 2000 Contaminated Soils Separation Technology. In: Mine Land reclamation and Ecological Restoration for the 21st Century. China Coal Industry. pp 444 - 463.

Brown, M. M.E. 1996 The Amelioration of Contaminated Mine waters by Wetlands. Ph. D. Thesis, Camborne School of Mines, University of Exeter.

Hallet, C.J., Froggatt, E.C., Sladen, P.J. and Wright, J.S. 1999 Wheal Jane - The Appraisal and Selection of the Long-Term Option for minewater Treatment. In:Mining and the Environment II (eds) Goldsack, D., Belzile, N., Yearwood, P. and Hall, G. Sudbury, Ontario, Canada. : 711 - 720.

Jenkin, L. E. T. 1998 Enhancement of Natural Recolonisation of Heathland on Metalliferous Mine Sites - the role of waste materials. 6th European Heathland Workshop. Bergen, Norway : 96 - 97.

Jenkin, L.E.T., Brown, M.M.E. and Watkins, D.C. 2000 Red River valley wetland enhancement project, Cornwall. In: Water in the Celtic world: managing resources for the 21st century. B. H. S. Occasional Paper 11: 347 - 355.

Littlepage, B., Schurman, S. and Maloney, D. 2000. Treating metals contaminated sludges with Envirobond. Mining Environmental Management 8 No.3: 22.

Mitchell, P.B. 1990. Reclaiming derelict metalliferous mining land with particular reference to Cornwall, UK. Land and Minerals Surveying 8: 7 -17

Mitchell, P.B. and Atkinson, K. 1991 The novel use of ion exchange materials as an aid to reclaiming derelict mining land. Minerals engineering 4: 1091 - 1113.

Pirrie, D., Power, M.R., Payne, A., Camm, G.S., and Wheeler, P.D. 2000 Impact of mining on sedimentation; The Camel and Gannel Estuaries, Cornwall. Geoscience in South West England. 10 (in press)19pp.

Spalding, A., Edwards, T, Sinkins, B. Purvis, O.W. and Stewart, J. 1996 Nature conservation value of metalliferous mining sites. In: The conservation value of metalliferous mine sites in Cornwall, eds. Johnson, N., Payton, P. and Spalding. A. : 31 -39.

Stoddern, T.J., Jenkin, L.E.T., Watkins, D.C. and Atkinson, K. 1999 Containment of mobile Metals in Natural Wetland systems. Studies of semi-natural wetland mesocosms with sub-surface flow. In: Mining and the Environment II. Sudbury, Ontario, Canada.: 703 - 710.

Wetzel,R.G. 1993 Constructed Wetlands: Scientific foundations are critical. Constructed Wetlands for water Quality Improvement. G.A. Moshirir (ed). Lewis Publishers: 3 - 7.

Whitbread-Abrutat, P.H., Atkinson, K. Paijitprapapon, A., and Charoenchaisri, P. 1997 Mobility of arsenic and other potentially harmful elements in metal mine wastes in Thailand. The International Conference on stratigraphy and Tectonic Evolution of Southeast Asia and the South Pacific. Bangkok, Thailand. : 693 - 713.

Geosciences for the Sustainability of Brazilian Mining

Arlei Benedito Macedo

Instituto de Geociências–Universidade de São Paulo,
Rua do Lago, 562 - CEP 05508-900 - São Paulo, SP, Brasil

INTRODUCTION

Since the arrival of Europeans in Brazil 500 years ago mining has left a mixed legacy: positive, in its importance to the national economy during almost this entire period, and negative, in its environmental impacts. These latter are the subject of this article, in which they will be exposed and evaluated, and proposals made for their control. Parts of this article are based on other syntheses published by the author (1, 2). Current impacts will be described, when possible, from direct experience, and the historical impacts from reports of scientists and travelers, and from what little that can be discerned as old impacts in areas of traditional mining.

THE ENVIRONMENTAL IMPACTS OF MINING IN THE BRAZILIAN HISTORY

Environmental impacts of mining are rarely mentioned in texts on the history of Brazil, even in those dedicated to the study of mining (3,4,5). In 1603 the first Regiment of the Mineral Lands of Brazil, already established in its article 46 the protection against the pollution of water streams by mining (6), the rules being repeated and enlarged in later codes. However, just in the decade of 1980 did effective protection begin. During almost all our history, and even now, in a large part of the mines, an environmental consciousness has not existed, without which the letter of the law is worthless.

From the point of view of mining technology, and particularly of the study and control of the environmental impacts caused by it, the history of mining in Brazil can be understood as superimposed models, with the predominance of one or the other in some historical periods, without however the most primitive models being annulled by the more developed ones. As in many other cases, such as of the land tenure policy (11), the history of mining in Brazil tends to repeat itself, with more complex models substituting the simpler ones in some points of the territory, while in others the cycle is just beginning.

These are the models:
- a: small mines, with inadequate technology;
- b: large mines, with inadequate technology;
- c: mines with adequate extraction technology, but without environmental control;
- d: mines with environmental control.

What is here considered adequate extraction technology includes, at least, ore deposit evaluation and mining project using geological and mining engineering principles, followed by the execution of the project with professional supervision. Until the eighties, environmental control of mining was almost nonexistent, and still is little and inefficiently applied, even when there exists an Environmental Impact Study.

The characteristics of the models are:

a: <u>Small mines, with inadequate technology</u>: Mining of easily extractable, usually alluvial deposits, with primitive techniques and little capital. They appear first during the colonization of virgin areas, where it is possible to find precious minerals, mainly gold, in modern sedimentary deposits. These, once mined, will return to minable grades after thousands of years. Deposits of this type were mined throughout the country, by individuals or small groups, using manual methods.

Environmental concerns are totally absent in such barbaric mines, but their impact is limited due to their small capacity of modification of the environment. Such is the case of the first attempts, in the 16th century, to mine gold in the coastal river basins in south and southeast of Brazil. The indirect impacts of that mining were more durable than the direct ones, as a consequence of the population centers they originated

That was the sole model in the 16th century and it prevailed in the 17th and 18th centuries until the discovery of the great gold and diamond deposits of Minas Gerais and Mato Grosso. It still continues, in the most distant reaches of the Amazon.

This model also reappears in periodically reactivated claim areas, where *"faiscadores"* or *"requeiros"* scratch out their survival from alluvium or tailing deposits, as in the case of old *"garimpo"* areas, like Diamantina or Poconé or new ones like Serra Pelada. Such independent microminers often work in the same area as companies or better-equipped *"garimpeiros"*.

Almost all mineral extraction for direct use in civil construction is also of small size, with inadequate mining technology. Sand is still manually extracted and transported in canoes or mule trains; manual crushing of rocks still exists, even supplying large cities like Manaus. Even when there is mechanization, in most cases there is no mineral evaluation, nor projects for mining or environmental control. Their environmental impacts are evident around all the inhabited points of the country.

From a legal viewpoint, these small mines almost never have mining titles, nor environmental licenses, unless they are close to large cities, in states where the environmental control is effective. The prevailing ideas among such miners is that rights on minerals before being mined do not exist, the ownership being legitimated by the work applied in their extraction. The same concept prevails among squatters, relating to land ownnership (11). These notions are totally adverse to the concepts of conservation of resources or of the environment. The primitive miner, being only interested in the product, and without bonds with the area, has no interest in mining rationally, nor in defending or reclaiming the environment.

The total impact of these mines, though ubiquitous, is limited, due to their small capacity to disturb the environment. Natural recovery is often sufficient to heal their small scars.

b - large mines, with inadequate technology: When larger mineral deposits were found, settlers set up great production structures, based on Native or African slave work until the 19th century, and later using free workers, and more recently, machines. Environmental concern continues nonexistent, and in most cases, adequate scientific and technological principles are not applied.

The technological ignorance of the colonizers was so great that many of the techniques used in Brazilian colonial mines were brought in from Africa by the slaves (5). Consequence, productivity and recovery were low, only gold and diamond being concentrated by simple gravimetric processes being recovered, while fine gold was lost and the deposits destroyed. In Burton's report of his trip to Minas Gerais in 1867 (12), he wrote: "The following sketch of gold-mining in the Brazil will show how little the Roman system has been changed since A.D. 50". Even today, in many cases he would not have much more to add, except for the greater use of mercury and the larger capacity for excavation.

The environmental impacts of mining in colonial and imperial Brazil, other than the silting of the stream beds, where not much mentioned in early scientific and technical reports, perhaps because they were considered natural and inevitable consequences of the activity. They left a bigger impression on travelers, like Courgy (21), who reporting a trip to Ouro Preto and Mariana in 1886, wrote: "As the men of then could not decide staying away from the light of the day, they didn't make underground galleries to cut the rich vein; they dug the land in a way of a funnel to have in the bottom the enough width to exploit the auriferous deposit, and when the depth became very considerable and it produced collapses, they abandoned everything to do new works in other places. The obtained gold was won, as can be seen, by means of rude works; this can be evaluated by the dimension of the embankments, whose vestiges the years were not capable to make disappear. Immense cuts done by the workers' hands can be seen in almost every direction".

Service in the mines was unhealthy, with slaves surviving from seven to twelve years of work, when they were not victims of accidents caused by incompetence and negligence, as those narrated by Eschwege (7) or Joaquim Felício dos Santos (8), that killed hundreds of workers at a time. Manumission, possible for those working in the final phases of diamond ore processing, who found a diamond larger than an octave (17 and a half carats) in size (13), was not even a dream for those who worked in the exhausting services of excavation or stream diversion.

The persistence of these primitive models is a characteristic of the Brazilian mining. The almost totally manual operation of the *garimpo* of Serra Pelada, in the eighties, with tens of thousands of workers, was very much like the excavations done by slaves of other centuries, the disregard for the working conditions and the lives of the miners remaining the same.

Even without machines, old excavations, made by groups of thousands of workers, as in Ouro Preto and Diamantina, left scars and silted watercourses, still recognizable today in the landscape, and in some cases the degradation begun by mining was further increased by erosion. Although such features may be hidden by vegetation in humid areas or obliterated by modern excavations, in dryer areas, like those around Diamantina, the impacts are more durable, with modern scars being added to the old ones to give some areas an aspect of almost continuous excavation, as around Guinda, Sopa and Extração.

Modern *garimpo*s increased the impacts greatly. The word *"garimpeiro"* comes from the habit of clandestine diamond miners working in the *"grimpas"* (tops) of the hills, thereby escaping the repression of the colonial government, which only allowed mining under strict conditions or under its direct administration (8). The derived terms *"garimpo"* and *"garimpagem"* are applied to mines and mining activity without previous evaluation of the deposit. In the Mining Code of 1967, permission to exercise *garimpagem* was individual, and the *garimpeiro* could only use equipments of small size and power. Law no. 7805, of 18.7.89, absurdly enlarged the permission for *"lavra garimpeira"* (mining by *garimpagem*) to encompass any mineral commodity, provided it was considered, at the discretion of the National Department of the Mineral Production (DNPM), to be possible of being mined *without* previous exploratory works, to establish the shape, size and composition of the deposit. Mineral extraction in these terms is a road straight for the destruction of the deposit.

140

Support for *garimpagem*, begun with Article 174 of the Federal Constitution of 1988, was immediately followed by a complementary Law and its regulation, in a hurry contrasting dramatically with the delay in the regulation of almost the entire Constitution. This is a dangerous setback, and one of the main reasons why it is almost impossible to opt for investment in legalized mining in Brazil.

In some cases of model "a", *garimpagem* is an individual activity, or almost. A *garimpeiro*, or a small group, mines a small area and lives off the sale of the product, whose price can be fair if it is bought by the government or on the free market. The situation gets complicated when the ones that indeed work depend on financing or authorization of "owners" of the areas, and must pay them a share of their production. Exploitation of the workers is greater when there is control of the sale of the product. Traditional (9) or modern (10) systems guarantee the dependence of the *garimpeiro* from his backer, which is larger when the backer is also "owner" of the area, transporter, owner of the landing strip, buyer of the production and vendor of supplies and women (the "Tapajós model")10. The illusion of wealth and the pressure of debts keep *garimpeiros* and women trapped in a web that few can escape.

In many cases the impacting power of the *garimpo* is the same or larger than that of organized mining. At the end of the eighties and beginning of the nineties there were five thousand dredges on the Madeira River. Many of them were true floating factories, with the capacity to extract and to process thousands of cubic meters daily, discharging tailings, mercury-rich effluents and lubricating oil into the water. Another case observed by this author was that of Poconé, where the renaissance of the *garimpo* in the 1980s brought in mechanized mining, which opened a hole of several square kilometers within of the urban perimeter, replacing streets with polluted ponds and tailings almost impossible to reclaim.

Garimpagem causes environmental problems even in metropolitan areas. In Curitiba, at the beginning of the eighties, road construction companies turned their machines, paralyzed by the economic crisis, to excavations as big as poorly done, mining gold ore, whose processing discharged mercury into water destined to public supply.

Companies duly legalized by the DNPM and even with respect to environmental control agencies readily ignore their already inadequate projects, and ambitiously attack the richer parts of the deposit, without any mining technology, destroying the environment, leaving behind a deposit unfit for any rational use. The environmental impacts remain for the society to reclaim, in the habitual privatization of the profits and "socialization" of the losses.

c - Mines with appropriate extraction technology, without environmental control: In these mines techniques of geology and mining engineering are applied for prospecting, evaluation, extraction and processing, but there is no control of the environmental impacts nor reclamation of degraded areas.

Mining technology was introduced late and only partially in Brazil. Although since the Alvará of 1608 the hiring of practical miners has been regulated, these positions were almost always been occupied by uneducated and untrained personnel (7), well-connected people seeking a sinecure. Only early in the 19th century did employees begin to exercise scientific and technical knowledge, such as the Brazilians Manuel Ferreira da Câmara Bittencourt Aguiar e Sá, responsible for the mechanization of the Extração mine and for the pioneer metallurgy of Morro do Pilar in Diamantina, and José Bonifácio de Andrada e Silva, who surveyed mines of São Paulo, and foreigners, like Friedrich Ludwig Wilhelm Varnhagen, who managed the Royal Iron Factory of São João de Ipanema and Wilhelm Ludwig von Eschwege, responsible for metallurgy in the Patriotic Factory, in Congonhas do Campo and for the underground mining of gold in the Passagem mine.

In 1824 the first successful foreign mining company, the Imperial Brazilian Mining Association, was established, applying modern techniques in Congo Soco, followed by the St. John Del Rey Mining Company, the most profitable of the time, in Morro Velho (5).

Geological and mining education had to wait until 1876 for the Escola de Minas de Ouro Preto (School of Mines), and until 1959 for the establishment of programs to graduate geologists. Up to this time little was known of the geology of the country, not withstanding the efforts of the national geologic services: Comissão Geológica do Império, founded in 1875; Serviço Geológico e Mineralógico do Brasil, 1907; Departamento Nacional da Produção Mineral, 1934, and some state services, the Comissão Geográfica e Geológica de São Paulo, founded in 1886, being the first.

The best mining and geological technology would be applied by the state companies, especially the Companhia Vale do Rio Doce (CVRD), founded in 1942, Petrobrás, in 1953, and Companhia de Pesquisa de Recursos Minerais (CPRM), in 1969, later transformed into the Geologic Survey of Brazil. Research and application of mining technology were stimulated by the Brazilian Institute of Mining (IBRAM), founded in 1976 through the initiative of private companies and by the Center for Mineral Technology (CETEM), 1978.

The First Ten-year Master Plan for the Development of Brazilian Mineral Resources - I PMD (1965-1974), together with new Mining Code in 1967 (Ordinance-law no. 227) and the establishment of fiscal incentives and of support of government organs for basic geologic surveys spurred a jump in the knowledge and development of deposits. This unfortunately did not happen with the II PMD (1981-1990), issued without adequate financing. The Long-term Plan for the Development of the Mineral Sector, established by Ordinance 918 (22/5/93), despite extensive consultation among the mineral community for its elaboration, which makes its proposals more realistic than those of the previous one, also suffers from the lack of political, financial and administrative support.

In summary, up to the end of the 1970s, Brazilian mining accompanied the efforts to modernize the country, by enlarging its production and equipment, but without any environmental concern. In many companies, even large ones, this attitude of contempt for the environment still prevails. The successive laws and regulations are not enforced, or, worse, they are enforced in the Brazilian fashion: forms are filled out, plans are formulated containing the bare minimum for their approval by the licensing agencies and not even this minimum is executed properly.

d: <u>Mines with environmental control</u>: Starting in the 1970s, the concern over environmental degradation spread throughout the country, influenced by the ecological movements in the developed countries and by the visible consequences of the disordered economic growth and the swelling of the great cities.

Even though environmental protection has been written into mining codes since 1603 (6), it has not being implemented. The DNPM only acted in favor of protection starting in the 1980s, forced by other sectors of society, and, even so, its performance has been irregular and insufficient.

Some of the historical marks in the legal protection of the environment in Brazil, and in particular of the control of mining, include:

- pioneer state laws, like no. 997 of São Paulo (31/5/76), establishing environmental protection at the state level;
 - federal laws: no. 6938 (31/8/81), altered by Laws 7804 (18/7/89) and 8028 (12/4/90), regulated by Ordinance no. 99.274 (6/6/90), which define the National Policy for the Environment.
- Law no. 7347 (24/7/85), regulating the Public Civil Action;
- Resolution no. 01 of the National Council for the Environment - CONAMA (23/1/86) requiring a previous Environmental Impact Study of potentially polluting enterprises;
- Federal Constitution of 1988. Brazil is probably the only country in which the obligation of reclaiming areas degraded by mining is incorporated into its Constitution (14), which in its Article 225 declares: *Everybody is entitled to the ecologically balanced environment,...* and specifies at the:

Par. 1st: *To assure the effectiveness of this right, it assigns to the Public Power.*

IV - *to demand, in the form of the law, for installation of work or activity potentially provoking of significant degradation of the environment, previous study of environmental impact...*

V - *to control the production, commercialization and employment of techniques, methods and substances that contain risks to life, the quality of life and the environment;* and in the

Par. 2nd: *That whoever exploits mineral resources is forced to reclaim the degraded environment, in agreement with a technical solution demanded by the competent public agency...* This paragraph was regulated by Ordinance no. 99.274 (6/6/90).

 - Resolutions nos. 9 and 10 (6/12/90) of the CONAMA, that regulate the environmental licensing of mining activities;
- Law of Environmental Crime, no. 9.605 (12/2/98).

Brazilian legislation, if properly applied, offers adequate protection to the environment. This was not and it is not the case, in the precise Portuguese-Brazilian tradition of beautiful words and ugly reality. Even so, since the seventies much has been done for the environmental control of mining. The main companies, led by CVRD, established their environmental services and started promoting a partial control of the impacts and the reclamation of the most visible damages, in some cases with good results, as in Carajás, in the Mineração Rio do Norte, in the areas of Alcoa in Poços de Caldas, in SAMITRI and in SAMARCO iron ore mines, in Mariana, in the protection and reclamation of the waters affected by phosphate mining in Araxá, in the experimental mining areas of oil shale of Petrobrás and in some quarries and sand pits in the larger metropolitan areas, this last advance due to progressive miners who founded the National Association of Producers of Aggregates for Construction.

During some time DNPM and IBRAM have sponsored research and courses on environmental technology applied to mining, working to convince mine owners and managers of the convenience of harmonizing mineral activity with the environment. States have set up structures of environmental control, enforcing legislation with varied results. In the developed states, and at discontinuous times, a serious work has been undertaken. In other cases control has been a tragic joke, with state environmental agencies controlled by *garimpo* owners, lack of funds and deviation of the few existing ones, even from international agencies, for uses more pleasant to the holders of power.

Starting with the Collor presidency (1991), and becoming worse since then, the dismantling of public services has strongly affected the patrimonial control of mining (DNPM) and environmental agencies at the federal, state and municipal levels. Even in states like São Paulo, experienced teams, with professionals trained at high cost, were dispersed by direct action of the government, or by indirect action, when many professionals sought early retirement, trying to guarantee rights threatened by reforms.

PRESENT SITUATION

The environmental control of mining and the reclamation of degraded areas are not priorities in Brazil. Even the CVRD, pioneer and leader in the application of environmental control and reclamation, when controlled by the Federal government, destined only US$ 20.744.500,00 to these activities in 1989. Compare this with the US$ 2,344 billion export revenue of the company and the US$ 734,5 million net profit in the same year. Or the twelve pages, of a total of 639, dedicated to the environment in a book produced for CVRD in commemoration of 50 years as a company, "Mining in Brazil and the CVRD" (5). Neither mining nor, consequently, its environmental control are priority for the Brazilian government, in none of its levels. The consequences of this disregard are:

1.   The division of mining into two segments: that of **organized mining**, that pays taxes, builds up an infrastructure used for the development of the area where it settles, pays wages, provides reasonably safe working conditions and assistance to workers, and is progressively adopting environmental control and reclamation of degraded areas and that of the *garimpagem*, with its pitiful working conditions, exploitation of workers, destruction of mineral reserves, environmental degradation and, frequently, association with crime.

2.   The persistence of rudimentary technological models of mining, leading to predatory extraction and environmental degradation, even in mines operated by organized and properly legalized companies.

3.   The impacts of mining affect the entire country. Although they degrade smaller areas than agriculture and

142

construction, mining can produce in Brazil (22, 23, 24, 25):

- deforestation, contributing to the reduction of plant and animal biodiversity and loss of the fertile layer of the soil;
- mining scars, sometimes of great extension and depth, causing more serious problems when they affect urban areas,as in the *garimpo* in Poconé and in some areas mined for construction materials;
- pollution of surface waters by sediments, in some cases for hundreds of kilometers of streams, as in the Rio Preto de Candeias, in Rondônia, polluted by cassiterite mining, or by toxic effluents, mainly mercury used for gold recovery;
- the tailings of all ores extracted prior to the mandatory environmental control, and of many not controlled after it,continue to pollute waters, soils and air, notably the acid drainage of tailing dumps of coal mines of southern Brazil and the pollution by lead sulfides, such as in the Ribeira River valley. In this case the pollution remains in sediments transported in the streams because of the naturally high pH of the waters. The acidification of the waters, at the bottom of the reservoirs planned for the area, will make thousands of tons of tailings with heavy metals soluble, with drastic consequences for the aquatic life of the area, including the largest fish breeding ground of the South Atlantic, the estuary-lagoon area of Iguape-Paranaguá.
- alteration of the amount and quality of groundwater, whose control and reclamation are much more expensive and complex than that of surface waters;
- destruction of dunes for construction sand in the Northeast and industrial sand in the Southeast
- air pollution by dust or gases from processing plants;
- social impacts, larger when they reach peoples of cultures very different from the contacting one, as has happened with the Ianomami indians of Roraima.

The degradation by mining can be important in urban areas, where each degraded square meter would be precious for residential or industrial use or for urban infrastructure. The obligation of reclamation of degraded areas is recent and still not totally applied, as it has been observed throughout Brazil and Bitar (20), has shown that, of the areas reclaimed in the São Paulo Metropolitan Area, only 5% were reclaimed by the concessionaires of mineral rights. The absence of a performance bond, a pecuniary warranty for the execution of the reclamation plan, allows the miner to escape his obligations by abandoning the area or by bankruptcy of the company.

On the other hand, the lack of municipal planning of land use eliminates the feasibility of mining areas that could first be mined and later used for urbanization. Thus construction minerals, mainly sand and clay, must be transported from distant areas, elevating freight costs, which account for the largest portion of their price.

The environmental and mining laws in Brazil are part of the problem, and not the solution. First, legislation is very complicated, with laws, ordinances, resolutions, orders, instructions, etc., etc., etc., appearing in great number, some altering others without being declared, in the traditional "the dispositions in opposite are revoked ", which are really revoked. At the same time, some fundamental legislative initiatives have not been implemented, such as:

- regulation of Article 23 of the Federal Constitution, that specifies the common competence of all levels of government for the protection of the environment and for registration, enforcement and fiscalization of the concessions for exploration and exploitation of mineral and hydric resources in their territories.
- establishment of real warranties, like performance bonds, which are the only efficient way to guarantee the execution of projects of environmental control and reclamation of degraded areas.
- mandatory inclusion of mining in Municipal Development Plans.
- regulation of mining in native reserves, where at present mining is both prohibited as legalized activity and tolerated as *garimpo*;

Second, the enforcement of the law is deficient. The National Department of Mineral Production remains without material and human resources and without political support to perform with minimal efficiency. A great part of its personnel struggles to do whatever is possible with what little is available. The computerization should be commended, which allows access to the mining law and statistics through the Internet (address *http://www.dnpm.gov*). Presently a bill seeking creation of a National Agency of Mining is being issued, substantially modifying the organization of the sector.

Other government functions relative to mining are in a worse state. The fiscalization of the mineral patrimony and of the environmental and working conditions of mining, also attribution of the DNPM, is extremely deficient. The geologic surveys are executed very slowly by CPRM, which is also affected by the scarcity and irregularity of funding.

In the environmental area, the fiscalization of mining is also an attribution of the environmental control agencies of the Ministry of the Environment and of the state governments. There are a few effective efforts of fiscalization, control, research and support to miners. The governments change, or the pressures of the polluters increase and the activities are discontinued, the teams dispersed, and funding vanished.

Even in the rare occasions in which the several agents execute their functions, it lacks coordination among their actions. The number of institutions that have or judge to have attributions over mining is frightening. In a study made by the Regional Council of Engineering - CREA-SP on sand mining, 18 different institutions were identified as related to mining in riverbeds in the São Paulo Metropolitan Area. It is less than the 26 reported for the State of Colorado (15) but it harms those who want to mine according to law.

In this insecure environment, part of the fiscalization actions are done by the State Attorneys and by NGOs, using the Law of Public Civil Action. In some cases, at least, this action seems to have larger effects of intimidating and increasing the costs for law abiding miners than in increasing the enforcement of the environmental law (15). The determination of the law of placing the informer NGO as a party in the lawsuit is not enforced, which impedes that be demanded the payment of costs and attorney's fees, in case of being the accusation groundless, or ten times the costs, in the case of unfairness. This favors the groundless accusation, which is aided by not being mandatory a technical inspection before the proposition of the lawsuit, even in cases in which a preliminary sentence is asked. This can be granted and a mine be paralyzed without crime. Until the situation is cleared, which can take much time in Brazil, the mine will remain shut. The contrary also happens. In cases of obvious ecological crime it is possible to a guilty and well advised company to drag a process for years, without any punishment (16).

In most cases the municipal governments, that have the attribution to administrate the mineral patrimony in relation to construction materials, besides their duties in land use regulation and environmental protection, are not equipped nor give importance to mining. Except for some state capitals and a few municipal governments, mining is neither regulated nor included in plans as a legitimate land use. Municipal action only happens to prohibit mining, in cases when the conflicts on land use become insoluble. The prohibitions are lifted up when it is noticed the obvious fact that is worse to have to buy construction materials in another city than to control their mining locally. Even so, the conflicts rarely lead to comprehensive legislation, being resolved case by case, according to the political and economic pressures.

Therefore it occurs in Brazil a contradictory situation. It is easy to do criminal mining, either of the point of view of the mineral patrimony that is appropriated without due concession, or of the environmental one. At the same time, for an operator mining according the regulations, everything is difficult, beginning by knowing which regulations to follow. He needs to obtain authorizations and licenses of federal, state and municipal organs, whose applications can wait years to be appreciated. Even a concession does not assure him safety. His claim can be invaded by *garimpeiros*, without the State guaranteeing his rights or even his life. He can suffer fines or closing, to incur in costs and to have difficulties for his defense, even operating conformable to law. His costs will consequently be larger than those of his unlawful competitors.

It is not surprising that did not happen here the expected "investment boom", after the fall of the prohibition of mining by foreign companies, even having Brazil great potential for large size and high grade deposits. The main reasons are insecurity in the ownership of the mines, restrictive legislation, bureaucratic difficulties and absence of reliable basic data in scales adequate for prospecting and exploration. For the national investors, scalded in so many economic plans, it is difficult to decide for a long term investment, especially with the current interest rates, that turn negative the liquid present values of almost all the investment analyses.

PROPOSALS

Mining is necessary to Brazil. More than this, it is important that mining works correctly, to meet the needs of a large and growing population, in an area lacking infrastructure and at the same time it continues to produce export surplus revenues. However, it is not desirable a disordered increase of production, without environmental control or reclamation.

Sustainable development should be sought in mining, even if it is difficult to apply this concept to nonrenewable resources. If it is applied the classic definition *("Sustainable development is development that meets the needs of the present without compromising the ability of future generations to meet their own needs")* (26), the sustainability in mining is impossible, because any use of a mineral commodity reduces its resource base (totality of the existent mineral substance), because the mineral substances are not renewable. The sustainability of the nonrenewable substances is reached when its consumption is the same or smaller than the generation of substitutes and, at the same time, the problem of the residues produced by its production and use is controlled (17).

The international guidelines do not help much. Mining is almost unknown in the Agenda 21. It is rarely mentioned, and almost always as a polluting agent. Almost nothing is told about its role in supplying raw materials to assure human life on Earth. It is repeated here, in the international level, and among environmentalists, the mixture of ignorance and disdain for mining that appears in the Brazilian politics. The Agenda specifies numerous institutional measures for sustainable development. Besides the lack of references to mining, it appears here the basic incoherence of the Agenda: while in many points the local initiative is praised, many detailed measures are proposed, turning it impossible to execute those details and to have some local initiative. It would be better to agree on basic principles and take measures adequate to each place.

The Ministry of the Environment developed a work program for application of the Agenda 21 principles to mining in Brazil, whose results are presented in the "Environmental Guidelines for the Mineral Sector" (22). These guidelines seem well done and applicable, since support is given to the technicians and companies disposed to follow them. In summary, these are the guidelines:

1 - Maintenance of agile, integrated and efficient legal, normative and institutional mechanisms for licensing, monitoring, and environmental fiscalization in the Mineral Sector.

2 - Internalization of modern concepts of Environmental Administration and of environmentally compatible technologies in the processes of extraction, processing and use of Mineral Resources.

3 - Maintenance of the knowledge base, formation and conscientization of human resources who make possible environmental planning and administration in the mineral Sector.

With the guidelines come a large number of programs, actions and recommendations. Some of them should be emphasized:

- Definition of competencies, unification and organization of the process of environmental licensing in the mineral sector
- establishment of real warranties for the rehabilitation of degraded areas;
- implementation of appropriate economic instruments to modern systems of environmental administration;
- to implement ways to support recycling;
- to encourage, to diffuse and to induce the use of new technologies for the environmental control and reclamation in mining;
- implementation of joint actions for the reduction of clandestineness;
- demonstrative projects for the regulation of the activities of *garimpo* and of extraction of sand;
- implementation of fiscalization campaigns in areas closed to mineral extraction.

These guidelines, programs and actions seem well elaborated and realistic, and their proposals contain much of

144

what it is consent among those who work seriously in the mineral sector, in government as well as in companies. However, it should be noted:

- their execution will demand a nonexistent coordination among the several government levels and between them and the private sector.
- the sources of the necessary resources are not declared for their implementation.
- it is worrying the treatment, in the same program, of *garimpo* and sand extraction. It reveals a disposition to deregulate the extraction of sand, which would be disastrous for the urban areas where it happens.
- there is an almost total absence of the academic community, both in the elaboration of the guidelines and in the forecast for participation in the implementation. Is the Brazilian University so remote from the mineral sector?

The sustainable development in mining does not depend exclusively on the mineral sector. The more efficient measures for the reduction of the exhaustion of resources and of the environmental impact of mining are those destined to decrease the use of mineral commodities. The production of more durable consumer goods, the reuse and the recycling, the rationalization of use, as well as the substitution by renewable resources, have larger effect than the actions taken inside the mineral sector for the sustainability. These internal actions are also important. The rational working of the deposits, that includes their correct evaluation, the control of environmental impacts and the reclamation of the disturbed areas, is fundamental for the best use of the resources and the integration of mining with other land uses and with the other sectors of economic activity in the search of sustainability.

How to arrive there?

It is necessary to exist a political will, in the highest level, to appropriate resources for the survey of the conditions of Brazilian mining and for the concrete planning of its development, followed by measures to implement it. This will does not exist now. The Executive and the Legislative alike show total disdain for the mineral production, expressed by the disastrous privatization of the Companhia Vale do Rio Doce and by the denial of resources for the agencies responsible for the mineral sector.

It is necessary an effort of all components of this sector to show the importance of mining and of its associated areas, as those of geologic and environmental survey. This demonstration will only be achieved if the isolation between the mineral sector and the society be broken. The chronicle of communication of the mineral sector is extremely negative. Mining is considered by the environmental agencies, NGOs and the general public as highly polluting, without its role being recognized as one of the most important bases of the economy, depending on mineral inputs the activities responsible for nearly a third of the Gross Domestic Product - GDP. This fraction used to be presented in the Mineral Yearbooks and Summaries of the DNPM. Not even there it is presented now. The dominant public perception is that mining is an activity of little importance and highly polluting, a perfect target for legal restrictions and budget cuts.

When a political will exists the results are immediate, as shown by the recent history of the Brazilian mineral sector. When there was a energetic effort of organization of the sector and some investment, as in the first Ten-year Master Plan (1965 -1975) there was an immediate answer in the discovery of deposits and in the attraction of investments for the sector, followed by an increase of production (18). Empty plans, without appropriate administrative measures nor financing, like the II Ten-year Master Plan or the Long Term Plan for the Development of the Mineral Sector, don't contribute to the improvement of the mineral production.

Even without a total reorganization of the sector, some concrete measures are necessary and possible:
- rationalization of the legislation and of its enforcement, allowing to the miner to know what should he do in relation to the mineral patrimony and to the environment, and facilitating his legalization, under the orientation of trained professionals of the control agencies, that also should have political support for efficient enforcement actions.
- enforcement and orientation activities to assure that environmental documents (EIS, Environmental Control Plans) contain reliable data, that can be used for regional-scale studies, and that the plans propose feasible activities. Particularly it would be necessary to enforce the need for a professionally conducted deposit evaluation, without which all mining and reclamation plans are fiction exercises.
- the establishment of fiscal incentives for the mining activity, compensating its disadvantages in relation to other investments, as the long term maturation and the risk, that is high for the low grade minerals.
- to establish real warranties of reclamation of the disturbed areas, as part of the licensing documents. The value foreseen for such costs should be object of a performance bond, a real warranty, like a bank letter, an insurance policy, a caution in money or a real state collateral. Part of the resources immobilized as warranty would be returned to the miner after each phase of reclamation be completed; the remaining would be returned only after final reclamation, which means an stabilization of physical processes and a self-sustainable revegetation. This is the only way to really assure reclamation; without a real warranty the door remains open for abandonment of the area, bankruptcy of the company and a rich and free owner.

the insertion of mining as a legitimate land use, in the municipal managing plans, after a study of the mineral potentiality
of the territory. Without this planning, at the same time mining destroys areas necessary for other activities, and areas with high mineral potential are made unfeasible for mining by permanent uses. This is particularly harmful for deposits of construction materials, of low value and great volume, which need to be mined close to the application. It is possible the sequential use of the land, mining being a temporary activity, followed by other (residential, industrial or recreational) after mine closure and final reclamation.

The University has important functions to fulfill for the development of the mineral sector and some of them are

145

its prerogatives. It should develop research, together with public and private institutions or by its own initiative, in this case the most valuable for its independence. The University personnel can consult for the mineral sector, provided it does not become their main activity, and not allowing that the dependence of contracts turns the University an uncritical appendix of the government or of the companies. Nevertheless, it is education the main and untransferable attribution of the University.

The author identified three attitudes in the activities of the environmental professionals (19): the knowledge of nature: search for the basic data on the Earth and of their relationships, without immediate economic objectives; the domain of nature: application of knowledge for the use of Earth resources; the integration with nature: with knowledge supporting the efforts for maintenance of humanity on Earth in sustainable bases.

According to the historical moment, each of these attitudes is privileged. The Greeks valued the knowledge, the industrial revolution demanded the domain; its consequences urge the integration with nature, under pain of destruction of humanity. It is not enough now to educate a scientist or a technician. Although these parts should be developed in the professional of the environment, it is fundamental to develop in him an attitude of respect to the human beings and to the other beings, animate and inanimate. The best way to develop this planetary conscience (ecologic? holistic? integrated?) is the involvement of the student, as early as possible, in actions for the defense, the rational use and the recovery of the environment.

The mining professional (geologist, mining engineer) cannot be just a technician. He must be a professional of the environment, being an obligation of the University to help him to develop his planetary conscience. At the same time, it should contribute to develop his political conscience, enabling him to defend his ideas. When there is a need to a political will to take measures for the development of mining in sustainable bases we cannot just hope that is spontaneously developed in government. The conscious professionals should act to disseminate the understanding of the importance of rational mining. It is necessary to break the isolation that mining and Geology have been keeping from the whole of the population. The concerned population will have conditions to demand a mineral production that meets its needs without destroying the environment. This is the fundamental, although not enough condition, for a change in the attitudes of the administration and of the productive sector in relation to mining.

REFERÊNCIAS BIBLIOGRÁFICAS

1 - Macedo, A.B., 1997 - Panorama da degradação da terra no Brasil. in REBOUÇAS. A., *Panoramas da Degradação do Ar, da Água Doce e da Terra no Brasil*. São Paulo, IEA-USP/CNPq, p. 114-151.

2 - Macedo, A.B., 1998 - *Recursos minerais não-metálicos*. Estudos Avançados, 12(33):67-87.

3 - Calógeras, J.P., 1904 - *As minas do Brasil e sua legislação*. Rio de Janeiro, Imprensa Nacional, 2 v.

4 - Guimarães, J.E.P., 1981 - *Epítome da História da Mineração*. São Paulo, Art/SECSP, 173 p.

5 - Monteiro, A.M.F.C. & Lamarão, S.T.N., 1992 - *A mineração no Brasil e a Companhia Vale do Rio Doce*. Rio de Janeiro, CVRD, 640 p.

6 - Herrmann, H., 1995 - *Mineração e Meio Ambiente: metamorfoses jurídico-institucionais*. Rio Claro, IGCE-UNESP, 355 p. (tese de doutorado inédita).

7 - Eschwege, W.L., 1979 - *Pluto Brasiliensis*. Belo Horizonte, Itatiaia/EDUSP, 2 v.

8 - Santos, J.F., 1976 - *Memórias do Distrito Diamantino*. Belo Horizonte, Itatiaia/EDUSP, 338 p.

9 - Machado F, A.M., 1985 - *O negro e o garimpo em Minas Gerais*. Belo Horizonte, Itatiaia/EDUSP, 141 p.

10 - Salomão, E.P., 1981 - Garimpos do Tapajós. *Ciências da Terra*, 1:38-45.

11 - Martins, J.S., 1998 - A vida privada nas áreas de expansão da sociedade brasileira. *História da vida privada no Brasil*. São Paulo, Companhia das Letras, v. 4, p. 659-726.

12 - Burton, R.F., 1869 - *Exploration of the highlands of Brazil, with a full account of the gold and diamond mines*. London, Tynsley, 2 v.

13 - Mawe, J., 1812 - *Travels in the interior of Brazil, particulary in the gold and diamond districts of that country*. London, Longman, 325 p.

14 - Brasil, 1995 - *Constituição - República Federativa do Brasil*. Brasília, CGSF, 292 p

15 - Macedo, A.B., 1995 - Case studies of excessive legal and administrative constrainsts to mining in the USA and Brazil. *First International Symposium on Mining and Development - Proceedings...* Campinas, UNICAMP, p. 185-191.

16 - Macedo, A.B., Mantovani, W. & Brighetti, G., 1997 - *Parecer técnico sobre o processo n 2766/89 da Comarca de Santos - Ação Civil Pública Ambiental movida pela Equipe Regional de Proteção ao Meio Ambiente da Baixada Santista, contra a Lello Empreendimentos Imobiliários Ltda*. São Paulo, IG-USP/CEPA, 1997, 14 p., mapas, fotos (inédito).

17 - Daly, H.E., 1991 - Sustainable growth research and exploration. National Geographic Society, 7(4) sp. apud LEMONS, J.F. & BERRY, D., 1996 - Sustainability in a Materials Society. *Nonrenewable Resources*, 5(4):277-284.

18 - Machado, I., 1989 - *Recursos Minerais - Política e Sociedade*. São Paulo, Pró-Minério/Edgard Blücher, 410 p.

19 - Macedo, A.B., 1997 - Recursos minerais e desenvolvimento sustentável. *Seminário Ciência e Desenvolvimento Sustentável*. São Paulo, IEA-USP, p. 130-132.

20 - Bitar, O.Y., 1997 - *Avaliação da recuperação de áreas degradadas por mineração na Região Metropolitana de São Paulo*. São Paulo, EPUSP, 184 p. (tese de doutorado, inédita).

21 - Courgy, E., 1997 - *Seis semanas nas minas de Ouro do Brasil*. Belo Horizonte, Fund. João Pinheiro, 132 p.

22 - MMA/ABC/PNUD, 1997 - *Diretrizes ambientais para o setor mineral*. Brasília, MMA/ABC/PNUD, 40 p.

23 - Ribeiro Filho, E., Moreschi, J.B. & Macedo, A.B. - Minerals and Mining in the Brazilian Economy. *Episodes*, Otawa, IUGS 11(3):215-221, 1988.

24 - Macedo, A.B. - Brazilian mining at a crossroads: "garimpeiros", companies and the environment. *AGID News*, Exeter, UK, Asssociation of Geologists for International Development, 1993, 74/75:16-19.

25 - Macedo, A.B. - Reclamation of mined lands in Brazil. *International Symposium on the Impact of Mining on the Environment - Problems and Solutions*, Nagpur, India, 1994; Proceedings... Balkema, Roterdam-New Delhi, p. 423-428.

26 - Comission on Environment and Development, 1987 - *Our Common Future*. Oxford, Oxford University Press, 340 p.

THEME 2 :
Waste Disposal and Eco-friendly Working

Environmental Protection and Management in the Spanish Reales Sitios: The Monte Del Pardo Royal Forest

Prieto[1], S. Soria[2], R.L. Fernández Suárez[1], J.A. Sáiz de Omeñaca[1], C. Cardiel[1], J. Martín, F. Tomé[2] and A. Muñoz[2]

[1]Escuela Técnica Superior de Ingenieros de Montes, Universidad Politécnica de Madrid, 28040-Madrid, Spain
[2]Patrimonio Nacional, Palacio Real, 28013-Madrid, Spain

INTRODUCTION

Since 1918 when the first National Parks were established in Spain up to the present, their common link is the purpose of preserving and protecting those natural areas showing paramount environmental values. Thus, those Parks now integrated in the State Network of National Parks have been catalogued in different ways such as *"Zona de Especial Protección para las Aves"* (ZEPA: Area of Special Protection for Birds, which has been admitted by UE directives), *Reserva de la Biosfera* (Biosphere Reserve) (MAB program), *"Zona Húmeda de Importancia Territorial"* (Wetlands of Regional Importance) (RAMSAR Agreement), *"Patrimonio de la Humanidad"* (UNESCO). The Council of Europe has awarded them all with the European Diploma, first class. Spanish National Parks play also an important role as referential elements for other protected natural areas, at least since 1975 when the Act for the Protection of Natural Areas was passed.

This legal text was later revoked and replaced in 1989 by a new one, the Act for the Conservation of Natural Areas, Flora and Wildlife: Its purpose is to achieve the coordination and development of the different conservationist policies emerging from the Spanish State and its component regional communities so as to widen the legal network protecting natural resources beyond the protected areas themselves. The new act established the Scheme for the Management of Natural Resources and the Guidelines for the Management of Natural Resources and instruments to achieve the conservationist goals desired; it also produced four different types of protection systems out of the many pre-existing ones i.e. 1) Parks (integrated in the Network of National Parks, directly administered by the State; 2) Nature Reserves, 3) Natural Monuments and 4) Protected Landscapes. This act also implied the transposition into the Spanish legal system of all EU directives on wildlife and flora protection, such as the one registered as 79/409 in connection with wild birds.

In 1993, the Spanish State Section of the European Federation of Natural and National Parks was created. This Section prepared, along its first year of life a database including the first finished catalogue of the Spanish Protected Natural Areas (among which the ones belonging to Patrimonio Nacional) which according to 1995 data mounted to 605: from this figure, 13 were National Parks, 81 Natural Parks, 75 Nature Reserves and 34 Natural Spots of Remarkable Interest (Parajes Naturales). The Act 23/1982, which regulates PN and also the Royal Decree 485/1987, developing such

Act, enumerate the assets embodied in the National Trust. They also establish the obligation to protect the environment and to ask for responsibilities in the assigned estates. In both ecological cases, they expressly mention Monte de El Pardo. As a result of the remarkable difference between this Park (that comprises a clearly limited independent area with strictly restricted access to nearly its whole extension and which enjoys an independent unitarian administration) and some other areas (which surrounding it) stretch as a supplemental green passage reaching Guadarrama mountain range, it was established, in 1985, what is called Parque Natural de la Cuenca Alta del Manzanares (the Natural Park of Manzanares River Upper Basin). This new entity was thought as a way preserving the whole environment of this really large area which also includes Monte de El Pardo. The fact that on August 7th, 1997, the Scheme for the Environmental Protection of Monte de El Pardo was promulgated supposed a detailed statement which described the management procedure to apply in order to preserve the soils, water resources, vegetation and flora, wild life, exploitation, etc; also imposed the need for devicing plans of actions, forest regeneration (reforestation) and improvement, control of erosive processes, endangered species, management and others. According to all these aspects, it can be considered that the protected natural areas under the administration of Patrimonio Nacional (National Trust) enjoy a legal framework as strict as that one ruling National Parks.

SPANISH FORESTS BELONGING TO PATRIMONIO NACIONAL

Patrimonio Nacional (PN) (Spanish National Trust) is a public law agency comprising the totality of the state assets devoted to the use and service of the Spanish Crown. They are, in addition, available to the Spanish people as a means of culture, research and education. An important part of these assets are woodlands and forests: Monte de El Pardo, within Madrid municipal limits comprising 15.840 ha. from which about 900 ha. are open to the public with free access; groves in Jardines de la Isla, el Príncipe and El Rebollo in Aranjuez stretching along several kilometers on Tajo river banks which mount about 30 ha., Bosque de la Herrería (497 ha.), El Cerrado (130 ha.) and Cuelgamuros (1.362 ha.) in San Lorenzo de El Escorial where the famous monument called Valle de los Caídos is built. Riofrío forest (625 ha.) in Segovia is also included in this list. They are, in full, about 18.000 ha covered by a large variety of ecosystems and landscapes which stretch on plains, river banks, slopes, medium and high mountain, etc. These woodlands and forests comprise formations of *Quercus ilex, Quercus pyrenaica, Quercus suber, Fraxinus angustifolia, Pinus sylvestris, Pinus pinaster, Pinus pinea, Juniperus thurifera*, etc. Wild life is rich, among others, the following species can be found: deer, buck, roe deer, wild boar, hare, rabbit, fox, imperial eagle, black vulture, night birds of prey, different duck species and so on.

The major functions of these spaces are connected with protection, landscape, social use and environment. They may be operated as natural or periurban parks, as the case demands. Among they all, Monte de El Pardo stands out because of its relevant importance. It is the one to be described in this paper since it shows very special characteristics and because it enjoys a management system which corresponds as a pattern with all the other national woodlands and parks.

Monte de El Pardo

Monte de El Pardo, owned by the State, belongs to Patrimonio Nacional (PN) as a result of being assaigned to the Head of the State's service. Thus, its management is a responsibility of the National Trust (PN) Board of Directors, a public law entity ruled by the Prime Minister's office.

The most ancient pieces of information about El Pardo date back from the first century A.D., under Roman ruling, and from them we can deduct that its original inhabitants

150

were shepherds dwelling in caves. Much later, during the Visigoth's period when chasing was regarded as one of the owner's exclusive privilege, Monte de El Pardo appears as one of the Crown properties (Bauer, 1985). This woodland is mentioned in the chronicle recording the reign of King Ramiro II of Leon who conquered Magerit (moorish name of Madrid area at that time) because of its luxuriant, green and leafy appearance and also for its hunting possibilities. In the Book of Chase (Libro de la Montería), written under the guidance of King Alphonso XI of Castile, El Pardo is described as a royal site for games where there was to point out the varied and rich fauna which could be chased as big and small game. In 1470, El Pardo is defined as a royal chasing site, what meant it was only devoted to the crown service. During the XVIth century, Emperor Charles V introduced a new system of chase in this woodland and Philip II divided it into quarters or sections (jwhich originally meant the 25 % of anything, in Spanish "cuarteles"). This word is still used nowadays in the Spanish forestry terminology giving name to one forest management unit.

Brown bears (Ursus arctos), very common in the area, disappeared by the end of this King's reign. The introduction of a new dynasty by Philip V of Borbon at the beginning of the XXVIIIth century gave way to a new hunting system based on game-beating ("ojeo"). Along the XIXth century and as a result of political and social affairs, surveillance declined and mismanagement led to important damage for this well-preserved ecosystem.

Later, and in order to favour its recovery, tillage was absolutely banned. The exploitation of cork oaks was nearly the only one allowed up to very recent times. Hunting was allowed only once a year. The recovery of cynegetic resources started in 1875, under the reign of Alphonso XII. Several game species were introduced by means of selected individuals (couples of bucks and wild boars) while surviving deer were also protected to let them reproduce freely. As a result of a legal regulation passed on March 7th 1.940, the estate was attached to the Head of State's use and service which is its present legal situation and has become a real asset for its preservation.

4. Different areas

The woodland covers an area of 15.840 ha. with a perimeter of 80 km delimited by means of a wall made of stone, brick and mortar. There are three clearly different areas if considered from the use, conservation and management viewpoints: a) public use, b) restricted public use, and c) area of exclusion (reserve) (Prieto et al., 1996).

Since it was opened to the public, the area of public use (about 900 ha.) experienced an important degradation, which could be appreciated in the form of soil compaction, littering and so on. This degradation was connected with a series of activities inappropriate for a suitable conservation.

The area of restricted public use comprises an extension of 35,34 ha. It allows visitors only on limited basis since it is a protected section of the park, which tries to achieve the preservation of wildlife in similar conditions to those ones in the exclusion area, but letting only authorized people to go in. This controlled access did not prevent the area from degradation continuance.

The reserve and special use areas (about 14.900 ha.) comprise the remaining surface of the park not included in any one of the previous sections. There is no public access to them at all, only the management staff and researchers are let in. They show highly relevant natural and scientific features since they have numerous peculiarities among which the ones connected with vegetation, fauna, geomorphology and landscape are to be denoted as well as the nearly unmodified ecosystems to which they belong. There are some other long-time established agricultural plots covering about 30 ha. The purpose of which is to provide supplementary food for game animals in the forest

151

during the season when grass production decreases. There is also a dam stretching about 550 ha. which entails vital importance for this forest wildlife, especially for certain bird species such as black and white storks, ducks, cormorants, gulls and others; in addition to some buildings, works, high voltage electric cables, water pipes and railroads.

5. Environment

Monte de El Pardo is located in an area between Madrid and the Central System of mountains of the Iberian Peninsula, within the Tajo river basin. Its main waterway is Manzanares river, which flows southward. Its climax height is 840 m and the lowest one is 590 m. The forest enjoys a moderate Mediterranean Climate affected by the continental character of central Spain (table 1). Summer drought and heat give way to rains and lower temperatures which are relatively more frequent along the other seasons (Nicolás & Casado, 1979; Comunidad de Madrid, 1984).

Table 1. Climatic variables of Monte de El Pardo.

Mean Annual temperature	10 – 13°C
Frost season mean length	5 - 7 months
Potential annual evatranspiration	700 – 900 mm
Mean Annual rainfall	400 – 800 mm
Dry season mean length	2 - 5 months

From the view point of lithology, the area is made up of granites and neis at the most important heights and detrital sediments coming from the breaking down of these rocks which adopt the form of sands. These may be light and rich in clay or thicker as they actually are in the recent, alluvial slopes. The main feature of these soils is their sandy nature which implies a poor water-retaining ability. They show a useful-water yield (i.e. moisture minus withering quotient) lower than 50 % of such equivalence. This being the reason that makes water a limiting factor for forest development in the area, so they are regarded as poor soils. Although these soils show a very small erosive nature resulting from their high contents of sand in addition to good permeability and standard granular structure of their surface horizon, many plots already show strong erosion giving rise to ravines and marlpits which let the lighter colours of the soil at sight (IGME, 1989).

The potential vegetation in Monte de El Pardo comprises large formations of holm oaks (*Quercus ilex*) on silex soils where *Juniperus oxycedrus* also thrive stretching over the meso-Mediterranean stratum (Izco, 1983). In this climatic holm oak forest which has been exposed to a secular and sever anthropozoogenic activity, the evolution has led to two large and different formations: *"dehesas"* (clear, uncrowded grazing forests) and *encinares* (more crowed oak groves). Supplementary and depending on the plots, we can find not dense formations of *Quercus ilex* and even intermingling with *Quercus suber* (cork oaks) individuals. On irregular soils, *Quercus ilex* can be found in association with holly trees (*Ilex aquifolium*); dwarf, bushy oaks and shrubs of *Cistus ladanifer* and *Retama sphaerocarpa* (on nearly all the plots). On water run-offs, ashes and poplars; and also *Pinus pinea* from reforestation works (Monte, 1984).

An Eurosiberian penetration in the area comprising willows, poplars and ashes (among which only the two last ones reach tree size) extends along the river banks, run-offs, floodable spots or over those ones with permanent edaphic moisture where they have access to sufficient water resources in order to counter-balance its summer deficit when the highest temperatures take place. The shrub stratum is not abundant. *Retama*

152

sphaerocarpa is the most common species together with *Santolina rosmarinifolia*, *Lavandula stoechas* and *Cistus ladanifer*.

After the destruction of oak forests and depending on the anthropozoogenic influence and the degree of soil alteration, bush and shrub formations usually came to occupy their place. These formations with heat and sun-loving nature behave as climax formation here in the form of chamaephyte and phanerophyte (disclimax or periclimax) usually thriving on light, erosion-weathered soils with poorly developed organic horizons.

Nearly all-vertebrate species typical of the Mediterranean forests are present together with others that were introduced because of their cynegetic value or accidentally. There are 26 mammal species, 119 bird species, 15 reptile species, 12 amphibian species, and 12 different types of fish. This large variety overrides the situation in other areas with similar characteristics becoming so and additional reserve, a focus for their irradiation to other territories. The presence in the park of some species at the top of the trophic chain is a clear sign of its high environmental quality though it is a fact that these populations are sometimes precarious, endemic and some of them are even at the risk of extinction such as the imperial eagle, the black vulture, the black stork, etc. The park is also an excellent stopover of winter place for some European migratory species and has been awarded the acknowledgement as Area of Special Protection for Birds (ZEPA) in 1988 by the EEC. It is to denote the fact that 28 out of the 32 bird species recorded in the Iberian Peninsula have been seen within this park limits. It is also a proved fact that 20 of these species usually nest within its limits. From the 130 couples of imperial eagles existing in the world, it is supposed that about 5 % to 10 % of them are nesting in Monte de El Pardo. On the other hand, there is a large number of small and medium predators such as: foxes, wildcats, and plenty of mustelids (marten, genets and badger). Different micromammals such as: rabbits, dormice, moles, country mice, water rats, shrewmice, etc. keep those predators alive.

Several natural agents (diseases, pests, fires, meteors, etc.) or some others connected with human activities (over-population of game species, acid rain, atmosphere and water pollution) affect the forest and endanger its vegetation.

Table 2. Inventory of cynegetic species in Monte de El Pardo.

Big name				
Species	Male	Female	Babies	Total
Deer	1.540	2.395	1.370	5.215
Buck	1.494	3.558	1.900	6.952
Boar				700
Small game				
Rabbit				50.000
Partridge				500
Fox				1.000

6. Cynegetic situation

From the cynegetic standpoint in Monte de El Pardo, deer and bucks are the only animal species that can be captured and this is done by means of selective methods. In order to manage everything concerning chase, a cynegetic scheme has been outlined according to the vegetation ability for self-reproduction, aiming at a progressive change to reach 1.500 deer and the same amount of bucks in a near future (Patrimonio Nacional, 1990; see Table 2).

7. Environmental Impacts in Monte de El Pardo

A series of installations or infrastructures are placed within the park limits which inevitably play effects on it (dam, small weirs, water reservoirs and pipelines, highways, railways and high voltage cables). They all are not connected with the park upkeeping and cause various environmental impacts on its landscape, vegetation and wildlife, particularly on birds. The environmental legislation nowadays in force makes it absolutely impossible the establishing of such installations, but they were made very long ago and on quite different legal conditions.

8. Thematic cartography

Map-drawings of the park performed in order to make its management easier use the scale 1: 10.000 and include all the previously described aspects in the following maps: topographic, of the different areas, hydrological, edaphic, of vegetation, of inventory division, transport ways, installations and infrastructures, phytosanitary, of fire risk and inventory by air-plotting.

9. Actions and outcome

Monte de El Pardo has performed different functions along the centuries among which the following ones are to denote: economic, agricultural, cynegetic and recreation ones. Nowadays, they are basically focused on the conservation of forests and wildlife as well as social use. Thus, its essential resources connected with the fulfillment of these multiple demands have been coordinated in order to achieve such goals. The concept *multiple-use demands* comprises protection, hydrologic, climatic, social, aesthetic, scenic and scientific aspects to which this park may contribute significantly. According to all these, the park management should tend to propose a series of provisions in order to reach a compromise able to fulfill the demands without disturbing the ability of the park for restoring its own resources. The purpose of the proposed Multiple-Purpose sustainable Management Scheme includes the following specific aims:
Ensuring the survival of the existing ecosystems.
Increasing the amount and quality of the existing renewable resources by means of suitable technics.

Allowing the neighbouring population to enjoy practical benefits (social use) and virtual ones (environmental, scenic).

This park management should take protection and conservation into consideration i.e., the static and dynamic aspects, which precisely define the idea of natural resources conservation. Thus, on the one hand we encourage integral environmental protection by establishing reserves in natural habitats which are typical of a given area and whose exploitation would become not only scarcely aggressive but economical as well. On the other hand and concerning the dynamic aspect, it should manage the resources aiming at the satisfaction of the people's needs trying to make visitors go into the forest safely. Irrespectively of those general principles governing the management of natural resources and specifically in the case of Monte de El Pardo, its outstanding character, which comes out of the following causes, must be considered:

It is the area in which the Head of State's official residence is sited, that means it must be mainly subject to this service in connection with this purpose, which implies a system of special regulations governing its definitions (assets), protection and administration (only in connection with this purpose).
Its nearness to Madrid makes it a popular recreation resort for its inhabitants, especially in spring and autumn.

154

It's remarkable environmental and ecological values. Since it is one of the best-preserved Mediterranean forests on plains, which is easily noticeable from the large amount of vertebrates dwelling in it and because it is the ideal place for birds to feed, rest and spend the winter.

Its scientific, research and educational interest, since it is an area showing little degradation and very few modifications.

According to the proposed Management Scheme and considering the previously mentioned peculiarities (particularly those ones connected with the Act 23/82 which regulates the National Trust, the operative goals to be taken into consideration when dealing with this park management should be the following:

Ensuring protection, conservation and integrity for its biodiversity and its characteristic ecosystems (soil, vegetation, wildlife, water and atmosphere) considered as exceptional or representative of its biological richness, leading to its restoration when necessary, in order to transmit them in the best possible conditions to the forth coming generations.

Preserving the scenery and the quality of the surface and ground waters in the park and also the quality of the incoming waters. Keeping air quality in order to help the reduction of the surrounding levels of pollution.

Establishing demarcations for areas of relevant interest for research and the study of the various environmental parameters. Promoting the orderly use of such demarcations for scientific research within the previously set division into smaller areas of the park above mentioned in this paper. Special focusing should be made on the protection and study of:

Populations of nesting, daily flying-by and seasonal (migratory) birds and of cynegetic fauna The reservoir. The thickets, streams, waterways and riverbanks.

The inner relationships of the typical Spanish ecosystems called *dehesa* (grazing forest) and the methods suitable for its management. The incidence of the different management schemes on wildlife.

Analysis of the number of visitors, their demands and social strata to which they belong.

Actions to be developed should consider the following criteria:

Conservation should override any other activity taking place in the park.

The natural systems, which make up the forest, should be given priority in an attempt to halt and/or produce an inversion in the regressive tendencies like erosion and ravines.

All those interventions, which could affect natural processes, should be performed in the shortest, minimal intensity and extent, provided they are consistent with the conservation of the park.

Concerning the area of public access, the main features of its use were considered. Certain measures leading to its recovery and upkeeping were proposed such as: cleaning, providing wastebins and containers to be periodically emptied, setting up of recreation spots, reforestation and protection along Manzanares river-banks, establishing training-workshops and vocational schools to offer environmental guidance. These interventions avoided degradation of the area by limiting human pressure on recreation areas and parking lots, but they did not lead to a better human behaviour towards the forest nor to a higher comprehension of the rustic areas.

The area of restricted public use comprises an extension of 35,34 ha placed eastwards between the areas of exclusion and public use. It is a protected section of the park, which tries to achieve the preservation of wildlife in a condition similar to that one in the area of exclusion but letting only authorized people, go in for observation. Nevertheless, degradation went on coming to a point at which cork oaks started to be threatened, especially because of soil compaction. Thus, in 1990 and after a general work of conservation, it was decided to devote the area only to scientific and cultural restricted groups of visitors who should ask for permission in advance and only on certain days. It is to denote the fact that very few scientific or cultural entities (like associations or schools) make use of this possibility, perhaps as a result of not being aware of it. This situation could be possibly modified by offering a scheme of guided visits to interested groups and even opening an ecological didactic trail to explain the different ecosystems, vegetation and wildlife present in the quarter.

REFERENCES

Bauer, E., 1985. Los Montes de España en la Historia. Ministerio de Agricultura, Pesca y Alimentación. Madrid. 610 pp.

Comunidad de Madrid, 1984. Guía de los montes de El Pardo y Viñuelas. Madrid. 221 pp.

IGME, 1989. Mapa Geológico de España 1:50.000. Instituto Geológico y Minero de Espana. Madrid. 71 pp.

Izco, J. 1983. Madrid verde. Comunidad de Madrid. Madrid. 516 pp.

Monte, J. P. 1984. Estudio de los diferentes ecotipos y fitocenosis del bosque mediterráneo en el Monte de El Pardo. Instituto Nacional de Investigaciones Agrarias. Madrid. 551 pp.

Nicolás, J. P. & Casado, I. G., 1979. Climatología básica de la provincia de Madrid. MOPU-COPLACO. Madrid. 261 pp.

Patrimonio Nacional, 1990. Estudio faunístico en terrenos del Patrimonio Nacional. Subdirección General del Patrimonio Arquitectónico. Servicio de Jardines, Parques y Montes. Madrid. 56 pp.

Prieto, A.; Soria S.; Muñoz, A.; Tomé, F., 1996. Plan de estado y protección medioambiental del Monte de El Pardo. Patrimonio Nacional. 330 pp.

Environmental Reclamation of an Old Municipal Solid Waste Dump: El Mazo, Torrelavega, Spain

P. Martinez Cedr'un[1], A. Zabala Ingelmo[2], J.A. Go'mez Peral[3] and J. Y Saiz De Omenaca

[1]DCITYMAC, Universidad de Cantabria, Spain
[2]Consejeria de Medio Ambiente y Ordenacion del Territorio, Gobierno de Cantabria, Spain
[3]Urbaser S.A.

INTRODUCTION

Urban wastes from Torrelavega (about 50,000 inhabitants) were wildly placed along 20 years at El Mazo rubbish dump. This fact gave rise to a strong struggle in the area. Finally, in 1999, this rubbish dump was closed by the environmental agency (Consejeria de Medio Ambiente del Gobierno de Cantabria) and afterwards its morphology was corrected, a major part of the area was devoted to new use and its decisive settlement was planned and started.

Environment

El Mazo is found in the central part of Cantabria coast. Its smooth orography enjoys warm-marine climatic conditions and its population density is over 250 inhabitants per km^2. The main environmental features of the spot are summarised in the next paragraphs.

Subsurface lithology

The dump is on siltstone clays combined with red mud altogether with micaceous and ferruginous sandstones whose grains range from medium to small size. These sandstones are compact and show whitish to reddish colour (facies Weald, Valanginiense Superior – Hauteriviense – Barremiense).

Triassiac clay facies made up of clay, siltstone, and multicoloured loams together with chalk and salts appears in the proximity of the dump basing on diapiric formations. Alluvial coverings along the riverbanks of Pas and Saja-Besaya rivers extend over large areas with a relatively important thickness. A brief summary about the geological surroundings can be found in IGME (1976).

Geomorphology

Water erosion of facies Weald materials showing little thickness (like clays and siltstones) led to a hilly orography on which the small hills are topped by sandstone levels which are usually more resistant to erosion but easily weathered into sand. Valky depths are broad and well covered by large amounts of detritus. The slopes vary between 5% and 20%. Natural stability at the moment is good but there may be the risk of earth slidings at those spots with deficient drainage. There may be superficial slidings on spots where the slopes are higher than the usual ones.

Water

The slope which was used for dumping was limited by La Tejera brook (on the East side) and by Cabo river (on the North side). The first one is a tributary of the last one and this runs into Saja-Besaya river. Facies Weald materials usually show little permeability. Nevertheless, sandstone and/or sand levels which are mingled in the area may determine small, not interconnected aquifer levels which are not really deep and shown limited horizontal continuity. They come out through small springs.
Some test performed in Cabo river bed showed a regolith made up of clays and permeable sandy mud 8-9 feet wide. From this up to at least 12 feet some facies Weald red clays appeared. They were practically impermeable.

Climate

Weather conditions establish a combined mesothermal, strongly maritime regime which adds to a subhumid one. Annual mean temperature is 58°F. In January (coldest month in the year) 48°F. and the warmest one (August) is 68°F. Absolute highest temperatures may even reach 92°F or more. Absolute minimal ones may be so low as 28°F. Rainfall reaches 63 in. yearly, autumn and winter being the seasons during which rains are more frequent and stronger. December is the most rainy month followed by November. Summer is the driest season.

Vegetation
According to its bioclimatic conditions, it can be said that the area belongs to the Eurosiberian region, being a part of the hilly – mountainous range of ash formations covering the northern part of the Iberian peninsula, from Galicia (in its western corner) up to its eastern end in the Basque country. It is also to denote the fact that these ash formations are usually mixed with individuals of the oak family. In the proximity of the area, it is very common to devote the land to meadows or prairies, which are usually highly productive on the basis of proper cultivation or intensive grazing, otherwise they become easily degraded. Eucalyptus globulus Labill. are also frequent in the area.

Others
Human settlements are nearly absent in the vicinity, but the road from Torrelavega to Solares runs close to it (200 yards) and it is an important traffic artery.

STARTING CONDITIONS
Urban solid residues were dumped on one slope forming a 90 feet high small hill with steep vertical slopes of 55° of inclination. The dumping are covered about 7.4 acres and the amount of residues was a little higher than 2° million cubic feet. Among other impacts, Cabo river course was modified by the dumps and it also prevented La Tejera Brook from running regularly which caused a damming of it waters; this gave rise to a large pond on the southern of the dump. The drainage of the pond took place through the residues. This fact increased the amount of leachate and polluted the waters as well.

CORRECTING PROCEDURE
Controlling and reducing the impacts on superficial and ground waters were urgent measures to be taken since they are very often the main polluting factors (Azurmendi & al., 1990; Saiz de Omenaca & al., 1993). Air and aesthetic aspects were to be considered also. In 1999 and as a first step, the environmental agency build a transferring unit on the spot in order to lead residues into the dump at Meruelo. This one is a daily controlled dump provided with a system for collecting and treating leachate. Gases are here reused by means of a cogeneration unit, which practically receives the whole amount of urban solid residues from Cantabria.

Water control
The subsoil being practically impermeable, there is no direct risk of ground water pollution. Thus, hydrogeologic delimitation is based on avoiding the connection within waters and the dumping unit. Once dumpings were stopped, the next step was to modify La Tejera brook course. In order to carry it out, a new river bed was open outside the dump, half-slope. It was given what it was thought to be the right size in order to let it give way to the expected floods. It was protected by means of stone fillings. In this way, the brook was prevented from increasing the pond located above the dumps. For draining rainwater from the pond, a trench was opened on the eastern side edge of the dumps up to the point at which water can be eliminated by means of gravity.

The collection of dumps leachate is done by a drain that is basically a trench deeping up to the impermeable shale lying beneath the lowest part of the dump. At the edge of this, residues are left once they have been reshaped as we are going to describe in the following item of this paper.

To be treated, leachates are taken into two different ponds for being stored and homogeinized. Each pond having 400-m3 capacity, size enough to contain them even on days of extremely abundant rain. They are able to store leachates on these exceptional conditions up to four days. Later, leachates are taken to Meruelo purifying unit, while other waters are only taken to this unit in case they shown polluting charges higher than legally accepted ones (R.D. 849/1986, del Dominio Publico Hidraulico).

Morphologic Reshaping and Impermeabilization
On its North face residues lay according to their dumping inclination which reached 55°. In order to avoid instability and after taking and putting the vegetal soil away, the slopes were reduced up to 30°, this was performed after considering the situation of the brooks and the projected new use.
In order to get impermeabilization, materials coming from the same spot were used. Once the residues were superficially compacted, a clay stratum was spread on them, this was artificially compacted by means of a low frequency vibratic static roller. The surface was let to be a little rough for allowing the vegetal soil to adhere better once replaced. Clay stratum minimal thickness is 2.7 feet, and that of vegetal soil 1.4 feet.
Water run-offs were left out of the impermeable area by means of trenches running along its perimeter to

prevent waters from spoiling the impermeabilization and coming into the residue mass. Thirteen exhaust wells have been open to let the gases out, they are connected to a control station. Biogases are removed by combustion.

ASSIGNEMENT FOR NEW USES: DUMPING PLACE FOR INERT MATERIALS

At the same time, a new container was deviced in the South slope to recive inert solid residues coming from Torrelavega, its size being 300.000 m^3 which represents about 12 years of practical use. It is to denote the fact that the clay used to make the former dump impermeable came from the spot where the new one was open.

This new dump is provided with a collection and distribution network for leachate which is absolutely independent of the old dump one. The main element of this new network is the ditch open by, the dump, and the starting point for dumping, which is located at the lowest part of the container. This ditch goes at least 1.7 feet into the impermeable clay. Collection and transportation are performed by means of a porous pipe placed along the ditch bottom on which gravel is spread up to a small container. From this point on, and up to the storing ponds the pipe becomes non porous.

Trenches to modify the course of rain waters were built on the eastern, western and southern faces of the hill, leading these waters away from the container towards their natural bed. These trenches are projected to be modified in the future in order to better exploitation as the amounts of dumpings increase.

In addition to all this, the banks of the brook running along the North face have been protected by means of stone fillings. The container impermeability has been secured with the help of two sheets, one is made with geotextile matter and the other one with high-density polystyrene. On the other hand a septic tank has also been build to achieve healthy conditions on the premises.

FINAL CONDITIONING

The most outstanding feature connected with this aspect is the preparation of the new soil surfaces to give way to new vegetation and also, the selection and planting of it.

As substratum, we used the vegetal soil previously taken out of the grounds, where the transferring unit was built and also from the spots where the morphologic reshaping was carried out together with the soils removed from the inert dump place and even all soils produced while preparing the land for construction. They are, in fact, materials which are original from the same spot and that should only be prepared (as a general rule) according to conventional techniques after going though the required analysis. It has been considered necessary to adjust pH in order to regulate the amount of available micronutrients and to provided organic matter to a maximum of 4% in order to make its texture better.

Vegetation is needed not only because of this protecting role (erosion control and slope stability) but also in order to achieve environmental and landscape balance for the area and its surroundings. As a result, the species to be planted were selected according to different criteria among which function, commercial availability, origin (whether they were or not common in the area), ability for adapting to the environmental and rusticity were the main ones. This last criterion entails special importance since, at it is widely known, substratum conditions over a urban solid residue dump are usually poor because of a wide amount of added factors : gases produced by the residues, limited availability of water and oxygen in the soil, reduced thickness of the covering stratum, little ability for cationic exchange, heat produced by residue break-down, texture and composition deficiency of substrata, etc.

In addition to these and among other aspects, vegetation has been planned to be distributed following the spontaneous tendencies in the area. At the same time, it has been projected to set the basis for reaching a biodiversity level similar to the natural ecosystems in the surroundings. Making the staff connected with the project conscientious about it and continued, realistic, practical surveillance after its finishing are also included in the project.

OUTCOMES AND CONCLUSIONS

Nowadays, urban solid residues are not dumped in the area. The former dump has ben reshaped and closed. The introduced vegetation (specially herbaceous cover) is thriving satisfactorily. Water currents in the area show acceptable pollution levels and the new dump container is ready to function. Everything seen to denote that in a not very long period of time a regrettable source of environmental degradation and social tension has been removed. At the same time, the starting point for the recovery of the land has been reached bringing it into a condition which is very close to the one enjoyed by its surroundings or even for devoting it to another useful use.

REFERENCES

Azurmendi, I.; Atxabal, K.; Ereno, I. & Saiz de Omenaca, J. (1990). Rehabilitacion de la circulacion hidrica en las antiguas explotaciones mineras de Malaespera (Bilbao, Vizcaya). IV Reunion Nacional de Geologia Ambiental y Ordenacion del Territorio, Com., pp. 33-38, Gijon.

IGME (1976). Mapa geologico de Espana E. 1:50.000 Torrelavega. Servicio de Publicaciones, Ministerio de Industria, 40 pp., map.

Saiz de Omenaca, J.; Ereno, I.; Atxabal, K. & Azurmendi, I. (1993). Reclamation of adverse effects at the Lezama-Leguizamon abandoned open pit mine (Bilbao, N.Spain). Environmental Geology .vol. 22, n.l,pp. 10-12

Geo-Environmental Impact of Granite Quarrying and Remedial Measures—A Case Study

R. Ravi Kumar[1], N. Shadakshara Swamy[1] and R.K. Somashekhar[2]

[1]*Department of Geology, Bangalore University, Bangalore 56, India*
[2]*Department of Environmental Science, Bangalore University, Bangalore 56, India*

INTRODUCTION

Bangalore, the capital of Karnataka is the fifth fastest growing city (due to rapid urbanisation, industrialisation, mining, etc.) in the World and, is encompassed by ornamental /dimensional stone quarries and crushing plants spread over the entire district. The rapid growing demand for granites has witnessed a large-scale increase in quarrying operations. Granite quarrying activities have attracted the attention of geologists and environmentalists to assess the environmental issues, as it is bound to alter the topography, nature and amount of overburden, which inturn decides the degree of land degradation (Rekha Ghosh and Saxena, 1995). The damages caused due to quarrying are not only confined to the immediate site, but also spread over a vast area around the operation sites (Mukhopadhyaya and Barve, 1993). In the present study an attempt has been made to evaluate the impact of mining on various aspects of environment, namely land, water, air, noise and flora.

STUDY AREA

Bangalore district forms a rolling topography with gentle slope towards south. The predominant lithology of the area includes light to dark grey granitic gneiss, peninsular gneiss, granite; metasediments, charnockites and migmatites often intruded by basic dykes. Soils have been classified into red gravely and non-gravely clay, lateritic gravely and non-gravely, colluvic-alluvial soils with salinity and shallow to deep black soils (National Bureau of Soil Survey & Land Use Planning).

QUARRYING TECHNIQUES

Based on the degree of mechanisation deployed for excavation, quarries have been categorised into manual, semi and mechanised types.

The excavation is being done manually using chisel and hammer, by flame torching and blasting (low-level explosives). The rocks are extracted using high level explosives (blasting); air compressors and, dimensional stones so obtained are crushed for aggregates to a desired size in the semi-mechanised quarries. In the mechanised quarries, heavy machineries such as poclaines, cranes, dumpers, air compressors, jack-hammer drills, jet burner techniques are employed, which inturn produce a lot of noise, dust and toxic elements.

METHODOLOGY

Geological map and Satellite (IRS-1C) imagery of Bangalore District were used to know lithologies and location of quarry sites respectively. Soil and water (surface and subsurface) samples were collected and analysed for different soil fertility and water quality parameters using Standard Methods (Jackson, 1973; Trivedy & Goel, 1987; APHA, 1992). **Envirotech APM – 410** High Volume Air Sampler was used to monitor air quality for suspended particulate matter (SPM), sulphur dioxide (SO_2) and oxides of nitrogen (NOx). The SPM ($\mu g/m^3$) is quantified using the equation,

$$\boxed{\text{SPM Concentration } (\mu g/m^3) = (Wf-Wi) \times 10 / V}$$

Where, Wf = Weight of exposed filter paper in g; Wi = Weight of filter paper in g; V = Volume of air sampled in m^3.

Spectronic-20 spectrophotometer was used for the estimation of sulphur dioxide and nitrogen dioxide Morris and Kratz (1977). Sound Level Meter 2031 CYGNET, [battery operated] capable of measuring 30 dB to 160 dB sound with ± 2dB accuracy was used to measure the noise levels.

RESULTS AND DISCUSSION

Land Degradation

Surface soil, a vital natural wealth produced and accumulated over thousands of years is stripped off and wasted in no time in the very first stage of quarrying activity. The wastes (aggregates) blasted materials (fly rocks), chemicals (explosives used for blasting) and dust (granite powder) generated from the activity spread and settle over the surrounding areas.

The NPK values (Table 1) showed steady increase with increasing distance from the operations. Most of the soil samples were low in nitrogen and potassium due to lower organic carbon content, resulted as a consequence of accumulation of dust. Available phosphorus in most of the samples was however, found to be high. The higher values of phosphorus may be due to alkalinity of soils, which in turn are derived from rocks being quarried. Organic Carbon plays an important role in the productivity and conditioning of soils (ICAR, 1997). The organic carbon in most of the samples varied from low to moderate, excepting where moderate to higher values prevailed (samples collected away from activity).

Surface and Sub-surface water

To evaluate the quality of water in and around quarries and crushing units, 54 water samples from the open/dug wells, tanks, ponds and tube wells were collected and analyzed for pH, turbidity, electrical conductivity, chlorides, calcium, magnesium, hardness, sulphates, dissolved oxygen and BOD (Table 2). The pH ranged from 5.62 to 9.30 with a mean value of 7.36 (groundwater) and 8.07 (surface water). The average "EC" value for surface and groundwater is 370 and 586 µmhos/cm respectively. The turbidity level in surface water exceeded the limits and varied from 1 to 155 NTU. This is perhaps due to intensive crushing and quarrying in the case of surface water. The average value of calcium and magnesium is well within the limits. Total hardness is not a pollution parameter but it helps to know whether water is fit for consumption or not. Water quality of 50% of the samples analyzed is found to be hard to very hard. The concentration of chloride, sulphates is also well within the limits except for a few samples. The dissolved oxygen ranged from 2.43 – 16.2 mg/l and 6.2 – 13.8 mg/l in groundwater and surface water samples respectively, and with maximum in surface water samples. The variations observed in surface water are perhaps due to the influence of dust on water system. The five-day biological oxygen demand (BOD) indicated lower values both in groundwater and surfacewater. Nearly 38 % of the samples showed relatively higher BOD values.

Air Quality

The status of ambient air quality with respect to SPM, sulphur dioxide and nitrogen dioxide is summarized in Table 3. The maximum concentration of SPM is found to be 14,013 µg/m³ at semi-mechanized quarry (due to large scale quarrying and stone crushing) and minimum being 78.3 µg/m³ in the mechanized quarry. The higher concentration of SPM is attributed to the high density of crushing plants and addition of dust, which are lifted by winds and vehicular movement. The average value of sulphur and nitrogen dioxide was found within the permissible limits as compared to the standard limits of 30 µg/m³ and 80 µg/m³ for sensitive, rural and residential areas (Table 3a). The higher concentration of sulphur dioxide is 32.5 µg/m³ (semi-mechanized quarry) and lower value being 13.5 µg/m³ (manual). Maximum concentration of oxides of nitrogen is recorded at mechanized quarry where large size equipments and machineries (petrol and diesel engines) are being used and, minimum concentration is at 200 m away from the quarry site (semi-mechanised). The maximum value of NOx and SO₂ can be attributed to semi and mechanised methods of quarrying, usage of diesel and petrol engines and, also to high frequency of vehicular movement.

Noise Level

The noise level ranged between 154.5 dB (A) in the mechanized (gas jet burner operation) to 48.9 dB (A) at 300 m away from the manual quarry (Table 4). The higher noise level was observed when more noise producing machineries are intermittently put into operation and, noise level reached the threshold level when all the machineries are in operation. The frequency of higher noise also depended on the frequency of passing of vehicles with air-horn, reaching a level of around 140-150 dB (A), culminating in discomfort experienced as actual pain. This Exposure could also cause significant negative impact on the psycho-behaviour of the individuals causing headache, irritation and even impairment of hearing (Yogamoorthi and Beena, 1996). Environmental Protection Agency (EPA) of US has recommended 85 dB (A) as warning not to be exceeded where the individual is exposed to noise for an eight hours work/day in a day environment.

Flora

The quarrying activities seem to have brought about changes in the ecology of the surrounding environment. There was a wide spread occurrence of many diseases on plants in the vicinity of quarries, possibly nurtured by continuous deposition of dust. The noticeable ones were grain smut; downy mildew of Sorghum and ragi; rice blast disease; tikka and leaf spot of groundnut and necrosis and chlorotic symptoms on leguminous plants. The agricultural crops showed stunted growth in several areas. This is again attributed to accumulation of dust which limpaired the physiological activities like photosynthesis, transpiration, respiration etc. The fine particles of dust also blocked the stomatal openings, through which exchange of gases and transpiration takes place.

REMEDIAL MEASURES

- Scientific method of quarrying should be adopted.
- Proper disposal of wastes can minimize Land degradation.
- Encroachment of agricultural land by quarries needs to be prevented.
- Augmenting soil erosion can reduce the deterioration of water quality.
- Planting trees can control the dust pollution caused by crushing units in particular. The plants mechanically stop smoke/dust from reaching us and also act as sinks for many noxious gases.
- Maintaining asphalt approach roads can control the dust.
- Usage of heavy machinery & movement of trucks will have to be regulated.
- The quarry owners should be asked to plant trees, in multiple rows.
- A room for collecting dust should be constructed to avoid further spread.
- The workers should be provided with filter masks, cotton ear plugs.
- Trees having thick fleshy leaves with flexible petioles posses the inherent capacity to withstand vibrations and thus reduce noise.
- Creating social awareness and inculcating proper civic sense among the workers / citizens.

CONCLUSION

The soil properties near the activity site have been affected to different degrees compared to that away from the non-activity site. The surface water quality is deteriorated due to intensive quarrying, rendering it not potable. There is not much influence of the activity on groundwater regime. The noise due to Jet burner in mechanized quarries has left impact not only on quarry workers but also nearby villagers. The SPM concentration is much higher and exceeded the prescribed limits at most of the sites. The plants around the sites are affected and showed stunted growth. Hence there is need to take up urgent action to mitigate the impact.

Table 1: Analytical Data of Soil

Parameters	Minimum	Maximum	Average
pH (Units)	4.5	8.4	6.00
Available Nitrogen (Kg/ha)	163	414	269
Available Phosphorus (Kg/acre)	17	84	39
Available Potassium (Kg/acre)	56	655	240
Organic Carbon (%)	0.0	2.82	0.9

Sample Size: 70

Table 2: Physico-chemical characteristics of Groundwater and Surface water

Parameters	Minimum	Maximum	Average	Std. Dev.	Co-efficient of Variation (%)
Groundwater					
pH (Units)	5.62	9.20	7.4	0.75	10
Turbidity (NTU)	1.0	155	16.5	33.2	201
Electical Conductance (μmhos/cm)	100	2200	586	496	85
Chloride (mg/l)	4.3	257	67.4	79	116
Calcium (mg.l)	8.8	119	33.3	15	45
Magnesium (mg/l)	0.98	63	18	9.0	50
Hardness (mg/l)	36	540	159	124	78
Sulphate (mg/l)	6.0	325	90	75	83
Dissolved Oxygen (mg/l)	2.4	16.0	8.16	3.2	39
B O D (mg/l)	0.0	13.0	3.53	3.8	108
Surface Water					
pH (Units)	7.2	9.3	8.07	0.64	8.0
Turbidity (NTU)	7.0	47	21.5	15.0	70
Electical Conductance (μmhos/cm)	100	1200	370	316	85
Chloride (mg/l)	17	61	34.0	14	41
Calcium (mg.l)	8.0	28	19	6.4	34
Magnesium (mg/l)	1.5	20	7.4	0.8	11
Hardness (mg/l)	30	164	96	5.5	6.0
Sulphate (mg/l)	14	223	95.5	43	45
Dissolved Oxygen (mg/l)	6.2	14	9.7	3.23	33
B O D (mg/l)	0.0	12.0	6.0	4.0	67

Table 3: Average Value of SPM, SO_2 and NO_2 ($\mu g/m^3$)

Sampling Units	Maximum	Minimum	Average
MANUAL TYPE QUARRY			
SPM	2672.14	111.70	1391.92
NO_2	19.5	13.5	16.5
SO_2	23.0	13.5	18.25
SEMI-MECHANIZED QUARRY			
SPM	14,013	286.66	13,724
NO_2	33.0	14.0	23.50
SO_2	32.5	18.0	25.3
MODERATE – MECHANIZED QUARRY			
SPM	493.3	78.3	285.8
NO_2	35.5	13.5	24.5
SO_2	24.0	15.0	19.5

Table 3a: National Ambient Air Quality Standards

Pollutant	Time Weighted	Concentration in ambient air		
		Industrial area	Residential, Rural & other areas	Sensitive area
Sulphur Dioxide ($\mu g/m^3$)	24 hourly / 8 hourly	120	80	30
Oxides of Nitrogen ($\mu g/m^3$)	24 hourly / 8 hourly	120	80	30
Suspended Particulate Matter ($\mu g/m^3$)	24 hourly / 8 hourly	500	200	100

Source: Central Pollution Control Board, New Delhi, 1994.

164

Table 4: Level of Noise at quarries

Distance (in metre)	Noise Level (dB(A))
Manual type of quarry	
At quarry site	59.7
100	57.1
200	51.0
>300	48.9
Semi-mechanised Quarry	
0	88.2
100	78.2
200	67.8
>300	57.4
Mechanised Quarry	
0	154.5
100	134.3
200	113.9
>300	102.5

REFERENCES:

APHA (1992). Standard Methods for the Examination of water and wastewater. APHA; Inc., 16th Edition, USA.

CPCB (1994). Standards for liquid effluents, gaseous emissions, automobile exhaust, noise and ambient air quality. June'95, Pollution Control Law Series PCL/4/95-96.

Environmental Protection Agency (1973). Water Quality Criteria. National Academy of Sciences, Washington, D.C., vol 1: 98pp.

Indian Council of Agricultural Research (1997). Hand Book of Agriculture. Krishi Anusadhan Bhavan, PUSA, New Delhi – 110 012.

Jackson, M. L. (1973). TextBook of Soil Chemical analysis. Prentice – Hall, Inc., Engle Wood Cliffs, Jersey.

Morris Kratz, (1977). Methods of Air sampling and analysis. Second edition APHA. Inter Society Committee Interdisciplinary Books and Periodicals for Professionals and layman.

Mukhopadyay, S. K. and Barve, S.P. (1993). Legislative measures to control land degradation due to mineral development – Special reference to Indian Scenario. Environmental Issues in Mineral Resource Development. Ed. – K.L. Rai. Gyan Publishing House. pp. 121-128.

National Bureau of Soil Survey and Land Use Planning (NBSS-LUP), Regional Centre, Bangalore.

Rekha Ghosh and Saxena, N. C. (1995). Small scale mining – Environmental impacts, Min-Env-95, Udaipur.

Trivedy, R.K. and Goel P.K. (1987). Practical methods in ecology and environmental science. Environmental publications, Karad (India).

Yoganmoorthi, A. and Beena, (1996). Studies on noise pollution level in the Pondichery town. South India. Poll. Res., 15(2): 155-158.

Management of Radioactive Waste and Environmental Impact of Mining and Processing of Uranium Ore

A.H. Khan, V.N. Jha, S. Jha and R. Kumar

Environmental Assessment Division, Bhabha Atomic Research Centre, Mumbai 400 085, India

INTRODUCTION

Uranium is the basic fuel for nuclear power. The first uranium ore deposit of economic importance was discovered at Jaduguda in the Singhbhum Thrust Belt in Bihar. After an initial phase of underground exploratory mining, the Uranium Corporation of India Ltd.(UCIL) commenced mining and processing of 1000 tonnes per day of ore at Jaduguda in 1967 to produce uranium concentrate commonly known as "yellow cake". Subsequently two other deposits at Bhatin and Narwapahar located 3 and 12 km northwest of Jaduguda were taken up for underground mining. The ore from these mines is also processed in the mill at Jaduguda which has an expanded capacity of 2090 tonnes/day [1]. The mineral concentrate obtained by tabling of the copper plant tailings which contain comparatively small values of uranium (typically 0.005 -0.009 % U_3O_8) is also processed in this mill. The mining and ore processing operations with emphasis on management of waste to control environmental impact are summarised in this paper. Environmental radiological surveillance has been an integral part of these operations since their inception. A summary of the environmental monitoring during the last few years is presented.

MINING OPERATIONS

The Jaduguda mine has been developed to a depth of 900 metres in three stages over the years. It has a horizontal strike length of about 800 metres. The central shaft serves as entry for men and material and as main ventilation intake route. Using latest mining machinery the ore is excavated from different locations and brought to a central location and hoisted to surface in a skip and discharged on to a conveyor system leading to the mill. Bhatin Mine, 3 km away, is a relatively small mine developed up to a depth of about 135 metres. It has entry through an adit which also serves as intake route for ventilation air and transport of the excavated ore to surface. The Narwapahar mine is one of the most modern mines in the country. Trackless mining with decline is used as one of the entries to excavate the ore up to a depth of about 140 metres. A vertical shaft has also been sunk to a depth of 355 metre for mining of deeper deposits.

ORE PROCESSING

The ore produced in the three mines is processed in the mill at Jaduguda to extract uranium. The initial process briefly comprises of crushing, screening ,wet grinding to the required size of -200 mesh and de-watering to control pulp density. This is followed by leaching with sulphuric acid in presence of an oxidising agent (MnO_2) in air agitated vessels. Temperature of 40 - 50 ° C is maintained by passing steam. The leachate is filtered and purified using ion exchange process. Uranium is precipitated in the form of magnesium diuranate (yellow cake) by addition of magnesia slurry to the pure liquor.

WASTE MANAGEMENT

Mine Wastes

Most uranium mines in the world produce low-grade ores containing 0.1 to 0.3% U_3O_8. Mines in India have lower grades than these. Only solid and liquid wastes from uranium mines and mill are of some concern.

(i) *Solid waste*: Depending on the availability of ore and recovery considerations in mining and milling operations each mine considers a cut off grade(% U_3O_8) below which it is uneconomical to process. For winning the ore excavation of some waste rock is inevitable. Due to very low radioactive content the solid wastes from mines do not pose serious problem. This waste rock is partly used as mine backfill material and the rest is used for land fill within the premises and for strengthening the tailings embankment.

(ii) *Liquid waste*: Large quantities of water are encountered in mines. This may contain dissolved uranium($U_{(nat)}$)and radium(^{226}Ra). Mine water is, therefore, collected, clarified and reused at appropriate stage in mill.

Mill Wastes

(i) *Mill tailings*: The bulk of the ore processed in mill in combination with the reagents used emerges as waste or 'tailings'. It comprises of the barren cake from filters containing all the undissolved radionuclides and the barren liquor from ion exchange columns having some dissolved activity. The tailings contain all the decay products of uranium originally present except for the parent mineral. Only about 15% of the activity present in the ore goes with the "yellow cake". After decay of the shorter lived component (^{234}Th) about 70% of the original activity of the ore remains with the tailings[2]. Radium-226 and Thorium-230 present in the tailings require long term management. Besides these, the tailings also contain the bulk of the chemical additives used during processing. Safe disposal of tailings is, therefore, an important aspect of the uranium ore processing.

(ii) *Run off water and other effluents*: The runoff water from ore yard are used in the process. Overflow from the magnetite (a byproduct) settling pit is sent to the Effluent Treatment Plant (ETP) for treatment. The effluents from storm water drains are treated for use as industrial water.

Tailings Treatment and Containment

The barren liquor from the ion exchange columns is treated with lime stone slurry initially to pH of 4.2 - 4.3 followed by addition of lime slurry to a pH of 10 - 10.5 and mixed with the barren cake slurry to a final pH of 9.5 - 10. At this pH the residual uranium, radium, other radionuclides and chemical pollutants like Mn get precipitated. The treated slurry is classified into coarse and fine fractions using a hydrocyclone. The coarse material forming nearly 50 % of the tailings is sent to mines as backfill. The fine tailings or 'slimes' are sent for permanent containment in to an engineered tailings pond. The slimes with precipitates settles down and clear liquid is decanted. A series of decantation wells and side channels are provided to lead the decanted liquid to an effluent treatment plant (ETP).

The tailings containment ponds are site specific. At Jaduguda the first and second stages of the tailings pond of about 33 and 14 ha surface area, respectively, are located adjacent to each other in a valley with hills on three sides and an engineered embankments on downstream side of natural drainage. These two containment pond are now nearly filled up. The third stage of the tailings pond, having an area of 30 ha, which is currently in use, is also located nearby in a similar setting. The underlying soil and the bedrock of these tailings ponds have very low permeability. The tailings pond area is fenced to prevent unauthorised access.

Consolidation of Tailings

A vegetation cover of non-edible grass and plant such as *Saccharum spontaneum* (kansh), *Typha latifolia* (cat-tail) and *Ipomoea carnia* (Amari) has been provided over the used-up portion of the first two tailings ponds[3]. This vegetation cover helps in suppressing generation and dispersal of dust and consolidates the tailings besides merging it with the local landscape.

Water Reclamation and Effluent Treatment

Though lime neutralization of tailings largely takes care of the dissolved pollutants in the process effluents, subsequent reduction of pH in the tailings pond over a period of time increases concentrations of some radionuclides and chemical constituents in the decanted effluents. Hence, these are further treated to meet regulatory discharge limits. In the ETP the tailings pond effluents are clarified , a part of which is sent to the plant for reuse in the milling process. The rest is treated first with $BaCl_2$ and then with lime to precipitate ^{226}Ra, Mn and other pollutants. It is clarified and the settled sludge carrying the $Ba(Ra)SO_4$, $Mn(OH)_2$ and other precipitates is sent to the tailings pond with the main tailings and clear effluent is discharged to environment after pH adjustment.

Environmental Surveillance

A comprehensive environmental surveillance is maintained around the mines, mill and the waste disposal facility to evaluate the effectiveness of control measures, assessment of personnel exposure, assessment of environmental impact and regulatory compliance.

Uranium tailings are low specific activity material. Presence of radium and other radionuclides, though in small quantities, make it a source of source of low levels of gamma radiation and environmental radon. The gamma radiation and radon monitoring data (median values) at the tailings ponds and nearby area are presented in Table- 1. The gamma radiation levels of 1.4 to 2.0 $\mu Gy.h^{-1}$ found at centre of the ponds reduce to 0.50 $\mu Gy.h^{-1}$ at the boundary and gradually reduce to the local background of 0.10 to 0.15 $\mu Gy.h^{-1}$ within a short distance. These radiation levels account for an annual exposure levels of about 1100 $\mu Gy.y^{-1}$ (1.1 $mSv.y^{-1}$). This is further confirmed from the readings of thermoluminiscent dosimeters (TLD) deployed at different locations to assess the cumulative radiation exposures in the region[4]. Similarly, due to large atmospheric dilution the radon levels of about 30.0 $Bq.m^{-3}$ observed above the tailings surface reduce to the local background value of about

Table-1: Gamma Radiation and Atmospheric Radon
At Tailings Pond and other Areas

Sample Locations	Gamma Radiation ($\mu Gy.h^{-1}$)	Atmospheric ^{222}Rn ($Bq.m^{-3}$)
Tailings pond II	2.00	32.5
Tailings Pond I	1.50	23.1
Tailings Pond III	1.45	9.6
Tailings Pond Boundary	0.48	9.7
Tuadungri (0.2 km, SE)	0.20	14.5
Dungridih (0.3 km,NE)	0.25	16.7
Chatikucha/Tilaitand (0.5 km,SSE)	0.13	9.5
Env.Survey Lab.(1.5 km, E)	0.14	11.0
> 2 km	0.12	11.8
> 5 km	0.11	9.4
Regional Background	0.10 - 0.15	10 – 20

10.0 $Bq.m^{-3}$ at about 0.5 km from the tailings pond. Radon concentration in the mine exhaust air averages around 6000 $Bq.m^{-3}$. It is released about 30 m above ground level. The atmospheric concentration reduces to the background value within the plant boundary.

The effectiveness of the ETP in controlling release of radioactivity in the aquatic environment is evaluated by measurement of the U(nat.) and ^{226}Ra in the inlet and outlet effluents. Decontamination efficiency of the ETP for the past few years is shown in Fig.1. The recipient surface water system down stream of uranium industry are monitored. Results of $U_{(nat)}$ and ^{226}Ra concentrations observed in the surface waters in public domain are presented in Table- 2. It may be noted that uranium and radium content of the Gara and Subarnarekha rivers down stream of UCIL operations are comparable to the respective background levels.

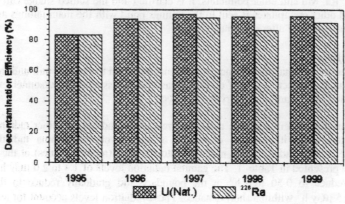

Fig.1 Decontamination Efficiency of ETP

Table- 2: Summary Of U$_{(nat)}$ and^{226}Ra Concentrations in Surface Waters

Sampling locations	No. of samples	U$_{(nat)}$ Conc. (µg.l^{-1})		^{226}Ra Conc. (mBq.l^{-1})	
		Range	Mean	Range	Mean
Gara River (NRW), U/S	20	<0.3 - 4.4	1.5	1 - 18	9
Gara River (NRW), D/S	20	0.6 - 13.4	3.7	5 - 74	18
Gara River (JAD), U/S	27	0.5 - 34.3	3.5	3 - 61	12
Gara River (JAD), D/S	36	0.5 - 54.9	14.8	5 - 283	44
Subarnarekha River, U/S	28	0.5 - 5.4	1.4	4 - 48	18
Subarnarekha River, D/S	30	0.5 - 11.9	5.0	7 - 55	20
DWC (Limit)			100		300

Analysis of ground water from around Jaduguda has been reported[5].The uranium and radium concentrations of 0.5 - 2.0 µg.l^{-1} and about 7 - 15 mBq.l^{-1}, respectively are of the order of the background levels and well within the derived water concentrations(DWC)or the drinking water limits.

CONCLUSIONS

The state of the art technology is used for the safe management of low specific activity radioactive waste at uranium mining and ore processing operations of UCIL. The in-built controls, treatment and surveillance have been effective in controlling the environmental releases of radioactivity. The impact of these operations on the local environment are only marginal and meet the regulatory requirements.

ACKNOWLEDGEMENT

The authors are grateful to Dr. V. Venkatraj, Director, Health, Safety and Environment Group and Dr. S. Sadasivan, Head, Environmental Assessment Division, BARC for the encouragement and support received. Thanks are due to the authorities of the UraniumCorporation of India Ltd for extending necessary facilities .

170

REFERENCES

1. Beri, K.K. (1998). Uranium Ore Processing and Environmental implications. Metal, Materials and Processes. Vol.10,No.1,pp. 99-108

2. IAEA (1992). Current Practices for the Management and Confinement of Uranium Mill Tailings. Technical Report Series No. 335. Iternational Atomic Energy Agency, Vienna.

3. Basu,S.K., Jha,V.N. and Khan,A.H. (2000). Uptake of Radionuclidesand trace metals by Plants Growing on or near Uranium Tailings. Proceedings of the Ninth National Symposium on Environment. Bangalore University, June 5-7.

4. Chougaonkar, M.P.,Mehta,N.K.,Srivastava,G.K., Khan,A.H.,and Nambi,K.S.V. (1996). Results of Environmental Radiation Survey around Uranium Mining Complex at Jaduguda using TLDs during 1984-94. Proceedings of the fifth National Symposium on Environment. Calutta. Feb. 28- March1.

5. Basu,S.K., Jha,V.N., Markose,P.M. and Khan, A.H.(1999). An Assessment of Ground Water Quality Around Jaduguda After Thirty Years of UCIL Operations. Proceedings of Twenty Fourth Conference of the Indian Association For Radiation Protection, Kakrapar, Surat, Jan. 20-22.

Monitor at and Near Hazardous Waste Disposal Sites

Ashok Dhariwal

Department of Civil Engineering, Jai Narain Vyas University, Jodhpur, Rajasthan, India

INTRODUCTION

Since the hazardous wastes are not useful to the industry producing them, they have typically been disposed of in the easiest and cheapest way. Waste water lagoons and ponds pose a threat to drinking water. Similar dangers exist for the second most common disposal technique i.e. non-secure landfill.

In determining the risk presented by a potentially hazardous waste site, it is necessary to perform site assessment. These site assessments must determine the source of contamination, provide a qualitative characterization of the transport mechanism by which the contaminants are migrating and determining the consequences likely to be associated with the contaminants and their effect on human health and the environment. Fig.1 shows all the environment media where pollutants could lodge, including air, surface water and sediment, soil and ground water. The environmental pathways of greatest concern obviously vary from site to site.

Monitoring activities have usually been carried out to help clarify the risks, if any associated with specific sites. While every data point that could be obtained may be of some interest, the challenge is to maximise the usefulness of monitoring data that are collected and analysed within cost and time constraints.

THE GENERAL APPROACH TO MONITORING

Monitoring Activities Involve Several Steps –

1. Deciding where and how to take representative samples of the media of interest and how to handle the samples enroute to the analytical laboratories.
2. Selecting analytical methods to investigate the presence or absence of either a finite or open ended list of chemicals.
3. Choosing a procedure and format for aggregating and presenting the results of the analysis and for setting forth the degree of confidence in data.
4. Determining how to interpret the monitoring data as to presence, quantity, transformation and migration of the chemicals of interest.

MONITORING TECHNIQUES

Initially, the monitoring process must determine the location, the quantities and the source of public contamination. This information might be found in industry and Government records. Once the existing records have been studied, additional information will usually be needed and must be obtained. Monitoring wells provide every warning of ground water contamination. Monitoring can reveal not only which pollutants are present and at what concentrations but also the Arial and vertical distribution of these pollutants. A visit to site is necessary to conduct actual field sampling is generally accomplished by drilling core holes in the site area. A well- designed pattern of core holes along with subsequent core analysis will help reveal and confirm the dimension and quantities of hazardous chemical present at the site. Core sampling is both expensive and may be hazardous to operators. To lessen the potential exposure of personnel to hazardous waste, remedial site operators can rely on geophysical techniques.

173

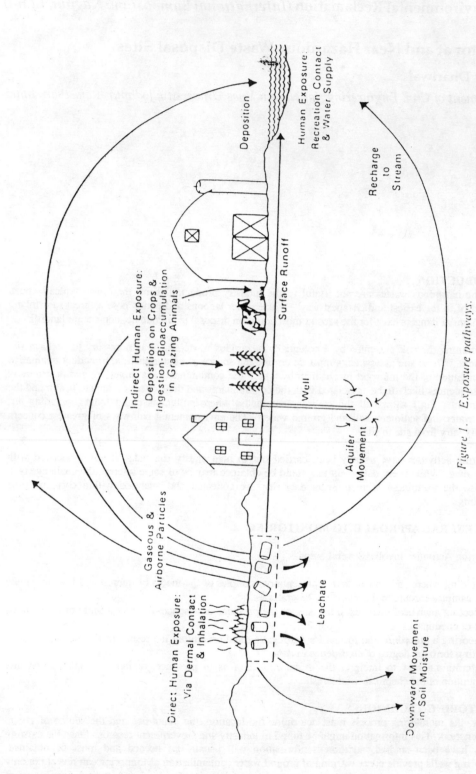

Figure 1. Exposure pathways.

Deposition

Human Exposure:
Recreation Contact
& Water Supply

Recharge
to
Stream

Indirect Human Exposure:
Deposition on Crops &
Ingestion-Bioaccumulation
in Grazing Animals

Surface Runoff

Well

Aquifer
Movement

Gaseous &
Airborne Particles

Direct: Human Exposure:
Via Dermal Contact
& Inhalation

Leachate

Downward Movement
in Soil Moisture

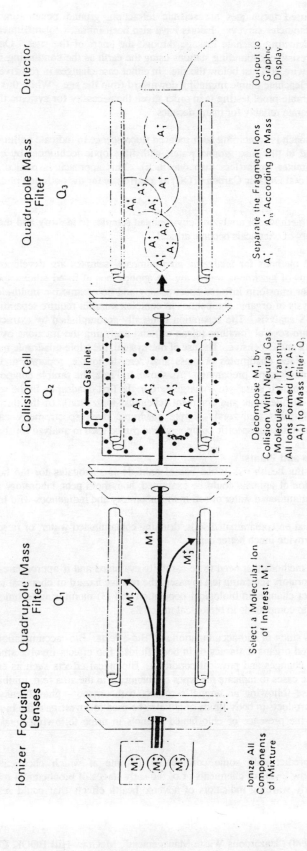

Ionizer Focusing Lenses

Quadrupole Mass Filter Q₁

Ionize All Components of Mixture

Select a Molecular Ion of Interest, M_1^+

Collision Cell Q₂

Gas Inlet

Decompose M_1^+ by Collision With Neutral Gas Molecules (●), Transmit All Ions Formed (A_1^+, A_2^+, ..., A_n^+) to Mass Filter Q₃

Quadrupole Mass Filter Q₃

Separate the Fragment Ions A_1^+, ..., A_n^+ According to Mass

Detector

Output to Graphic Display

Figure 2 Triple-stage quadrupole mass spectrometer schematic.

175

Geophysical Techniques

The most commonly used techniques are seismic refraction, ground penetrating radar, electrical conductivity and magnetometer surveys. Efforts have also been initiated to instrument new sites with electrode systems for detecting leachate leakage through the liners of the sites. One approach is to surround the site with resistively sounding stations using the earth as the conducting medium. Another approach is to embed a wire grid just below the site. In either case changes in resistively measurements would signal a possible leachate plume migrating downward from the site. While this technique seems very appealing, considerable proof testing is in order given the necessity for systems that will minimize false alarms and will operate reliably for many decades.

Another promising approach is combining laser induced fluorescence to indicate pollutant contamination of an aqueous body and in this case ground water, with fibre optic techniques by entering the earth through very small diameter wells (less than one inch). This approach is particularly attractive in attempting to measure Total Organic Carbon (TOC) as a surrogate for the pollution plume.

Analytical Methods

The need for improved methods for analyzing environment samples to identify and measure very small amounts of a wide variety of chemicals become critical.

A number of advanced methods for analyzing environmental samples are developed. Of particular relevance to the problems of hazardous wastes are the applications of fused silica capillary columns in GC/Ms systems. Fourier transform infrared spectroscopy and non-extractive multi-elemental analysis. MS techniques for analysis of organic compounds in complex samples require separation of the sample components prior to MS analysis. The separation is usually accomplished by extracting the samples, separating the extract into several fractions (clean up) and analyzing the fractions by GC / MS These steps are time consuming and expensive. The use of one system, the triple quadruple mass spectrometer, uses a combination of three quadruples or mass analyzers, to ionize, separate and analyze sample components with minimum sample preparation. As shown in Fig.2, the sample components are ionized and separated according to their mass – to – charge ratio in the first quadruple. In the second quadrupole these ions collide with an inert gas and fragment (chemical ionization). In the third quadrupole the fragments are identical (mass analyzers). Triple quadruple mass spectrometry can provide rapid screening of complex mixtures for specific compound and can be used to analyze for these compounds.

Monitoring In Animals and Plants

One approach to assessing health risk is the use of animals as surrogates for the human population. Three different population of animals could be examined, household pets, laboratory animals raised in homes or exposed to contaminated water or soil in the laboratory and indigenous wild life.

Laboratory animals raised on contaminated soils, drinking contaminated water, or raised in homes with contaminated air may provide much better data.

The types of biological methods that need to be carefully evaluated and if appropriate, standardized for field use include (1) laboratory screening test to assess the relative hazard of chemical mixtures, (2) field survey methods to detect changes in biological population and (3) monitoring techniques to detect the presence of some specific compounds in biological materials.

Bio monitoring includes study of Bio-accumulation and Bio-effects. Bio- accumulation can be taken in whole body or in selected organs & tissues or in body fluids. Bio effects involve study of bio-system responses, pathological changes and physical condition. Biological effects such as enzyme inhibition can also be used in some cases to indicate the types of chemicals in the area (e.g. inhibition of red blood cell acetyl cholinesterase following an animal's contact with orqano - phosphorus chemical). The detection of metabolic products in body fluids might also be used to investigate the biological uptake of some compounds (e.g. the presence of chlorinated phenols in urine following uptake of chlorinated hydrocarbons).

Monitoring helps to predict with some confidence, the rate at which chemicals under people. Measurements of very low levels of chemicals or of the early stages of biochemical reactions in people or animals serve as early warning indicators of adverse health effects that could result from further exposures.

REFERENCE

SChartes A. Wentz. (1998) "Hazardous Waste Management", McGraw-Hill BOOK COMPANY, New York.

Joseph H. Highland. (1999). "Hazardous Waste Disposal Assessing the problem", ANN ARBOR SCIENCE. New Delhi.

Schweitzer G.E. "Risk Assessment at Hazardous Waste Sites, AMERICAN CHEMICAL SOCIETY SYMPOSIUM Series 204.

Solid Waste Management Network of Jodhpur City

Ashok Dhariwal

Department of Civil Engineering, Jai Narain Vyas University, Jodhpur, Rajasthan, India

SYNOPSIS

Solid waste management has a greater importance in the view of transportation system and operation has increased in size and complexity. The conventional approach in optimal transportation management by using individual experience is losing its place in the view of complexity of solid waste management. During the incidence of epidemics, conventional method of collection, transportation and disposal is not adequate to eradicate the spreading of communicable diseases like plague. Identification of minimal spanning tree is an essential tool to achieve the above objective. An attempt has been made in this paper to identify the normal shortest path and emergency optimum path for integrated solid waste management of Jodhpur city by using optimization techniques such as transportation problem, shortest route algorithm and minimal spanning tree.

INTRODUCTION

With the influx of industrial revolution coupled with ever accelerating urbanization, increasing consumerism and population exodus, the problem of solid waste is inflated significantly. The most obvious environmental damage is causal by improper collection of solid waste as aesthetic, ground water contamination due to leaches from dumping sites and air pollution due to gases released from dumping sites. It also causes public health problems indirectly by encouraging the growth of population of flies (which can transmit fever, cholera, diarrhoea, tuberculosis, anthrax etc.),rats (which can spread the plague, murrain typhus fever, rabies etc.), cockroaches, mosquitoes (which can cause malaria, yellow fever, dengue, mosquito borne diseases, filarisis etc.) and other pests. Respirative disease is a possible consequence form inhaling dust from refuse. In the view of the above problems, proper collection and disposal within a stipulated time with optimum cost is essential. In the view of complexity, present solid waste collection and disposal, the existing practice in collecting, transporting and disposal of solid waste may not be adequate for reducing transportation and operating cost and also for eradication of environmental damage. In the view of the above, an attempt has been made to optimize the solid waste management in Jodhpur City by using advance optimization techniques.

METHODOLOGY

Identification Of Optimum Paths From Transfer Points To Disposal Sites

In Jodhpur city 75 collection (dustbins) are scattered in entire city. In order to facilitate quick Collection and disposal with a minimum cost, 33-transfer points are selected covering the entire city. The distance among transfer points and distance from transfer points to disposal site are measured using road map of Jodhpur city. There are large numbers of alternate routes from each transfer points to disposal sites. In order to select optimum route from each transfer points to disposal sites shortest route algorithm of a network is adopted. Input parameters for algorithms are transfer point nodes and distance among the transfer points and disposal sites. The result of the algorithm is presented in Table 1 to 4.

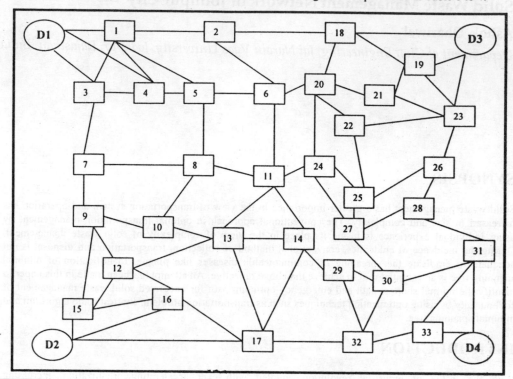

Fig.1 Network Map of Study Area

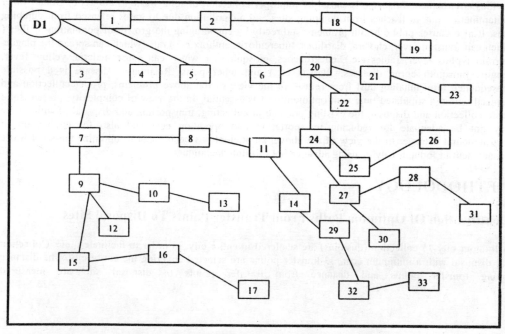

Fig. 2. Minimal Spanning tree for disposal site 1

178

Table 1 Shortest distance to various transfer points from disposal site 1

Transfer point	Distance Kms	Path
1.	0.70	DI – 2
2.	1.40	DI – 1 – 2
3.	0.45	DI – 3
4.	0.80	DI – 4
5.	1.20	DI – 1 –5
6.	1.70	DI – 1 – 5 – 6
7.	0.85	DI – 3 – 7
8.	1.65	DI – 3 – 7 – 8
9.	1.25	DI – 3 – 7 – 9
10.	1.85	DI – 3 – 7 – 9 – 10
11.	2.15	DI – 3 – 7 – 8 – 11
12.	2.45	DI – 3 – 7 – 9 – 12
13.	2.15	DI – 3 – 7 – 9 – 10 – 13
14.	2.65	DI – 3 – 7 – 8 – 11 – 14
15.	2.10	DI – 3 – 7 – 9 – 15
16.	2.70	DI – 3 – 7 – 9 – 15 – 16
17.	3.15	DI – 3 – 7 – 8 – 11 – 14 – 17
18.	2.40	DI – 1 – 2 – 18
19.	3.20	DI – 1 – 2 – 18 – 19
20.	2.40	DI – 1 – 5 – 6 – 20
21.	3.10	DI – 1 – 5 – 6 – 20 – 21
22.	2.95	DI – 1 – 5 – 6 – 20 – 22
23.	3.50	DI – 1 – 5 – 6 – 20 – 21 – 23
24.	2.80	DI – 1 – 5 – 6 – 20 – 24
25.	3.20	DI – 1 – 5 – 6 – 20 – 24 – 25
26.	3.90	DI – 1 – 5 – 6 – 20 – 21 – 23 – 26
27.	3.20	DI – 1 – 5 – 6 – 20 – 24 – 27
28.	3.70	DI – 1 – 5 – 6 – 20 – 24 – 27 – 28
29.	3.90	DI – 1 – 5 – 6 – 20 – 24 – 27 – 30 – 29
30.	3.50	DI – 1 – 5 – 6 – 20 – 24 – 27 – 30
31.	4.20	DI – 1 – 5 – 6 – 20 – 24 – 27 – 28 – 31
32.	3.95	DI – 1 – 5 – 6 – 20 – 24 – 27 – 30 – 32
33.	4.55	DI – 1 – 5 – 6 – 20 – 24 – 27 – 30 – 32 – 33

Table 2 Shortest distances to various transfer points from disposal site 2

Transfer point	Distance Kms	Path
1.	2.5	D2 – 15 – 9 – 7 – 3 – 1
2.	3.2	D2 – 15 – 9 – 7 – 3 – 1 – 2
3.	2.1	D2 – 15 – 9 – 7 – 3
4.	2.5	D2 – 15 – 9 – 7 – 3 – 4
5.	2.9	D2 – 15 – 12 – 13 – 10 – 8 – 5
6.	2.9	D2 – 15 – 12 – 13 – 11 – 6
7.	1.7	D2 – 15 – 9 – 7
8.	2.4	D2 – 15 – 12 – 13 – 10 – 8
9.	1.3	D2 – 15 – 9
10.	1.9	D2 – 15 – 12 – 13 – 10
11.	0.2	D2 – 15 – 12 – 13 – 11
12.	0.9	D2 – 15 – 12
13.	1.5	D2 – 15 – 12 – 13
14.	2.0	D2 – 17 – 14
15.	0.5	D2 – 15
16.	1.0	D2 – 16
17.	1.5	D2 – 17
18.	3.7	D2 – 15 – 12 – 13 – 11 – 24 – 20 – 18
19.	4.3	D2 – 15 – 12 – 13 – 11 – 24 – 20 – 21 – 19
20.	3.3	D2 – 15 – 12 – 13 – 11 – 24 – 20
21.	4.0	D2 – 15 – 12 – 13 – 11 – 24 – 20 – 21
22.	3.7	D2 – 15 – 12 – 13 – 11 – 24 – 25 – 22
23.	4.3	D2 – 15 – 12 – 13 – 11 – 24 – 25 – 22 – 23
24.	2.9	D2 – 15 – 12 – 13 – 11 – 24
25.	3.3	D2 – 15 – 12 – 13 – 11 – 24 – 25
26.	4.2	D2 – 15 – 12 – 13 – 11 – 24 – 27 – 28 – 26
27.	3.3	D2 – 15 – 12 – 13 – 11 – 24 – 27
28.	3.8	D2 – 15 – 12 – 13 – 11 – 24 – 27 - 28
29.	3.1	D2 – 17 – 32 – 29
30.	3.1	D2 – 17 – 32 – 30
31.	3.6	D2 – 17 – 32 – 33 – 31
32.	2.6	D2 – 17 – 32
33.	3.2	D2 – 17 – 32 -- 33

179

Table 3 Shortest distances to various transfer points from disposal site 3

Transfer point	Distance Kms	Path
1.	2.95	D3 – 18 – 2 – 1
2.	2.25	D3 – 18 – 2
3.	3.35	D3 – 18 – 2 – 1 – 3
4.	3.20	D3 – 18 – 2 – 1 – 4
5.	2.75	D3 – 19 – 21 – 20 – 6 – 5
6.	2.25	D3 – 19 – 21 – 20 – 6
7.	3.75	D3 – 18 – 2 – 1 – 3 – 7
8.	3.25	D3 – 19 – 21 – 20 – 6 – 5 – 8
9.	4.15	D3 – 18 – 2 – 1 – 3 – 7 – 9
10.	3.60	D3 – 19 – 21 – 20 – 24 – 11 – 13 – 10
11.	2.75	D3 – 19 – 21 – 20 – 24 – 11
12.	3.90	D3 – 19 – 21 – 20 – 24 – 11 – 13 – 12
13.	3.30	D3 – 19 – 21 – 20 – 24 – 11 – 13
14.	3.25	D3 – 3 – 7 – 8 – 11 - 14
15.	4.40	D3 – 19 – 21 – 20 – 24 – 11 – 15
16.	4.35	D3 – 19 – 21 – 20 – 24 – 11 – 14 -- 17 – 16
17.	3.75	D3 – 19 – 21 – 20 – 24 – 11 – 14 – 17
18.	1.25	D3 – 18
19.	0.55	D3 – 19
20.	1.55	D3 – 19 – 21 – 20
21.	0.85	D3 – 19 – 21
22.	1.40	D3 – 23 – 22
23.	0.80	D3 – 23
24.	1.95	D3 – 19 – 21 – 20 – 24
25.	1.75	D3 – 23 – 22 – 25
26.	1.20	D3 – 23 – 26
27.	2.05	D3 – 23 – 22 – 25 – 27
28.	1.60	D3 – 23 – 26 – 28
29.	2.75	D3 – 23 – 22 – 25 – 27 – 30 – 29
30.	2.35	D3 – 23 – 22 – 25 – 27 – 30
31.	2.10	D3 – 23 – 26 – 28 – 31
32.	2.80	D3 – 23 – 22 – 25 – 27 – 30 – 32
33.	2.55	D3 – 23 – 26 – 28 – 31 - 33

Table 4 Shortest distances to various transfer points from disposal site 4

Transfer point	Distance Kms	Path
1.	4.30	D4 – 31 – 28 – 27 – 24 – 20 – 6 – 2 – 1
2.	3.60	D4 – 31 – 28 – 27 – 24 – 20 – 6 – 2
3.	4.70	D4 – 31 – 28 – 27 – 24 – 20 – 6 – 2 – 1 – 3
4.	4.40	D4 – 31 – 28 – 27 – 24 – 20 – 6 – 5 – 4
5.	3.80	D4 – 31 – 28 – 27 – 24 – 20 – 6 – 5
6.	3.30	D4 – 31 – 28 – 27 – 24 – 20 – 6
7.	4.40	D4 – 31 – 28 – 27 – 24 – 11 – 8 – 7
8.	3.50	D4 – 31 – 28 – 27 – 24 – 11 – 8
9.	4.30	D4 – 32 – 17 – 14 – 13 – 10 – 9
10.	3.70	D4 – 32 – 17 – 14 – 13 – 10
11.	3.00	D4 – 31 – 28 – 27 – 24 – 11
12.	3.40	D4 – 32 – 17 – 16 – 12
13.	3.40	D4 – 32 – 17 – 14 – 13
14.	2.80	D4 – 32 – 17 – 14
15.	3.50	D4 – 32 – 17 – 16 – 15
16.	2.90	D4 – 32 – 17 – 16
17.	2.30	D4 – 32 – 17
18.	3.00	D4 – 31 – 28 – 27 – 24 – 20 – 18
19.	2.80	D4 – 31 – 28 – 26 – 23 – 21 – 19
20.	2.60	D4 – 31 – 28 – 27 – 24 – 20
21.	2.50	D4 – 31 – 28 – 26 – 23 – 21
22.	2.45	D4 – 31 – 28 – 27 – 25 – 22
23.	2.10	D4 – 31 – 28 – 26 – 23
24.	2.20	D4 – 31 – 28 – 27 – 24
25.	2.10	D4 – 31 – 28 – 27 – 25
26.	1.70	D4 – 31 – 28 – 26
27.	1.80	D4 – 31 – 28 – 27
28.	1.30	D4 – 31 – 28
29.	1.70	D4 – 31 – 29
30.	1.60	D4 – 31 – 30
31.	0.80	D4 – 31
32.	1.20	D4 – 32
33.	0.70	D4 – 33

Shortest route algorithm determines the shortest path from one specified node to all other nodes. The algorithm assumes that the distance C_{ij} separating nodes j and I is nonnegative. The algorithm is iterative in nature. If a network consists of n nodes, n-1 iterations will be required to reach the optimum solution. The algorithm consists of two parts. A labeling procedure is used to determine the shortest distance from one specified node to each of the other nodes. The second part of the algorithm is a backtracking procedure used to define the actual shortest distance route to each node.

Labeling Procedure

The label assigned to the node n has the form (d_n, p), where
d_n = Distances from the origin to node n using a specified route
p = The node preceding node on the route.

180

At the algorithm progress, the labels will be classified as either temporary labels or permanent labels. Nodes are initially assigned temporary labels. The temporary labels may change as the algorithm progresses. When the shortest route to a given node is identified, the node is assigned a permanent label.

Step 1 : Consider all nodes connected directly to the origin. Assign temporary labels to each where the distance component equals the distance from the origin. The predecessor part of the label is the origin.

Step 2. Permanently labels the node whose distance from the origin is the minimum. (Break any ties arbitrarily). If all the nodes are all permanently labeled, perform the backtracking procedure.

Step 3. Identify all unlabeled or temporarily labeled nodes that can be connected directly with the latest permanent labeled node to the node of interest. If the node of interest is unlabeled, assign a temporary label. If the node already has a temporary label, update the label only if the newly calculated distance is less than that for the previous label. Updating must also include the predecessor part of the label. Go to step 2.

Backtracking Procedure

To find the shortest route from the origin to any node, identify the predecessor part of its permanent label, and move back ward to that node. Identify the predecessor part of the permanent label for this node, and move backward to that node. Continue in this manner until arriving at the origin. The sequence of nodes traced forms the shortest route from the origin to the node in question.

Allocation Of Transfer Points To Disposal Sites

In order to minimize the transportation cost, the allocation of transfer points to each disposal sites was done by using transportation problem. Transportation cost from each transfer point to disposal site was estimated based on the distance among transfer points assuming that the transportation cost is directly proportional to the distance. The basic objective of transportation problem is to minimize the transportation cost. Transportation problems are linear programs that have special structure, their constraints can be divided into two sets (Supply and Demand). When the problem has this transportation structure, using specialized versions of the simplex method that are more efficient than general simplex method that can solve it. When a problem is formulated as transportation problems, two types of entities are implied, points of supply (Sources) and points of demand (Destinations). Each source i have a supply capacity of S_i and each destination j has specified demand D_j. The cost of supply a unit product from source I to source J is C_{ij}. The problem is to determine how much of demand at each destination J that should be supplied by each sources I to minimize the cost. Defining the variables X_{ij} to be the number of units of product supplied or transported from source I to destination J, the transportation problem have the following form

Minimize $Z = \sum\limits_{i=1}^{M} \sum\limits_{j=1}^{N} C_{ij} X_{ij}$

Subjected to

$\sum\limits_{i=1}^{M} <=$ for every source i = 1 to M

$\sum\limits_{j=1}^{N} <=$ for every source j = 1 to N

Identification Of Shortest Of The Optimum Route

During the emergencies like outbreak of plague and other type of communicable disease and strikes of municipal employees, a shortest route for the entire network is needed for effective collection and disposal. Minimal spanning tree, which is optimum tree, was developed which covers the entire network. This path is useful to collect the solid waste in the entire area in order to control the spreading of diseases during epidemics.

A classic network problem requires selecting area for a network, which provide a route between all parts of nodes and minimize the total distance (cost or time) to do so. Another way of viewing this problem is that it involves selecting then network that forms a tree that spans (connects) the nodes of interest. This type of problem has wide applicability, especially to transportation and telecommunication networks, planning of communicity in the process of drawing up blue prints for sewer service, in this case the planners knows the locations of the users (nodes) and wishes to provide sewer service to all users at a minimum cost of construction. The minimum spanning tree problem requires identifying the network of sewers that reaches all users and minimizes the total amount of sewer pipe.

Minimal Spanning Algorithm

Step 1. : Select any node and identify the least distant (cost or time) node.Connect the two.

(Break any ties arbitrarily).

Step 2 : Identify the unconnected node that is least distant from any connected node, and connects the two.

(Break any ties arbitrarily).

Step 3 : Repeat step 2 until all nodes have been connected.

RESULTS AND DISCUSSION

Fig 1 shows the network of transfer points and disposal sites. Table 1 to 4 show the optimum paths of various transfer points to disposal sites 1 to 4 respectively. If these paths are operated for collection and disposal of solid waste, the operating cost is optimum. The above tables also show the shortest distance from transfer points to disposal sites. The optimum paths of each collection points were connected through other transfer points to disposal sites. The table also reveals that there are 12 longest in shortest paths that cover larger collection/transfer points in the entire town. These paths can be operated during unavoidable circumstances such as absentees of large number of labor and repairing of the vehicles etc., in order to maintain the hygienic conditions in the town.

Allocation Of Transfer Points To Disposal Sites

Various transfer points were allocated to different disposal sites based on minimization of transportation cost, which are shown in Table 5. The table shows the amount of solid waste to be disposed per day for each disposal site. These data is useful to estimate the life period of disposal site which is helpful in developing the strategy to identification of new disposal sites before the expiry of the existing disposal sites.

The Figure 2 shows the minimum spanning tree, which is optimum path for the entire network. Optimum paths of transfer points to disposal site can operate during normal conditions to minimize the transportation cost only whereas minimal spanning tree (a single tree network) can operate during the epidemics and emergencies such as strikes of municipal employees and bandhs in order to maintain the hygienic conditions in the short time and also prevent breakout of diseases. This network is also helpful in deciding the scheduling of vehicles and total distance traveled per day by each vehicle.

Minimal spanning tree for disposal site 1 is shown in Fig 2.

CONCLUSIONS

Based on the above models, the following conclusions have been drawn.

1. Optimum paths developed in network model are useful in reducing the transportation cost and also for effective solid waste management.

2. Transportation models in integrated solid waste management are a strategy for effective disposal of solid waste based on the cost of transportation.

3. Minimal spanning tree model is a vital tool in solid waste management during epidemics.

4. All the above models are essential instruments in developing the strategies for integrated solid waste management.

Table 5 Distance to various disposal sites from various transfer points

Transfer point	D1	D2	D3	D4	Estimated waste per day (qtis)
1.	0.70	2.50	0.70	0.70	057.60
2.	1.40	3.20	1.40	1.40	035.20
3.	0.45	2.10	0.45	0.45	046.40
4.	0.80	2.50	0.80	0.80	038.40
5.	1.20	2.90	1.20	1.20	027.20
6.	1.70	2.90	1.70	1.70	059.20
7.	0.85	1.70	0.85	0.85	041.60
8.	1.65	2.40	1.65	1.65	038.40
9.	1.25	1.30	1.25	1.25	080.00
10.	1.85	1.90	1.85	1.85	042.60
11.	2.15	0.20	2.15	2.15	072.40
12.	2.45	0.90	2.45	2.45	028.90
13.	2.15	1.50	2.15	2.15	048.00
14.	2.65	2.00	2.65	2.65	037.20
15.	2.10	0.50	2.10	2.10	049.20
16.	2.70	1.00	2.70	2.70	027.30
17.	3.15	1.50	3.15	3.15	120.40
18.	2.40	3.70	2.40	2.40	060.20
19.	3.20	4.30	3.20	3.20	027.60
20.	2.40	3.30	2.40	2.40	099.70
21.	3.10	4.00	3.10	3.10	013.00
22.	2.95	3.70	2.95	2.95	027.20
23.	3.50	4.30	3.50	3.50	014.30
24.	2.80	2.90	2.80	2.80	070.40
25.	3.20	3.30	3.20	3.20	035.40
26.	3.90	4.20	3.90	3.90	140.20
27.	3.20	3.30	3.20	3.20	070.40
28.	3.70	3.80	3.70	3.70	030.70
29.	3.90	3.10	3.90	3.90	120.60
30.	3.50	3.10	3.50	3.50	038.00
31.	4.20	3.60	4.20	4.20	124.60
32.	3.95	2.60	3.95	3.95	146.80
33.	4.55	3.20	4.55	4.55	040.00

Table 6 Optimal allocation of collection centres to each disposal sites and quantity allocation from various transfer points to disposal sites.

Transfer point	D1	D2	D3	D4
1.	058			
2.	035			
3.	046			
4.	038			
5.	027			
6.	059			
7.	042			
8.	038			
9.	080			
10.	043			
11.		072		
12.		029		
13.		048		
14.		037		
15.		050		
16.		027		
17.		120		
18.			060	
19.			028	
20.			100	
21.			130	
22.			027	
23.			015	
24.			070	
25.			035	
26.			140	
27.				070
28.				031
29.				121
30.				038
31.				125
32.				147
33.				050

REFERENCES

1. Budnick, S.F.,(1996) " Principles of operations research for management", Richard, D.Irwin Inc.Illionis

2. Georage, Technobanoglous, Hillary, T. Rolt, E.,(1997) " Solid Waste Engineering Principles and Management Issues",McGraw Hill, New York.

34. 3.Radha Krishna Murthy, B and N. Kumaraswamy,(1998) " A Rational approach to Integrated Solid Waste Management-Transportation problem with modified cost coefficients", Environmental pollution control journal, Vol. 2, No.1, PP 55-57, New Delhi, India.

3. Ravindran, A et al, (1987) " Operations research principles and practice", John Wiley and Sons, New York.

4. Sasisubramanian, R et al, (1997), " Computers aided optimization of transportation models for solid waste management systems using linear programming", J. of the Institution of Public Health Engineers, No.1, PP 17-23, India.

Land Application of Municipal Solid Wastes—Need for Regulations

K. Jeevan Rao

Department of Soil Science & Agricultural Chemistry, Agricultural College, Acharya N.G. Ranga Agricultural University, Mahanandi, Nandyal (R.S.) 518 502 (A.P.), India

INTRODUCTION

Indian urban population is increasing due to various reasons, such as the concentration of industries and better employment opportunities in the urban areas. As per the census data the number of cities having a population of more than 1,00,000 have increased from 216 to 296 in the last decade (1981-91) with a decadal growth rate of 46.87%. Consequently, municipal agencies face difficulties in providing various public services such as education, water supply and public sanitation including collection and disposal of urban solid wastes (USW) to the desired level. Of these services municipal solid waste (MSW) management gets the lowest priority, mainly because disruptions and deficiencies in it do not directly and immediately affect public life and cause public reaction. About 255 million people are estimated to reside in Indian urban areas and about Rs. 19200 million (roughly US $576 million) are spent every year on municipal solid waste management. Inspite of such a heavy expenditure the present level of service in many urban areas is low and may be a potential threat to the public health in particular and environment in general[1].

In general, " solid waste" is the term used internationally to describe non liquid waste materials arising from domestic, trade commercial industrial, agriculture, mining activities and from the public services. Their disposal creates lot of environmental and health problems in the cities. So proper management and safe disposal of solid waste in cities in needed to create a clean and better living condition for the urban people.

Municipal bodies have to manage the solid wastes arising from residential, commercial and institutional activities along with waste from street sweepings. Normally the municipal bodies handle all the waste, deposited in the community bins located at different places in the city. The municipal solid waste is transported to processing/disposal facilities. Majority of the municipalities does not weigh their solid waste vehicles but estimate the quantities on the basis of the number of trips made by the vehicles. Since the density of waste is considerably less as compared to the material for which these vehicles are designed to carry, such data on quantity cannot be relied upon.

MUNICIPAL SOLID WASTE GENERATION

The per capita waste quantity tends to increase with the passage of time due to various factors like increased commercial activities, standard of living, etc. (Table 1). Increase in per capita waste quantity is also known to occur at a slightly lesser rate than the increase in GDP/capita. This increase is estimated to occur in India at a rate of 1-1.33% per year. In majority of Indian cities, the per capita production of refuse ranges from 0.15 to 0.35 kg day^{-1}. The overall average garbage generation per capita is estimated at 0.33 kg day^{-1}. The data on per capita production of refuse quantity from other countries is found to vary from 0.5 to 1.5 kg day^{-1}. The maximum production of 2.5 kg day^{-1} is reported from USA[2]. Normally, the per capita value does not change appreciably in locations with comparable climatic conditions and living and food habits. But the change in standard of living and density of population per unit area has a perceivable influence on the quantity and quality of garbage generated from urban areas[3].

Table 1: Solid waste generated in India's major cities

Cities	Solid waste generated (tonnes)
Mumbai	6050
Delhi	4000
Calcutta	3500
Bangalore	2000
Chennai	2000 (USW); 500 (debris)
Lucknow	1500
Hyderabad and Secunderabad	1300
Ahmedabad	1280
Surat	1000

Source: The Hindu, Survey of the Environment.

Such a large quantity of waste requires proper collection, transportation and disposal. For this, knowledge about the composition of wastes and how they should be collected and disposed off is needed [4,5].

The increasing cost of chemical fertilizers and their limited availability, coupled with the concern for efficient utilization of energy and natural resources, have generated an interest in alternative uses and utilization of urban and industrial wastes. Use of organic fertilizer instead of chemical fertilizer can result in a two-thirds energy saving[6].

The manurial value of composts and its use as a soil conditioner has been subjected to detailed study by several investigators because of its advantages over inorganic fertilizers and its long term beneficial effects [4,5].

A comparison of wastes from Indian city with that of European city is given in table 2. As personal income rises, paper wastes increase (4% Indian city and 27% in North European city): kitchen wastes decline; metals and glass increase, total weight generated rises and the density of wastes declines.

Table 2: Waste from Indian city and North European city – a comparison[7, 8]

Item	Indian	N. European
Paper	4%	27%
Vegetable/putrescible matter	75%	30%
Dust etc. under 10mm	12%	16%
Metals	0.4%	7%
Glass	0.4%	11%
Textiles	3%	3%
Plastics	0.7%	3%
Others, stores, ceramics etc.	7%	3%
Weight/person/day	414g	845g
Weight/dwelling/day	2.5kg (6 persons)	2.5kg (3)
Density kg/cu.m	570	132

ENVIRONMENTAL POLLUTION DUE TO MUNICIPAL SOLID WASTES

Presently most of the waste is disposed in low-lying areas without taking proper precautions. This is observed to result in breeding of rodents, flies and other domestic pests and also tends to pollute the surface as well as ground-waters. Hence, it is necessary to adopt sanitary land filling techniques so that such problems are avoided[4]. Improper and unscientific disposal of urban solid wastes may pollute air, water and soil in several ways[9].

Atmospheric Pollution

One of the critical public health hazard from a sanitary land fill is due to the gasses emanating from the decomposition of the organic matter under partially anaerobic conditions of the gases generated (CO_2, CO, CH_2, H_2 and H_2S), CH_4 and H_2 can be explosive. CO_2 can cause acidic reactions when dissolved in water and can be the cause for corrosive problems. Hydrogen sulphide is a known corrosive, malodorus and if produced in high quantity can be irritating and toxic to man and animals. Cases are on record in which methane has escaped from landfills and caused severe damage. Horizontal movement and distribution of the gases in dangerous concentrations beyond the confines of the fill upto 300 meter away has been reported[9].

Water Pollution

The land disposal of solid wastes creates an important source of ground-water pollution . Most landfills throughout the world are simply refuse dumps. Only a few if at all can be regarded as sanitary landfills, indicating that they were designed and constructed according to engineering specifications. Leachate from a landfill can pollute ground-water if water moves through the fill material. The possible sources of water include precipitation, surface water infiltration, percolating water from adjacent land and ground-water in contact with the fill.

Entry of pollutants into shallow aquifers occurs by percolation from ground surface, through wells, from surface waters and by saline water intrusion. The extent of pollution in groundwater from a point source decreases as pollutants move away from the source until a harmless or very low concentration level is reached. Because each constituent of a pollution source may have a different attenuation rate. The distance to which pollutants travel will vary with each quality component.

The important aspect of ground-water pollution is the fact that it may persist underground for years, decades or even centuries. This is in marked contrast to surface water pollution. Reclaiming polluted groundwater is therefore much more difficult, time consuming and expensive than reclaiming polluted surface water. Under ground pollution control is achieved primarily by regulating the pollution source and secondarily by physically entrapping and when feasible removing the polluted water from the underground.

Soil Pollution

Soil gets polluted in a number of ways which are grouped together and broadly described as mechanical or gravitational. A major difficulty in predicting potential hazards associated with land disposal of organic solid waste is the inherent variability in the composition of wastes. The unleached composts may be high in available boron. Boron caused severe toxicity when shredded refuse was incorporated. Thus potential phytotoxicity exists for B, Zn, Cu and Ni. Not only is refuse heterogeneous in total content of an element, but the chemical forms of the elements also vary[4,5].

RATIONALE FOR LAND APPLICATION

Land application refers to addition of wastes to soil for the purposes of improving crop growth, either through addition of nutrients or improvement of soil physical or chemical properties, either in the short or longer term. Materials used for land application should be studied and documented to be safe and effective for use in agriculture, with no overriding constraint to their use. This is distinctly different from land treatment, which suggests that materials are being applied to land primarily in order to dispose of (treat) them.

The key to land application is the idea of safe, beneficial reuse, which is the major criteria used by many environmental agencies to approve materials for land application. The environmental issue has been partially addressed by the adoption of the 40 CFR 503 ruling in USA which specifies acceptable concentrations and annual and cumulative loadings of nine contaminant metals applied to crop land in municipal biosolids[10]. While this ruling has been applied no other types of by prdoucts, it is not clear that the specifications apply to inorganic types of wastes.   Some further study is necessary to demonstrate the safety of these loadings for other types of wastes. The European Union (EU) has adopted much more stringent standards than those of United States Environmental Protection Agency (USEPA), and a considerable debate is in progress on this topic.

Environmental Aspects of Wastes Applications

For many in the public and regulatory sectors, the primary question regarding land application of wastes centers on public safety and human health. Without a clear demonstration that levels of contaminants added in a given by product applications will have acceptably minimal effects on human and ecological health, such application should not be made.

Unfortunately, even within the scientific community, no universal consensus exists as to what materials and rates are safe for use in agriculture. Fundamental questions regarding plant uptake of elements, the impacts of contaminants on microbial communities in soils, and the solubility and rate of movement of contaminants in soils are still under active investigation. A significant body of research was recently summarized by the USEPA in formulating rules for biosolid use, and was used to set limits for safe use of that material. However, European countries have (in many cases) more stringent rules[11]. For many types of potentially land-applied materials there are no available environmental guidelines.

In the United States, state environmental agencies are largely responsible for implementation of federal regulations regarding land application. Where no such regulations exist, state agencies develop their own guidelines or consider materials on a case by case basis. State department of agriculture are involved in registering byproducts as fertilizers or lime substitutes, and sometimes do not effectively interact with state environmental agencies to ensure safe use of these materials. There are examples of quite hazardous materials being land applied where state agencies have failed to coordinate[12].

Given the fact that some industries may view land application of marginally useful or environmentally questionable materials as a low cost alternative to land filling, it behoves land owners to ask by product producers and state agencies about materials being applied, including information on potential benefits, analysis of contaminants, and rates of application [14].

United States Government Regulations Concerning Land Application of MSW- An Overview

The USEPA regulates the handling, disposal and land application of wastes under Congressional authority, and through rule making associated with legislative acts. Nearly all such regulation deals with industrial and municipal by products, with agricultural by products being expressly exempt (This exemption may change as manure applications are becoming increasingly linked with water quality degradation). State departments of agriculture and/or environmental affairs implement federal regulations, and may modify them in some cases. The European Union (EU) has also adopted resolutions, implemented by member countries, regulating handling and use of wastes[14].

Municipal solid wastes, pulp and paper by products (except highly alkaline residues or high dioxin sludges), and many other more less inert by products are classified nonhazardous. Biosolids are exempt, being covered under special rules of 40CFR 503. Phosphogypsum has been regulated by USEPA due to its radionuclide content in a ruling that is still under litigation. Animal manures and most agricultural byproducts are currently exempt and unrestricted.

Biosolid use on land is regulated separately under authority of the Clean Water Act in 40CFR 503, revised in 1993 in a careful review of available data on hazards associated with land application of this material [10]. Processing guidelines were established to reduce disease vectors in sludge by specifying treatments such as alkali treatment, sterilization, or composting. Sludges treated in this manner contain minimum numbers of pathogenic microbes, and are considered Class A. Class A sludges can be land applied without permit or restriction, although records must be kept on amounts applied to individual fields. Class B sludges, not treated to decrease disease vectors, must be applied by permit and more carefully managed and monitored. Management practices such as injection or incorporation of sludge into soil are also specified.

Maximum safe metal levels were a point of much debate in codifying 40 CFR 503. The approach used was to assess risk to human populations based on pathways such as direct soil injection (children), contamination of food (vegetable and animal) by plant uptake, and ground water pollution by leaching

[13]. The pathway with the greatest risk (in terms of cancer incidence or toxic reaction risk of an assumed highly exposed individual) for a given metal was chosen to limit

Table 3: USEPA and EU limits for metals in biosolids and loading rates for land application of biosolids annual and lifetime)[14]

Element	USEPA			EU		
	Metal[1] concentrations	Annual loading	Cumulative loading	Metal[4] concentrations	Annual loading	Cumulative loading
	------ mg kg^{-1}------	---kgha^{-1} -------		----- mg kg^{-1}------	---kgha^{-1}	--------
As	41-75	2	41	----	(0.7)[5]	(100)[5]
Cd	39-85	1.9	39 (21)[6]	20-40	0.15	2-6
Cu	1500-4300	7.5	1500	1000-1750	12	100-200
Cr	[3000][2]	150	3000	----	(15)	(800)
Pb	300-840	15	300	750-1200	15	100-600
Hg	17-57	0.85	17	16-25	0.1	2-3
Mo	[18-75][3]	[0.9][3]	[18][3]	----	(0.2)	(8)
Ni	[420][2]	21	420	300-400	3	60-150
Se	[100][2]	5	100	----	(0.15)	(6)
Zn	2800-7500	140	2800	2500-4000	30	300-600

From USEPA 40 CFR 503; Smith (1996).
1. Lower value is for high quality unrestricted use; higher value is ceiling for application.
2. Limits for these metals are tentative
3. Limits on Mo are being removed by USEPA
4. Lower value for soils at pH < 6, higher value for pH>7
5. Values in parenthesis are UK limits; no EU values specified for these metals
6. USEPA is considering reducing Cd loading to 21kg ha^{-1} at USDA's request.

Loading of that metal during land application. The results were maximum metal concentrations in sludges for high quality sludges that can be used without restriction, and ceiling levels for other sludges above which land application is not permitted; annual and cumulative (lifetime) loading rates (kg ha^{-1}) are also specified for a given site (Table 3). USEPA regulations are formulated based on a maximum 50Mg ha^{-1} application rate for high quality sludge, and a 1000 Mg ha^{-1} lifetime application[14]

The 40 CFR 503 limits are considerably higher than previous regulations, and are, in general greater by a factor of 5-15 for most metals than EU soil-loading rates, which were arrived at in a more conservative estimation of risk (EU Directive 86/278/EEC; Smith 1996. Interestingly, metal concentrations in sludges do not differ significantly between the two, but maximum rate of application in Mg ha^{-1} allowed (computed as [loading in kg ha^{-1}] / [concentration in mg kg^{-1}] x 1000) differ significantly. Using the lower metal concentration limits for EU sludges and corresponding annual metal loading rates, only about 5-10 Mg ha^{-1}yr^{-1} of sludge can be applied, compared to 50Mg ha^{-1}yr^{-1} for the USEPA limits. Depending on N and P needs of the crop on a given site and the available nutrients in the sludge, more than 10Mg ha^{-1} may be needed to meet annual nutrient needs, therefore requiring supplemental inorganic fertilizer be added. Rates based on N or P required would seldom exceed the 50Mg ha^{-1} limit for metals based on USEPA values.

Metal limits in 40 CFR 503 are often applied to the utilization of other by products for land application. This is probably justified for other organic by products such as MSW compost or pulp and paper sludges, which have a similar matrix and impact on soil chemistry. However, for inorganic materials such as gypsum, fly ash, and by product dusts with a largely inorganic matrix, this could lead to increases in leachable or plant-available metals above safe levels, due to the fact that metals in organic matrix are generally less soluble[14].

CONCLUSIONS

Land application of Municipal Solid Wastes improves Crop growth, through the addition of nutrients or improvement of soil physical or chemical properties. In India. no consensus exists as to what materials and rates are safe for use in agriculture. Fundamental questions regarding plant uptake of elements, the impacts of contaminants on microbial communities in soils and the solubility and rate of movement of contaminants in soils are still under active investigations. However United States and European countries have set more stringent rules. In India for many types of potentially land applied materials there are no available environmental guidelines. The approach for developing guidelines should be based on assess risk to human populations based on pathways such as direct injection, contamination of food by plant uptake, and ground water pollution by leaching. The pathway with greatest risk for a given metal should be chosen as a limit to load that metal during land application. The maximum metal concentrations in MSW that can be sued without restriction and ceiling levels for other wastes above which land application is not permitted, annual and cumulative loading rates are to be developed for a given site, under Indian conditions, Rates based on Nitrogen and Phosphorus required would seldom exceed the limit for metals based on USEPA and EU values. Since there is no legal requirement in India for disposal of MSW on agricultural land it is reasonably understandable that the Indian agencies may adopt the USEPA and EU guidelines as an initial attempt to regulate metal accumulation in soil, until local research is able to modify these standards. In India there is a need to regulate the land applications of MSW through rule making associated with legislative acts.

REFERENCES

1.   Shekdar, A.V. (1999). Municipal Solid Waste Management. The Indian perspective. J. Indian Association for Environ. Management. 26:100-108.

2.   Bhide, A.D. (1984). Urban Solid Waste Management – An Assessment. In National Environmental Engineering Research Institute, Silver Jubilee Commoration, Nehru Marg, Nagpur. 219-225.

3.   NEERI. National Environmental Engineering Research Institute. (1983). Solid Waste Management – A course manual, Nehru Marg, Nagpur.

4.   Jeevan Rao, K. (1994). Environmental Pollution from Urban Solid Wastes – Challenges and Strategies. National Symposium an Environmental Pollution, A.V. College, Hyderabad.

5.   Jeeva Rao, K (1998). Urban Solid Waste Management with reference to Hyderabad. In Environmental impact assessment and Management (ed) B.B. Hosetti and A. Kumar, daya publications New Delhi. 148-171.

6.   Golueke, C.G. (1977). Biological reclamation of Solid Wastes. Rodale Press Emmans. 1-3.

7.   Trivedi, P.R and Gurdeep .R. (1992). Disposal of Solid Waste. In Solid Waste Pollution. Akashdeep Publishing House, New Delhi. 116-186.

8.   Trivedi, P.R and Gurdeep. R. (1992). Management of Solid Wastes. In Management of Pollution Control. Akashdeep Publishing House, New Delhi.

9.   Liptak, B.G. (1974). Land Pollution. In Environmental Engineers Hand Book Vol. 3. USA.

10.   USEPA. United States Environmental Protection Agency. (1993). Part 503. Standards for the use of disposal of sewage sludge. Fed .Reg. 58:9387-9404

11.   Smith, S.R .(1996). Agricultural recycling of sewage sludge and the environment. CAB International, Walling Ford, U.K.

12.   Wilson, D. (1997). Fear in the fields, part 1 and 2. Seattle times, July 3.

13   Ryan, J.A. (1994). Utilization of risk assessment in development of limits for land application of municipal sewage sludge. In clapp, C.E., Largon, W.E., and Dowdy, R.H.(ed). Sewage sludge: Land utilization and the environment. Soil Science society of America, Madison. WI. 55-65.

14.   Miller, D.M. and Miller, W.P. (2000). Land Application of wastes - In Hand Book of Soil Science.(ed). M.E. Sumner . CRC. Press, U.S.A. 217-245.

Utilisation of Fly Ash, The Solid Waste Generated from Thermal Power Plant

Mitali Sarkar, Mahadeb Das, Sucharita Manna and Pradip Acharya

Department of Chemistry, University of Kalyani, Kalyani 741 235, West Bengal, India

1. INTRODUCTION :

More than 70% of total electric power produced in India is obtained from thermal power stations, which use mainly coal as fuel (1). Fly ash is one of the major culprits to degrade the environment, which constitutes about 70% of the total amount of residue generated in coal fire plant. More than 150 tones of fly ash are produced annually worldwide (2). Such huge amount of fly ash is being disposed of either by dumping in open beds, by land filling or mixing it with water and discharging the ash laden water in impounding pond, lagoons, rivers or sea. Moreover, the recent research is directed towards use of fly ash laden soil for crop production (3). Each of the above practice is connected with a distinct environmental problem and merits careful consideration. In an aim to reduce the degradation of the environment, fly ash has been utilised as an effective adsorbent to remove some pollutants like Phenol, surfactant (Sodium Lauryl Sulphate) and dye (Crystal Violet).

2. POSSIBLE SOURCES AND ENVIRONMENTAL IMPACTS OF THE STUDIED POLLUTANTS :

Phenol : Phenol an its derivatives are present in effluents from a number of industries viz., synthetic rubber, oil refineries, steel plant, plastic, coke over plant etc. Phenols impart taste and odour to water and are toxic to fish and other aquatic life even in very low concentration.

Sodium Lauryl Sulphate : Surfactants (Surface active agents), particularly the anionics are used in each industry either as such or as an active component of the synthetic detergents. Not only do they cause contamination problem to the aqueous environment they also affect other treatment processes due to foam formation.

Methyl Violet : Dyes are used in various industries viz., textile, paint, pigment, leather etc. Methyl Violet is the important member of triphenyl methane dye. Color affects aesthetic nature of water and inhibits sunlight penetration into the stream and therefore reduces photosynthetic action.

3. METHODOLOGY :

The Figure 1 is presented to give an idea of the method carried out in the experimental project in order to set up positive potential utilization of fly ash from thermal power plant.

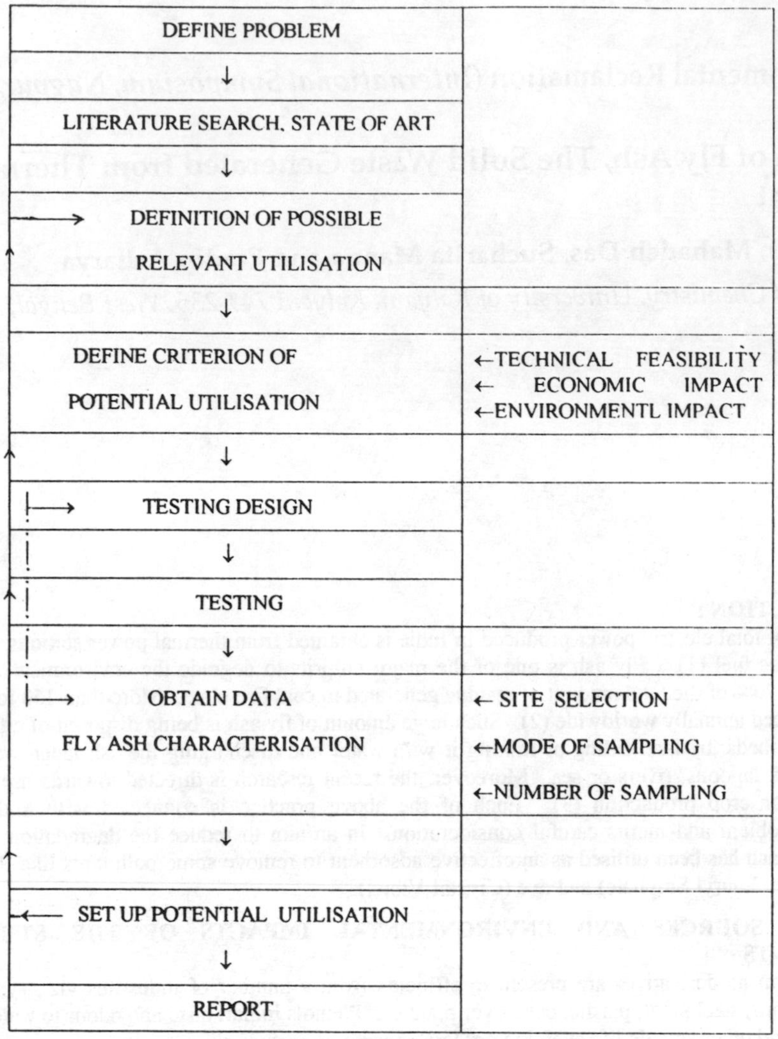

DEFINE PROBLEM	
↓	
LITERATURE SEARCH, STATE OF ART	
↓	
→ DEFINITION OF POSSIBLE RELEVANT UTILISATION	
↓	
DEFINE CRITERION OF POTENTIAL UTILISATION	←TECHNICAL FEASIBILITY ← ECONOMIC IMPACT ←ENVIRONMENTL IMPACT
↓	
→ TESTING DESIGN	
↓	
TESTING	
↓	
→ OBTAIN DATA FLY ASH CHARACTERISATION	← SITE SELECTION ←MODE OF SAMPLING ←NUMBER OF SAMPLING
↓	
← SET UP POTENTIAL UTILISATION	
↓	
REPORT	

Figure 1 : A general flow chart of the proposed methodoloogy.

4. EXPERIMENTAL :

All the reagents were of AnalR grade. Standard solution of adsorbate (500 mg/l) was prepared by dissolving it in minimum volume of acetone and diluting the solution with distiled water to prepare the test solution. Fly ash was collected from Bandel Thermal Power station, Triveni, West Bengal, dried, Sieved and used without any pretreatment.

Procedure :

Batch study was performed by shaking 1.0 g of fly ash with 50 ml of aqueous solute solution of appropiate concentration, temperature and pH in glass bottles in a shaking incubato r. At the end of the predetermined interval, the absorbent was removed and solute concentration was measured spectrophotometrically at the respective wavelength.

5. RESULTS AND DISCUSSION :

The chemical analysis of fly ash (4) includes loss on ignition at 800° C, silica, aluminium oxide, ferric oxide, calcium oxide and magnesium oxide. Physical properties such as specific gravity and surface area are also determined (Table)

Table 1 : Physico-chemical characteristics of fly ash

Constituent	Weight percetage
SiO_2	72.90
Fe_2O_3	03.70
Al_2O_3	14.51
CaO	02.00
MgO	00.80
Loss on ignition	08.99
Density	03.24 g/cm^3
Surface area	12.97 cm^2
pH zpc	04.80

Adsorption occurs due to the accumulation of solute from aqueous solution to the surface of adsorbent. The time at which equilibrium is attained is known as equilibrium time and the concentration at this time is the equilibrium concentration. The adsorption equilibrium and the adsorption time depend on several factors such as nature of solute and sorbent, solution pH and temperature. The optimum operational conditions for different pollutants are given in Table2.

Table 2 : Optimum adsorption condition for the studied pollutants.

Pollutant	Equilibrium Time (min)	pH	Dose of fly ash (mg/l)
Phenol	300	7.9	20
Sod.Lau.Sulph	120	7.0	10
Crystal Violet	135	6.5	10

Moreover, it is observed that low solute concentration, smaller particle size and higher temperature favours adsorption and in each case permissible limit of the solute (5) is attained.

Equilibrium adsorption data was fitted to linearised Langmuir adsorption isotherm model, expressed as follows:

$$C/q = 1/Qb + C/Q$$

Where, q indicates the amount of solute adsorbed at equilibrium, C the equilibrium concentration of the adsorbate, Q and b are isotherm constants.

The isotherm constants were obtained from the slope and intercept of linearised Langmuir adsorption isotherm plot (Table 3)
In order to examine the feasibility as well as its nature of the process, the thermodynamic parameters were evaluated from the following relations,

$$\Delta G = -RT\ln K,$$
$$\Delta H = \Delta G + T\Delta S,$$
$$\ln K = -\Delta H/RT + \Delta S/R \text{ and}$$
$$K = C/C'$$

Where, ΔG, ΔH and ΔS are the change of free energy, enthalpy and entrophy respectively, K is the equibrium constant, C' is the concentration of solute on fly ash at equilibrium.

Table 3 : Isotherm constants and thermodynamic parmeters.

Pollutant	Q(mg/g)	b (mg/l)	-ΔG (kJ/mole)	ΔH (kJ/mole)	ΔS (J/mole)
Phenol	15.87	0.0139	12.83	19.85	107.97
Sod.Lau.Sulph	25.03	0.1979	6.94	7.22	45.98
Crystal Violet	80.31	0.2120	5.01	33.27	125.66

The negative value of free energy (Table 3) indicates that the process is spontaneous and favorable. The process is endothermic (positive ΔH) in nature and is aided by increased randomness (Positive ΔS).

193

6. CONCLUSION :

The process is found to be cheap and simple. It can therefore be used to remove the pollutants effectively from wastewater.

7. REFERENCES :

1. Sachder, A.K., (1992) Beneficiation of power grade cals, its relevance to future coal use in India. Energy Management, 2, 15.

2. Piekos, R., Paslawska, S., (1992). Leaching characteristics of fluoride from coal fly ash. Fluoride, 31, 188.

3. Iverma, K.C., Scenario of utilisation of fly ash in India. (1994). Irrigation and power, 51, 69.

4. WHO, Guidelines for Drinking Water Quality, Vol. 1, (1984). World Health organisation, Geneva p.85.

5. APHA, AWWA and WPCF, (1990). Standard Methods for Examinations of Water and Waste Water, APHA, Washington, 19th Edn

Geoenvironmental Reclamation (*International Symposium, Nagpur, India*)

Environment Oriented Development of Mining Areas in Himalaya

A.K. Soni[1] and A.K. Dube[2]

[1]*Scientist, Central Mining Research Institute, Regional Centre, Roorkee, U.P., India*
[2]*Ex. Director, Central Mining Research Institute, Dhanbad, Bihar, India*

ABSTRACT

"Environment Oriented Development" is the key of success of overall mining operation, which commence from exploration and ends with reclamation. This paper summarises the various aspects of environment and ecology with respect to mining of minerals in eco-sensitive Himalayan region and finally arrives at the conclusion that in tough and rugged terrain conditions extraction of minerals should be limited, selective and controlled. To keep the environmental degradation under control -

❑ No single element of mining or environment alone can itself minimise the negative impacts of development intervention i.e. mining.
❑ Short cuts should be avoided as far as possible.
❑ Integrated strategy on watershed basis should be the approach for planning.
❑ Best Practise Mining (BMP) should be practised as far as possible.

1.0   INTRODUCTION

Environment is the basic dimension of development in eco-sensitive regions, especially in the Himalayas. In hilly areas it is imperative that development must take into account the fact that environment in the hills has a great bearing on environment in the plains. Himalayan eco-systems are fragile too owing to its less resistance and resilience. Soil conservation, land use, water resources management, energy resources from water, forest or mineral are all interdependent. Each one of them are of vital importance in integrated planning in Himalaya e.g. soil conservation without forestry and land use planning will not help in retention of soil moisture nor help in solving the basic needs of water, fuel and fodder. Hence, in sensitive areas the mining and environment should go together and in specific term development should be **"Environment Oriented Development"**. This basic principle is the key of success of overall mining operation, which commence from exploration and ends with reclamation.

The basic objectives of environment oriented development which is nothing but sustainable development of mineral bearing areas in Himalaya are -

• protection of environment and ecology
• protection of biological diversity and maintain essential ecological processes and life support systems.
• enhancement of overall standard of living by following a path of economic development that safeguards the welfare of future generations

Here, we would like to bring forward "Best Mining Practise (BMP)" and "Integrated Approach (IS)" concept so as to maintain maximum performance within the limit of an achievable and acceptable environmental protection level in fragile /sensitive areas. The inherent mistake of treating hill development as an extension of plain development should not be repeated again and again.

2.0   WHAT ARE THE ENVIRONMENTAL ISSUES OF CONCERNS?

Given below are the checklist of points which planners should use to analyse and help choose the optimal mine design/layout in mountains.

- Ambient air quality and air quality at work sites ;
- Water quality Vs hydrological regime of site and areas nearby ;
- Noise, vibrations and fly rocks;
- Soil and water conservation;
- Land degradation including landslides and impact on aesthetic environment (visual impacts);
- Prevention and control of acid mine drainage (if, any);
- Flora and fauna;
- Subsidence (if, any);
- Waste handling and management including tailings;
- Socio-economic issues;
- Reclamation and rehabilitation;
- Wild life (endangered species), archaeology and heritage protection.

3.0   DESCRIPTION OF AFFECTED ENVIRONMENT

The goal of the mining is to conduct mineral winning and mineral benefaction activities in an environmentally sound and safe manner. Possible impacts of mining operation on the existing environment and community may or may not be significant. An essential starting point to describe the affected environment is to have a fundamental knowledge of local environmental conditions, knowledge of accepted best environmental management practices and the possible consequences to the environment resulting from the mining activities. The description of affected environment commonly refereed as *Environmental Impact Assessment (EIA)* or preparation of *Environmental Impact Statement (EIS)* shall be done under two major heads, namely –

(a) Environmental Performance Assessment by monitoring various parameters in the field
(b) Analysis of data

Such description contains information about environmental setting of study area, which forms the basis to know whether the mining process is causing any effect on physical, biological and social system. It also informs about the status of environment before and after mining. Big scale mining projects require detail EIA/EMP studies and small-scale mining projects usually require review of environmental factors. Though it is difficult to concentrate the complete procedure for description of affected environment in few pages however, briefly this can be accomplished in the following manner.

Environmental Performance Assessment: Environmental performance assessments are based on detailed environmental monitoring and are the measures of the success of strategies implemented. The diversity of climate, ecosystem, land uses and topography greatly influences the design of environmental monitoring programmes and thereby performance assessment. In environmentally sensitive regions these should be based on actual monitoring program data. It also enables to review and improve the management plans or measures. The environmental performance needs to be monitored periodically against the objectives set out in the Environmental Management Plan (EMP)

- to detect short term and long term trends.
- to find out the causes of environmental degradation.
- to improve environmental practices and procedures.
- to demonstrate community/government that the operation complies with environmental quality/standards.
- to help/assist in regional or micro level planning.
- to know the pre-mining and post mining status.

The environment performance assessment is though a very typical and broad area. For Best Mining Practises (BMP) it encompasses aspects related with air, water, land, biology, noise, vibration and eco- system protection.

Air Monitoring: Ambient air monitoring and air monitoring at the work sites are two different areas in respect of air environment assessment. Dust and fumes (blasting) are two by-products, which deteriorates the overall air environment in any mineral extraction project. Air pollution and its monitoring at the project site, where actual operation is going on needs to address the following requirements -

- location of monitoring site
- frequency of observation
- equipment to be used for monitoring and data collection
- appropriate quality control procedures to ensure reliability of results

Both laboratory and field analysis of collected data should be done to obtain the most representative results of air monitoring. *High Volume Air Samplers and Portable Dust Samplers* may be used for sampling of air. *Radiometric and Gravimetric Analysis* should be undertaken.

Water Monitoring: In order to monitor water quality, it is necessary to know the characteristics of natural water available in the area and what might be introduced to it as a result of mining operations. Likely impact of mining on water regime may be either of the following types in mountain areas.

- Groundwater contamination
- Contamination of water channel in the vicinity of mine due to run off from mine site.
- Siltation of water channels

Though, chances of groundwater contamination exist very less in such type areas because of the very nature of mining process and obvious topographical reasons. Moreover, ground water table lies at a considerable depth and is highly fluctuating according to the hill profile. It is observed in some limestone mining areas of Sirmour district in Himachal Pradesh that natural spring occurring in the mining areas are dried. The causes for this may be due to change in water table level or hydrological regime of the area. Hence, it is almost certain that impact of drying of springs or change in water channel courses may occur as a result of mining.

To detect the pollutants present in mine water, 35 standard parameters as laid down under *Water Prevention and Control of Pollution Act* must be evaluated. Thereafter, quality is monitored by analysis / measurement in laboratory, which enable to know the *Key Indicators* of water pollution. Special attention must be paid to sampling method and for preservation and handling of water samples prior to analysis. It should be done according to the laid down procedure of standards. Water pollution parameters such as pH, BOD, COD turbidity, DO and coliform measurements must be done with high precision for the discharge of mine water into mountain water channels i.e. rivers stream, nallah etc. for protection of aquatic life forms.

Water is since a major transport medium for contaminants, a judicious water monitoring programme needs to address the following requirements for best practise mining in eco-sensitive areas.

- Water sampling points and its location should be such so that it truly represents the water quality of the area.
- Sampling frequency should be in tune with hydrological variability and season requirements.
- Standard sample collection, preservation techniques (before analysis) should be adopted.
- Standard analytical methods of chemical analysis should be used.
- Appropriate and high quality control procedures
- Evaluate the test results and review monitoring progress and/or practises.

Land Degradation Assessment: Land degradation due to mining is manifested by various physical impacts. Land monitoring for its evaluation essentially relates to the land management components of an Environment Management Plan (EMP) and is the most important component in mountainous terrain because of the fact that degraded land look like scars which is an eyesore. Best practice as regards to land include quantification of land degradation i.e. change detection analysis over a time period. A land monitoring programme needs to address the following:

- Identify areas to be monitored for each of the key issues
- Specify appropriate method and frequency of monitoring for each aspect
- Evaluate and review results/observation and adjust monitoring progress and or practises.

Geographic Information System (GIS) and *Remote Sensing Techniques* can be used for land degradation assessment as the mountainous terrain are not easily accessible for field data collection. Baseline data on soil and vegetation can be collected from various information sources or can be derived from field survey. Land maps prepared in this manner can assist in

197

- Quantification of land degradation over time period
- Management measures for erosion control and landslips / slides
- Topsoil/subsoil management
- Protection of specific landscape features/aesthetic features

Land Management based on *Land Capability Classification Approach* (Soni, 1997b) is an important tool, for sustainable land development in mineral rich area.

Restoration of Derelict Land: Mining land is basically a derelict land, which has loosened its organic content and water holding capacity. Its restoration to original or near original shape can be done by appropriate treatment and after care. Though, restoration by plantation is the most adequate and suitable method for making the derelict land green but social compulsion and industrial need of the area in mountains demands for alternative use of mining area for miscellaneous purposes such as development of area as recreational centre, tourist resort, playground, cattle grazing ground (Kanji house) etc. Restoration of derelict land of mining areas in mountain require following measures in a phase wise manner.

Phase -I
- collection of complete information or micro-climatic conditions of the area which include rainfall, temperature humidity, snowfall etc.
- collection of contour map of the area.
- collection of physico-chemical characteristics of soil such as pH, organic/humus content, K, N and P content, soil depth, texture.
- collection of measures required to check soil erosion and water conservation.
- collection of vegetation records and information about species diversity, community structure and site specific soil -plant- animal inter relationship of the area prior to the mining activities.

Phase - II
- Immediate measures to check soil and water conservation.
- Improvement of land quality by additives, by fertilisers or by manure etc.
- Choice of appropriate local species. Selection should follow basic ecological approach. Give preference to local species and those plant species, which are capable of growing on barren rocky areas deficient in moisture, nutrients and organic matter.
- Other land management measures taken locally can be adopted. The practices adopted should be such, so that it helps in improvement of land quality on long run.

Phase -III
- Ground to be levelled, compacted and terraced.
- Wherever permissible tillage operation can be used.
- Mechanical means of removing stone may be adopted if, soil: stone ratio is more.
- Nitrogenous fertilisers are always needed for derelict land at the initial stage of land quality improvement. Legumes, grown on mine soils are likely to increase the amount of soil nitrogen as mine soils are low in organic matter and nitrogen. To acquire organic matter the tree leaves may be allowed to decompose and *earthworms* can be added to accelerate the process of decomposition.
- Vegetation grown on mined land utilise the water and nutrients of mine soil for their survival. Plants must have about 20 different nutrient elements to grow and develop properly. Not all plant uses all elements but six elements viz. Nitrogen, Potassium, Sulphur, Calcium and Magnesium are the most needed elements. Therefore, selection of suitable fertilisers, containing the essential elements must be done. If, there is a deficiency of one nutrient needed for plant growth the addition of some other nutrient will not increase or support the plant growth. Therefore, only deficient nutrient is to be added.
- Second stage of land management according to land categories (Class I to VIII) should be applied.

Restored area should be protected from grazing and proper care should be taken for fire protection and other hazards.

Noise, Vibration, Air Over-Pressure and Fly Rocks: Blasting is an important component of mining process. The environmental hazards associated with blasting includes noise, vibration, air over-pressure and fly rocks. Topography of mountains is a responsive or an automatic multiplier factor of noise, vibration and fly rocks environmental problems.

When blasting is carried out it is accompanied by a loud noise called air blast. It is an atmospheric pressure wave consisting of high frequency sound that is audible and low frequency sound or concussion that is sub-audible and can not be heard. Air blast over pressure is measured in Decibels (dB) or in Pounds per sq. inch (psi). In environmentally conscious places air blast is seldom a problem in normal blasting operations. In most of the places, it causes psychological fear rather than any probable damage. The detrimental effect of

198

air blast are- cracks in the window pans of buildings. Noise, vibrations and fly rocks components are well addressed in any mining projects/reports and environment management plan. Number of measures to prevent these hazards are suggested but for best practise mining some cost-effective measures should be augmented into practice and these includes -

- Proper Blast Design which include placement of explosive, blast pattern shape, point of initiation, sub-drilling, burden, spacing, stemming column length, etc.
- Use of delay detonators and correct selection of delay intervals
- Adequate change per delay and proper loading during of explosives
- Direction of initiation
- Selection of appropriate time for blasting
- *Muffle Blasting* to control fly rocks
- Low density explosive in loose and fractured rock mass
- Angular holes in conformity with the slope of the bench
- Adoption of *Controlled Blasting Techniques* (if, need be)

Biological Environment Assessment: The check list required for assessment of biological environment is a long one, which interalia includes flora and fauna, endangered species wildlife population and sex ratio, species density, pattern of their multiplication/breeding, growth and mortality rate, migration pattern, disease pattern, aquatic life and its pattern etc. These are to be identified during the EIA process. Since, it is not practical to monitor entire systems, *Indicator Species*, processes, group or communities are usually selected. Their selection should be based on their efficacy or sensitiveness in the system. Due to complexities and dynamics of biological systems, which nature has offered, studies needs to done over several seasons / years. It will define the key indicators that will be used in the assessment process. A biological environment assessment program must be studied in conjunction with ecosystem preservation plan, as they are inter-wovened. A biological monitoring program needs to address the following requirements:

- Define community and species dynamics.
- Select appropriate indicators for direct toxicity or bio-accumulative measurements.
- Consider variation due to space and time.
- Evaluate direct impact on biological communities.
- Use widely accepted and standardised methods where possible.
- Collect adequate data for appropriate statistical analysis with special attention on short and long term effect, local and regional effects, individual species and broad community impacts.
- Evaluate and review test results.
- Adjust monitoring program and/or practices according to requirement of area.

Eco-system Protection: It is well realised that some effects of mining can be predicted easily because they occur immediately but other effects happen over a longer period and are more difficult to predict and measure. Eco-system and its protection is one such long term effect. Best practise mining should include steps to identify these long-term effects and incorporate them in the management plan. Long term impact include -

- Vegetation changes caused by alteration in ground water table.
- Change in the wild life and bird migration pattern.
- Siltation of water channels and change in their courses.
- Gradual effect of air pollution on vegetation's health and gradual shifting of one form of vegetation into another.
- Acid mine drainage.
- Instability of post mining land forms after mine closure.

4.0   INTEGRATED APPROACH / STRATEGY

To facilitate area specific micro level planning for management of mineral bearing resource areas, it is the best option to apply integrated approach/strategy on watershed basis. Watershed as a unit of environmental planning and management constitute a technical and ecological entity which is self contained, physically composite and functionally coherent for land, water and air resource planning and management. It has compatibility with the existing block/village level planning. According to the multilevel planning policy of Govt. of India at national, state, district and lower area levels natural resource data management is done on watershed basis considering each watershed as a constituent unit for planning. Hence, ecologically fragile areas of hilly region containing mineral resources of economical importance, the study of environmental parameters and other related factors should be done on a watershed basis because of the fact that -

* This approach provides the production which is resource centred and environmentally friendly and helps in promoting sustainable development.
* Integrated strategy on a watershed basis for land utilisation, soil and water conservation and utilisation or recycling of waste can be rationally articulated for comprehensive short term and long term planning and their effective implementation.

5.0 BEST MINING PRACTICE

Best mining practise or Best Practise Mining (BMP) does not connote to any designed/formulated method but implies to the continuous improvement of existing practises so that the negative impacts are minimal. To follow best practice mining, the developmental planning require some prior considerations as given below -

* Watershed must be a unit of environmental management in ecologically fragile areas. Since, Himalayan region is characterised by sharply changing features from watershed to watershed a *"Watershed Development & Management Plan"* for total life span of the project should from an essential constituent of planning.

* While designing and implementing a mining project (as a part of developmental activity for the region) at both macro and micro planning level explicit considerations should be given to the characteristic features of the mountainous region which are termed as *"Mountain Specificities"* and includes inaccessibility, fragility, marginality, diversity and niche.

• Guidelines for grant of mineral rights should be framed beforehand, keeping in view this type area and its delicacy. These guidelines should be drawn in consultation with technical experts and representatives of the mine owners so that they are practical in implementation.

• Small scale mines which account for considerable percentage of the production needs special attention. To develop best practice mining in small mines due consideration for environmental and mining problems on site specific basis must be given on case to case basis. It basically includes –

* creation of green barriers along haulage roads and along the periphery of outer limit of the quarry,
* projected parapet walls all along the approach roads with necessary super elevations,
* mobile environmental monitoring unit for a group of mines and an internal wing of environmental management for various mining operation,
* community nursery,
* use of *hydroseeder* for reclamation of high angled slopes,
* progressive restoration and scientific reclamation practices,
* use of rippers as alternate to blasting,
* *controlled blasting* and use of *sequential blasting machine* to control fly rocks and ground vibrations,
* gravel/sand packed pits to check water pollution,
* improved design of check walls/check dams with filtering arrangements,
* environmental friendly and scientific methods of transportation for a cluster of small mines,
* environmental friendly equipment, machine and accessories like continuous miners,
* scientific methods of waste disposal by *back-filling* or by *fill construction in lifts*.

6.0 CONCLUSIONS

On the basis of description given above it is inferred that to harness maximum from existing mining practises **Integrated Strategy** must be adopted and **BMP** should be practised as it causes least harmful effects on environment and improve performance of overall mining operation. It was observed that in hills leases are granted all along the road sides or on the hill slopes (nearby the roads) to private small mine owners. Such mines does not follow scientific practices, as a result of which ugly scars are created on hill slopes which remains visible for road traffic. At the same time unscientific practices also lead to public debate on environment ground.

In brief, it is concluded that mining of minerals in tough and rugged terrain conditions like Himalaya is not at all a simple issue. Research and past precedence have led to the conclusions that the scientifically planned strategy should be the only approach with no short cuts as far as possible. Mining in fragile and sensitive hill areas should be limited, selective and controlled and very large-scale mechanisation has limited scope on account of the environmental sensitivity.

7.0   REFERENCES

Babu, C.R., Jha, P.K., Nair, S. (1990), "Biological Reclamation of Derelict Lands: Problems and Prospects", Environmental Management of Mining Operations, ed. B.B. Dhar, Ashish Publishing House, New Delhi, pp. 327-335.

Banerjee, P.K. (1993), "Environmental Problems Related to Mining in Hilly Region", Proceedings of Seminar on Environmental Policy issues in Mineral Industry, MGMI, Calcutta, pp. 89-95.

EPA (1995), Best Practice Environmental Management in Mining –Various Modules, Environmental Protection Agency, Australia.

ICIMOD (1983), " Mountain Development-Challenges and Opportunities", Proceedings, First International Symposium, Kathmandu, (country statements and selected papers), p. 122.

Moddie, A.D. (1980)," Environment Oriented Hill Development", Studies in Himalayan Ecology and Development Strategies, ed. Tejvir Singh, English Book Store, New Delhi, pp. 193-198.

Noetstaller,R.(1993),"Small Scale Mining : Practises, Policies, Perspectives, In Small Scale Mining - A Global Overview, Ed.-Ajoy K. Ghose, Oxford IBH & Company, new Delhi, pp.3-10

Paithankar, A.G. (1993), "Large Scale Hill top Mining with Concern for the Environment", Proceedings, Innovative Mine Design for the 21st Century, eds. Barvden and Archibald, A.A. Balkema.

Poore, D (1980), "Developing Tropical Rain Forest Ecologically", ed. Tej Vir Singh, English Book Store, New Delhi, pp. 104-111.

Rai, K.L., "Geological and Geo-environmental Aspects of Mineral Resources Development in Lesser Himalayan Tracts of Sikkim and U.P. Himalayas, India", Proceedings, Asian Mining - 1993, Calcutta, India.

Saxena, N.C.(1995)," Environmental Management Plan (EMP) Preparation for Mining Projects - An Approach", The Indian Mining and Engineering Journal, Vol. 34, No.5, May, pp. 35-43.

Shapkota, P.(1991)," Small Scale Mining in Bhutan - A Perspective, Small Scale Mining - A Global Overview, ed. Ajoy K. Ghose, Oxford IBH Company Private Ltd., New Delhi, pp. 221-239.

Soni A.K., Dube A.K. and Srivastava S.S. (1995), Theoretical Study of Some of the Factors Affecting Environment in a Limestone Quarry of Himachal Pradesh, Journal of Mines, Metals & Fuels, Vol. XLIII, No.5.pp 111-113.

Soni A.K., (1995), Environmental Study of a Limestone Mining in Himalayas, Journal of Mining Research, Vol. 3, Nos. 3&4, January- March, pp.1-8

Soni A. K. (1997a), Fragile/Sensitive Ecosystems, Proceedings of National Seminar on "Eco-Friendly Mining in Hilly Region and its Socio-economic Impacts (HILMIN'97), June, MEAI Himalayan Chapter, Shimla, pp. 224-231.

Soni A. K. (1997b), Integrated Strategy for Development and exploitation of natural Mineral Resources of the Ecologically Fragile Areas, Ph.D. Thesis, Indian School of Mines, Dhanbad, p. 238.

****

Environmental Evaluation on Leaching of Trace Elements from Coalashes from Thermal Power Stations of Eastern India

Gurdeep Singh

Professor and Head, Centre of Mining Environment
Indian School of Mines, Dhanbad 826 004, India

INTRODUCTION

Electricity is an essential need of any industrial society and no nation can progress without adequate supply of power. Growth in its demand during the past decades has been phenomenal and has outstripped all projections. In India also, there has been impressive increase in the power generation from a low capacity of 1330 MW in 1947 at Independence to about 81,000 MW at end of March 1995 (Trehan et al. 1996). However, despite this substantial growth there still remains a wide gap between demand and supply of power which is expected to worsen in the years and decades to come. We already are experiencing shortage of nearly 8% of the average demand and 16.5% of peak demand. Moreover, with the quantum jump expected in demand for power in the future due to rapid industrialization and changing life styles of populace as a result of economic liberation, shortage shall further increase unless immediate steps are taken to increase power production. It has been estimated that 1,40,000 MW of additional power would be required to meet such demand by the end of 10th five year plan i.e. year 2007 A.D. (Trehan et al. 1996).

India has vast reserve of coal and it is expected that this shall remain as prime source of energy in the early part of 21st century. Presently, thermal power stations account for about 70% of installed generation capacity of 81000 MW (Sampath et al. 1996). In a pulverised coal fired thermal power plant about 40-55% ash is produced. About 20% falls down due to gravity and is removed as bottom ash and the remaining fly ash. this fly ash is collected by mechanical and/or electrostatic precipitator. As per the available estimates the production of coal ash in India including both fly ash and bottom ash is about 70 million tonnes per annum which is likely to touch 155 million tonnes per annum by 2020 A.D. Thus, the utilisation of coal ashes has drawn considerable concern and attention of scientists, technologists, environmental groups, government, regulators etc.

Studies of trace elements and the elements present in coal ash are distributed into the fractions of the coal ashes based on volatilisation temperature (Bachor et al. 1981). It is found that elements appear to partition into three main classes.

1.   Elements that are not volatilised and reported equally in both fly ash and bottom ash. These elements include Al, Ba, Ca, Ce, Co, Cu, Fe, Hf, K, La, Mg, Mn, Rb, Se, Si, Sm, Sr, Th, Ta and Ti.
2.   Elements that are volatilized on combustion and preferentially get adsorbed on the fly ash as flue gas cools down. These include As, Cd, Ga, Mo, Pb, Sb, Se and Zn.
3.   Elements that remain almost entirely in the volatilised state tend to escape to the atmosphere as vapours. These are Hg, Cl and Br.

Coal ash is a alumino silicate glass consisting of the oxides of Si, Al, Fe and Ca with minor amounts of Mg, Na, K, Zn and S and various trace elements. The concentration associated with the ash may be either adsorbed on the surface of particle or incorporated into matrix (Natusch et. Al. 1974). A mechanism that appears to be common for all ashes during their formation is the condensation of metal and metalloid vapours on refractory core materials. As the ash particles and gas stream exist from the combustion chamber and proceed upto the flue gas, this results in locally higher concentration of many trace elements at the surface of ash particles and accounts for the generally higher concentration of these elements as particle size decreases (Markowski & Filbly, 1985). The association between trace elements and major elements/minerals may be an important factor in determining the leachate composition of water in contact with ashes.

It is recognised that the health hazards and environmental impacts from coal fired thermal power stations result from the mobilization of toxic elements from ash. The large amount of ash that accumulates at thermal power plants, its possible reuse and the dispersion and mobilization of toxic elements from it, require greater attention (Palit et.al., 1991). Mobilization of various elements from the ash into the environment depends on climate, soils, indigenous vegetation and agricultural practices (Page et. Al., 1979). Present study on environmental characterisation is in continuation with earlier studies, to evaluate leaching of trace elements from coal ashes from a few Thermal Power Stations situated in eastern India are presented in this paper (Gurdeep Singh & Sanjay Kr., 1996, 1999, 2000; Gurdeep Singh & Kumari Vibha, 1999). The possible water contamination is also envisaged through the leachate analysis from ash pond disposal sites in real life situation.

EXPERIMENTAL METHODOLOGY
The ashes collected/sampled were analysed for their leaching characteristics using open column percolation experiments. The columns of the fly ashes, bottom ashes and pond ash were packed in April 1996 and leachates were collected twice a week in a times period of about three years. Every time the leachates were analysed for trace elements. Standard sampling and leaching analysis methods were followed in this study on environmental evaluation of coal ash. These are briefly described below.

SAMPLING
Coal ash samples from Thermal Power Stations were collected on five different days from each of the power plant over a week and a final homogenised sample was prepared while appropriately mixing various portions. Leachate samples were periodically collected twice a month from the final discharge of the ash pond which receives fly ash (80%) and bottom (20%) in slurry form from the plant.

LEACHING CHEMISTRY
Open column percolation leaching experiments were carried out on the coal ash samples to ascertain its leachate chemistry as briefly described below.

OPEN PERCOLATION COLUMN EXPERIMENTS
In these experiments, deionized water is percolated through a packed column of coal ash (fly ash, bottom ash and pond ash packed separately in different columns) in the presence of oxygen at a rate which depends on the natural permeability of the material. The open columns for leaching experiments were made of PVC pipe four inches in diameter and two feet in length. The column setup involved packing the coal ash material at optimum moisture and density conditions as determined by the Proctor test (ASTM Standards, 1990). The coal ash material was packed into the column in two inch lifts with a 2" x 2" wooden rod, about 4 feet long. Each packed layer was scarified, by lightly scraping the top of the packed layer with a long thin rod to ensure proper interlocking of the material. The top six inches of the column was left unpacked to allow for the addition and maintenance of the leaching medium. About 200 ml of leaching medium (de-ionized water) was added at the top of the column once every alternate day to maintain sufficient supply of water to the packed coal ash material. The top end of the column was exposed to the atmosphere and the bottom end was connected to quarter inch tubing. The columns discharged the leachates through this tubing into the 250 ml polypropylene beakers. The leachates were collected in these beakers and analysed.

ELEMENTAL ANALYSIS OF LEACHATES
The leachate samples were filtered and acidified with 2 ml of nitric acid and then preserved in polypropylene sampling bottles. The samples were kept in a refrigerator until further analysis. Sodium and potassium were determined using Systronics flames photometer. Concentration levels of trace elements were evaluated using Atomic Absorption Spectrophotometer (AAS) GBC-902. Working/standards solutions were prepared according to instructions given in the operation manual of the GBC-902 AAS.

Optimized operating conditions such as lamp current, wave length, slit width, sensitivity, flame type etc. as specified in the manual, were used for analysis of a particular elements.
Merck-AAS standards were used for standardisation and calibration of AAS. Three standards and a

blank of the concerned element were used to cover the range 0.1-0.8 Abs. The calibration was performed by using the blank solution to zero the instrument. The standards were then analysed with the lowest concentrations first and the blank was run between standards to ensure that baseline (zero point) has not changed. Samples were then analysed and their absorbance recorded. The calibration was performed in the concentration mode in which the concentration of sample was recorded.

RESULTS AND DISCUSSIONS

Analysis of twenty two elements were carried out from each of the leachate samples collected from open column experiments and the observations are summarised in Table 1 to 3 for Flyash, Bottom ash and Pond ash, respectively. It is noticed from the observations that the concentration of thirteen elements, namely, chromium, nickel, cobalt, cadmium, selenium, aluminum, silver, arsenic, boron, barium, vanadium, antimony and molybdenum were below the detection limit (.001 mg/l) in the entire study period. Among the other nine elements only calcium and magnesium were observed in the leachates throughout the study period while the concentration of other elements showed a decreasing trend to below detection limit (.001 mg/l). In the leachates from actual ash ponds, lead and manganese were found absent but iron, calcium, magnesium, sodium, potassium, copper and zinc were present throughout the study period.

A comparison of the concentration levels observed with respect to nine significant leachable elements in the leachates of fly ash, bottom ash, pond ash and also leachates from actual ash pond disposal site with the permissible limits as per IS:2490 is presented in Table 4 which indicates that the concentration of all the elements during the entire study period was below the permissible limits.

It can be inferred that no significant leaching occurs and toxicity is manageable with respect to trace elements both in the ash pond disposal site as well as in the open column leaching experiments. Further, analysis results of leachates from open column percolation experiments resemble closely with those of actual ash pond leachates. The physical set up of the open columns more closely resembles with because the flow of the leaching medium is influenced by gravity alone and the solid to liquid ratio is more close to the field situation. Hence, open column leaching experiments may be used in predicting the long term leaching behaviour that can be observed in the field. Coal ash leachates as generated from open percolation column leaching experiments and those from ash pond disposal site closely resemble and as such do not pose any significant environmental impacts in the disposal system. Overall, coal ash would not seem to pose any environmental problem during its utilization and/or disposal.

CONCLUDING REMARKS

On the basis of the study of the leaching of trace elements from coal ashes, following conclusions can be drawn:

1. In the study period of about three (3) years there was practically no leaching of thirteen elements namely, chromium, nickel, cobalt, cadmium, selenium, aluminium, silver, arsenic, boron, barium, vanadium, antimony and molybdenum from all the ash samples.
2. Out of the nine elements found in the leachates only calcium and magnesium were found to be leaching in the entire period. The leaching of other seven elements namely, iron, lead copper, zinc, manganese, sodium and potassium was intermittent. The leaching of sodium and potassium practically stopped after 35 and 40 days, respectively.
3. The concentration of the elements in the leachates was invariably well below the permissible limits for discharge of effluents as per IS:2490 and also for drinking water as per IS:10500.

Overall, the coal ashes evaluated in this study were found to be environmentally benign and can be engineered for their bulk utilization particularly for mined out areas reclamation and for soil amendment for good vegetation.

The Centre of Mining Environment at ISM Dhanbad is currently engaged in evolving low technology high volume field demonstration to show that coal ash particularly fly ash can be disposed and utilized as fill material in an environmentally acceptable way in reclamation of abandoned mines (Bradley C. Paul, Gurdep Singh et al. 1995, 1998). Use of fly ash as backfill material for reclamation of mined out sites provide benefits such as easy availability, cheaper to transport because empty coal carriers returning from the power plant can "back haul" it to the mine site. From the standpoint of the power plant, this is essentially a waste material which requires large costs of handling and a disposal to comply with environmental regulations. From the environmental point, this waste material will go back to the same place where it was mined and use of this material serves as extra benefit to power plants. Studies are also in progress to use fly ash for agriculture development.

REFERENCES
Atomatic Absorption Spectophotometer Operating Manual, 1990. AAS, GBC-902, Australia

Annual Book of ASTM Standards, 1990, vol. 11, 10.

Bahor, M.P., Mclaren, R.J., Niece and Pedersen, H.C., 1981, Coal Ash Disposal Manual, Second Edition Electric Power Research Institute, Final Report (CS-2049)

Bradley C. Paul, Gurdeep Singh, Steven Esling, Chaturvedula and Pandal, H. 1998. The impact of scrubber sludge on ground water at an abandoned mine site. Environmental Monitoring and Assessment, 50:1-13.

Bradley C. Paul and Gurdeep Singh, 1995. Environmental evaluation of the feasibility of disposal and utilization of coal combustion residues in abandoned mine sites. Proceedings, First World Mining Environment Congress. Dec. 11-14, 1995. N. Delhi Oxford & IBH Publ. Co. Pvt. Ltd., New Delhi. Pp. 1015-1030.

Bradley C. Paul, Gurdeep Singh and Chaturvedula, S. 1995. Use of FGD by products to control subsidence from underground mines: groundwater impacts. Proceedings International Ash Utilization Symposium October 23-25, 1995. Lexington, Ky. USA.

Gurdeep Singh and Sanjay Kumar Gambhir, 1996. Environmental evaluation of fly ash in its disposal environment. Proceedings, International Symposium on Coal – Science Technology, Industry Business & Environment, Nov. 18-19, 1996, editors. Kotur S. Narsimhan & Samir Sen. Allied Publishers Ltd., New Delhi. Pp.547-556.

Gurdeep Singh and Kumari Vibha 1999. Environmental assessment of fly ash in its disposal environmental at FCI, Ltd. Sindri, Poll. Res. 18(3) 339-343.

Gurdeep Singh and Kumar Sanjay, 1999. Environmental evaluation of coal ash from Chandrapura Thermal Power Station of Damodar Valley Corp. Indian J. Environmental Protection 19\8(12) 884-888.

Gurdeep Singh and Kumar Sanjay, 2000 Environmental evaluation on leaching of trace elements from coal ashes: a case study of Chandrapura Thermal Power Station. Journal of Environmental studies and Policy 2(2); 135-142.

Markowski, G.R. and Filbly, R., 1985. Trace element concentration as a function of particle size in fly ash from a pulverised coal utility boiler. Environmental Science & Technology, vol. 19, pp. 796-800.

Natusch, D.F.S., Wallace, J.R. and Evans, C.A., 1974. Toxic trace elements; preferential concentration in respirable particles. Science, Vol. 183 no. 4121, pp. 203-204.

Page, A.L., Elseewi, A.A. and Straughan, I.R., 1979, Physical and chemical properties of fly ash from coal fired power plants, Res. Review, vol. 71, pp. 83-120.

Palit, A., Gopal, R., Dube, S.K. and Mondal, P.K. 1991, Characterisation and utilisation of coal ash in the context of Super Thermal Power Stations. Proceedings International Conference, on Environmental Impact of Coal Utilisation, 26-27, pp. 154-155.

Sampath, R. "Fly Ash – its pollution and potential for utilisation", Seminar on Fly Ash Utilization, 26-27 March, 1996, New Delhi.

Trehan. A., Krishnamurthy, R., Kumar. A. "NTPC'S experience in ash utilization" Seminar on Fly Ash Utilization, 26-27 March, 1996, New Delhi.

Table 1   Summary of Leachate Analysis of Fly Ashes

Open Column leachate experiments				
	FA # 1		FA # 2	
Element	Range	Average	Range	Average
Iron	BDL-3.60	2.80	BDL-3.08	2.50
Lead	BDL-0.090	0.068	BDL-0.10	0.05
Calcium	10-30	20	25-40	30
Mangnesium	10-25	17	16-35	20
Copper	BDL-0.090	0.08	BDL-0.09	0.05
Zinc	BDL-0.10	0.08	BDL-1.10	0.80
Manganese	BDL-0.10	0.085	BDL-0.10	0.075
Sodium	BDL-20	8	BDL-20	10
Potassium	BDL-20	10	BDL-40	20
Chronium	BDL	BDL	BDL	BDL
Nickel	BDL	BDL	BDL	BDL
Cobalt	BDL	BDL	BDL	BDL
Cadmium	BDL	BDL	BDL	BDL
Selenium	BDL	BDL	BDL	BDL
Aluminium	BDL	BDL	BDL	BDL
Silver	BDL	BDL	BDL	BDL
Arsenic	BDL	BDL	BDL	BDL
Boron	BDL	BDL	BDL	BDL
Barium	BDL	BDL	BDL	BDL
Vanadium	BDL	BDL	BDL	BDL
Antimony	BDL	BDL	BDL	BDL
Molybdenum	BDL	BDL	BDL	BDL

Table 2: Summary of Leachate Analysis of Bottom Ashes

Open Column leachate experiments				
	FA # 1		FA # 2	
Element	Range	Average	Range	Average
Iron	BDL-3.10	2.3	BDL-3.20	2.50
Lead	BDL-0.09	0.070	BDL-0.09	0.070
Calcium	30-55	40	18-48	30
Magnesium	12-40	25	10-32	22
Copper	BDL-0.056	0.050	BDL-0.060	0.050
Zinc	BDL-1.071	1.060	BDL1.092	1.080
Manganese	BDL	0.08	BDL-0.080	0.070
Sodium	BDL	10	BDL-20	15
Potassium	BDL	15	BDL-20	12
Chromium	BDL	BDL	BDL	BDL
Nickel	BDL	BDL	BDL	BDL
Cobalt	BDL	BDL	BDL	BDL
Cadmium	BDL	BDL	BDL	BDL
Selenium	BDL	BDL	BDL	BDL
Aluminium	BDL	BDL	BDL	BDL
Silver	BDL	BDL	BDL	BDL
Arsenic	BDL	BDL	BDL	BDL
Boron	BDL	BDL	BDL	BDL
Barium	BDL	BDL	BDL	BDL
Vanadium	BDL	BDL	BDL	BDL
Antimony	BDL	BDL	BDL	BDL
Molybdenum	BDL	BDL	BDL	BDL

Table 3: Summary of Leachate Analysis of Pond Ashes

Open Column leachate experiments				
	FA # 1		FA # 2	
Element	Range	Average	Range	Average
Iron	BDL-3.22	2.5	1.02-3.50	2.60
Lead	BDL-0.06	0.050	BDL-0.09	0.04
Calcium	18-60	40	18-60	38
Magnesium	10-20	15	10-20	12
Copper	BDL-0.050	0.050	0.011-.050	0.04
Zinc	BDL-1.030	1.00	0.93-1.030	1.00
Manganese	BDL-0.060	0.050	BDL-0.65	0.05
Sodium	BDL-20	10	5-30	10
Potassium	BDL-20	15	10-30	16
Chromium	BDL	BDL	BDL	BDL
Nickel	BDL	BDL	BDL	BDL
Cobalt	BDL	BDL	BDL	BDL
Cadmium	BDL	BDL	BDL	BDL
Selenium	BDL	BDL	BDL	BDL
Aluminium	BDL	BDL	BDL	BDL
Silver	BDL	BDL	BDL	BDL
Arsenic	BDL	BDL	BDL	BDL
Boron	BDL	BDL	BDL	BDL
Barium	BDL	BDL	BDL	BDL
Vanadium	BDL	BDL	BDL	BDL
Antimony	BDL	BDL	BDL	BDL
Molybdenum	BDL	BDL	BDL	BDL

Table4:Comparison of Elemental Analysis pf Samples with Permissible limits

Leachats	Elements	Observed range mg/l	Permissible Limits		
			Discharge in inland surface water (IS:2490)	On land for irrigation IS:2490	Drinking waterIS:10500
FA#1	Iron	BDL-3.60	--	--	--
	Lead	BDL-0.90	0.1	--	0.1
	Calcium	10-30	--	--	75
	Mangnesium	10-25	-	--	30

	Copper	BDL-0.090	3	--	0.05
	Zinc	BDL-1.10	5	--	--
	Manganese	BDL-0.10	--	--	0.1
	Sodium	BDL-20	--	60	--
	Potassium	BDL-20	--	--	--
FA#2	Iron	BDL-3.08	--	--	--
	Lead	BDL-0.10	0.1	-	0.1
	Calcium	25-40	--	--	30
	Mangnesium	16-35	--	--	30
	Copper	BDL-0.09	3	--	0.05
	Zinc	BDL-1.10	5	--	--
	Manganese	BDL-0.10	5	--	--
	Sodium	BDL-20	--	60	--
	Potassium	BDL-40	--	--	--
BA#1	Iron	BDL-3.10	--	--	--
	Lead	BDL-0.09	0.1	--	0.1
	Calcium	30-50	--	--	75
	Mangnesium	12-40	--	--	30
	Copper	BDL-0.056	3	--	0.05
	Zinc	BDL-0.071	5	--	--
	Manganese	BDL-0.10	--	--	0.1
	Sodium	BDL-20	--	60	--
	Potassium	BDL-20	--	--	--
BA#2	Iron	BDL-3.20	--	--	--
	Lead	BDL-0.09	0.1	--	--
	Calcium	18-48	--	--	75
	Mangnesium	10-32	--	--	30
	Copper	BDL-0.060	3	--	0.05
	Zinc	BDL-1.092	5	--	--
	Manganese	BDL-0.080	--	--	--
	Sodium	BDL-20	--	60	--
	Potassium	BDL-20	--	--	--
PA#1	Iron	BDL-3.22	--	--	--
	Lead	BDL-0.06	0.1	--	0.1
	Calcium	18-60	--	--	75
	Mangnesium	10-20	--	--	30
	Copper	BDL-0.050	3	--	0.05
	Zinc	BDL-1.030	5	--	--
	Manganese	BDL-0.060	--	--	0.1
	Sodium	BDL-20	--	60	--
	Potassium	BDL-20	--	--	--
FA#2	Iron	1.02-3.50			
	Lead	BDL-0.09	0.1	--	--
	Calcium	18-60			
	Mangnesium	10-20			
	Copper	0.011-.050	3.0	--	
	Zinc	0.93-1.030	5.0	--	
	Manganese	BDL-0.65			
	Sodium	5-30	--	60	
	Potassium	10-30			
AP (Disposal site)	Iron	1.02-2.94	--	--	--
	Lead	--	0.1	--	0.1
	Calcium	18-46	--	--	75
	Mangnesium	10-19	--	--	30
	Copper	0.1-.047	3	--	0.05
	Zinc	.93-1.015	5	--	--
	Manganese	--	--	--	--
	Sodium	5-10	--	60	--
	Potassium	8-18	--	--	--

Concentration in mg/1
BDL – below detection limits, i.e. 0.001 mg/1.

Environmental Mitigation—WCL'S World Bank Projects

S.K. Mitra[1] and D.K. Roy[2]

[1]*Chief General Manager, World Bank Projects W.C.L., Nagpur, India*
[2]*General Manager (Environment), W.C.L., Nagpur, India*

1.   INTRODUCTION

Environmental awareness in our country has been there since time immemorial. The environmental attributes of Earth, Water, Fire, Air and Space (Ether) have been revered as gods and still continue to be. The water of the ganges and the leaves of Tulsi are considered most favourite offerings for worship of God. Plants, animals (Flora and Fauna) have been worshipped and still are. But ironically, at the same time, our country appears to be one of the forerunners in destruction of plant and wild life. This seems to be an unexplainable contradiction.

With the pressure on development and welfare of ever increasing population of our country, the coal production targets had to be commensurate, since coal is the prime energy source. The production target had to be raised from 250 MTY in the terminal year of VIIIth plan to 314 MTY at the end of IXth five-year plan period. This would call for a massive financial input of the order of Rs. 13,700 crores. With absence of Govt. Budgetary support, this gap of requirement of fund & internal resource availability could only be met with borrowings. The World Bank was perceived as the most viable lending agency, for such huge sum.

Western Coalfields Ltd., to sustain desirable levels of production and profitability, decided to take loans from World Bank and its funding agencies like JEXIM, IBRD etc for purchase of heavy equipments needed in the industry. Remarkably enough, these loans are attached with some unusual strings (which can rather be called umbilical chord, because the strings relate to the every existence of the Nature's attributes). These are – commitments for environmental safeguards to mitigate devastations caused by the heavy giant-like equipments. This safeguard has been named Coal Sector Social & Environmental Mitigation Project (CSESMP) which would enable Coal India Ltd. to strengthen its capacity to deal more effectively with environmental & social issues through implementation of high priority Environmental and Social Mitigation programme.

WORLD BANK FUNDED PROJECTS

Western Coalfields Ltd. has five projects, all Open Cast Mines, which have been covered by World Bank funding. They are :

1.	Durgapur Open Cast Project	}
2.	Padmapur Open Cast Project	} In Chandrapur District
3.	Sasti Open Cast Project	}
4.	Niljai Open Cast Project	} In Yeotmal District
5	Umrer Open Cast Project	} In Nagpur District

The projects have been selected on the strength of their inherent profitability and production capacity/ productivity. The funding for the five projects, for procurement of equipments amount to Rs. 410.00 crores. But for getting this funding the precondition is to restore that umbilical chord of environmental and social mitigation for which a part of the loans is earmarked. Though the environmental part is financially insignificant, being just 4% of the loan amount, it nevertheless occupies greater importance, since this is a pre-conditional activity.

ENVIRONMENTAL MITIGATION – A PRO-ACTIVE APPROACH

There is a distinct thrust on the above mentioned five projects as regards environmental mitigation activities with a vision of making these project the models for other projects to follow. These thrust areas will be discussed in the following points vis-à-vis the achievements in these aspects.

The World Bank stipulates the following activities to precede the actual requisition of the loan amounts. The activities are listed as scheduled activities with firm implementation dates.

1. For Air Pollution Control:
   (a) Mobile water sprinklers on roads
   (b) Dust extractors in drills
   (c) Plantation on plain land, avenue and over-burden dumps
   (d) Dust suppression activities in industrial roads and structures
2. For Water Pollution Control:
   (a) Treatment plants for mine discharge water
   (b) Treatment plants for effluent from workshops
   (c) Treatment plants for discharge of Domestic sewage
   (d) Collection and treatment of over-burden dump seepage water
3. For Land Degradation Control:
   (a) Stability of over-burden dumps
       i) biologically   -   by plantation
       ii) technically   -   by benching, sloping, gully plugging,
                                           garland drains, crack filling etc
   (b) Catch Drains and Sedimentation Ponds for arresting siltation from OB dumps on fertile land down stream.
   (c) Plain land and Avenue plantation.
4. Effluent / Chemical Hazard Control:
   (a) Disposal system of recovered oil and grease from workshops
   (b) Settling tanks in coal handling plants
   What is more important is a number of studies covering all above activities before their implementation to safeguard against arbitrary implementation was undertaken through reputed agencies.
   These studies cover –
   (a) Studies on disposal of oil and grease
   (b) Study on Bio contamination of soil from domestic effluents
   (c) Study on over-burden dump safety and stability
   (d) Study on suitability of top soil in reclaiming overburden dumps etc

One aspect of significance is, the organisation and manning for the above activities, which obviously could not be done with the existing set up with their pre-occupation and existing assignments.

Formation of an Environmental Cadre of Executives and also training modules for the persons opting in this cadre, followed by refresher training modules were done at the CIL corporate level. To increase the overall capability, environmental training was imparted to executives of various other disciplines engaged in mining operation.

ENVIRONMENTAL MITIGATION – ACTIONS

The above listed environmental items could not have been tackled just by persons and the funds. For the success of the activities, action plans or schedules had to be prepared and agreed.

With all the above ground work, the result came systematically in planned manner. The achievements of the various items in these projects are as below:

i) Mine discharge sedimentation pond - Since mining activities require pumping out of sub soil water to keep the working area dry, substantial quantities of water are discharged on the surface. Fortunately there is no polluting activity involved in this; and the major pollutant is only suspended particulate matter. Hence the treatment scheme comprises of only sedimentation.

Durgapur & Padmapur projects have sedimentation pond of capacity 1012m3 which is a masonry of size 2x38x(2x15)x0.9. The water, after settlement, is carried through open drainage for gainful use in residential colony after treatment.

Sasti project have sedimentation pond of capacity 798m3 which is a masonry of size 38x(2x15)x0.7. The water, after settlement, is carried through open drainage for gainful use in residential colony after treatment.

Niljai project have sedimentation pond of capacity 1080m3 which is a masonry of size 40x(2x20)x0.8. The water, after settlement, is carried through open drainage for gainful use in residential colony after treatment.

Umrer project have old abandoned mine sump of capacity 2,50,000m3 and it now serves as sedimentation pond.

ii) Overburden seepage sedimentation tank - The seepage water from the dumps (due to rains) carries with it soil particles which, being non-nutritious, is not desirable to be carried & deposited on down stream side fields. Hence the sedimentation pond of capacity has been created.

This collects water for settlement of suspended solids. The tank has been used as a water body and aqua-park by developing garden, boating, benches, fountains etc and is now a high point of eco-friendly mining.

iii) Workshop Effluent Treatment Plant - A 150 KL plant complete with pre-sedimentation and sedimentation tanks. Oil and grease traps, flash mixers, Clarifloculator, sludge collection system, sludge drying bed, clear water tank etc. The plant emits zero discharge and treated water is fully recycled for use in workshop only. The oil and grease are fully recovered and collected in drums. The collected oil and grease are used for lubrication of minor machinaries like coal tubs, rope pulleys etc of underground mines.

There are sludge drying beds and sludge farms which encourage the sludge to turn nutritious material by natural vegetation that grow over it.

iv) Domestic Effluent Treatment Plant - A modern Domestic Effluent Treatment Plant (DETP) catering to the combined residential colony of Durgapur and Padmapur Projects has been constructed. The plant consists of two oxidation ponds with 4 Nos. fixed type mechanical surface aerators, followed by two polishing ponds.

The treated effluent will be used for irrigation of plantation and green cover all around.

In the Sasti Project the domestic effluent collected through a network of sewers in an oxidation pond with two mechanical surface aerators followed by polishing pond. The capacity is 1 MLD.

In Umrer too a similar system has been provided with IMLD capacity.

In Niljai Project, extended aeration system comprising of aeration tank, classifier, sludge drying bed, etc. with related facilities have been constructed.

The capacity is 0.5 MLD.

Dust Suppression:

The conventional mobile water sprinklers have been deployed for periodic sprinkling over haul roads etc. Besides one 28 KL automatic water sprinkler for each of the five projects has been procured which has proved very effective and convenient.

Besides, automatic mist type spray nozzles have been fixed at dust generation points.

A common difficulty faced in all projects is that in dry seasons, the sprinkling fails to produce desirable results as water evaporates very fast and surfaces become dry quickly. To get maximum effective suppression with minimum quantity of water consumption as well as with just adequate movement of the sprinklers, studies are being taken up to use chemicals which can retard the evaporation loss, besides creating bond between dust particles.

Environmental Management encompasses an important issue which is social environment.

For any business to survive & flourish, there has to be acceptability of the venture by local & neighbouring population. This can only happen if the people around the project or the people who get directly of indirectly affected by the establishment of the project have an inviting attitude towards the venture rather than an opposing stance. This may only come if the message goes that the project proponent will cause benefit to them in enhancing their prosperity.

In this direction, WCL has reached out to the close-by villages around the above named projects helping them on the road to better living and developing capability to become independent and confident members of the society. Trainings like tailoring, mechanic, goat-keeping, be-keeping, and other vocations have been imparted to all willing persons. In Umrer, the ladies of a village started making food items, packaging & marketing in the city of Nagpur. Even big departmental stores are now keeping the items and the first month's returns itself is encouraging by any standard.

CONCLUSION

It is our endeavour that these five projects will set the trend for all other projects regarding environmental & social mitigation activities and in near future, a new trend of comraderie between industry & people will dawn.

Waste Dump Reclamation and its Managements

T.N. Singh and V.K. Singh

Department of Mining Engineering, Institute of Technology, Banaras Hindu University, Varanasi 221, U.P., India

INTRODUCTION:

Today the globe is engaged in seemingly endless discussion about the sustainable development while the mining industry stands the cross roads involved in the problems of growth, sustainability and the imperative need of an environmental agenda. One of the issues of primary concern for mining industry especially open cast mines is of dumps and their reclamation. It is essential for a safe and healthy working environment in mines. Today the race for growth and productivity has led to faster depletion of mineral resources, a major part of which goes to open cast mining. Estimates indicate that for extracting about 20 billion tons of mineral raw material, the mining industry has to excavate more than 2000 billion tonnes of earth and rock mass. Globally of which about 90% is waste or 'residuals' which needs to be disposed off in an ecologically sustainable manner. Restricting to Indian mining industry, the major open cast projects of seventh and eighth five year plans will produce more than 20,000 million cubic meter of overburden which has to be handled by the various projects (Khandelwal and Majumdar, 1991). In fact, land degradation is inevitable in mining particularly in open cast and original soil ecosystem, structure, texture and horizonations are dratstically destroyed (Sooper and Seaker, 1984; Juwarkar et al. 1989; Juwarkar et al., 1992). The number of spoil dumps is increasing due to stock piling of overburden which do not have supportive and nutritive capacity, devoid of organic matter and microbial activities (Juwarkar et. al., 1992; Visser, 1985).

Reclamation:

With the new emerging trends in the advancement of technology, our mining industry has been equally affected by this increasing trend of automation. Due to advancement in technology it has been possible to exploit deeper and submarginal grade ores which could not be exploited earlier. Since the quantum of waste has increased almost exponentially, its disposal has created a problem because of its irretrievable damage to the environment. Waste dumps should be properly reclaimed and planning for it is a multi disciplinary task (Prabhakar and Kamraj, 1976). Thus, mine planning and reclamation planning should go hand in hand, and the surrounding environment can be improved by adopting reclamation techniques (Rao and Balakrishnan, 1976)

The reclamation of mine waste dumps situated on land is a major issue of concern and should be practised in order to support the following points:

- ❖ To mitigate the impact on general environment (Ghosh et al., 1984; Johnson and Bradshaw, 1978)
- ❖ To deploy the dumps into a productive use (Norton, 1990)
- ❖ To avoid the damage of sliding of dumps in case of dumps saturated with rain water (Akerss and Muter, 1974)
- ❖ To maintain the standards of aesthetic and scenic beauty (Ghosh, 1991)
- ❖ To assist in the procedure of ecosystem development (Hussain, 1990; Kleiman and Layton, 1979; Larsen and Vimmer, 1990)
- ❖ For the successful rehabilitation of mine spoils (Chatterjee, 1988)

In India, reclamation means afforestation of the waste dumps and of the tailing dams. There are also a few instances of land being put to use for other purposes such as agriculture (orchards, meadows or grazing land), recreation (recreation zones, parks or public open spaces), construction, biomass plantation for energy, water use (fishing, swimming boating and yatching) and wild life habitats.

Reclamation Techniques:

The various procedures recommended for reclamation of spoil and overburden, extraction of coal, reshaping of spoil, replacement of stock piles soil material, fertilizing, mulching and planting suitable mixture of plant species.

Top Soil Replacement

Topsoil should be taken into proper consideration as an ecosystem and not simply a physical resource. A number of scientists have recommended covering of the spoil with top soil of variable depth for preserving the productivity, natural habitat, protection of soil and water resources. Various soil properties like water holding capacity, nutrient supply capability, buffering capacity and plant root depth are affected by the thickness of topsoil. These soil properties control the plant growth, plant composition and nutritional quality of the vegetation.

213

Enhanced plant survival and growth, reduced run off and erosion, increased infiltration are facilitated by spread of 5cm of topsoil. It was observed that this topsoil prevents dispersion of spoils and thus prevents scaling and crushing. Dumps with high initial concentration of Ca and K and low concentrations of Fe require lime and fertilization. Hodder (1977) found that the top 5 cm soil layer has sufficient number of seeds for revegetation of the areas beyond their original densities. It was reported that many viable seeds and propagules from a number of species were present in the forest top soil associated with surface mining in Tennessee and practically all these soils and soil emerging from these propagules survived on topical mine soils and were well established by one growing season. Barth (1984) reported that the thickness of top soil for maximum production was 50 cm, 74 cm and indeterminate and no top layer required for generic, sodic, acidic and oil like spoils respectively. Most exact guidelines for soil replacement are based on spoil properties like texture, electrical conductivity and sodium adsorption ratio.

In the areas with undesirable sub soil characteristics, the soil profile can be segregated into "A" and "B" horizons and replaced in natural order over the regraded spoil material. With the top soil application, the microbial community speeds up the attainment of ecosystem stability in reducing the time required to establish the nutrient cycling soil aggregate formation, symbiotic relationship with plants and related function. The lower activity of microbial community was measured in spoil. However, when the topsoil covers the spoil, the microbial activity increases to 62% at a soil depth of 6 cm and upto 70% for soil depth of 14 cm. But further increase in soil depth from 14 to 150 causes no consistent change in microbial activity. It was suggested that P concentration could be enhanced in several crops by returning topsoil or adding manure. The effect of topsoil and sub soil replacement upon various parameters of crop quality after four years were evaluated.

When the spoil was covered with 20 or 60 cm thick topsoil, higher P concentrations were observed. Greater run off and erosion was observed when smooth spoil had steep slopes. Hence the replaced topsoil are quickly eroded from steep slopes in absence of vegetation. The addition of straw reduced the erosion and run off from the respread topsoil. If the reclaimed land is used for agricultural purposes, then the topsoil thickness should be more than 30cm. Compacted overburden exhibits better water holding capacity as compared with coarse and medium textured overburden. Thus compaction causes an increase in the water content at 0.3 bar (field capacity) excluding silty clay.

Chemical Amendment:
The spoil materials which are sodic and acidic in nature require chemical amendment so as to speed up the process of reclamation and to reduce the cost associated with top soil replacement. The sodic spoils of Northern Great Plains (India) require following chemicals associated with chemical reclamation:
(i)   Ca or Mg salt
(ii)   Enough water to transport the Ca ions to the cation exchange sites to displace the sodium ions and
(iii)   Sufficient hydraulic conductivity so that Na ions are transported out of the root zone.
It was reported that under the climate prevailing in most of the mine areas of Northern Great Plains addition of gypsum reduces exchangeable Na content by 30 to 50% in the upper 30cm of the material within few years after treatment. The Na replaced must be leached below the root zone, which is achieved by the use of mulches and follow in combination with gypsum. In an experiment, topsoil with straw and topsoil with gypsum was applied. They observed that the treatment with 5 cm of topsoil gives the best grass growth.

The benefit of top soil spreading and gypsum incorporation in mine spoils of various qualities were compared and it was concluded that application of gypsum is inferior to soil placement when sufficient amount of top soil is available for respreading. The nutrient in soil can be restored by periodically adding organic manure and fertilizers. In acidic spoils, 1% Fe rich pyrite generate lime requirement of 40 tonnes/hectare for pH control to 15 cm depth and 100 – 400 tonnes/ hectare of lime may be needed in high pyrite spoils for pH control to 45 cm depth to avoid the die back of vegetation after land reclamation.

Mulching:
Different organic mulches like stray, hay, crop residues were applied on spoils in a number of experiments in North Dakota, but no consistent effect upon exchangeable Na or other cations or upon salt, pH or other properties of the spoil were observed. Favourable results were reported during the application of waste materials like fly ash, sewage, sludge dredge spoil and organic waste on reclamation. An increase in the vegetable productivity of strip mine spoil was observed through the application of sawdust and green manure. Applying 5 cm modifies the microclimate or less of saw mill residues on drastically disturbed area. It was found that increased level of nitric acid, extractable Cu, P,

214

Zn, and pH in acidic mine spoil in west Virginia is a result of lime, garbage, mulch and sewage sludge but its effects were generally confined to 0-7.5 cm layer (Singh, 1994).

Schneider et al. (1981) suggested the use of sewage sludge for reclamation of coalmine spoils. Hay harvested from grasslands was used as a mulch and seed source for revegetation of mine spoils. Prairie hay provides seeds best adapted to the climate and soils for the mine spoil located in the prairie region. The effect of spoils shaping and water harvesting methods on the establishment s of perennial plants on semi arid coal mine spoil in New Mexico were evaluated by Scholl and Aldon (1979). Surface sealing materials, ground paraffin and a Silicon water repellent with a latex soil binder were used and it was reported that shaping of spoil surface for harvesting water, improved the growth of all perennial plants while sealing the surface area has little beneficial effect on plant growth.

Revegetation:
It is the most crucial step of reclamation. A proper decision is taken on the choice of species and method of revegetation after collecting and analysing the information on climate, physical characters of overburden, expected water availability in the root zone and anticipated use of the area after reclamation. Information on the mechanism and the course of natural vegetation is of profound assistance in planning a speedy recovery of the land.

A knowledge of physico chemical properties of the overburden material is a prerequisite to any revegetation plan. It was estimated that it may take upto 2160 years of natural succession on overburden for N pool to grow enough to support a stable, self sustaining plant soil system. The physico-chemical characters crucial to prediction of plant growth potential for overburden include textures, pH, electric conductivity, infiltration rate, water holding capacity, soluble Ca, Mg, Na, B, cation exchange capacity, exchangeable cations, gypsum and Ca carbonate equivalents and carbonate and bicarbonate, sulphate, chloride and nitrate content of saturation extract. It was felt that a large pool of organic N and a high rate of ammonification are necessary to sustain vegetation and to prevent N immobilization. In some situation, addition of P and seeding and transporting leguminous species can also stabilise steep slopes (greater than 60%)

Direct spot seeding at a proper rate is a better and cost effective method of revegetation for oak, pine black walnut and black locust as compared to planting of bare root seedlings on overburden. Mono and mixed stands of grasses and legumes were established successfully by direct seeding through drilling or broadcast with judicious inputs of N and P and proper timing with respect to rainfall regime and temperature conditions. It was observed that reclaimed native grasslands once established were equally stable from erosion so long as the land is not heavily grazed. An increase in topsoil depth from 5 – 60 cm increases plant production by 20 to 28% as a result of increased N uptake, which was observed from a modeling exercise. Addition of low N soil organic amendment reduces the primary production by 20% due to increased N uptake by soil microbes. While application of N fertility caused a six fold increase in primary production and when combined with irrigation, production increased by 100% compared to fertilizers alone.

Some recent technological advancements and innovations has led to evolution of newer techniques like:
Central drainage system through two concentric pipes, with outer pipe perforated with course gravel to control erosion and acid mine drainage for external dumps (Fig. 1). The top of the dump is being given inward gradient 0.5% for proper drainage and contour furrows are maintained. Erosion is being controlled by erecting bunds, masonary chutes, check dams etc. over the dump area. Polluted water through pipes and drainage ditches is collected on a collecting pond where it is treated with chemicals. These water than pass through settling pond and then reused for other purposes.
Concurrent reclamation of internal dumps with advancement of mining area (Fig. 2) will minimize the area of exposure and help in restoration of the degraded ecosystems (Fig. 3) (Bradshaw, 1978).

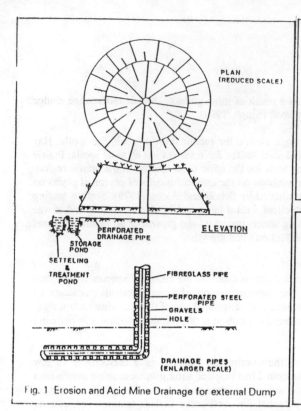

Fig. 1 Erosion and Acid Mine Drainage for external Dump

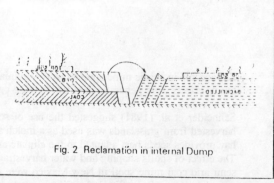

Fig. 2 Reclamation in internal Dump

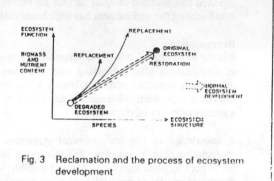

Fig. 3 Reclamation and the process of ecosystem development

Besides this, various institutes have undertaken considerable research related to land reclamation. The various problems of mine land, reviewing of previous reclamation techniques, waste characteristics has led to new dimensions in the field of reclamation (Mitchell, P.B, 1990; Alkinion et al., 1991). The various techniques evolved can be extended to other forms of toxic pollution, other than that arising from metalliferrous mine waste and could be used as an end of pipe technology to mine pollution problems from tailing disposal areas and dumps.

These in-situ techniques involve the treatment of contaminated land by the addition of ion-exchange materials, preferably with the minimum disturbance of any naturally vegetated or regenerated areas. The term amendment is used generically for the materials added to contaminated land, as it describes the spoil modifying nature of the added material (Atkinson et. al., 1990). The amendments are studied for a variety of characteristics like:
1. Theoretical cation exchange capacity (CEC)
2. Acid neutralisation capacity
3. Cation exchange characteristics from ideal metal sulphate solutions
4. Stability in acid media
5. Thermal stability

In application of the amendments it is however important to recognise that no specialised equipment is required, the amendment is ploughed upto a depth equivalent to the rooting layer. Apart from the introduction of this amendment, no other material including top soil, need to be brought onto the site native species to reseed naturally, to speed up the level of highest previous use.

Existing reclamation schemes for mining activities are either in the form of a reclamation bond or a reclamation fund (Indrapravish and Clark, 1994). The difference between these two forms is that a bond is primarily used for land reclamation, not for other environmental problems and is specific to a given mine or mining area, whereas a fund can be used for swider purpose, i.e. to solve other types of pollution problems or for reclamation of former unclaimed mine area with no responsible owner.

Most countries implement different measures depending on the size of the mine. Malaysia has different funds and management for large scale and small scale mines. The former is called mine Rehabilitation Fund and latter is called the Common Rehabilitation Fund (Otto, 1991)

CONCLUSION:
It is essential to have a proper reclamation plane before the mine operation. The topsoil should be preserved safely and utilized after reclamation, so that soil fertility can be useful for growth of plants.

The reclamation area should be used for recreative purpose like school, playground etc. priority should be given to minimise the local impact through a continuation of recent improvements in tip design and tipping practices including progressive restoration. Each site has individual requirements, which can be satisfied by site specific prepared by design engineers. Thus, it is essential to have action oriented long term research efforts.

ACKNOWLEDGEMENT:

Authors acknowledge their sincere thanks to CSIR, New Delhi and AICTE, New Delhi for financial assistance.

REFERENCES:

Atkinson, K., Edwards, R.P., Mitchel, P.B and Waller C.P. (1990) Roles of industrial minerals in reducing the impact of metalliferrous waste in Cornwal. Trans IMM 99: A158 – A172.

Akers, D.J. and Muter, R.B. (1974) Gob pile stabilisation and reclamation. Proc. of 4th Mineral Waste Utilisation Symp., Chicago, Illinois: 229-238.

Barth, R.C. (1984) Soil depth requirements to reestablish perennial grasses on surface mined areas in the North Great Plain, Mineral and Resources, vol. 27, no. 1:1-20.

Bradshaw, A.D. (1990) Ecological approaches to the handlin gof mine waste. J. of Mine and Mineral world, June '90: 22-28.

Chatterjee, P.C. (1988). Mine spoil and their rehabilitation, Mining and Environment in India. Ed. Koshi and Bhattacharya, HR. Publisher, Nainital: 394-409.

Ghosh, R. (1991). Reclaiming waste dumps lands of Jharia coal fields Eastern India. Int. J. of Surface Min., vol. 5, no. 4: 185-190.

Ghosh, S.K., Baliya, B.P. and Malik, G.P. (1984) Environmental control and reclamation. Coal Min. in India, CMPDIL Pub.: 303-307.

Hussain, A. (1990). Biological plant colonization on the lateritic waste dumps of bauxite mine – A case study. J. Mines, Metals and Fuels, Dec. 1990: 391-395.

Intarapravish, D. and Clark, A.L (1994) Performance guarantee schemes in the mineral industry for sustainable development.

Juwarkar, A.S., Malhotra, A.S., Kumar, A. and Juwarkar Asha (1989). Reclamation of mined land and minc spoil using sewage sludge and soil. National seminar on protection of environment and ecology by mining industry 1: 397-409.

Juwarkar, A.S., Juwarkar, Asha; Pande, V.S. and Bal, A.S. (1992). Restoration of manganese mine spoil dump productivity using pressmud, pp 827-830. In R.K. Singhal, A.K. Mahrotra and J.L. Collins (ed): environmental issues and management of waste energy and mineral production, Balkema, Rotterdam, Brookfield.

Johnson, M.S. and Bradshaw, A.D. (1978) Prevention of heavy metal pollution from mine waste by vegetative stabilisation. Trans. Inst. Min. Metal., 86A: 47-55.

Kleiman, L.H. and Layton, D.E. (1979). Reclamation technique and vegetation response of Decker coal. Symp. on Surface Coal Min. and Reclam., Kentucky, Oct. 23-25: 255-259.

Larsen, M.M. and Vimmer, J.P. (1990) Effect of mine spoil and seeded herbaceous species on survival of planted trees. Int. J. of Surface Min., vol.4, and no.2: 120-129.

Mitchell, P.B. (1990) Reclaiming metalliferrous mining land- with particular reference to Cornwall, U.K., Land and Minerals surveying 8: 7-17.

Mitchell, P.B. and Atkinson, K. (1991) The novel use of ion exchange materials as an aid to reclaiming direlict mining land. Minerals Engineering 4: 1091-1113

Niortaon, J. (1990) Use of mining wastes for reclamation. Mine and Quarry Environment, Vol. 4, no. 2: 10-14.

Otto, J. (1991) Model state mining legislation, Malaysia, United Nations

Prabhakara, S.K. and Kamraj, V. (1976) Advance planning on refuge dumping and reclamation and recultivation in mines areas. Nat. Sem. On Opencast Min., April 9-11, Neyvelli, India:133-138.

Rao, S.K. and Balakrishnan, M.S. (1976) Reclamation and afforestation on programme in the spoil heap and refilled area of Neyvelli opencast mine. Nat. Sem, on opencast Min. April, 9-11, Neyveli, India: 27-32.

Singh, T.N., Goyal, M. and Kumar, P. (1994) Impact of waste lead –zinc mines, Int. Symp. on mineral beneficiation, Recent trends beyond 2000 A.D., Nagpur (India)

Sopper, W.E and Seaker, E.M, (1984) Strip mine reclamation with municipal sludge. Project summary EPA 600/82-84/035, Municipal environmental Res. Lab. USEPA Cincinnati, O.H. Stevenson, F.J. 1982. Humus Chemistry. John Wiley & Sons, New York.

Schneider, K.R., Wittweer, R.F. and Carpenter, S.B., (1981) Trees respond to sewage sludges in reforestation of acid spoil, Proc. Symp. on surface mining hydrology, sedimentology and Reclamation, Univ. of Kentucky, Lexington, Kentucky, pp. 291-296.

Scholl, D.G. and Aldon, E.F., (1979) Water harvesting to establish perennial plants on semiarid coal mine, Wali, M.K. (ed) Ecology and coal resource development, vol. 2, Pergamon press, NY: 1724-1728.

Visser, S. (1985). Management of microbial process in surface mined land reclamation in Western Canada, pp. 203-341. In R.L. Tate and D.A Klein (ed). Soil reclamation processes. Marcel Dekker, New York.

Eco-friendly Iron Ore Mining at N.M.D.C.

D. Rajasekhar[1], M.H. Sheriff[2] and G.S. Naidu[3]

[1]*Dy. Manager (Env. Engg.),* [2]*Sr. Manager (Env.) and* [3]*Dy. General Manager (Env.)*
NMDC Ltd., Hyderabad, India

PREAMBLE

India ranks 5th amongst the major iron ore production and 4th in the list of major iron ore exporting countries. There are proved reserves of 10.6 billion tonnes of hematite and 3.1 billion tonnes of magnetite ores in our country. With increase in activities, during the years to come, it is necessary to adopt environmental friendly mining practices with a pragmatic approach for creating sustainable development.

National Mineral Development Corporation (NMDC) Ltd., a public sector company incorporated in the year 1958 under the Ministry of Steel, GOI, is a pioneer organization in developing various mineral deposits in India and abroad. The corporation has to its credit three highly mechanised iron ore mines (2 at Bailadila in M.P and 1 at Donimalai in Karnataka) with annual production of 17 million tonnes of ROM ore and the only mechanised diamond mine at Panna in M.P in the country with an annual production of 41,000 carats of diamonds. Technical breakthroughs like developing ferrite powder, pigment grade ferric oxide and ultra pure ferric oxide from the "blue dust" (a form of powdery iron ore) are some of the credits of its
UNIDO recognized R & D unit, which is coveted as the centre of excellence. The corporation is a glorified "mini-ratna" with annual net profit earnings of well above Rs.200 crores apart from invaluable earnings of foreign exchange by exporting iron ore to Japan, S.Korea, China, Pakistan and other countries.

CORPORATE ENVIRONMENTAL POLICY

Corporate Environmental policy is for safe and scientific mining wherein environmental attributes are not seen in isolation. Eco-governance is duly embedded into the MIS activities of the Corporation.

THE 3-E's CORRELATION (EXCAVATION, ECONOMICS AND ECOLOGY)

Beattle EIA is adopted to identify every environmental hazards with respect to each operation involved at NMDC. This is as below.

EXCAVATION

Controlled blasting with milli-second delay detonators, optimising explosive charge per hole, adequate spacing are adopted. Shovel-dumper combination is used to transport the excavated ore upto the Crushing plant, which is situated on the hill top. The lead distance between the different ore faces and the crushing plant is 1-3 Km.

Controlled blasting eliminates ground vibration and air blast problems. Blasting is adopted only during afternoon periods when greatest instability in atmosphere prevails. This leads to greater dispersion of air pollutants and thereby less GLC. Dumpers are adequately maintained to prevent heavy automobile pollution in the mining area. Haul roads are regularly wetted along with a chemical compound called "Syntron" for reducing water consumption and increasing adsorption capacities of loose soil to the ground.

Waste rock as shale, phyllite, BHQ is dumped at strategic places devoid of thick vegetation, steep slopes and water courses. At Bailadila as the total area is falling under reserve forest, there is no option except to select such places where the crown density of forests is less than 0.2.

ORE CRUSHING

Dumpers unload the ROM ore into the Primary crusher at "Dumper Platform". Gyratory crushers are used in the primary crusher while cone crushers are used to crush the ore further at Secondary and tertiary crushers. The crushed ore is transported to the Screening plant through closed conveyors and eco-friendly tunnel which generates electricity when the transmission rate exceeds 1600 tph.
Atomised mist spray of water takes care of fugitive emissions during dumper unloading operations.

SCREENING PLANT

The crushed ore is washed at the Screening plant basically to improve the Fe%. Quantity of gangue minerals like silica and alumina gets reduced. About 0.7 cum of water is required for washing every tonne of iron ore. Slimes to an extent of about 15% by volume is generated which contains lateritic material as well as micro fines of iron ore. Maximum recovery of fines is being achieved by adoption of hydrocyclones, slow speed classifiers and thickeners.

To take care of water pollution due to screening operations, a tailing dam is construction and the effluents from led into it for proper clarification. Desilting to an extent of 4-10 lakh cum is envisaged every year from of these tailing dams and the material is either dispatched by rail or stored at a place away from the catchments of natural water courses.

LOADING YARD

Crushed of iron ore gets commercial market in three forms viz., lumps (10-150mm), calibrated lump ore (6-30 mm) and fines (-10mm). These products are dispatched by electrified BG line in Box-N wagons for exports and also for use at various steel plants in the country.
To control water pollution at the stock yard, an interconnected drainage network is established and the final discharge passes through a sedimentation basin. Thus only clarified water is let out to meet the natural drainage.

ALLIED INFRASTRUCTURAL FACILITIES

Service centre, Auto workshops are other areas wherein certain air, water and noise pollution are expected. Water pollution mainly in the form of form oils and grease gets treated through a ETP. Air pollution is controlled by good maintenance of all vehicles and house keeping. Noise pollution is controlled through development of green belts and by providing ear muffs to employees at places unavoidable of reduction in noise levels.

TOWNSHIPS

Well designed oxidation ponds in the form of facultative stabilisation ponds is established for treatment of domestic sewage. This treatment is selected due to abundance of insolation (>2500 langleys) and availability of land. Also, it is very cheap and rarely requires maintenance.

Thus it may be seen that every activity has embedded environmental protection into it. In addition to the above, the following methods are also in vogue for protection of ecology and environment.
1. All fine ore dumps are provided toe walls in the form of chain link mesh with boulders to protection erosion of fines.
2. All waste rock dumps are provided with retaining walls/ toe walls for controlling flow of scree due to precipitation run off. Storm water and garland drains are provided to stream line the runoff.
3. Wind breaks are created by scientific afforestation so as to avoid ground contact. More than 22 lakh saplings have been planted over an area of about 1,400 ha. The survival rate is healthy at +90%.
4. Fully covered RCC diversion channels over an length of 1.9 km is provided at the fine ore dump of Bailadila Dep-5 project for avoiding contact with runoff.
5. Medicinal plant nursery is established at Bailadila Dep-5 project in addition to full fledged nurseries at every project for the purpose of supply/ protection of saplings before they are planted. This is an humble exercise to increase the undergrowth.
6. Noise control is taken care of, by mechanising the complete operations and thus reducing the contact time of employees directly exposed to such noise prone areas. Further, ear muffs are

220

provided as an occupational safety measure.

7. Wild life corridors are created for avoiding outer migration and man-animal conflicts.

8. Many areas have good "Climax community". Afforestation outside these areas not only protect this natural vegetation but also, acts as an extension as the Importance value of the region is duly regarded.

9. Check dams are constructed at strategic places basically to reduce the velocity of water flow thus helping in arresting silt loads to a good extent. Regular desilting of these check dams is an ongoing practice.

10. Creation of parks with musical fountains, well laid out roads with avenue plantation and display of sign boards on environmental protection creates awareness amongst the employees as well as the local tribal people, without whose support and help, it is near impossible to maintain fragile eco-balance.

11. Concurrent reclamation and rehabilitation of mined out areas and waste rock dumps to the extent possible.

MONITORING MECHANISM

Environmental engineers and scientist with adequate exposure are recruited and posted at Head office and projects for looking after the environmental aspects. Horticulturists, Mining Engineers, Hydro-geologists and Geologists guide these core people for the cause.

Regular environmental quality data generation is done at all the projects and the data is assimilated for providing betterment of results. Annual Audits are also done for temporal assessments. Use of satellite data viz., IRS-1c-LISS III and PAN, Arc Info/ Erdas/Easypace softwares are used for assessing the density of plantation done in the Mining lease areas and outside, land use planning for avoiding visual intrusion, etc. Exploring these areas into the water and air pollution aspects are planned in the near future.

BUDGETS

Necessary budgets are provided for environmental protection both as capital and recurring investments. Capital investments include those items which can be included in the "Gross block" while the revenue items include the routine measures for supplementing to the cause of sustainable development. These budgets are never diverted for any other cause. During the previous year 1999-2000, about Rs 10 crores has been spent only on environmental protection at 4 major production projects of NMDC.

CONCLUSION

The authors realised that there are three major factors that are responsible for eco-governance.

I. Creative, proactive and benevolent apex management.

II. Support and awareness amongst all employees and the surrounding population.

III. Need based thorough examinations on pollution instead of orthodox generalised theories and providing amelioration instantly.

ACKNOWLEDGMENTS

The authors are deeply indebted for the constant inspiration given by Mr.P.R.Tripathi, CMD, Shri S.K.Agrawal, Director(Production), Shri V.Satyanarayana, Director(Technical), Shri V.Rajagopal, Director(Finance) and Shri Murli Manohar, GM(I,E&TS) of NMDC, Hyderabad.

NOTE

Any part or fact of this paper is purely the view of the authors DRS, MHS & GSN and not necessarily that of NMDC.

Remediation of a Former Sewage Treatment Works and Return to Public Open-space, Exeter, UK

John R. Merefield[1] and David P. Roche[2]

[1]*School of Engineering & Computer Science, University of Exeter, Harrison Building, North Park Road, Exeter, EX4 4QF, United Kingdom*
[2]*GeoConsulting, 19 Richmond Road, Exeter, Devon EX4 4JA, UK*

INTRODUCTION

It was planned to develop the part of the Belle Isle Nursery site, Trews Weir, Exeter, United Kingdom for use as a public open space (Figure 1). In view of part of its present-day usc as a nursery and former uses which included a sewage works which could have contaminated the site, Exeter City Council decided that a contaminated land survey was necessary.

The main aim of this survey was to ascertain whether the site was or was not chemically contaminated. Accordingly, the University of Exeter was contacted as they were based locally and had successfully carried out similar investigations in the south west of England. As it seemed likely that the land contained some contaminated areas, the investigating team called in supporting consultants with experience in remediation of such 'brownfield' sites.

METHODOLOGY

Site Walkover

After an initial meeting with officers of Exeter City Council at Belle Isle Nursery a site walkover was conducted during May 1993. This enabled planning for the site investigation in order to meet the conditions laid down for the identification of potentially contaminated land. In the United Kingdom guidance is issued by Her Majesty's Government in the form of reports from the Interdepartmental Committee on the Redevelopment of Contaminated Land (ICRCL 1983; BSI 1988).

At this early stage, it was recommended that eight shallow (1 meter depth) trial pits be dug to permit sub-surface soil sampling. A ninth sample was to be taken from the soil mound at the SE end of the site. Heavy metal analyses including lead and cadmium were to be carried out on all the samples and an organic screening on three more selected to represent the site as a whole.

Sampling

For administrative reasons, the decision to proceed with the survey was delayed until October 1995. However, the sampling plan was to be retained as before with the addition of one sample from a new soil heap. Exeter City Council dug the pits by JCB and sampling was conducted. During the course of this work, an additional pit was dug, as local knowledge confirmed this to be central to the former sewage operations. Seventeen samples were taken from the trial pits near surface and sub-surface and two samples from the two spoil mounds (Table 1), by means of a screw soil auger. They were sealed in labelled plastic bags for subsequent analysis of heavy metals. The three samples for organic-screening were collected separately in acid/organic solvent washed glass bottles and stored in a cool bag prior to analysis.

Table 1. Description of the trial pit soil samples from Belle Isle Nursery.

Sample no.	Depth. metres	Comments
TW 01a	0.2	brown clay-rich soil
01O**	0.2	
01b*	0.8	gravel horizon 0.5m
TW 02a*	0.1	crumbly soil
02b	0.7	large stone (limestone) w end of trial pit dark bands, red brick remains
TW 03a*	0.1	clay-rich surface soil
03b	1.0	pebbles/cobbles with light red clay
TW 04a*	0.1	evidence of bonfires, dark humus-rich layer down to 0.5m., red zone >0.5m depth
04b	0.6	red clay-rich zone
TW 05b	0.5	dark humus-rich surface soil down to 0.3m red oxidised zone >0.3m, red clay bricks, wall remains
05O**	0.5	
TW 06a	0.2	dark black humus horizon, fine
06b	0.6	red-stained pebbles & red clay sub-soil
TW 07a*	0.1	dark black humus top-soil
07b	0.5	red clay-rich sub-surface zone
TW 08a	0.1	dark black fine humus topsoil
08b*	0.6	red clay, bottles, New Red Sandstone cobble
TW 09	top	tip site, brown humus, leaves, plastic bags
TW 10	2m from top	recent above-ground level mound of light brown soil with pebbles & cobbles
TW 11b	0.5	site of sewage settling area
11O**	0.5	
11c	1.0	thin light brown surface soil & red oxidised clay-rich zone below

denotes sampled but not subjected to analysis
**sampled for organic screening only*

Analyses

Soil samples were oven-dried overnight at 50°C and ground in a TEMA tungsten carbide disc mill for 4 minutes. Five grams of the powder were mixed with 4 drops of 2% aqueous polyvinyl-pyrrolidone in an agate pestle and mortar, backed with boric acid and pressed at 3 tons for 3 minutes to provide a pellet for X-ray fluorescence spectrometry (XRFS). The sample was then analysed for heavy metals using a Philips PW1400 sequential X-ray spectrometer with reference to an international set of calibration standards.

Approximately 30 g (fresh weight) of soil were transferred to a 100 ml glass bottle. Deuterated (d8)-napthalene was added to the soil as an internal standard to a final concentration of 4.7 ppm (mg/kg) and mixed thoroughly to ensure homogeneous distribution throughout the soil matrix. Fifteen ml pentane and 5 ml acetone were added and the sample sonicated for 2 hours to maximise contact between solvent and soil particles. After sample acidification with 2 ml 10% nitric acid, clearing by filtration, cleaning through activated flurosil and elution with a 95:5 mixture of pentane : toluene, a small volume of 1-bromonapthalene was added as a marker to enable accurate determination of the final volume of the extract. Extracts were analysed by gas chromatography with mass spectrometry (GC-MS) using a Hewlett Packard 5890 series gas chromatograph linked to an HP 5972 mass selective detector.

RESULTS

Trial Pit Results
Results from the initial heavy metal analyses are given in Table 2.

Table 2. Heavy metal analyses of soils from Belle Isle Nursery.

Sample no.		Lead Pb	Arsenic As	Zinc Zn	Copper Cu	Nickel Ni	Chromium Cr	Cadmium Cd
TW	01a	303	n.d.	189	66	49	75	n.d.
	02b	1054	n.d.	412	79	51	72	n.d.
	03b	160	n.d.	139	44	45	66	n.d.
	04b	91	n.d.	57	29	48	82	n.d.
	05b	1259	n.d.	258	54	157	195	n.d.
	06a	2318	n.d.	1209	285	56	214	n.d.
	06b	178	n.d.	231	41	40	67	n.d.
	07b	310	n.d.	181	72	45	80	n.d.
	08a	1520	n.d.	742	261	70	168	n.d.
	09	86	2	153	41	44	62	n.d.
	10	184	n.d.	137	38	30	57	n.d.
	11b	25	17	67	24	37	67	n.d.
	11c	70	6	99	29	59	103	n.d.

All results given in parts per million (ppm)
n.d.: not detected

Concentrations for lead proved particularly high in near-surface samples, 6a (exceeding 2000 ppm Pb) and 8a (1520 ppm Pb), and in sub-surface samples 2b and 5b (Table 2). Thus all three areas at Belle Isle Nursery were contaminated with lead. As ICRCL trigger levels for lead are 2000 ppm threshold for parks and open space, these values gave cause for concern. Similar guidance from the Greater London Council (GLC-Kelly values) classes lead concentrations between 500-1000 ppm Pb as slightly contaminated and 1000-2000 ppm Pb as contaminated. Importantly, highest chromium values correlated with highest lead values. This strongly suggested residual material from the former sewage works to be the likely source of this contamination.

Results of the organic screening showed surface sample TW01O, from 0.2m depth yielded a wide range of organic contaminants, some at relatively high concentrations. Of 76 compounds isolated, 51 were identified and 22 gave match qualities of greater than 90%. The majority of those identified were polycyclic aromatic hydrocarbons (PAHs) or their derivatives (25 compounds), or derivatives of the bicyclic molecule napthalene (7 compounds). Biphenyls, furans and thiophene derivatives were also prominent. Very few aliphatic (long-chain) compounds were identified. The three most prominent peaks were identified as phenanthrene, fluoranthene and pyrene and present in concentrations significantly above that of the standard (approx. 7 ppm, or mg/kg, wet weight) in the original soil sample. The other compounds identified were present in the high parts per billion (ppb) to low parts per million (ppm) range. The predominance of bicyclic and polycyclic aromatics and near absence of long-chain aliphatic hydrocarbons suggested that the contamination did not result from spillage or seepage of motor or other lubrication oils. Contamination from creosote or other wood preservatives which may have been used on the site during its time as a plant nursery appeared more likely. Subsurface samples TW05O and TW11O were much less contaminated with organics, both in the number of compounds isolated and their relative concentrations. Nevertheless, there was some evidence of contamination at site 5 with low levels of aromatic hydrocarbons including the PAH pyrene. Fourteen compounds were isolated from sample TW11O including alcohol, 2 ketones and a complex organic acid additive. These compounds can be of biogenic origin and could have related to the former use of the site as a sewage treatment works.

Soil Auger Survey

Budgetary constraints dictated a phased approach to sampling and analysis with each outcome dictating the next step. Initially, thirty near-surface (to a 30cm depth) soil samples were taken in July 1996, by means of a screw soil auger. The samples were from strategic locations across the area destined for new end-use as a public open space (Fig. 1). An additional set of 5 samples from soil stored at nearby Trood Lane landfill site and intended for cover at Belle Isle, were also analysed to assess its status. The five (3 from brown silty soil & 2 from red sandy soil) were taken for lead analysis. Two more were sampled in July for organic screening.

Soils were pH tested in the field at Belle Isle using a Solomat portable pH/Eh metre and equal volumes of soil and de-ionised water. They were returned to the laboratory for repeat analysis at the end of the day.

Additionally, a further set of 23 soil samples were collected from the site during August for lead analysis along margins where soil remediation would necessitate construction of retaining walls, and to confirm the most elevated levels of contamination. Lead (Pb) was selected as the key indicator of metal contamination at this site, based on the initial analyses, and enabled the more detailed analyses to be targeted with reduced costs. An additional soil sample was collected at the Trood Lane site for quantitative analysis of benzene, toluene and xylene.

The lead results gave a range from 61 ppm to 2437 ppm Pb. The highest result was similar to that from the initial survey 2 at 2318 ppm Pb, whilst lead concentrations at 2006, 1848 and 1747 further confirmed the contamination at the southern half of Belle Isle site. Samples analysed from the central section of the site better defined the hotspot there, the new values giving from 900 to 1397 ppm Pb. Attention was paid to the area supporting 3 lines of trees south of the E-W aligned path crossing the centre of the site. Concentrations of lead in the soils there proved relatively lower (271 to 860 ppm Pb) except for one site away from the trees (1493 ppm Pb) at the junction with a riverside path. Attention was also paid to the north of the site as the desk study had uncovered previous structures there related to the site's former use as a sewage works. Values here proved relatively low, from 99 to 207 ppm Pb.

In order to plot the area of contamination, these Pb values plus the 30 obtained from the second soil survey were contour plotted using Kriging interpolation based on a linear variogram model. Results from this are illustrated in Figure 2. Areas above 1000 ppm Pb and 2000 ppm Pb are shaded to indicate the key sectors requiring remediation. The three sites tested for soil pH, representative of low, intermediate and high levels of lead, proved neutral giving a range from pH 7.33-7.60. Samples from Trood Lane identified for remediation at Belle Isle gave low lead values from 17-28 ppm Pb. The organic screening of the topsoil indicated traces of long aliphatic side chain benzene derivatives, which might have been derived from fuel oil. As a precautionary measure, therefore, a further sample was tested for quantitative analysis of benzene, toluene and xylene which all proved below .02 milligrams per kilogram (parts per million).

GEO RECLAMATION: REMEDIATION STRATEGY
The nature and extent of the contamination identified at the site gave rise to two possible options for remediation:

- Removal of the 'hotspot' areas by excavation (to about 1m depth) and disposal off-site at a suitably licensed landfill, and replacement by importation of clean soils.
- Cover over 'hot spot' areas by importation and placement of clean soils (up to 1m depth) to a slightly raised/domed profile.

Covering over the hotspot areas was chosen in this case, as it would prove the most cost effective and appropriate remedial treatment for the site. Without the costs of excavation, removal and disposal, it was possible to maximise the area of cover treatment and thereby minimise future risk to public users. The importation of clean soil cover provided an effective separation between the contamination in the ground and the main potential targets at risk, namely people and especially children playing and gardeners working on the ground.

Key elements of the works now comprised:
- consultations and approvals from regulatory authorities
- a Remediation Plan, defining methods, procedures and safety precautions
- further sampling and testing to narrow the limits of excavation and to check new cover materials if won on site
- CDM requirements including Health and Safety File
- Engineering design, specification, quantification and cost estimates.

CONCLUSIONS
Taking the initial results of the inorganic and organic analyses together, soils of the Belle Isle Nursery site, Exeter, UK were proven to be contaminated. The selective survey indicated this pollution to be just above the UK ICRCL threshold guidelines and it was a matter of professional judgement as to whether action was needed. However, the site proved unsuitable for direct development as a public open-space amenity without prior remedial action. A subsequent more detailed soil survey was thus undertaken based on Pb analysis to accurately define the limits and to assess the scale of this contamination. The feasibility of remedial action at Belle Isle could then be assessed by specialists with reference to the scope, health & safety aspects, quality assurance and cost implications.

Key elements of the survey comprised:
- A desk-top survey of previous use
- A site walkover
- Examination using trial pits
- Organic screening of trial pit samples
- Inorganic analysis for heavy metals
- Additional sampling and analyses to delimit 'hotspots' for remediation

Key elements of the Remediation comprised:
- Use of lead contour plots to mark out areas requiring remediation
- Covering of the 'hot spots' by importation and placement of clean soils (up to 1m depth) to a slightly raised/domed profile
- Analysis of imported soils before and after remediation

ACKNOWLEDGEMENTS

Grateful thanks are due to Exeter City Council for their permission to publish this paper and to English Partnerships for the funding required to carry out the remediation works.

REFERENCES

ICRCL. Interdepartmental Committee on the Redevelopment of Contaminated Land (1987). Guidance on the assessment and redevelopment of contaminated land. (ICRCL) 59/83: Second Edition, July, 18pp.

BSI. British Standards Institution (1988). Code of practice for the identification of potentially contaminated land and its investigation. Draft for Development. DD 175:1988, 28pp.

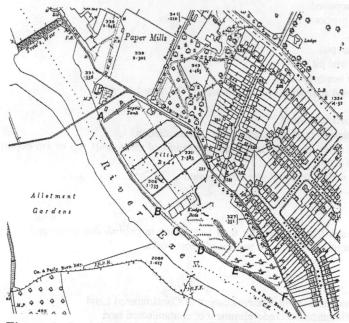

Figure 1. Location Map, Belle Isle Site Exeter, UK.

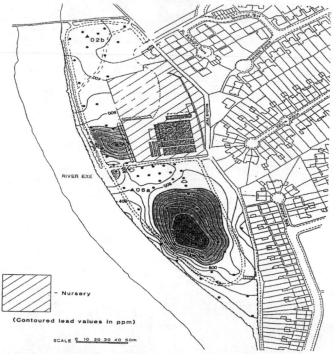

- Nursery

(Contoured lead values in ppm)

SCALE 0 10 20 30 40 50m

**Figure 2. Soil sampling sites at Belle Isle, Exeter
with Pb contours delineating the metal contamination.**

Environmental Management of Manganese Mine Spoil Dumps

D.K. Sahani[1] and Asha Juwarkar[2]

[1]MOIL, *Nagpur (India)*
[2]NEERI, *Nagpur, India*

Introduction

Mining is one of the major contributor for the creation of wasteland in the country. In India, around 0.8 million hectares of land is under mining affecting the local environment drastically. Mining specially opencast, drastically destroys the land ecosystem, structure, and microbial community. Several microbial processes such as nitrogen and carbon cycling, humification and soil aggregation are practically non-functional and hence do not support biomass development. It is also responsible for contamination of water bodies, stream, groundwater and adjoining lands leading to severe environmental problems.

Revegetation and reclamation of mine land offers a great challenge as more often they do not have suitable surface soil to provide bedding layer to encourage and support the biomass development, while in other cases they do not support plant growth due to presence of toxic metals. Gradual increase in such landscapes due to extensive mining activities may endanger not only the agro-forest productivity of the country but also the ecosystem and ecological balance. Hence, R&D is urgently warranted for the reclamation of degraded land in open cast mines through plantation with a view of restoring its productivity and fertility as also the ecological cycles in the rhizosphere with identification of appropriate plant species for plantation on such land.

The Integrated Biotechnological Approach (IBA) was used for the reclamation of manganese mine spoil dumps at Dongribuzurg and Gumgaon mines for field trial and technology was extended on large scale at Tirodi mine. It envisages the use of organic wastes of diversified origin (industrial, domestic, etc.) as an ameliorative material enables the development of supportive and nutritive capacity of degraded lands. The introduction of specialized strains of endomycorrhizal fungi promotes stress tolerance in the plants, while the *Rhizobium* and *Azotobacter* help to meet the requirements of nitrogen for plant growth, thereby promoting self sustainable land ecosystem.

Methodology

Manganese Ore India Limited (MOIL) undertook a collaborative project National Environmental Engineering Research Institute (NEERI), Nagpur to develop and implement a scientifically planned IBA at Gumgaon, Tirodi and Dongribuzurg manganese mines for solving the problems of mine spoil dumps by restoration and revegetation through biological reclamation process. Based on extensive laboratory study carried out to ascertain the improvements in spoil quality to register proper plant growth by using an appropriate blend of soil, spoil, organic matter and site specific biofertilizer strains, the IBA was extended at field level to reclaim manganese mine spoil dumps at above mentioned locations. For field trial, IBA was implemented on an area of 10 hectare each at manganese mine sites viz Gumgaon and Dongribuzurg. For plantation around 2500 plants of high economical and ecological importance were planted. While at Tirodi mine, IBA was demonstrated on an area of 10 hectares and 16,000 plant saplings were planted.

Selection of Suitable Ameliorative Materials for Manganese Mine Spoil Dumps

Depending upon the suitability, economic feasibility and easy availability near mine sites, organic amendments like pressmud (a waste material from sugar factory) and farm yard manure (FYM) were selected as ameliorative material before using these ameliorants were analysed for their physico-chemical and microbiological properties (Black, 1965).

Isolation and Identification of Nitrogen Fixing Bacteria and VA Mycorrhizae

The symbiotic nitrogen fixers for leguminous plants were isolated from the root nodules of leguminous plants viz. shishum and cassia and was identified as *Bradyrhizobium japonicum* on the basis of morphological , biochemical and cultural characteristics. The nitrogen fixing strains for nonleguminous trees viz. *Azotobacter* was isolated from rhizosphere of nonleguminous plants viz. shiwan and teak by Serial Dilution Pour Plate method and specific media. It was identified as *Azotobacter chroococcum* on the basis of morphological , biochemical and cultural characteristics. Vesicular arbuscular mycorrhizal fungi were isolated from rhizospheric soil samples were collected from plants growing near mine sites by Wet Sieving and Decanting Method. The isolated spores were identified as *Glomus* and *Gigaspora* sp. on the basis of their size, shape, colour, stock attachment and wall thickness. The mass inoculum of *Rhizobium*, *Azotobacter* and VAM were prepared in the laboratory. The mass inoculum preparation of *Rhizobium* and *Azotobacter* was done in 20 litre capacity fermenter (Bioengineering Alt Segenrainstrasse Ch. 8363 Wald. Made in Switzerland) with 18 litre working capacity by using Yeast extract mannitol medium and Gensen's medium respectively. For large scale production of VAM spores, *Bryophyllum* plant was used as a host plant for multiplication of VAM spores. The plastic bag containing soil and sand were planted with host plant inoculated with VAM spores. After 3 months the entire content of the bag having germinated spores fungal hyphae and mycorrhizal roots was used as VAM inoculum.

Selection of Plant Species

Plant species were selected on the basis of their ecological, economical importance and availability in the mining area and growth performance in the pot studies conducted with different blends of spoil and organic wastes. On this basis different plant species like teak (*Tectona grandis*), neem (*Azadirachta indica*), shishum (*Dalbergia sissoo*), shiwan (*Gmelina arborea*), cassia (*Cassia seamea*), mango (*Mangifera indica*), awla (*Embelica officinalis*), bamboo (*Dendrocalamus strictus*), etc. were selected. These plant species with selected blends of organic wastes were planted on different mine spoil dumps and were inoculated with suitable site specific strains of biofertilizers.

Methodology Used for Plantation at different Sites

Pitting technique was adopted for plantation on slopes of mine spoil dumps. Pits at the slope of dump were of 0.6 m x 0.6 m x 0.6m dimension and at elevated level were 1m x 1m x 1m size dimensions. Each pit was filled with 4 parts of spoil + 1 part of soil. The most responsive treatment selected for plantation at Gumgaon and Dongribuzurg mine sites was 4 part of spoil + 1 part of soil + pressmud @ 100 t/ha and for Tirodi mine treatment used was 4 part of spoil + 1 part of soil + FYM @ 50 t/ha. The VAM spores (10 g) having approximately 30 spores) were applied to each pit by mixing with the bedding material to enhance the nitrogen fixation, development of profuse root system in plants, solubilization and mobilization of nutrients. The saplings of selected plant species i.e. leguminous plants viz. shishum, cassia, etc. were treated with respective site specific cultures of *Bradyrhizobium japonicum* while the non-leguminous plants viz. teak, shiwan, etc. were treated with *Azotobacter* sp. by root inoculation method before planting in pits at the rate of 10^5 cells per sapling for leguminous and nonleguminous plant species.

Improvement in various physico-chemical properties of mine spoil and microbial population such as bacteria, fungi, actinomycete, *Rhizobium*, *Azotobacter* and VAM spores was determined by suitable methods.

Results and Discussions

Physico-chemical Characteristics of Manganese Mine Spoil Dumps at Various Sites

The manganese mine spoil dumps selected for R&D at Dongribuzurg, Gumgaon and Tirodi was devoid of vegetation because of poor physico-chemical properties. The bulk density of mine spoil at three different sites was high and it ranged from 1.39-2.0 g/cc and had low water holding capacity in the range of 5.20-29.31%, while the porosity varied from 16.2-49.28%. The mine spoil had pH in the range of 5.71-7.8 and contained low soluble salts (ECe - 0.14-0.78 mS/cm). Organic carbon content of mine spoil was low which was in the range of 0.102-0.77 %. Nutrients such nitrogen, phosphate and potash were also low. Although heavy metals like Cr, Pb, Cu, Mn and Zn were present in spoil but they were not of serious concern. Characteristics of manganese mine spoils from various sites is given in **Table 1**.

Microbiological Status of Barren Manganese Mine Spoil Dumps at Various Sites

Manganese mine spoil from different sites were characterized with respect to microbial parameters and results indicated that mine spoil had very low profile of microflora due to surface mining activity (**Table 2**). The top soil which is often rich in organic matter and harbor a highly diverse and active microbial community is disrupted causing a drastic alteration of bacterial, fungal population and of other microbial groups, with a concomitant decrease in microbial activity and biomass. The total count of bacteria was in the range of 38-54 CFU/g whereas fungi were in the range of 9-25 CFU/g. Actinomycetes which are usually dominant in organic amended soil was absent in Dongribuzurg and Gumgaon mine, whereas at Tirodi its counts were very low. *Rhizobium*, *Azotobacter* and VAM were absent in all the three mine sites. The limited microbial growth was attributed to low moisture status of spoil, extreme temperature of spoil, adverse physical and chemical factors and above all low levels of organic matter. These constraints could be resolved by use of organic waste amendments for establishment of productive rhizospheric microbial population.

Physico-chemical Characteristics of Ameliorative Material used in the Study

Pressmud was used as an organic amendment for revegetation at Dongribuzurg and Gumgaon mine sites. It was found to be a good source of organic carbon (37-38.2%) and also rich in plant nutrients. It also had good water holding capacity (80.20-84.68%). A ton of pressmud provides 14.8 kg nitrogen, 57.8 kg phosphate and 21.9 kg potash besides micronutrients. Presence of sugar in the pressmud help in fast reestablishment of microflora. At Tirodi mine, FYM was used as an organic amendment. FYM is a readily available organic amendment of domestic origin. It has high organic matter (39.6-42.2%) with appreciable water holding capacity (71.5-73.3%). Besides improving the physical properties of mine spoil like water holding capacity, porosity, etc. required to develop good structure, it also provides good nutrient source in terms of N,P and K.

Effect of Ameliorative Material and Use of Mycorrhizae-biofertilizers on Survival of Plants at Manganese Mine Spoil Dumps

On mere spoil dump plant showed less than 18% survival due to low nutritive and supportive capacity of the dump. Use of ameliorative material like pressmud at the rate of 100 t/ha at Dongribuzurg and Gumgaon sites and FYM at the rate of 50 t/ha at Tirodi site along with endomycorrhizae and respective biofertilizers enhanced the plants survival rate on dumps. On an average 90 to 100% survival of plants was observed on mine spoil dumps at the three sites under study. The higher survival rates of plants was due to the development of better rhizosphere and use of encomycorrhizal which enabled the plant to tolerant stress conditions.

Effect of Mycorrhizae-biofertilizer Inoculation and Use of Ameliorative Material on Plant Growth at Manganese Mine Spoil Dumps

Growth performance of plants grown on manganese mine spoil dump after addition of pressmud/ FYM as ameliorative material alongwith top soil boosted the plant growth many fold. Inoculation of leguminous plants with endomycorrhizae-*Rhizobium* and nonleguminous plants with endomycorrhizae-*Azotobacter* further enhanced the plant growth. Further mycorrhization of plants with respective inoculum of biofertilizers enabled plants to develop stress tolerance as well as promoted profuse root development. This resulted in high nitrogen fixation which is exhibited through excellent plant growth. The growth performance of different plant species in terms of their height planted on manganese mine spoil dumps is given in **Table 3**.

Improvement in Physico-chemical Properties and Nutrient Status of Manganese Mine Spoil Dumps

Changes in physico-chemical and nutrient characteristics of manganese mine spoils from three different sites were evaluated at the time of plantation and then after 3 years of plantation and results are depicted in **Table 4**. The bulk density of manganese mine spoil was high i.e. 1.39-1.82 g/cc. After 3 years of plantation the bulk density of mine spoil decreased in the range of 0.72-1.58 g/cc due to pressmud/ FYM amendment. The maximum water holding capacity of mine spoil improved from an initial range of 5.20-29.31% to 39.5-62.28 %. The porosity of mine spoil also increased from an initial range of 16.2-49.28% to 43.4-60.44%. Similarly the organic carbon content of mine spoil increased from 0.102-0.77% to 1.52-3.89% within 3 years of time period. Improvements in nutrient status of mine spoil with respect to N,P and K was also noticed.

Effect of Mycorrhizae-biofertilizer Inoculation and Use of Ameliorative Material on Microflora Development in Manganese Mine Spoil Dumps

Improvement in microbiological status of rhizospheric mine spoil was determined at the time of plantation and then after 3 years of time interval. The microbial count as observed after amending the spoil with top soil, pressmud/ FYM indicated that these organic materials improved the organic matter content of mine spoil by providing nutrients for microbial growth which subsequently increased the rate of microbial development. There was pronounced improvement in the total bacterial count in the rhizosphere spoil samples from all the three mine sites. The improvement in bacterial count was in the range of 9.1×10^8 to 7×10^9 CFU/g from an initial low count in the range of 6×10^3 - 9×10^3 CFU/g (**Table 5**). The fungal and actinomycetal population also increased by many folds. The nitrogen fixing strains of *Rhizobium* and *Azotobacter* also increased in the range of 3.4×10^7-5×10^8 CFU/g. The counts of VAM spores also improved in the range of 37-48 spores/10 g of spoil after 3 years of plantation. The increase in VAM spore count was owing to firm anchorage of plants due to supportive material (pressmud/FYM) and favourable conditions for germination and subsequent infection of roots of plant species by VAM spores. Results showing improved counts of various microbial communities were found very much similar to the microbial population comprising of bacteria, fungi, actinomycetes reported in good productive soil which is in the range of $1-34 \times 10^6$, $1-9 \times 10^5$, $1-36 \times 10^4$ CFU/g respectively (Babich and stotzky, 1977). These observations confirmed that the microbial activity in mine spoil dumps was re-established in minimum 3 years to optimum level generally found in good productive soil by using proper blend of organic material and specialised cultures of biofertilizers.

Conclusion

Demonstration or Implementation of Integrated Biotechnological Approach for bioremediation of manganese mine spoil dumps has led to the development of supportive and nutritive rhizosphere in barren mine spoil through appropriate blending of mine spoil with organic amendments like pressmud/

231

FYM, inoculation of plants with specialized cultures of nitrogen fixing microorganisms and strains of mycorrhizal fungi for profuse root development and contributing to stress tolerance in plants.

Demonstration of IBA has turned barren mine spoil into lush green forest and contributed to various environmental benefits viz. oxygen production, soil conservation, development of top soil and provides carbondioxide sinks. The demonstration of technology has opened an avenue to implement the technology for different mining sectors and also for different types of wastelands.

References

Babich, H. and Stotzky, G. (1977). Sensitivity of various bacteria, including actinomycetes, and fungi to cadmium and the influence of pH on sensitivity. *App. Env. Microbiol.*, 33:681-688

Black, C.A. (1965) Methods of Soil Analysis Part I&II No. 9 in Series. American Society of Agronomy Inc., Publisher, Wisconsin, USA.

Table 1 : **Physico-chemical Characteristics of Mn Mine Spoil at Various Sites under MOIL**

Parameters	Dongribuzurg	Gumgaon	Tirodi
Physical			
Bulk density, g/cc	1.65-1.82	1.6-2.0	1.39-1.61
Maximum water holding capacity, %	13.8-17.8	5.20-8.7	17.29-29.31
Porosity, %	16.2-21.3	16.2-20.0	32.14-49.28
Chemical			
pH	7.5-7.8	6.9-7.4	5.71-6.56
EC, mS/cm	0.24-0.78	0.32-0.59	0.14-0.26
Na, mg/l	0.70-1.46	0.6-1.4	0.21-0.70
K, meq/l	0.1-0.6	0.4-1.6	0.042-0.168
Ca+Mg, meq/l	1.6-5.72	2.2-3.2	0.85-2.20
HCO_3, meq/l	0.9-3.1	1.0-3.1	1.8-2.1
Cl, meq/l	1.3-3.8	1.1-2.7	5.6-7.2
Nutrients			
Organic Carbon, %	0.102-0.103	0.104-0.108	0.39-0.77
Nitrogen, mg/100g			
Total	3.2-4.5	4.1-9.3	37.0-69.0
Available	0.092-1.2	1.2-3.4	0.28-0.40
Phosphate, mg/100g			
Total	4.7-6.2	4.6-7.3	37.0-68.0
Available	1.4-1.8	0.92-3.2	9.0-17.8
Potash, mg/100g			
Total	10.8-13.2	8.6-14.8	250-704
Available	2.8-4.2	1.8-3.8	172-417
Heavy Metals			
Cr, mg/kg	4.2-9.8	-	0.826
Pb, mg/kg	9-18	15.9-25.0	2.120
Cu, mg/kg	24.3-32.5	23.2-28.8	0.728
Mn, mg/kg	1872-2350	5180-8780	9.280
Zn, mg/kg	19.6-24.6	-	1.760

Table 2 : **Microbial Population in Barren Mn Mine Spoil Dumps at Various Sites under MOIL**

Microbial groups	Dongribuzurg	Gumgaon	Tirodi
Total bacterial count (CFU/g)	41	38	54
Fungi (CFU/g)	25	23	9
Actinomycetes (CFU/g)	Nil	Nil	7
Rhizobium (CFU/g)	Nil	Nil	Nil
Azotobacter (CFU/g)	Nil	Nil	Nil
VAM/10 g	Nil	Nil	Nil

Table 3 : **Growth Response of Plants Grown on Manganese Mine Spoil Dumps**

Plant species	Average plant height in cm	
	During plantation	After 30 months
Teak	30	514
Neem	41	579
Shishum	38	465
Awla	32	416
Mango	40	373
Cassia	36	426
Shiwan	38	392

Table 4 :   **Improvement in Physico-chemical Properties and Nutrient Status of Manganese Mine Spoil Dump at Various Site due to IBA after 3 Years of Plantation**

Parameters	Dongribuzurg	Gumgaon	Tirodi
Physical			
Bulk density, g/cc	1.50-1.58	1.30-1.42	0.72-1.08
Maximum water holding capacity, %	43.6-46.5	39.5-42.4	56.42-62.28
Porosity, %	43.4-47.8	45.8-49.5	52.02-60.44
Chemical			
PH	6.7-7.1	6.3-6.8	7.88-8.00
EC, mS/cm	1.2-1.52	0.92-1.0	1.18-2.42
Na, mg/l	1.9-2.9	1.4-2.0	0.70-0.98
K, meq/l	8.1-10.0	6.2-7.0	0.20-0.42
Ca+Mg, meq/l	1.2-2.1	0.7-0.9	0.41-0.53
HCO_3, meq/l	6.8-9.2	1.6-3.5	3.3-4.2
Cl, meq/l	1.9-4.1	3.1-4.9	6.8-10.5
Nutrients			
Organic Carbon, %	1.8-2.4	1.9-2.2	1.52-3.89
Nitrogen, mg/100g			
Total	260.8-278.0	248.6-268.2	200-440
Available	28.4-32.2	32.8-42.2	21.2-48.2
Phosphate, mg/100g			
Total	77.8-110.2	98.2-110.2	230-380
Available	6.5-7.2	6.2-6.8	46.0-98.0
Potash, mg/100g			
Total	248.8-268.2	240.9-260.8	420-1280
Available	8.6-8.9	8.2-8.8	38.0-58.4
Heavy Metals			
Cr, mg/kg	7.2-11.2	0.6-1.2	0.302-0.385
Pb, mg/kg	13.4-21.4	12.8-26.2	0.920-1.356
Cu, mg/kg	28.2-33.2	24.6-29.6	0.532-0.928
Mn, mg/kg	2428-3209	5458-8726	5.320-7.832
Zn, mg/kg	58.6-68.5	41.9-59.2	0.740-0.918

Table 5 :   **Microbial Population in Manganese Mine Spoil Dumps at Various Sites Treated with Ameliorative Material and Biofertilizer**

Microbial groups	Dongribuzurg		Gumgaon		Tirodi	
	Control	Treated	ɔntrol	Treated	ntrol	Treated
Total bacterial count (CFU/g)	$6x10^3$	$7x10^9$	$9x10^3$	$5x10^9$	$7.3x10^3$	$9.1x10^8$
Fungi (CFU/g)	$7.2x10^2$	$2.8x10^4$	$2.2x10^3$	$9x10^5$	$7.0x10^2$	$8.0x10^5$
Actinomycetes (CFU/g)	$2.5x10^2$	$3.2x10^6$	$1x10^2$	$3x10^6$	$5.0x10^2$	$9.0x10^5$
Rhizobium (CFU/g)	$3x10^1$	$7x10^7$	$5x10^2$	$5x10^8$	$2.3x10^1$	$7.6x10^7$
Azotobacter (CFU/g)	$4x10^2$	$3.4x10^7$	$3x10^2$	$4.1x10^7$	$7.1x10^1$	$3.6x10^7$
VAM/10 g	15	39	17	48	14	37

Impact of Coal Mining Industry on Environment with Possible Solutions—A Case Study

Chandrani, D. Prasad[1] and A.G. Paithankar[2]

[1]*Lecturer, Mining Dept., R.K.N.E.C., Nagpur, India*
[2]*Retd. Head of Department (Mining), V.R.C.E., Nagpur, India*

INTRODUCTION

The opencast project block is situated on a flat to moderate plains R.L varying between 264 to 280m. Part of the block is below high flood level. The mine is having total mineable reserves of 23mt. upto a depth of 100m. Balance reserve as on date is about 14mt. Average stripping ratio of the mine is 1:3.7. Effective thickness of coalseam is 14m. Working is carried out by dragline as well as shovel- dumper combination .The overburden consist of black cotton soil 10m thick and sandstone of medium hardness upto coalseam, having average total thickness of 62m.

ENVIRONMENTAL IMPACT OF MINE --Coal deposits are depleting assets and generally occurs in isolated area, under cover of thick forest, river, nallah, etc. And are required to be opened up at the place of occurrence only. This causes disturbance to the existing forest or agricultural land, which ultimately result into environmental imbalances, polluting air, water, and soil and acoustic levels.

The various impacts observed in this mine are air, water, and noise & land degradation.

There is no damage to forestland; also no environmental impact due to blasting .One of the socio economic impact is rehabilitation.

AIR POLLUTION

Air is considered to be polluted, when pollutants are present in such a concentration that, they are harmful to man, animal, plants and material property, causes harm or reduces the sense of well being of the population leaving in the region.

Sources Of Air Pollution

1. Drilling and blasting --- drilling raises dust .Use of explosives introduces oxides of nitrogen as well as % of carbon monoxide in air.
2. Transportation of O.B/coal in & out of the quarry roads: - Dust is raised due to the movement of HEMM on haul road. Air of high velocity raises coal/O.B dust from roads and carrying trucks too. Exhaust of HEMM produces sulphur dioxide and carbon monooxide.
3. Coal handling plant: --dust is raised at transfer and discharge points of conveyor.
4. Overburden dump. -- whenever there is a high wind, particles of O.B are raised in air. It
   Remains suspended for long-time especially fine particles because of very low terminal settling velocity.

The Main Pollutants found are of two types: -----

1. Particulate Matter (PM):--includes sand or coaldust, fumes,liquid droplets of grease,etc. Subset of particulate matter called SPM (suspended particulate matter) comprises particles with size less than 100 microns and is likely to remain suspended for a long time. PM chemically reacts with the molecule of respiratory system and brings about adverse chemical changes. This reduces the lung capacity of humans.

   Gaseous Pollutants: --------It includes sulphur dioxide, oxides of nitrogen and carbon monoxide SO_2, NO_x contributes the erosion of rock materials, damage to buildings etc. Due to acid rains SO_2 causes brancho-constriction in asthmatics at relatively low concentration (0.25 to 0.5 ppm).

Thus air pollution makes a man sick. It causes burning of eyes, nose and causes trouble in breathing, longterm injury to lungs and breathing passages. It spreads diseases like bronchitis, pneumokoniosis etc that's why precaution must be taken to reduce air pollution within specified standard limits.

Mitigation Measures Adopted includes: ———

1. Green Barrier i.e., plantation have been developed all along the haulroad, quarries and around the colony, industrial area at interval of 2m, to arrest the travel of dust farther which may cause soil infertility.
2. Water spraying is done at regular intervals on haulroads, O.B dumps, coal stockyard, near CHP to suppress dust.
3. 1km length fixed sprinklers system has also been provided along roads.
4. Black topping of roads to eliminate dust generated by HEMM movement.
5. Plantation developed on overburden dumps to prevent soil erosion.
6. To avoid spillage of coal tarpouline cover is used on carrying trucks.
7. Dust extracters are used on drills.
8. Regular maintenance of quarries roads as well as surface roads.
9. Regular maintenance and overhauling of HEMM.

To keep the concentration of pollutants always within safe limits and to be abreast of changing conditions, air quality monitoring is carried out. Samples are collected on fortnight basis each of 24hrs duration.

Methodology of sampling and analysis

Samples are collected with APM -451 respirable dust sampler to monitor ambient air quality with respect to SPM, RPM (Respirable particulate matter), SO^2 and oxides of nitrogen.

SPM: ----------Ambient air quality laden with suspended particulate enters the respirable dust sampler through the inlet pipe of sampler by means of a high flow rate blower (1.1 to 1.5 cum/min. As the air passes through the cyclone, coarse, non-respirable dust (size > 10 micron) is separated from the air stream by centrifugal forces acting on the solid particle. These separated particles fall through the cyclone conical hopper and collect in the sampling bottle placed at bottom. The fine dust forming the respirable dust fraction (size < 10 micron) of the total suspended particulates passes through the cyclone and is carried by the air stream to the glass micron fibre filter paper. The filter and the carrier air exhausted from the system through the blower retain the respirable dust. The mass concentration (microgm/cum) of SPM in the ambient air is computed by measuring the mass of collected particulates and the volume of air sampled.

NOx: ---------Determination of oxides of nitrogen is based on the procedure of "Jacobs and Hochheiser method". In this method the air is collected daily in the field and analysed in the laboratory using spectronic 20D+ spectrophotometer. Nitrogen dioxide is collected by bubbling air through a sodium hydroxide solution to form a stable solution of sodium nitrite. The nitrite ion produced during sampling is determined colorimetrically (with the help of spectrophotometer measuring absorbance at 540nm) by reacting the exposed absorbing reagent with phosphoric acid, sulphanilamide and N (1-napthyl) ethylenediamine dihydrochloride. Converting it to sulphuric acid with hydrogen peroxide before analysis eliminates the interference of sulphur dioxide.

SO2:---------Determination of SO2 is based on the procedure of " West and Geake " method. Sulphur dioxide from the air stream is absorbed in a sodium tetrachloromercurate solution of dichloro-sulphito-mercurate. The amount of SO2 is then determined or estimated by the colour produced when P- Rosaline hydrochloride is added to the solution. The colour is estimated by a reading of absorbence at 560nm in the spectrophotometer.

WATER POLLUTION– Water pollution means such contamination of water or such alteration of the physical, chemical or biological properties of water or such discharge of any sewage or trade effluent or of any other liquid, gaseous or solid substance into water (either directly or indirectly) as may or is likely to create a nuisance or render such water harmful or injurious to public health or safety to domestic, commercial, industrial, agricultural or other legitimate uses or to the life and health of animal or plants or of aquatic organisms.

Sources of water pollution

1. Mine water discharge, which contains overburden, coal, oil matter, toxic substances like sulphide mineral, gases etc.
2. Industrial effluent containing grease, oil and mud of washed vehicles.
3. Domestic liquid effluents.

Before being discharged into river it must be treated, to project aquatic life and human life from diseases.

Measures to Control Water Pollution are as follows: ---------

1. Mine Water Discharge: -- The total mine water discharge being collected into settling tank of size (38*30*0.7) m of capacity 798 kl to settle the suspended solid particles. Then the water is being used for dust suppression on haulroads, for fire fighting purposes and for plantation on overburden

dump within the premises. Settled and dried mud can be used for cultivation because of high % of nitrogen in it.

2. Industrial Effluents: -------- the effluent is being treated by " EFFLUENT TREATMENT PLANT (ETP)" near workshop. After washing of HEMM water is treated and is reused for the same. The collected oil and grease is disposed to nearby underground mine for lubrication of haulage, etc., .ETP gives 1-barrel oil/month for 4 dumpers wash a day.

Effluent Treatment plant -- Workshop effluent is allowed to settle in settling tank for mud particles. The water is then carried to oil and grease trap. Oil and grease being lighter floats on water, and is removed

And collected in oil and grease tank. Water from oil and grease trap is taken to flash fixing chamber where alum is added. Daily 4 dumpers and 1 dozer is washed, 1 dumper requires 20 litres of water for washing. each dumper or dozer produces 5 cum of effluent /day. Alum required per dumper is 1.5kg in mixing chamber. Then water is carried to clarifloculater. Here sludge remains at bottom due to flocculation action and water from top is channelised and collected is treated water tank. Water in this tank is now free from oil, grease and mud. It is then stored in storage tank and is reused for washing of HEMM. Sludge from cariflocculater is collected in sludge bed tank and is dried in sunlight, to prevent disintegration of sludge and finally disposed off. (Fig. No. -1).

3. Domestic and service building effluents are collected and biological treatment is given by means of extended aeration. This will ensure that after the treatment effluent confirms to IS 4764.

4. Vegetative plantation has been done on overburden dumps to prevent excessive suspended soil particle in surface run-off for stabilisation.

Pollutant Received by Water can be grouped as: ----

1. Physical -- colour, odour, suspended solids, turbidity, taste and temperature.
2. Chemical --Total dissolved solids (TDS), alkalinity, hardness, metals (toxic and non-toxic), organic, fluorides, nutrients.
3. Biological-- bacteria,viruses, etc,.

To keep all these within permissible limits and to be abreast of changing condition, samples are collected on monthly basis in plastic zuricanes from prefixed location and are transported to the laboratory for analysis. pH , temperature, conductivity ,etc,. Are measured with the help of portable water analysis kit. Presence of metals is determined using SQ 118 photometer.

NOISE POLLUTION -- Acoustic noise is usually defined as unwanted sound.

Sources of noise: ------

1. Drilling and blasting.
2. Loading and transportation of coal/overburden.
3. Crushing of coal in CHP.
4. Frequent movement of HEMM in workshop.

For shorter duration noise level exceeds, although it does not even cover one-third of excessive level situation. Even at moderate and low level, noise is undesirable because of its interference with speech and sound. The human ear is considered one of the most sensitive instruments able to detect the sound over a wide varying range of levels. Human exposure to excessive noise and loud sound causes fatigue in the ear may also result in loss of speech or hearing, the most damaging effect. Though noise generation is inevitable, it can be reduced to safe limits by various mitigatory measures.

Measures Adopted to Control Noise Pollution are as follows----

1.The creation of green belts around the quarry area, CHP and housing colony which serves as an acoustic barrier.

2.Systamatic & controlled blasting is carried out.

3.Noise absorbing pads are used in foundation of vibrating machine and also in cabins of HEMM to minimise the effect of noise. Absorbing pads being made up of fibreglass and mineral wool.

Sound absorbing matter works by converting acoustic energy in the sound wave into heat, due to friction generation as the air molecules in the sound wave passes through the materials.

4.Silencers/ mufflers are also used in HEMM.

5.Earplugs / Muffs are used by employees to reduce noise level exposure.

Besides these precautions noise level is continuously monitored quarterly with the help of integrated sound level meter with LEQ(Energy Equivalent Sound level). Noise level monitoring helps us to identify any further increase in noise at any place to be dissipated within limits. Noise level at certain places i.e., at workshop, bunkers & near crusher has exceeded the limits.

LAND DEGRADATION---

Opencast mining creates land degradation directly by excavation and then by overburden dumping, while indirect effects causes siltation in surrounding water bodies, lowering of water table & other related effects. Excavation lowers the original topographic height of land, which disturbs soil profile & finally the soil quality. Land degradation is minimised by suitable land reclamation, dump slope stability, vegetation & afforestation. Effect of land degradation is well explained in Fig. No. 2.

A) Land Reclamation: -------- Land Reclamation is the process of reforming and regrading the waste land approximately to the original topography & bringing about permanent self sustaining vegetation.

It involves four steps: -----

1. Storage of top soil.
2. Levelling, dressing & backfilling.
3 & 4. Replacement of topsoil & Ecological Restoration.

B) Dump slope stability-- A huge amount of O.B is being utilised in the formation of embankment 30 m wide, height being 6m above high flood level of river, around the quarry, service buildings & colony area. Since these area are under high flood level of 272m R.L. low lying area is also filled with O.B to raise the level above H.F.L. The final void (8.26 ha) at the end of mine life will be filled up with water body to a depth of 35m. Dump slope if not stabilised properly causes slidings of debris, which may block haulroads, damage powerlines, cover exposed coal in working area rendering mining uneconomical & other effects. Thus external dump must be stabilised and prevented from soil erosion.

<u>Various measures adopted include –</u>

1. Plugging of gullies on dumps.
2. Dump slopes have been maintained at 28 dig. and plant species of bamboo, eucalyptus, etc. have been planted.
3. Coir matting.
4. Quality inspection of O.B dumps is carried out.
5. Catch drains are provided around O.B dumps.
6. Top surface drainage has also been provided.
7. Sealing of fissures/ cracks to prevent infiltration of water.

Rehabilitation of nearby village with a population of 179 families has been carried out and is compensated properly with money & jobs. Cost impact of reclamation and environment control per tonne of coal raised is RS.2.50.

CONCLUSION

The case study of this mine reveals that management has adopted effective measures, to curb the adverse impact of mining on environment, during its production, handling, transportation and usage. Analysis of monitoring data indicates concentrations of certain parameters particularly dust are very close to permissible limits, which warrants need to more careful monitoring control to prevent there possibility of crossing the limits. Simply determining the concentrations of SPM & RPM is not adequate. It is also necessary to consider the shape, size and geochemistry of dust particles & thus should be made mandatory.

Energy is the most important factor for sustainable economic development and coal is a major source of energy for human beings. About 70% of India's commercial energy need is met by coal based power plants. Coal is also useful in other industries like Iron & steel industry, cement industry, paper mills, etc. therefore it is necessary to understand environmental impacts of mining .we must strike a balance between mining and environment protection.

Protection of environment is of paramount importance because cost of living in degraded environment is much more, than the cost, that will be incurred for undertaking environmental reforms.

238

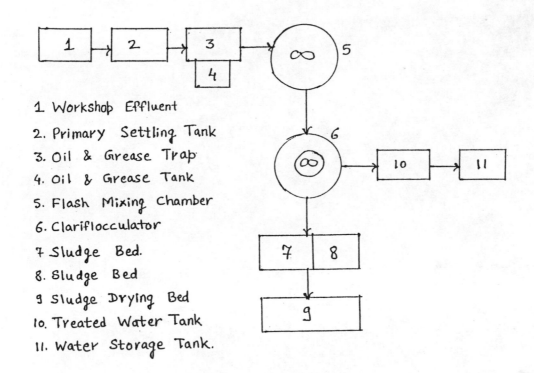

1. Workshop Effluent
2. Primary Settling Tank
3. Oil & Grease Trap
4. Oil & Grease Tank
5. Flash Mixing Chamber
6. Clariflocculator
7. Sludge Bed.
8. Sludge Bed
9. Sludge Drying Bed
10. Treated Water Tank
11. Water Storage Tank.

Fig-1. FLOWSHEET OF EFFLUENT TREATMENT PLANT

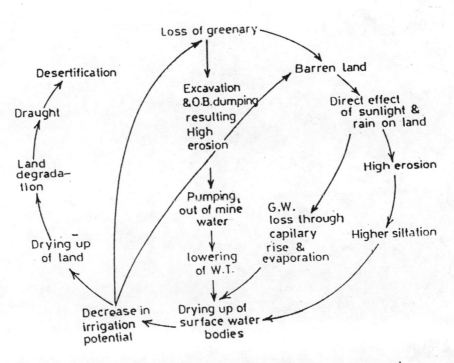

Fig. 2   Land degration cycle due to mining and
related activities

THEME 3 :
Water Pollution

THEME 3:
Water Pollution

Natural Aeration in Reclamation of Surface Waters: A Case Study of an Iron Ore Mine

Y.G. Kale[1], A.G. Paithankar[2], Shanta Satyanarayan[3] and D.K. Ghosh[4]

[1]*Asstt. Controller of Mines, Indian Bureau of Mines, Nagpur, India*
[2]*Prof. & Head (Retd.), Mining Dept. VRCE and Prof. Emeritus, RKNEC, Nagpur, India*
[3]*Sr. Asstt. Director, National Environmental Engineering Research Institute, Nagpur, India*
[4]*Chief Controller of Mines, [4]Indian Bureau of Mines, Nagpur, India*

ABSTRACT

Environmental pollution consequent to any developmental activity is inevitable and mining activity is no exception to it. Water pollution is one of the major environmental problems arising out of mining activities. Impact of mining activity on surface water quality in and around Kudremukh iron ore mine, which is India's largest opencast mine in metalliferous sector, has been studied.

The study was conducted in 1993-94 for all the seasons. Initially seven sampling sites were selected, three on river Bhadra which passes through the leasehold area and one each on four tributaries of the river. All the samples were analysed for physico-chemical parameters including heavy metals. Except iron and manganese concentrations, all other parameters were well within the prescribed limits. Maximum values of iron and manganese concentrations were found to be 1.38 and 0.38 mg/l as ferrous iron and manganous manganese respectively. In the downstream side of the active mining zone in the summer season being a critical season, additional five water samples were collected at locations one km apart. The results indicated that ferrous iron and manganous manganese concentrations were reduced to 0.204 and 0.050 mg/l respectively. Thus, it may be concluded that ferrous iron (Fe^{+2}) and manganous manganese (Mn^{+2}), which are in soluble form, are oxidised to ferric iron (Fe^{+3}) and manganic manganese (Mn^{+4}) respectively due to the natural aeration. This is mainly due to the hilly terrain, which results in turbulence of water flow and increased exposure of river to the atmosphere leading to self-purification of the stream.

Field observations were further confirmed by laboratory experiments on 'Tray Aerator' where iron and manganese concentrations were reduced to 'not detectable' level. The present paper discusses in detail the results of the study.

INTRODUCTION

Kudremukh iron ore mine, India's largest opencast mine in metalliferous sector, designed to produce 22.6 million tonnes of ROM ore to generate 7.5 million tones of iron ore concentrate per annum. It is nestled high in the picturesque mountains of Western Ghats in South India. The unique features of the Kudremukh mine are: that it is the only mine in India where low grade iron ore is mined.[1] The ore is mostly magnetite in composition. It is the first long distance slurry pumping operation in India, which radically changed the entire economics of this project.[2,3]

The method adopted for mining is opencast with fully mechanized means. The bench height is 14 m. The blast holes are drilled with the help of electrically operated drill machines of 312 mm diameter. The site mixed slurry is being used to blast the holes. The blasted re is loaded in dumpers having capacity of 120 tones in conjunction with 10.7 m^3 capacity shovels.

243

DRAINAGE SYSTEM

The Kudremukh mine area receives a heavy rainfall of average 7000 mm per annum and is a source of several perennial rivers. This also forms the catchment area of river Bhadra which is the main river of the region. The river Bhadra originates at a distance of 15 km northwest of the mine. The perennial tributaries of river Bhadra are Lakya, Kunya, Sitabhoomi, which flow down to Bhadra at northern half of the leasehold, and Kochige and Kudremukh flow down to Bhadra at southern half of the leasehold.

Due to mining activity, soil erosion and wash off of solid waste in the form of top soil, waste rock and ore was anticipated. Therefore, to prevent the flow of ore and soil, two rock filter dams have been provided across the valleys, where the direct discharge of water from a mine gets filtered through specially prepared vertical filter beds, embedded into the dam walls made of sized stones. The rock walls and the filter beds arrest most of the suspended solids leaving the discharge water cleaner. As a result of production of 7.5 million tones as finished concentrate, about 13 to 14 million tones of material per annum is generated as tailings. Therefore, for the disposal of tailings, a major 'earth fill' dam has been constructed across the river Lakya, a tributary of river Bhadra. This dam has a capacity of 260 million tonnes to store the tailings. To allow overflow of surplus monsoon water, the dam is also provided with the tunnel-spillway at the upstream side of the reservoir.

EXPERIMENTAL METHODS AND RESULTS

The study was conducted in 1993-94 and all the three seasons were covered. Seven sampling sites were selected for surface water, one each in four tributaries and three in river Bhadra. All the samples were analyzed for physico-chemical parameters including heavy metals as per the standard methods.[4]

Water Quality Evaluation

The colour of the water was turbid in monsoon, slightly turbid in winter and clear in summer. Water was odourless in all the seasons. Temperature ranged from 23 ° to 30° C. Throughout the stretch of the river Bhadra and its tributaries pH value did not show much variation and ranged from 7.2 to 7.6. Turbidity was higher in monsoon season but well within the limit of 10 NTU[5] and ranged from 3.0 to 3.26 NTU. Total dissolved solids ranged from 144 to 254 mg/l. Water quality is quite satisfactory in respect of dissolved oxygen varying between 7.2 to 8.4 mg/l, indicating good oxygenation. Total suspended solids in river Bhadra were ranging from 30 to 68 mg/l.

Organic pollution load in terms of COD and BOD, at all sampling stations was well within the limits[6] of and it ranged between 14-54 mg/l and 3.0-6.4 mg/l respectively. Fluoride values throughout the stretch of the river surveyed were well within the prescribed limits.[5]

Iron and manganese concentration were higher in the downstream location at Station R7 i.e. 1.38 and 0.38 mg/l respectively. At the upstream side of active mining zone the iron concentration was lower than the prescribed limit but it showed a gradual increase towards downstream. (Figure-1). Manganese concentration was on the higher side right from the upstream and it showed increasing trend towards downstream. (Figure-2) Zinc concentration was below the tolerance limit in all samples.[5] Concentration of lead was found in some of the samples during summer season ranging between 0.009-0.017 mg/l, much below the tolerance limit.[5] Metals like copper, chromium, nickel, cadmium, mercury and arsenic were not detected in any of the samples.

Natural Aeration

Kudremukh has a natural advantage of high altitude location. So it was thought proper to explore the possibility of natural aeration further downstream of mining zone beyond station R7 that may help in reducing the ferrous iron and manganous manganese by the process of oxidation. Therefore, five additional sampling stations were identified outside the mining lease

244

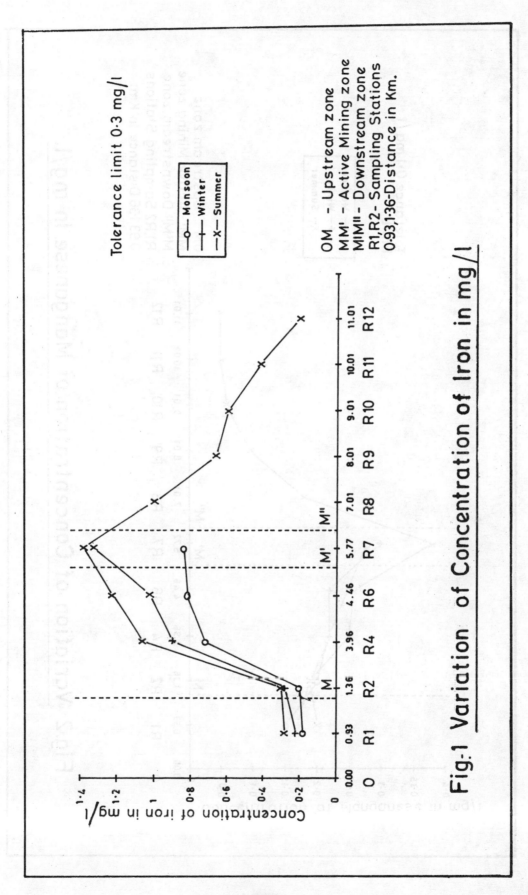

Fig.1 Variation of Concentration of Iron in mg/l

Tolerance limit 0·3 mg/l

Monsoon
Winter
X— Summer

OM - Upstream zone
MM' - Active Mining zone
MM" - Downstream zone
R1,R2- Sampling Stations
0·93,1·36-Distance in Km.

Concentration of iron in mg/l

O   R1   R2   M   R4   R6   M' M" R8   R9 R10 R11 R12

0.00 0.93 1.36 3.96 4.46 5.77 7.01 8.01 9.01 10.01 11.01

245

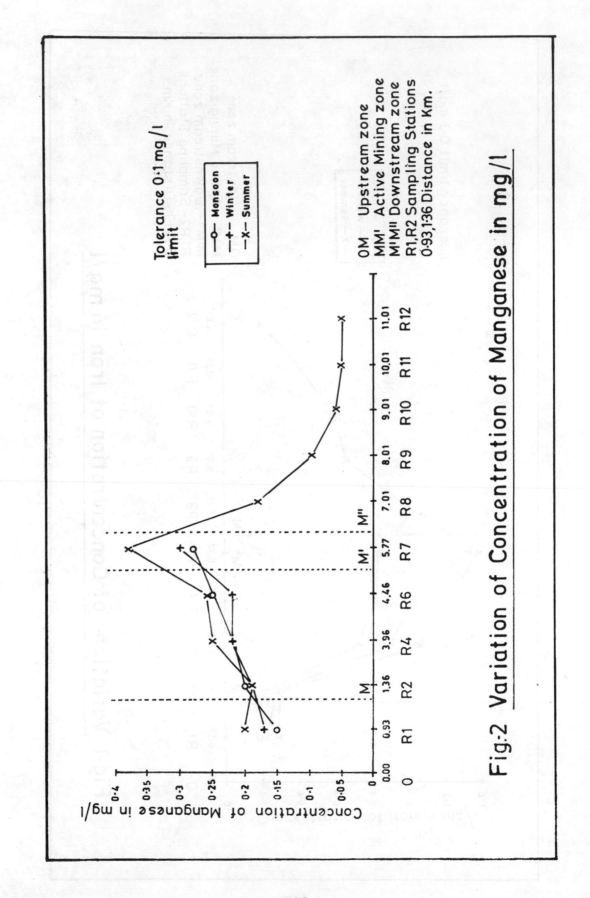

Fig.2 Variation of Concentration of Manganese in mg/l

further along the downstream stretch of river Bhadra, spaced 1 km apart in summer season of 1995, the summer being the critical season as dissolved solids are highest. The analysis of results shows that the ferrous iron concentration was reduced to 0.204 mg/l and manganous manganese concentration to 0.050 mg/l at a distance of about 5 km from the boundary of the active mining zone. Thus it could be concluded that the ferrous iron (Fe^{+2}) and manganous manganese (Mn^{+2}), which are in soluble form, are oxidised to ferric iron (Fe^{+3}) and manganic manganese (Mn^{+4}) (insoluble form) respectively due to the natural aeration.[7] This is mainly because the flow of the water is all along through hilly terrains, which results in turbulence of water flow. High speed wind also helps in aeration of water flow. The chemical reactions, which have taken place, are:

$$Fe^{++} + 3H_2O \rightarrow Fe(OH)_3 + 3H^+ \quad \text{---- For iron and}$$

$$Mn^{++} + 2H_2O \rightarrow MnO_2 + 4H^+ \quad \text{---- For manganese}$$

Laboratory Experiment

The above field observations were confirmed by conducting laboratory experiments on 'Tray Aerator'. Water is discharged through a riser pipe and distributed on to a series of trays from which water falls through small openings to the bottom. Coarse media such as stones ranging from 50 to 150 mm in diameter, were placed in the trays to increase the efficiency. With this media good turbulence was created and large water surface was exposed to the atmosphere. Five trays of size 0.6x0.55x0.01 m were used with a spacing of 0.3 m between each tray in structure of 1.8 m height. Openings were of the size 6-8 mm diameter. The operation continued for about 10 minutes. The outlet water was analyzed and it was found that both the ferrous iron and manganous manganese concentrates were reduced to 'not detectable' level.

The amount of oxygen needed to oxidise the iron and manganese is given below[7,8] which substantiate the above results:

$$4Fe^{++} + O_2 + 10H_2O \rightarrow 4Fe(OH)_3 + 8H^+$$

$$4 \times 56 \text{ mg } Fe^{++} \leftrightarrow 2 \times 16 \text{ mg } O_2$$

$$1 \text{ mg } Fe^{++} \leftrightarrow 0.14 \text{ mg } O_2 \quad \text{----------(i)}$$

$$6Mn^{++} + O_2 + 6H_2O \rightarrow 2Mn_3O_4 + 12H^+$$

$$2Mn_3O_4 + 2O_2 \rightarrow 6Mn_3O_2 \text{ (Black)}$$

$$6Mn^{++} + 3O_2 + 6H_2O \rightarrow 6MnO_2 + 12H^+$$

$$6 \times 55 \text{ mg } Mn^{++} \leftrightarrow 3 \times 2 \times 16 \text{ mg } O_2$$

$$1 \text{ mg } Mn^{++} \leftrightarrow 0.29 \text{ mg } O_2 \quad \text{---------(ii)}$$

Therefore, from (i) and (ii) it is computed that 0.14 mg and 0.29 mg of oxygen is needed to convert 1 mg of ferrous iron to ferric iron and 1 mg of manganous manganese to manganic manganese, respectively.

CONCLUSIONS

From the above study, it is concluded that the impact of iron ore mining at Kudremukh on the river water quality is marginal. Dissolved oxygen in the river stretch studied indicates good oxygenation. Suspended solids were below the permissible limit. Major pollution problem is the increase of iron and manganese concentration above permissible limit. However, the iron and manganese concentration is reduced to a level, which is well below the prescribed limit within a stretch of 5 km on the downstream side of station R7 due to natural aeration. This appears to be mainly because of turbulent flow owing to hilly terrain and high-speed wind to which the water flow is exposed.

Therefore, it is suggested that wherever topography permits the mine water should be discharged in such a manner so as provide proper turbulence and consequent aeration.

ACKNOWLEDEMENT

The authors are thankful to Shri A.N. Bose, Controller General, Indian Bureau of Mines for kind permission to present this paper. The authors are also thankful to M/s Kudremukh Iron Ore Company Ltd. for permission to conduct this study. The views expressed in this paper are those of authors and not necessarily of the organization to which they belong.

References

1. Report of the Expert Group on classification of minerals with regard to their optimum industrial use. Department of Mines, Government of India, December 1989 pp36

2. Kudremukh makes great strides (1995) Minerals & Metals review, January 1995 pp60

3. Kudremukh Iron Ore;(1978) A report; Mining Magazine, London, January 1978 pp26-27

4. Manual on Water and Waste water analysis, National Environmental Engineering Research Institute, 1988

5. Indian Standards. Specifications for drinking water, Indian Standards; 10500-1983

6. General Standards for discharge of effluent. Environment (Protection) Rules, 1986 Government of India.

7. Manual on Water Supply and Treatment; Third Edition.(1991) Central Public Health and Environmental Engineering Organisation, Government of India, March 1991, pp284-285

8. Joshi N.S., Vaidya M.V. & Eli Dahi (1998) Up and downflow, filterartion for iron removal from groundwater in hand pump connection, Copenhagen pp25

#

Fluoride Contamination in Groundwater in Birbhum District of West Bengal—A Geoenvironmental Hazard

Mitali Sarkar*, Sumit Chakrabarty and Aparna Banerjee

Department of Chemistry, University of Kalyani, Kalyani 741 235, West Bengal, India

1. INTRODUCTION :

Environment and ground water are now the two major concerns facing the international community. The irrational exploitation and utilization of natural resources in the course of industrialization and urbanization has resulted in environmental pollution and ecological degradation. Fluoride contamination in drinking water is a worldwide problem and 29 countries are reported to be affected with fluorosis, the fluoride related disease. The problem in India, too, is known for quite a long time. It is estimated that in India 64 million people are affected with fluoride toxicity arising from groundwater contamination (1), while 67 million people is at risk (2). Many attempts have been made to tackle the problem; still it remains a threat.

2. OCCURRENCE OF FLUORIDE IN GROUND WATER :

Fluorine is a normal constituent of most soils and rocks. It occurs in calcium granite (500 mg/l), in alkaline rocks (1200 Mg/l), in shells (750mg/l) and in sandstone (270 mg/l). The main sources of fluoride in natural drinking watger are Fluorite (CaF_2), Fluorapatite [$3Ca_3(PO_4)_2$-CaF_2], Cryolite (Na_3AlF_6), Magnesium Fluoride (MgF_2) and replacement of ions of crystal lattice of micas and many other minerals.

As rain water percolates through the soils, it comes in contact with rocks and minerals present in the aquifer materials. Due to the presence of acid in the soil, dissolution of the fluoride from country rocks occurs. While most of the soluble fluoride is leached out from the soils and coprecipitated with $CaCO_3$, only part of fluorine enters the clay minerals, and remains as find suspended particles in water. The concentration of fluoride ions in natural water depends upon several factors like:

(a) distribution of weathered fluoride bearing materials,
(b) accessibility of circulating water to these minerals,
(c) extent of fresh water exchange in aquifer,
(d) evaporation and evapotranspiration,
(e) formation of ion pairs such as $CaSO_4$, $CaHCO_3$ etc.,
(f) Complexing of fluoride ion with aluminium, beryllium, ferric ion and series of mixed fluoride hydroxide complex with boron.

3. FACTORS AFFECTING FLUORIDE CONTENT IN SOIL AND WATER

There are various factors affecting fluoride content in soil and water like soil reaction (saline and alkaline), exchangeable Na content, electrical conductivity, organic matter content etc. Report indicates that fluorine concentration increases with salinity of irrigation water (3)

Besides this sewage water used for irrigation purposes, various industrial wastes, soil near brick fields, mining areas have been found to be potential sources of fluoride in soils. Such accumulated fluoride in soils either showed toxicity or reaches to the ground water in its toxic concentrations which ultimately affect plant growth, animal and human health (4).

4. FLUORIDE AND HUMAN HEALTH :

Fluoride concentration in drinking water has a physiological significance for human (Table I) and some animals. Deficiency of fluoride in potable water causes over calcification of teeth and bones and dental carries. Continuous high intake of fluoride will result in mottled enamel of teeth, skeletal fluorosis and sometimes-severe osteocleosis. Moreover excess fluoride affects functioning of calcium, the essential element for blood clotting and every muscular contraction. Excessive accumulation of the calcium fluoride in the renal system causes formation of stones in kidney and eventual renal failure (5).

Table 1 : Fluoride content in drinking water and physiological effect of human body :

Physiological effect	Fluoride Content (mg/l)
Dental caries	>1.0
Mottled teeth	2.0
Osteocleosis	5.0 to 8.0
Crippling and thyroid change	20.0 to 80.0
Retardation	100.0
Kidney disorder	125.0
Death	>2500.0

5. GEOLOGICAL BACKGROUND :

Rampurhat and Nalhati blocks, the studied area lie in the northern part of Birbhum district. The area is bound by east longitudes 87°45' and 87°52' and north latitudes 24°09' and 24°20'. The village Nasipur lies on the Rajmahal trap consisting of volcanic eruption of basaltic composition. This lava flows cover an extensive area of 4000 square kilometer and extend upto a maximum depth of 450 bgl. Drilling data of Geological Survey of India reveals existence of five cycles of lava flows with intermittent sedimentary beds known as trappens. Thickness of the individual lava flows ranging from 15 to 90m and that of intertrappeans varies from 1.5 to 6m which is composed of siltstone, shale and volcanic tuffs (6).

6. RESULTS AND DISCUSSION :

The groundwater samples collected from Rampurhat and Nalhati block of Birbhum district were analysed for their water quality parameters following standard procedure (7). The analytical results showed fluoride concentration ranging from 5.0 to 16.0 mg/l in the villages Nasipur, Bhabanandapur and Rampurhat. The highest concentration found is 16.0 mg/l, which is ten times higher, the permissible limit of fluoride as recommended by WHO (8) and Indian Standard (9). People in these areas are worsely affected with fluoride toxicity. Many have died, most are horribly disfigured, crippled for life. The village Nasipur today is called "disabled village". The analytical investigation further revealed a negative correlation between fluoride and hardness, while a positive correlation between fluoride and sodium. Samples rich in fluoride are soft in nature. There is a positive between soil pH and fluoride concentration. Jaglan studied the impact of irrigation on the elevated fluoride in ground water (10.)

7. AVAILABLE REMIDIAL OPTIONS :

No anthropogenic source is detected for the elevated fluoride concentration in the ground water in this region. It is geogenic in nature. Both long term and short-term remedial methods are needed to tackle the problem. The long term approach involves hydrogeological study of the area to explore the causative factors for spreading fluoride vertically as well as laterally and prediction of safer zone for trapping fluoride free groundwater. The long-term measure is a capital intensive venture and need significant time to implement. As a shor-term (interim) measure, for supply of safe drinking water to the people, alum treatment method is a convenient method (11). Practically no treatment is available for advanced fluorosis.

8. SUGGESTION :

Each source of drinking water in city and village should be analysed for the presence of toxic concentration of ions. Public awareness should be generated.

9. REFERENCES :

1. Fluoride Research and Rural Development Foundation Report, (1999). Ministry of Rural Developments, Government of India, India.
2. Jacks, G., Bhattacharya, P., High-fluoride groundwaters in India, (2000). Proc. Conf. Ground water, June, Denmark.,
3. Barbiero, L., van Vliet-Lanoe, B.(1998). The alkali soils of the middle Niger valley : origins, formation and present evolution. Geoderma 84,323.
4. Sarkar, M., Das, D.K., (1990) Status of fluoride in water, soil and plant in some areas of West Bengal. Proc. Conf. Industry and environment, Karad, Maharashtra, India.

5. Indian Council of Medical Research, (1975). Special report series no.44.
6. Sengupta, A., Studies on sporadic fluoride contamination in drinking water at Nalhati - I block, district Birbhum, (1998). Proc. Conf. Ground water management, December, Calcutta, India.
7. APHA, AWWA and WPCF, Standard Methods for Examination of Water and wastewater, (1999).APHA, Washington, 19th edition.
8. International standards of drinking water, (1971). WHO, Geneva.
9. Indian Standard, Drinking water specification, (1991) UDC 628.
10. Jaglan, M.S. (1996). Irrigation development and its environmental consequences in arid regions of India, Environmental. Management, 20,323.
11. Sharma, S., Joshi, J.D. (2000) Fluoride reduction in water, Res. J. Chem. Env. 4,69.

Impact of Overburden Geochemistry on the Quality of Mine Effluent—A Case Study

A. Jamal[1], S. Siddharth[1] and R.K. Tiwary[2]

[1]*Department of Mining Engineering , I.T., B.H.U., Varanasi 221 005, U.P. India*
[2]*Central Mining Research Institute, Dhanbad 826 001, Bihar, India*

INTRODUCTION

The objective of this research paper is to provide geo-chemical information about coal and associated overburden, to enable mine professional to manage spoil in most favourable way in order to minimize the impact on water environment in and around the mining area.

In this study, samples of coal, associated rocks (i.e. overburden) and mine effluent from different location were collected from the West Chirimiri Opencast Project. This mine is located in Sarguja district of Madhya Pradesh.

The slides of rock samples were prepared for identification of minerals (microscopic study). For determination of geo-chemical parameters, rock samples were powdered and digested with acids. The water samples were also analyzed. The variable parameters (Temperature, pH value, TDS, Conductivity etc.) were analyzed by portable water analysis kit. The hydrological parameters were correlated with geochemical parameters of overburden. The results are discussed in subsequent paragraphs.

CHARACTERISTICS OF OVERBURDEN

A detailed analysis of coal-associated rocks was done. The rocks in the study area are mostly sandstone, and in lesser amount, shale and clay. On the surface also, the most available rock is sandstone. The chemical behavior of sandstone were studied under the following heading –

1. Petrological character.
2. Geochemical character.

1 - Petrological Character of Sandstone

The microscopic investigation of sandstones occurring over the surface were made in order to determine the mineral constituents. Model analysis has been performed to know the acidic or alkaline nature of sandstone and the results are given in tabular form in Table No.1.

Minerals constituents of sandstone may be divided into reactive and non-reactive component.
The reactive component is most important and it affects the water quality to a greater extent. Among reactive component the percentage of feldspar ranges from 5.2 & 13.3 with an average of 9.31.

The reactive component is the most important and it affects the water quality to a greater extent. Among reactive components the percentage of feldspar ranges from 5.2 to 13.8 with an average of 9.6 %. Whereas, the percentage of Biotite ranges from Nil to 2.1 and Muscovite ranges from 0.8 to 4.9 with an average of 0.86 and 3.2 percent respectively. The Biotite is relatively less stable than muscovite and hence occurence of Biotite is less than Muscovite.

The matrix is usually are argillaceous. The high percentage of matrix (Avg. 36.9) are responsible for friable and soft nature of sandstone.

Table - 1 Petrological Characters of Sandstone

Sample No.	Reactive Component (%)				Non reactive Component (%)		
	Feldspar	Biotite	Muscovite	Cement/ Matrix	Quartz	Rock fragment	Heavy minerals
SSt-1	7.8	2.1	2.82	40.5	42.6	4.2	Nil
SSt-3	9.3	1.4	4.2	29.6	53.8	1.6	0.4
SSt-4	8.1	1.3	4.0	31.2	51.6	3.6	0.2
SSt-5	5.2	1.0	4.9	37.0	48.6	3.1	0.1
SSt-6	13.3	Nil	0.8	29.8	52.8	2.5	0.3
SSt-7	10.7	0.9	1.8	27.1	59.4	Nil	Nil
SSt-8	13.8	Nil	3.5	31.4	50.1	Nil	1.0
SSt-10	8.6	0.2	3.6	35.8	51.2	1.2	Nil
Average	9.6	0.86	3.2	36.9	51.2	2.02	0.25

Among non reactive component, the percentage of Quartz ranges from 42.6 to 59.4 with an average of 51.2. There are few grains of rock fragments and heavy minerals, which are relatively stable, and non reactive component of the sandstone.

The reactive components in each sandstone are relatively high and are responsible for degradation of water quality, when the latter comes in contact with the sandstone. The high content of reactive component increases the susceptibility of water pollution. The non-reactive component is also a threat for water pollution. It may pose the silting problems of nallas and streams and is a source of suspended solid in mine water also. The modal analysis of sandstone suggest that the given rock types in the area is arkose.

2. Geochemical Character of Sandstone

The geo-chemistry of arkose were analyzed. The results are shown in Table 2. It may be observed from this table, that the percentage of reactive oxides (CaO, MgO, Na_2O, K_2O, Fe_2O) in the sandstone is 12.45 (average). The percentage of non-reactive component (i.e. Silica) is 68.20. These reactive oxides are responsible for the degradation of water quality in mine.

Table 2 : Geo-chemical Analysis of Sandstone

SiO_2	Al_2O_3	CaO	MgO	Na_2O	K_2O	Fe_2O_3	TiO_2	MnO	Ignition loss (H_2O^+)
68.20	10.41	2.45	4.62	1.26	2.39	1.73	0.88	0.016	3.84

Total 10 samples were analyzed

GEO-ENVIRONMENTAL CHARACTERS

To determine the geo-environmental character of coal and associated overburden leading to the production of acidity and alkalinity. Acid production and neutralization potential of coal and overburden samples were determined. The results are summarized in Table 3 and 4.

Table 3 : Acid Potential (Kg of $CaCO_3$/ton)

Sample No.	% of sulphide sulphur	Acid potential (Kg of $CaCO_3$/ton)
Sh-1	1.528	47.75
Sh-2	1.508	47.13
Sh-3	1.526	47.69
Sh-4	1.512	47.25
C-1	1.598	49.93
C-2	1.469	45.91
C-3	1.453	45.41
C-4	1.038	32.43
C-5	0.994	31.06

Table 4 : Neutralization Potential of Sandstone (Kg of CaCO₃/ton)

SSt₁	44.88
SSt₂	47.52
SSt₃	45.91
SSt₄	46.03
SSt₅	43.62
SSt₆	39.04
SSt₇	41.06
SSt₈	42.02
SSt₉	43.14

It may be observed from table 4, that the neutralization potential of arkose associated with coal is high. It ranges from 39.04 to 47.52 kg of CaCO₃/ton. However in case of shale, the acid production potential is high and almost in the range as in coal. At this AP, the pH value of sump water should be in severe acidic range.

The neutralization potential of associated rocks shown in Table 4 indicates that there is ample alkalinity in arkosic sandstone to neutralize acidic water produced by coal and shale. Because of this alkalinity, the pH value of sump water is high but still under acidic range.

HYDRO-CHEMISTRY OF MINE WATER

The hydro-chemical characteristics of sump waters collected both from coal and overburden faces have been given in Table 5. It may be observed from this table that the water of this mine is acidic in nature.

Table 5 : Hydro-chemical Characteristics of Mine Water

Parameters	Mine Water		
	Coal face	Sandstone face	Main sump
Temperature (⁰C)	31.8 – 32.2	31.5 – 32.0	32.2 – 32.8
pH value	3.72 – 4.12	8.63 – 8.96	4.90 – 5.12
TDS (mg/L)	1050 – 1120	330 – 380	1270 – 1330
SO₄ (mg/L)	670 – 730	19 – 33	190 – 217
Fe (mg/L)	0.7140 – 0.8002	0.218 – 0.436	0.512 – 0.542
Ca (mg/L)	122.6 – 127.8	34.5 – 37.8	98.14 – 101.56
Mg (mg/L)	86.24 – 88.18	24.6 – 25.64	72.52 – 73.62
Na (mg/L)	49.0 – 51.46	29.4 –30.22	64.06 – 65.16
K (mg/L)	55.1 – 56.2	13.4 – 14.6	51.42 – 52.18

It may be observed from table 5 that when water comes in contact with coal, acidity of mine water increases and pH value decreases. The pH value of water occurring near coal working face is acidic in nature. It ranges from 3.72 to 4.12 whereas nature of mine water in sandstone benches, is alkaline in nature. The pH value ranges from 8.63 to 8.96. Along with this, concentration of sulfate and other parameters is also high is the water near the Coalface. The concentration of sulfate ranges from 670 mg/L to 730 mg/L. The concentration of iron is also in high and its concentration ranges from 0.7140 to 0.8002 mg/L. It may also be observed from table 5, that acid water has high solubility and thereby concentration of TDS is more in comparison to alkaline water. The concentration of TDS ranges from 1050 mg/L to 1120 mg/L in acidic water and 330 mg/L to 380 mg/L in alkaline water respectively.

IMPACT OF OVERBURDEN ON MINE WATER QUALITY

A laboratory investigation has also been made in order to know the contribution of each lithounits of the area in polluting the quality of mine drainage. The total dissolved solid released by rocks of unit weight is defined as pollutant productions potential. It may be observed from Table 5, that coal and shale is more vulnerable to release TDS in high concentration. At similar experimental condition, release of TDS by shale and coal is more than arkosic sandstone.

Table 6: Physico-chemical Characteristics of Leached Water

Rock type	Temp ⁰C	pH value	TDS
Sandstone	33.3 – 33.6	7.72 – 7.94	268.0
Shale	34.2 – 34.8	3.72 – 3.96	642.0
Coal	34.6 – 35.0	4.02 – 4.14	498.0

During its passage from the points of seepage to the discharge, water comes into contact with a variety of rocks in a mine. The water and the rocks undergo reciprocal changes. The interaction of water with

Various rock units may lead to desirable and undesirable changes in hydrochemistry. The relation of rock types with R-pH value is shown in figure1.

R-pH of varoius rock type

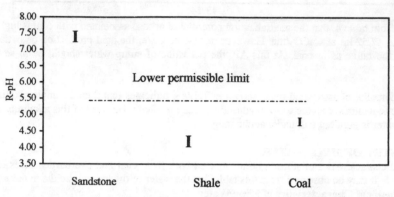

It may be observed from figure 1 that the R-pH value of sandstone ranges from 7.72 to 7.94 whereas sump water in sandstone bench varies from 7.72 to 7.96. The R-pH of shale ranges from 3.72 to 3.96 and of coal from 4.02 to 4.14, whereas sump water in Coal + Shale bench ranges from 3.72 to 4.12. The R-pH values of shale and coal are in acidic range, whereas the pH value of sump water is slightly less acidic as compared to the pH value of leached water from coal and shale. This is perhaps due to mixing of overburden (i.e. sandstone) leached water with coal sump water. CONCLUSIONS

Field observation and laboratory investigation conclude the following point:
1. The quality of mine water is governed by rocks occurring in the area.
2. The sandstone associated with coal as overburden is responsible for increasing the pH value of water and it is a source of iron also.
3. The shale associated with coal, as overburden is responsible for the acid drainage. The pH of leached water from shale ranges from 3.72 to 3.96. The main cause of acidity in the mine is pyrite.
4. The neutralization potential of sandstone is responsible for increase in the pH value of acidic water.

ACKNOWLEDGEMENT
The financial assistance from CSIR, New Delhi is dully acknowledged.

REFERENCES
Jamal, A., Dhar, B.B. and Tiwari, R.K. (1997). Prediction of water quality drainage in coal mines – A case study. VI Int. Mine Water Association Congress, Bled, Slovenia.

Jamal, A., Dhar, B.B., Siddharth, S. and Tiwary, R.K. (1998). Impact of overburden spoils on water quality in coal mines. In Proceedings of National Symposium on "Environmental Geochemistry", June 26-27, Deptt. of Applied Geochemistry, Osmania University, Hyderabad.

Shay, T.O., Hossner, L.R. and Dixon, J.B. (1990). A modified Hydrogen Peroxide Oxidation Method for Determination of Potential Acidity in Pyrite Overburden. J. Environ. Qual. 19, 778-782.

Skousen, J., Renton, J., Brown, H., Evans, P., Leavitt, B., Brady, K., Cohen, L. and Ziemkiewicz, P. (1997), Neutralization potential of Overburden samples containing Siderite, J. Environ. Quality, 26 : 673-681.

Sobek, A.A., Schuller, W.A., Freeman J.R. and Smith, R.M. (1978), Field and laboratory methods applicable to overburdens and mine soil. EPA-600/2-78-054. U.S. Govt. Print office, Washington, D.C.

Contaminant Transport in Saturated Aquifers

T.R. Nayak and R.K. Jaiswal

National Institute of Hydrology, Regional Centre, 278, Manorma Colony, Sagar 470 001, M.P., India

ABSTRACT

In water resources development and management, one of the main problems is that of water quality. As greater development and use of groundwater continues, combined with the reuse of water, quality suffers unless consideration is given to protect it. Special attention should be devoted to the pollution of groundwater in aquifers due to their very slow velocity. In order to protect the groundwater resources from deterioration, knowledge of pollutant movement and level of concentration with time is essential.

In the present study a model has been developed to predict the pollutant movement in adsorbing and non-adsorbing saturated porous media. The model is based on Galerkin finite element method using eight-nodded isoparametric parabolic elements. The present model gives better results than the four-nodded finite element model and it is close to the analytical solution. It is also observed that the propagation of concentration are considerably retarded by clay strata in saturated aquifers.

1.0 INTRODUCTION

In order to meet the growing demand of an ever-expanding world population, available natural resources are being exploited at an increasing rate. This exploitation involves large scale development of nuclear plants, coal fired plants, coal mines, oil refineries and oil wells; and an ever accelerating expansion of municipal waste treatment plants, agricultural farms and cattle feed lots saturated with organic and inorganic wastes and frequent dosage of fertilizers, pesticides and herbicides. Percolating rainwater contains carbon dioxide derived from the atmosphere, which increases the solvent action of the water. Sodium and calcium are commonly added cations and bicarbonate, carbonate and sulphates are corresponding anions. Important sources of Chlorides are from sewage, connate water and intruded seawater. All these activities, through accidental system leaks or natural return flows of their diluted wastes, present a threat to the ground and surface water environment. Special attention needs to be devoted to the pollution of ground water in aquifers. Due to their very slow velocity, when groundwater pollution occurs, the restoration to the original, non-polluted state is more difficult.

Migration of contaminants in groundwater has become an area of increasing research interest in recent decades. Modelling studies, laboratory experiments and field experiments have been conducted by research workers with the ultimate goal of understanding the behavior of contaminants in groundwater and for predicting the future contamination level. Various investigators have attempted numerical solutions of the convective-dispersion equation governing the movement of pollutants in a porous media. Guymen et al (1970) and Nalluswami et al (1972) used the finite element method based on the variational principle for the solution of the dispersion problem in a rectilinear flow field. Smith et al (1973) compared the variational approach with the Galerkin method and concluded that the latter was more versatile. Pinder (1973) also used the Galerkin method to solve the two dimensional groundwater flow and dispersion equations in cartesian coordinates and applied the model to simulate the movement of chromium contaminated groundwater in Long Island, New York. Prakash (1976) solved the convective-dispersion equation for the transport of a radioactive tracer through an adsorbing porous in the cylindrical polar coordinate system. Daus and Frind (1985) developed a more general model that formulated in a 'natural' coordinate system sonsistent with the principal directions of hydraulic conductivity. David and Stollenwerk (1987) presented a summary table that references the various kinds of models studied and their application in predicting chemical concentration in ground water.

In the present study a model has been developed for the movement of pollutant in unconfined saturated aquifer. One-dimensional flow field for concentration/pollutant movement is considered. The pollution concentration usually gets modified while moving with groundwater due to the effect of mechanical dispersion, convection, diffusion and adsorption. A computer program has been developed for Galerkin finite element model using eight nodded isoparametric parabolic elements. A major advantage in using this technique is its versatility in handling system with irregular geometry. The seepage velocities at Gaussian points are obtained by using finite element method.

2.0 MODEL DEVELOPMENT

The movement of pollutant in groundwater is mainly due to advection and hydrodynamic dispersion. The transport of dissolved solid at the mean velocity of groundwater is called advection. Hydrodynamic dispersion results from mechanical dispersion and molecular diffusion. Fick's law plays an important role in obtaining the governing equation for advective-dispersive mass transport. The other law used is the conservation of mass principle.

2.1 Basic Governing Equations

The governing differential equation of the solute transport in saturated homogeneous isotropic porous media including the effect of adsorption and radioactive decay can be written as (Bachmat and Bear, 1961):

$$\partial/\partial t \, [nC+(1-n)S] - \partial/\partial x_i[nD_{ij}(\partial C/\partial x_j)-u_iC] + \lambda[nC+(1-n)S] - qC^* = 0 \qquad \ldots.1.0$$

in which,

C = the concentration of the dispersing mass in liquid phase, ML^{-3}
S = the concentration of the dispersing mass in solid phase, ML^{-3}
N = the porosity of the porous medium, $L^3 L^{-3}$
D_{ij} = the hydrodynamic dispersion tensor, $L^2 T^{-1}$
U_i = the concept of the Darcy velocity vector in the ith direction, LT^{-1}
λ = the radioactive decay coefficient, T^{-1}
q = the volumetric fluid injection rate at the source per unit volume, T^{-1}
C^* = the concentration of the source fluid, ML^{-3}

<pre>
t   =   time instant, T
x_i, x_j =   spatial coordinates, L
</pre>

In the above equation, the first term represents the rate of change of total dissolved and adsorbed mass; the second term denotes the dispersion and advection; the third term is the mass change due to the decay; and finally the term qC^* represents the injection or withdrawal rate. In this equation, change of mass as a result of volume changes due to variation of pressure is neglected.

Equation 1.0 is a single equation in the two state variables, C and S hence, an additional relationship between C and S may be required. An adsorption isotherm is an expression that relates the quantity of an adsorbed component to its quantity (expressed as concentration) in the fluid phase, at constant temperature. Assuming linear equilibrium adsorption, $S = k_d C$, and substituting S in equation 1.0, we get

$$_i n.\partial C/\partial t + (1-n)k_d\, \partial C/\partial t - \partial/\partial x_i[nD_{ij}\, \partial C/\partial x_j - u_i C] + \lambda[nC + (1-n)\, k_d\, C] - qC^* = 0 \quad ....2.0$$

$$\partial C/\partial t[1+\{(1-n)/n)\}k_d] - \partial/\partial x_i[D_{ij}\, \partial C/\partial x_j - (u_i/n)C] + \lambda C[1+\{(1-n)/n\}k_d] - qC^*/n = 0 \quad ....3.0$$

put, $R = [1 + \{(1-n)/n)\}k_d]$ ⁄ (R ≥1), then eq. 3.0 may be written as:

$$R\, \partial C/\partial t - \partial/\partial x_i[D_{ij}\, \partial C/\partial x_j - (u_i/n)C] + \lambda RC - qC^*/n = 0 \quad ....4.0$$

R is called the coefficient of retardation or the retardation factor. It can be assumed to be a constant for a given homogeneous region.

2.2 Initial and Boundary Conditions

The initial condition used for transport equation may be given as:

$C(x,y,0) = C_0(x,y)$ on the domain

Where, C_0 is the initial concentration in the domain.

Two types of boundary conditions generally used in field applications of contaminant transport models are: Dirichlet type where the concentration (C) is specified and Newmann type where flux is specified.

3.0 FINITE ELEMENT MODEL

The Galerkin technique has been used to determine the approximate solution for transport equation 4.0 under the prescribed initial and boundary conditions. Employing the Galerkin weighted residual approach the transport equation becomes:

$$\iint_\Omega N_i\big[R\, \partial C/\partial t - (D_{xx}\, \partial^2 C/\partial x^2 + D_{xy}\, \partial^2 C/\partial x\partial y + D_{yx}\, \partial^2 C/\partial y\partial x + + D_{yy}\, \partial^2 C/\partial y^2) +$$
$$1/n(\partial/\partial x\, (u_x C) + \partial/\partial y\, (u_y C)) + \lambda RC - (q/n)C^*\big]\, dxdy = 0 \quad ....5.0$$

i − 1 to …...n. (No. of nodes in the finite element grid)

The equation is first integrated by parts to reduce the order of equation and shape functions are introduced to depict the spatial variation of solute concentration. The final equation can be written in matrix form as:

$$[P]\{C\} + [K]\{C\} = \{F\} \qquad\qquad \ldots 6.0$$

in which, $[P]$ and $[K]$ are $(n \times n)$ square matrices and $\{F\}$ is $(n \times 1)$ column vector. The matrices $[P]$, $[K]$ and vector $\{F\}$ are given by:

$$p_{ij} = \iint_{\Omega}^{e} R\, N_i\, N_j\, dxdy \qquad\qquad \ldots 7.0$$

$$k_{ij} = \iint_{\Omega}^{e} \Big[Dxx\, (\partial N_i/\partial x)\, (\partial N_j/\partial x) + Dxy\, (\partial N_i/\partial x)\, (\partial N_j/\partial y) + Dyx\, (\partial N_i/\partial y)\, (\partial N_j/\partial x) +$$
$$Dyy\, (\partial N_i/\partial y)\, (\partial N_j/\partial y) + (u_x/n)\, N_i(\partial N_j/\partial x) + (u_y/n)\, N_i(\partial N_j/\partial y) + \lambda R N_i N_j \Big] dxdy$$
$$\ldots 8.0$$

$$f_i = \iint_{\Omega}^{e} (q/n)C^* N_i dxdy + \int_{S}^{e} \Big[\{Dxx(\partial C/\partial x) + Dxy(\partial C/\partial y)\} n_x + \{ Dyx(\partial C/\partial x)$$
$$+ Dyy(\partial C/\partial x)\} n_y \Big] N_i dS \qquad\qquad i,j = 1,2, \ldots,n \qquad \ldots 9.0$$

4.0 RESULTS

In the preset study, Galerkin finite element method is used to predict the movement of contaminant/pollutant front in one-dimensional saturated aquifer. Eight-nodded isoparametric parabolic element was taken to get better approximation of mass transport. The rectangular finite element mesh as shown in Fig. 1 has been used, whose dimensions are the same as those used by Wang and Anderson (1982) in order to compare the results. The boundary conditions for $t>0$ have been chosen to be $C(0,t) = C_o = 10.0 \text{ gm/m}^3$ and $C(100,t) = 0$. The right boundary conditions at $x = 100$ km is the approximation to the boundary condition $C(\infty,t) = 0$. No transport boundary conditions are imposed along y-direction, thereby making the solute transport problem one-dimensional. The initial condition is $C(x,0) = 0$ for all x. The longitudinal flow velocity $u_x = 0.1$ m/day and the longitudinal dispersion coefficient $Dxx = 1.0 \text{ m}^2$/day are assumed. The transverse flow velocity u_y and transverse dispersion coefficient Dyy have been taken as one tenth of u_x and Dxx respectively.

4.1 Transport of Pollutant in Non-Adsorbing Saturated Porous Media

The solute distribution along x-direction of the mesh, for two different time steps, using the analytical solution given by Wang and Anderson (1982) and the Finite Element Model with four-nodded and eight-nodded Galerkin solution method has been presented in Fig. 2. It has been observed that for given Dxx, u_x, n and R the quadratic polynomial basis functions give more or less the same solution as analytical solution. The movement of concentration with time is presented in Fig. 3.

4.2 Pollutant Movement in Adsorbing Saturated Porous Media

Mansell et al (1977) described the non-linear reversal adsorption-dispersion relationships of Phosphorous transport through water saturated and unsaturated cores from surface and sub-surface horizons of Oldsmar fine sand. Physical and Chemical properties and physical parameters for soil cores of Oldsmar fine sand have been given in Table 1. The movement of Phosphorous for no adsorption and with adsorption has been given in Fig. 4. The adsorption coefficient R = 1.27 has been used.

Table 1

Physical Properties		Physical Parameters	
Sand	97.0 %	Dispersion coeffi.	39.36 cm²/hr.
Silt	1.6 %	Pore water velocity	25.0 cm²/hr.
Clay	1.4 %	Porosity	0.46
Organic Matter	0.0 %	Distribution coeffi.	0.23

Gupta and Greenkorn (1974) have given dispersion and adsorption parameters for Ottawa washed sand mixed with 0.0 % to 7.5 % Kaolin clay. The other relevant parameters of the Ottawa washed sand have been presented in Table 2. The natural sand generally contains about 2 – 2.5 % clay and sandy loam about 7 –10 % clay. The model has been applied to the transport of Phosphate in the saturated porous media of Ottawa washed sand. The results of Phosphate movement in the above mentioned media has been presented in Fig. 5.

Table 2

Packing % clay	Total Porosity	Permea-bility	Pore water velocity	Dispersion coefficient	Distribution coefficient
2.5	0.352	0.212	1.55	1.882	0.068
7.5	0.365	0.013	0.74	1.557	0.189

5.0 CONCLUSION

The results obtained by the present model have been compared with the results given by Wang and Anderson (1982). It has been found that the solution given by this model is more or less same as the analytical solution, hence the quadratic shape functions give better approximation for solute transport in saturated porous media. The model has also been applied to the movement of pollutant in adsorbing porous media and it has been found that the effect of adsorption is to retard the overall rate of propagation of concentration with time. The adsorption of phosphate ions increases with the increase of clay content in the porous media. Hence, it can be concluded that in a formation containing fair amount of silt and clay, adsorption may significantly reduce the concentration of accidentally released pollutants.

REFERENCES

1. Bachmat, Y. and Bear, J. (1964). The general equation of hydrodynamic dispersion in homogeneous isotropic porous mediums. *J. Geophysical Res.* 69(12): 2561-2567.
2. Daus, A.D. and Frind, O.E.(1985). An alternating direction Galerkin technique for simulation of contaminant transport in complex groundwater systems. *Water Resour. Res.* 21(5): 653-664.
3. David, B.G. and Stollenwerk, K.G. (1987). Chemical reactions simulated by groundwater quality models. *Water Resour. Bull.* 23(4): 601-615.
4. Gupta, S.P. and Greenkorn, R.A. (1974). Determination of dispersion in non-linear adsorption parameters for flow in porous media. *Water Resour. Res.* 10(4): 839-846.
5. Guymen, G.L., Scott, V.H. and Herrmaqnn, L.R. (1970). A general numerical solution of two dimensional diffusion-convection equation by the finite element method. *Water Resour. Res.* 6(6). 1611-1617.
6. Mansell, R.S., Selium, H.M., Kanchansut, P., Davidson, J.M. and Fiskell, G.A. (1977). Experimental and simulated transport of phosphorus through sandy soils. *Water Resour. Res.* 13(1): 189-194.

7. Nalluswami, M., Longenbaugh, R.A. and Sunanda, D.K. (1972). Finite element method for the hydrodynamic dispersion equation with mixed partial derivatives. *Water Resour. Res.* 8(5) : 1247-1250.

8. Pinder, G.F. (1973). A Galerkin finite element simulation of groundwater contamination on Long Island, N.Y. *Water Resour. Res.* 9(6): 1657-1669.

9. Smith, I.M., Farraday, R.V. and O'Connor, B.A. (1973). Rayleigh Ritz and Galerkin finite elements for diffusion-convection problems. *Water Resour. Res.* 9(3): 593-606.

10. Wang, H.F. and Anderson, M.P. (1982). Introduction to groundwater modelling: Finite element methods, W.H. Freeman and Co., Sanfrancisco.

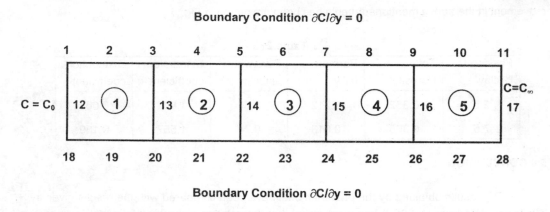

Fig. 1   Finite Element Mesh for One Dimensional Solute Transport Problem

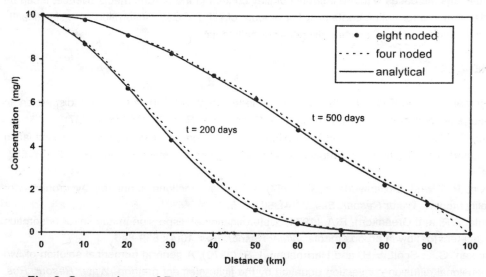

Fig. 2 Comparison of Results with 4-Noded and Analytical Solution

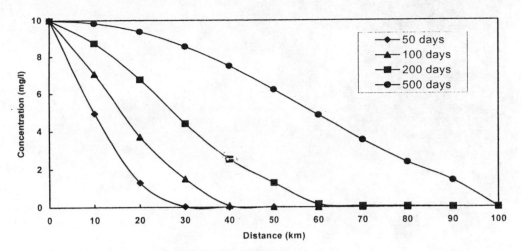

Fig. 3 Propagation of Concentration Front with Time

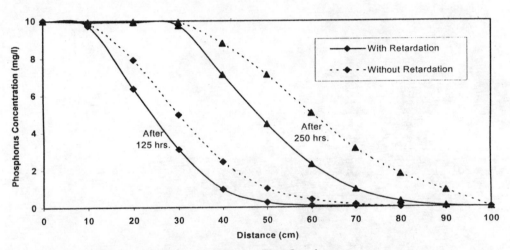

Fig. 4 Phosphorus Movement in Oldsmar Fine Sand

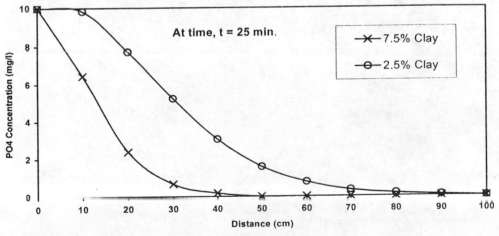

Fig. 5 Effect of Clay Contents in Phosphate Transport

An Economically Viable Treatment Process for CD (II) Removal from Water and Wastewater

Y.C. Sharma

Pollution Control and Research Laboratories, Metallurgy and Materials Engineering Department, National Institute of Foundry and Forge Technology, Hatia, Ranchi 834 003, Bihar, India

1.0   INTRODUCTION

Metals are important for industrial development of a nation. The metallic species can be classified as essential or non-essential depending upon its requirement to fauna, flora or human beings. Cadmium is a non-essential depending upon its requirement to fauna, flora human beings.   Cadmium is a non-essential toxic heavy metal. Its occurrence in nature is rare (1,2) but it is toxic to plants, animals and humans at high exposure, chronic exposure to cadmium results in its steady accumulation in the body of fauna and human beings. It is a potent anzyme inhibitor and damages liver and kidney (3).   Precipitation, ion-exchange, solvent extraction, etc. are the well documented methods of water and wastewater treatment rich is cadmium (4-6) but all these processes are expensive and inhibit their large scale application to developing nations.   Activated carbon adsorption is a popular choice for removal of metallic species from industries effluents in developed nations but cost of this treatment too restricts its large scale application to developing nations like India.   The present communication has been addressed to the application of china clay, a local geological for reclamation of water and wastewater rich in Cd(ii). Effect of various important parameters on removal of cadmium by adsorption on china clay has been studied.

2.0   EXPERIMENTAL

China clay is a clay mineral available in plenty in Patherghatt, Bihar (India) but for the experiments it was obtained from the Institute's laboratories. The sample was ground and sieved with appropriate sieves to maintain an average particle size of 100 m. The average particle size was measured by particle size analyser, model HIAC-320 (ROYCO Inst. Div., USA) and the surface charge by lazer zee meter, Model – 500 (Penkem Inc. N.Y., USA) and the surface charge by lazer zee meter, Model – 500 (Penkem Inc. N.Y., USA).   The surface area of the powdered adsorbent was determined by a 'three point' N_2 gas adsorption method using Quantasorb Surface Area Analyser Model-QSM (Quantachrome Corp., USA) and the porosity by Mercury Porosimeter.   The chemical analysis of china clay was carried out by using Indian Standard Methods IS :1527 (1960). Batch adsorption experiments were carried out by shaking. 1.0 g adsorbent with 50-ml aqueous solutions of cadmium of desired concentrations, and pH in different glass bottles in a thermostat shaker at 100 rpm.

After predetermined intervals of time, the sample bottles were taken out and the reaction mixtures were centrifuged. The progress of adsorption was examined by

determining the amount of cadmium left in aliquot by Atomic Absorption Spectrophotometer (GBC, Australia). The pH of the solutions was maintained by 0.1 HCI/NaOH solutions. The ionic strength of the solutions was maintained by 0.01 M $NaCIO_4$. All the experimental were carried out at 10 M particle size.

3.0 RESULTS AND DISCUSSION

The results obtained for the present system have been discussed under following headings:

3.1 Characterization of China Clay: The analysis of china clay (Table 1) shows that SiO_2 and Al_2O_3 are present in it as major constituents and oxides of other metallic species are present in traces.

Table 1: L Chemical and physical characteristics of china clay

Constituents	% by weight
SiO_2	46.22
Al_2O_3	38.40
CaO	0.86
Fe_2O_3	0.68
MgO	0.37
Loss on ignition	13.47
Mean particle diameter (μm)	100
Surface area (m^2g^{-1})	13.52
Density (gcm^{-3})	2.69
Porosity	0.33

3.2 Effect of Contact Time and Concentration on Removal of Cd (II) by Adsorption on China Clay

This study shows that in initial stages the removal of Cd (II) from the solution increases and acquires saturation in 80 min. Removal of Cd (II) decreased from 80.3 to 41.0% (Table 2) on increasing the Cd(II) concentration from 0.5×10^{-4} M to 2.0×10^{-4}M at 6.5 pH, 0.01M $NaCIO_4$ ionic strength and 30°C temperature. The finding that higher removal is achieved in low concentration range is of industrial importance.

Table 2 Effect of contact time and concentration on removal of cadmium by china clay.

Time (min.)	% removal at 0.5×10^{-4} M	% removal at 1.0×10^{-4} M	% removal at 1.5×10^{-4} M	% removal at 2.0×10^{-4} M
10	28.3	18.3	14.9	9.4
20	43.8	36.0	24.7	19.0
30	56.0	42.2	31.6	24.8
40	63.4	51.0	38.3	30.5
50	68.6	56.7	43.2	39.0
60	74.5	61.1	47.5	37.3
70	77.8	62.8	50.0	39.6
80	80.0	64.0	50.7	41.0
90	80.0	64.0	50.7	41.0
100	80.0	64.0	50.7	41.0
110	80.0	64.0	50.7	41.0
120	80.0	64.0	50.7	41.0

3.3   KINETIC MODELLING
Legergren's rearranged model (8) :

$$\text{Log } (q_e - q) = \log q_e - (K_{ad}/2.303).t \quad (1)$$

Where q_e and q (both mgg^{-1}) are the amounts of cadmium removed at equilibrium and at any time t respectively, K_{ad} (min^{-1}) is the rate constant of removal of cadmium by adsorption on china clay, was used for kinetic modelling of the process. The straight line plots of log (q_e-q) VS t (Figure 1) at different temperatures indicate validity of the above model. The values of rate constant at different temperature were determined by these plots and were found to be 4.7x102 min1, 4.5x10^{-2} min^{-1} and 3.4x10^{-2} at 30,40 and 50°C temperature respectively. This is clear that lower temperature values favour the removal in the present system.

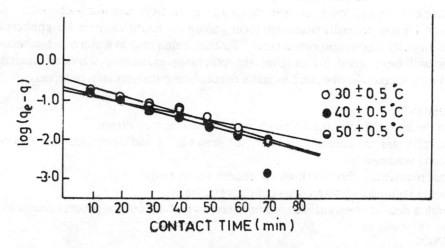

Figure 1. Lagergren Plot for Cd (II) removal at different temperatures.

3.4   INTRAPARTICLE DIFFUSION STUDY
There exists a possibility of intraparticle diffusion in a rapidly stirred batch reactor and in many adsorption processes this is a rate limiting step. This possibility was tested by determining the value of coefficient of intraparticle diffusion (9) :

$$D = 3.03 \; r^2 / t_{1/2} \quad (2)$$

Where D is a coefficient of intraparticle diffusion, r (cm) is the radii of adsorbent particles, and $t_{1/2}$ (min$^{1/2}$) is the time of half adsorption. The value of D was found to be 3.25x10^{-10} cm^2 sec^{-1} and this indicates intraparticle diffusion to be rate controlling step (10)

The rate constant of intraparticle diffusion, Kp was determined by the graphical method. The amount of cadmium adsorbed was plotted against $t_{1/2}$ (Figure 2). This graph shows double nature : curved in initial stages and linear at final stages. The initial curved portion is indicative of boundary layer diffusion effects (11) and the final linear portion is indicative of intraparticle diffusion effect. The value of rate constant of intraparticle diffusion, Kp at 30°C was calculated from slopes of linear portion and was found to be 2.0x10^{-2} mgg^{-1} min$^{-1/2}$

267

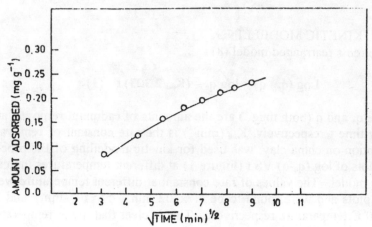

Figure 2 Interparticle diffusion Plot for Cd (II) removal

Thus china clay can be used for removal of cadmium from cadmium rich water and wastewater. It is a naturally occurring local geological and its large-scale application will incur only transportation cost on user. Further, china clay is a non-toxic substance and thus will be a good choice over the otherwise expensive activated charcoal, activated carbon, etc. Further studies and a detailed cost analysis is in progress.

CONCLUSION

Based on the above studies the following conclusions may be drawn :

1. China clay can be successfully used for removal of cadmium from Cd (II) rich waters and wastewaters.
2. Higher removal is observed in lower concentration range.
3. Removal is higher at tower values of temperature.
4. Though a detailed cost analysis has not been carried out but the process seems to be economically viable

REFERENCES

- Wood, J.M. (1974) Biological Cycles for toxic elements in the environment. Science 183 : 1044-1052.
- Nriagu, J.O. (1988). A Silent epidemic of environmental metal poisoning, Environ Pollut 50:139-161
- Kannam, K. (1995). Fundamentals of Environmental Pollution, First Edition, S Chand & Co. Ltd, ND, India.
- Westall, J.C., Chen, H. Zhang, W. and Brownawell (1999). Sorption of linear alkylenezene sulfonates on sediment materials. Env. Sci. Tech. 33:3310-3118.
- Chu, K.H., Hasim, M.A., Shang, S.M., Sammel, V.B. (1997). Biosorption of cadmium by algal biomass : adsorption and desorption characteristics. Water Sc. Tech. 35:115-122.
- Sharma Y.C. (1995) Economic treatment of Cd(II) rich hazardous waste. J. Colloid Interface Sc. 173:66-71.
- Indian Standard Methods of Chemical Analysis of Fire Clay and Silica Refractory Materials (1960), IS:1527.
- Tiwari, P.H., Compbell, A.B. and Lee, W. (1972). Adsorption of Co^{2+} by oxides from aqueous solution. Can J. Chem. 50: 1642-1648.
- Michelsen, L.D., Gideon, P.G., Pace, E.G. and Kutal, L.H. (1975), U.S.D.I. office of water Research and Technology Bull. No.74
- Crank. J. (1965), The mathematics of diffusion, Clarendon Press, London.
- McKay, G., Otterburn, M.S. and Sweeny, A.G. (1980), The removal of colour from effluents using various adsorbents-III, Silica: Rate Processes, Water Res. 14:15-20.

Evaluation of Heavy Metal Pollution Index for Surface and Spring Water Near Limestone Mining Area of Lower Himalayas

Bably Prasad

Scientist, Central Mining Research Institute, Dhanbad 826 001, India

1. INTRODUCTION

Lime stone mining in the ecologically fragile Lower Himalayas has been a contentious issue since many years. Sirmour district of Himachal Pradesh is one area in the region very rich in lime stone deposits of different grades and the Mushroom growth of mines alongwith other industries here has exacerbated the environmental and ecological devastations by removing vegetation cover, modifying landforms, creating air and water pollution and ultimately threatening biodiversity. The major sources of water and drinking water in these area are spring and some perenial rivers like Giri, Tons and Yamuna. Extensive mining in this area may damage the springs as well as deteriorate the quality of spring and river water. Heavy metals, one of the important water quality parameters, may also get enriched in water due to mining activities. Thus monitoring of heavy metals in spring and surface water used for drinking purpose assumes great significance from the point of view of human health. In this context monitoring of heavy metal pollution in spring and surface waters of the limestone mining belt of Sirmour district is of paramount importance.

The pollution parameters monitored for the assessment of the quality of any system gives an idea of the pollution with reference to that particular parameter only. Quality Indices are useful in getting a composite influence of all parameters of overall pollution. Quality Indices make use of a series of judgements into a reproducible form and compile all the pollution parameters into some easy approach. Several methods have been proposed to develop Quality Indices for estimation of characteristics of surface water with water quality parameters (Horton, 1965; Lohani and Todino, 1984; Tiwary and Mishra, 1985; Joung et al., 1979; Landwehr, 1979). In recent years much attention has been paid towards the evaluation of heavy metal pollution in ground and surface water with development of Heavy Metal Pollution Index(HPI) (Mohan et.al., 1996; Reddy, 1995; Prasad and Jaiprakas, 1998).

In the present paper, application of weighted arithmetic average mean method of indexing to assess the overall pollution due to heavy metals in spring and surface water near lime stone mining area of Sirmour district has been done. The concentration of seven heavy metals iron, manganese, lead, copper, cadmium, chromium and zinc has been evaluated for nine important springs and eight important locations for surface water of Giri and Tons rivers in pre and post monsoon seasons of an year.

2. INDEXING APPROACH

The HPI represents the total quality of water with respect to heavy metals. The proposed HPI is based on weighted arithmetic quality mean method and is developed on two basic steps. First, by establishing a rating scale for each selected parameters giving weightage to selected parameter and second, by selecting the pollution parameter on which Index is to be based. Rating system is an arbitrary value between zero to one and its selection depends upon the importance of individual quality considerations in a comparative way or it can be assessed by making values inversely proportional to the recommended standard for the corresponding parameter (Horton, 1965; Mohan et al., 1996). In the present formula, unit weightage (W_i) is taken as value inversely proportional to the recommended standard (S_i) of the corresponding parameter (Reddy, 1995). Iron, manganese, lead, copper, cadmium, chromium and zinc have been monitored for the model Index application. The Heavy Metal Pollution Index (HPI) model proposed is given by (Mohan et al., 1996)

$$HPI = \frac{\sum\limits_{i=1}^{n} W_i Q_i}{\sum\limits_{i=1}^{n} W_i} \qquad \text{---------------} \qquad (1)$$

where Q_i is the sub index of the ith parameter. W_i is the unit weightage of ith parameter and n is the number of parameters considered.

The sub index (Q_i) of the parameter is calculated by

$$Q_I = \sum\limits_{i=1}^{n} \frac{(M_i(-) I_i)}{(S_i - I_i)} \qquad \text{-----------------} \qquad (2)$$

where, M_i is the monitored value of heavy metal of ith parameter, I_i is the ideal value of ith parameter, S_i is the standard value of ith parameter. The sign (-) indicates the numerical difference of the two values, ignoring the algebraic sign.

Generally, pollution Indices are estimated for any specific use of the water. The proposed Index is intended for the purpose of drinking water. The critical pollution index value for drinking water is 100.

3. EXPERIMENTAL

Sirmour is the southern most district of Himachal Pradesh State of India and covers an area of 2825 sq km in the lesser Himalayas and Siwalik ranges. The two rivers, namely, Giri and Tons flow through the limestone-mining belt of the district and carries the discharges of mining activities. In the mining area, there are several springs on which the local population depends for drinking water. Due to blasting in the mining zone there is always a threat to these natural springs in terms of their damage as well as quality deterioration.

Taking all these into consideration four representative sampling locations have been selected for Tons and four locations for Giri river. Giri river is the main water resource of the district which passes through limestone mining belt. At some places it receives run off water from the mines and also through the drainage flowing in the mining areas. Sampling locations have been selected for the assessment of water quality of the Giri river at the down stream of each mining area. Similarly for Tons river, sampling points have been selected at the down stream of drainage because these drainage discharge mining wastes from different mining areas into it. The nine sampling locations have been taken for spring that cover to assess the water quality of entire area as they are the only sources of drinking water supply in hilly area. Sampling was done in pre-monsoon (May-June) and post-monsoon (October-November) seasons of the year to evaluate variation in annual concentration of heavy metals. All the samples have been acid digested, prepared and analysed by Atomic Absorption Spectrophotometer (AAS) and Inductively Coupled Plasma Spectrophotometer (ICP) according to standard methods (Arnold, et al., 1992).

4. RESULT AND DISCUSSION

From the results it has been observed that concentration of heavy metals such as Fe, Mn, Pb, Cu, Cd, Cr and Zn in Giri river and Tons river water is well below the permissible limit of drinking water standards. The concentration of iron has been found maximum at four sampling points out of total eight points in Giri and Tons rivers. Where as chromium has been found only in 2 sampling locations of Tons river. In all the nine spring water samples concentration of iron has been found to be maximum and chromium has not been detected at all in the samples.

The variation in the concentration of heavy metals in the spring and river water in two different seasons has been found to be very insignificant. On the whole both the surface and spring water samples in and around the mining belt have been found to be safe from the heavy metal pollution. The monitored data have been used to evaluate the HPI and to assess the validity of the proposed Index model.

The HPI has been determined for surface and spring water by taking the average value of heavy metals of both the seasons using equation (1). In Table-1 and 2 detailed calculation of pollution Index with unit weightage (W_i) and standard permissible value (S_i) are presented for surface water and spring water, respectively. The pollution Index calculated with average values of all metals including all sampling points of surface and spring water for both the season comes out to be 9.3663 and 7.2796 respectively, which are well below the critical Index value 100. The calculated Index values indicate that in general the surface and spring water are not contaminated with respect to heavy metals pollution. The HPI has been calculated separately for Giri river, Tons river and all spring water sampling points. The values are cited in Table-3A and 3B. This enables us to assess the quality of water at each river and spring water sampling points, which can also be used to compare with each other. Percentage deviation with mean values is also calculated for each sampling point. At all the places, HPI calculated is below the critical index value of 100, but the Index value of Giri river is more than that of Tons river, indicating that Giri river is more polluted than Tons river. Four sampling points of spring water are showing Index value lower than the mean value and the percentage deviation is at negative side which indicate better water quality with respect to heavy metals. The rest of the five sampling points are showing index value more than the mean value. The method used to calculate HPI has been found to be very useful to study and compare variations of overall pollution level, that include many parameters together and also it is very useful to assess overall pollution level with respect to heavy metals.

5. CONCLUSION:

The Heavy Metal Pollution Index model used here, has been proved to be a very useful tool in evaluating the over all pollution level of surface and spring water in terms of heavy metals. The HPI calculated for surface and spring water of Sirmour district's limestone-mining area has been found to be below the Index limit of 100. This shows that the surface and spring water are not polluted with respect to heavy metals inspite of the prolific growth of mining and other allied activities in the zone.

ACKNOWLEDGEMENT

The authors are thankful to Scientists of Environmental Management Group of Central Mining Research Institute, Dhanbad for collection of samples.

REFERNECES

- Arnold E.G., Lemore, S.C. and Andrew, D.E.: American Public Health Association, Inc. 1992, *'Standard Methods for the Examination of Water and Waste water'*, APHA,18th Edition.
- Horton, R.K.: 1965, *'An Index Systems for Rating Water Quality'*, J. Water Poll. Cont. Fed, 3, 300.
- Joung, H.M., Miller, W.W., Mahammah, C.N. and Gultjens, J.C.A..: 1979, *'A Generalisez Water Quality Index Based on Multivariate Factor Analusis'*, J. Environ. Quality, 8, 95.
- Landwehr, T.M.: 1979, 'A Stastical View of a Class of Water Quality Indices', Wat. Resour. Res., 15, 460.
- Lohani, B.N. and Todno, M..: 1984, J. Environmental Engineering Division (ASCE), 110 1163.
- Mohan, S.V., Nithila, P., Reddy, S.J.: 1996, *'Estimation of Heavy Metal in Drinking Water and Development of Heavy Metal Pollution Index'*, J. Environ. Sci. Health., A31(2), 283.
- Nishidia, N., Miyai, M., Tada, and F.,Suzuki, S.: 1982, 'Computation of Index of Polltion by Heavy Metals in River Water' Env. Poll., 4, 241.
- Prasad, B. and Jairakas K.C.: 1998 'Evaluation of Heavy Metals in Ground Water Near Mining Area and Development of Heavy Metal Pollution Index', J. Environ. Sci. Health. (Accepted for publication).
- Reddy, S.J.: 1995, *'Encyelopedia of Environmental Pollution and Control'*. vol.1, Enviro. Media, Karlla, India, p-342.
- Tiwary, T.N. and Mishra, M.: 1995/85,: 'A Preliminary Assignment of Water Quality Index to Major Indian Rivers' J. Env. Prol., 5, 276.

Table - 1

HPI calculation for surface water of Sirmour lime stone mining area.

Heavy Metal	Mean Value (ppb) (M_i)	Standard permissive value (ppb) (S_i)	Height desirable value (ppb) (I_i)	Unit weightage W_i	Sub index Q_i	Wi x Qi
Fe	79.5	1000	100	0.001	2.2777	0.0022
Mn	15.57	300	100	0.0033	42.125	0.1390
Pb	5.0	50	----	0.020	10.00	0.200
Cu	9.625	1000	50	0.001	4.25	0.0042
Cd	1.0625	10	----	0.100	10.62	1.0625
Cr	0.700	10	----	0.100	7.00	0.700
Zn	16.5	15000	5000	0.00006	49.83	0.0029

$$\Sigma\ W_i = 0.22536$$

$$\Sigma\ W_iQ_i = 2.1108$$

$$HPI = 9.3663$$

Table - 2

HPI calculation for sprin water o Sirmour lim stone minin area.Heavy Metals	Mean value (M_i) (ppb)	Standard permissibl e value (S_i) (ppb)	Height desirable value (I_i) (ppb)	Unit weightage (W_i)	Sub index (Q_i)	Wi x Qi
Fe	271.16	1000	100	0.001	19.017	0.0190
Mn	19.61	300	100	0.0033	40.195	0.1326
Pb	7.888	50	-	0.02	15.776	0.3155
Cu	10.722	1000	50	0.001	4.134	0.0041
Cd	1.166	10	-	0.1	11.66	1.166
Cr	0.000	10	-	0.0	000	0.000
Zn	24.166	15000	5000	0.00006	49.758	0.0029

$$\Sigma\ W_i = 0.2253 \qquad \Sigma\ W_iQ_i = 1.6401$$

$$HPI = 7.2796$$

272

TABLE 3A

Heavy Metal Pollution Index of individual surface water

Surface water	HPI value	% deviation
Giri river	10.3177	+ 6.93
Tons river	8.9794	- 6.93

Mean = 9.6485

Table 3B

Heavy Metal Pollution Index of spring water at various sampling location:

Sampling point	HPI	% deviation
S1	8.4407	+ 33.15
S2	4.9665	- 21.65
S3	8.3945	+ 32.42
S4	6.6143	+ 4.33
S5	0.7672	- 87.89
S6	3.6355	-42.65
S7	11.6236	+ 83.36
S8	9.5988	+ 51.41
S9	3.0120	- 52.48

Mean = 6.3392

Environmentally Friendly Technologies for the Pulp and Paper Industry

J.D. Dhake

Laxminarayan Institute of Technology, Nagpur University, Nagpur, India

ABSTRACT:

Worldwide awareness about environmental protection has forced pulp & paper industry to take review of existing technologies in terms of environmental regulation to modify the existing technologies or develop altogether new technologies. The innovations taking place in the pulp & paper industry have been extensively reviewed [1]. Most of innovations are either chemical or biological.

This paper reviews some of the developments during last two decades which have taken place in relation to "pulping" and bleaching processes to make them capable of meeting the challenges of twenty-first century in terms of the stringent environmental regulations, ever growing demand and energy usage.

Pulp and Paper industry uses lignocellulose material derived from forest, agriculture and recycled products, Major raw materials are wood bamboo, rice straw, wheat straw, baggage and waste paper, rags etc.

In the manufacture of paper major processes involved are: -

i) <u>PULPING</u> :- Separation of cellulose fibers from Lignin and other constituents of raw material.

❖ <u>Chemical Pulping</u> :- This involves digesting the raw material with alkali (NaOH) or (NaOH-Na_2S) at the temperature of 165^0 C. and at about 10 atmospheric pressure.

❖ <u>Mechanical Pulping</u>:- Lignin is separated by grinding wood. Mechanical pulps containing degraded lignin is used for news print.

❖ <u>Semichemical Pulping</u>:- uses combination of chemical and mechanical energy for separation of fibbers.

ii) WASHING: - Pulp produced by above process is thoroughly washed and screened.

iii) BLEACHING: - The brown color of the pulp due to the presence of residual lignin and other resinous compounds is bleached by using chlorine, hypochlorites, chlorine dioxide to the reasonable brightness.

iv) STOCK PREPARATION: - Bleached pulp is refined in beater, refiners to separate the fibers from fiber bundles. The chemicals such as rosin, alum, dyes, loading and filling materials, other additives are added to pulp water slurry.

v) PAPER MAKING: - The pulp slurry containing only about 0.5% fibers and 99.5% water is run on four drinier / cylinder mould machine (moving wire mesh). The wet web of paper on the wire mesh is picked to press section, pressed and dried in drier section.

vi) CHEMICAL RECOVERY: - The black liquor and washings generated in washing section is concentrated and burnt in the furnace to recover heat and chemicals (alkali). The NaOH and Na_2S thus recovered is again recycled in the process of pulping step.

275

ENVIRONMENTAL PROBLEMS ASSOCIATED WITH CONVENTIONAL PULPING AND PAPER MAKING SYSTEMS.

❖ The large use of water and discharge of coloured effluent having high BOD, COD, Suspended solids.

❖ Presence of highly toxic chlorinated by-products collectively known as AOX adsorbable organic halides) in bleaching effluent. eg. 2,3,8,- tetrachloro-dibenzo-p-dioxin (TCDD) and 2,3,7,8, tetrachloro-dibenzo-furan (TCDF).

❖ Control of obnoxious odour due to the presence of sulfur compounds,

❖ Emission of SOx, NOx, (which leads to acid deposition) and other particulate material detrimental to the health of people living in the vicinity of paper mills.

❖ Disposal of huge quantity of solid waste generated in the processes.

❖ Denudation of forests.

NEED OF ECOFRIENDLY TECHNOLOGIES FOR THE PULP AND PAPER INDUSTRY

The pulp and paper industry is facing increasing pressure in terms of environmental regulations, energy uses and profit margins. Therefore new or modified technologies will be necessary to meet the challenges of twenty first century.

Most of the innovations, modifications taking place in the industry can be classified in two broad categories: - I) Chemical and ii) biological as well as combination of the two. These developments have taken place during last two decades.

In order to keep up with the increased demand for pulp and paper and to simultaneously meet increasingly stringent environmental regulations, the industry is looking towards improvements in the conventional technologies or all together new technologies,

Some of the developments in environmental friendly pulping, bleaching technologies are discussed in following paragraphs.

ORGANOSOLV PULPING :-

This is the pulping process using the organic solvents (in place of NaOH and Na$_2$S) to aid in the removal of lignin from wood.

Of the many systems tried; ALCEL Process using Ethanol and Water; Organocell process using Methanol + NaOH in Methanol /Water have been found to be most potential and are at commercial scale up stage.

Environmental advantages of Organosolv Process are: -
❖ Since cooking medium is only alcohol and water odours, emission of SO2 leading to acid rain will not be present.
❖ Coproduct lignin is less degraded and is used as a substitute for Phenol-formaldehyde resin, which is otherwise made from non-renewable petrochemicals.
❖ Discharge of spent chlorinated bleaching chemicals is reduced.
❖ Organosolv mill can have much smaller size than the size of the Kraft mill needed to be economic. Small mills can be scattered within a geographic region and thereby dissipate the environmental and socioeconomic impacts of any large single operation.

BIOLOGICAL PULPING

The concept of biopulping is based on the knowledge of bio-degradation of cellulose, lignin and hermicellulose by micro organisms. Biopulping was developed to provide environmentally clean pulping process. This process has potential to overcome some problems associated with convential pulping.

BIO MECHANICAL PULPING.

Several white rot fungi are capable of degrading lignin in wood chips. The pretreatment of wood chips followed by mechanical treatment is known as biomechanical pulping. Some of the white rot fungi capable of degrading lignin in pretreatment are *P.Crysosporium,Dichomitus.squalens,Poria medullapanis ,H. setulosa, C.subvermispora, P.tremellosa, P.;Subserialis and P.brevispora*. Of these *P.Chrysosporium* and *C.subvermispora* are most studied and are found to be most effective.

VARIABLES IN BIO MECHANICAL PULPING:

The most important variables affecting the biopulping are wood chip sterilization, inoculum, addition of nutrients to chips, aeration, wood species, wood ageing, wood chip movement during incubation.

Lignin degradation fungi can alter cell walls of wood in short period after inoculation. These changes are identified through the use of histological and ultrastructural techniques. Results demonstrated that lignin modification, rather than removal, is involved during biopulping.

Biopulping is an environmentally friendly technology that substantially increases will through put or reduces electrical energy consumption to the extent of 50% at the same throughput, results in stronger paper and lowers the environmental impact of pulping. Fungal pretreatment is found to be effective in depitching of pine wood chips, for improving dissolving grade pulp.

BIO-CHEMICAL PULPING:

Pretreatment; of wood chips with fungi followed by conventional before chemical treatment is known as biochemical pulping.

Biosulphite/ Biokraft pulping :- Fungal treatment with *C. Subvermispore* followed by conventional sulphite pulping improved the cooking of wood chips leading to kappa number as much as 30 % lower than control pulp (after two weeks of pretreatment time). This means less energy consumption at the same yield.

Pretreatment of wood chips with *P. chrysosporium* followed by conventional Kraft pulping resulted in a decrease in the refining energy needed for Kraft pulping, increased yield at similar kappa number and increased tensile, bustring and tearing strength of paper.

Fungal treatment causes softening and swelling of wood cells. This results in improved chemical penetration during pulping and thus in turn could result in easily bleachable, low kappa number pulps, reduced cooking time and temperatures, reduced pulping chemicals load, reduced effluent waste loads and so on.

ELEMENTAL CHLORINE FREE (ECF) AND TOTALLY CHLORINE FREE (TCF) BLEACHING OF PULP.

Because of regulatory authorities applying much stricter controls, industry was faced to develop technology that can address the environmental issues related to bleaching. This led to the development of ECF and TCF bleaching processes, which reduces the formation of AOX in effluents.

ECF BLEACHING:-

In ECF bleaching chlorine is replaced by chlorine dioxide ClO_2 has 2.5 times more potential oxidizing power than chlorine, which is evident from following reaction-

$$Cl_2 + 2e^- \longrightarrow 2Cl^-$$
$$ClO_2 + 2 H_2O + 5 e^- \longrightarrow Cl^- + 4OH^-$$

This practically means that at the same oxidant equivalence, ClO_2 introduce about a fifth of the chlorine in to bleaching reaction. Complete replacement of molecular chlorine with ClO_2 reported to have lowered AOX from 6.7 to 1.7 Kg/Ton of pulp, the color and BOD of the effluent were also found to be lower.

TCF BLEACHING: -

TCF bleaching uses chemicals (that do not contain chlorine) such as oxygen, ozone, Hydrogen Peroxide, Peroxy acetic acid, Peroxysulfuric acid etc. All these chemicals are oxidizing agents that generate acidic groups in the residual lignin and finally bring about the cleavage of β-aryl ether bonds leading to the formation of acids and CO_2.

Development of ozone bleaching technology has been quite rapid over the last ten years. The first commercial scale operations commenced in 1992 and at present there are number of mills throughput the world applying ozone-bleaching technology.

CLOSED CYCLE BLEACH PLANT: -

In closed cycle TCF bleach plants using ozone or peroxide as bleaching agents, mills are recycling the filtrates from the alkaline stages. The capital and operating costs for a new TCF closed cycle bleached Kraft pulp mills are lower than those for a new conventional ECF pulp mill. The great saving in capital cost comes from elimination of the Waste Water Treatment Plant

BIO- BLEACHING OF PULPS:-

Bleaching process aims at removing residual lignin in the pulp, which could no be removed in the pulping process. This residual lignin has to be removed selectively without degrading cellulose. In conventional bleaching this is achieved by using chlorination of pulp followed by hypochlorite treatment. This leads to the effluents containing highly toxic AOX compounds.

Biobleaching aims at removing lignin by the use of microorganisms rather than by chlorine compounds.

In last decades many laboratories have been engaged in research progrmme to identify the organisms that most effectively biodegrade lignin. In one of such programmes, screening of fungi, isolated from 2068 samples of decaying lignocellulose, for phenol oxidase activity on wood agar medium lead to the isolation of a potential and selective lignin degrader strain IZU-154. This isolate effectively delignified and brightened hard wood kraft pulp under solid state fermentation condition (upto 25% consistency).

Using delignifiecation with IZU-154 as a pretreatment to bleaching with chlorination, alkali extraction and chorine dioxide allowed the target brightness of 88% ISO to be reached with 72% less chlorine, 79% less NaOH, and 63% less C/O2. The COD in efflent was reduced by 49%, color by 78% as compared to control bleach liquor.

Another approach being developed for biobleaching of pulp is the use of hemicellulase enzymes (Xylanase) which degrade the covalent bond between xylan and lignin thus releasing lignin, which can then be more readily removed during bleaching. Xylanase also caused partial depolymerisation of the hemicellulose in the fiber wall, which facilitates removal of the residual lignin during bleaching.

Present status of biobleaching is at pilot scale. However in Canada, Xylanase is being used at commercial bleaching of pulp since 1994-1995.

REFERENCE : -

1. Raymond A.Young and Mnsood Akhtar "Environmentally Friendly Technologies for Pulp and Paper Industry"; John Wiley & Sons Inc. (1998).

Development by Environmental Management Planning at Dongri Buzurg Mine of Manganese Ore (India) Ltd.

D.L. Chaudhary[1], A.K. Nag[2] and G.G. Manekar[3]

[1]*Sr. Dy. General Manager (Prod.),* [2]*Dy. General Manager (Mines) and*
[3]*Sr. Manager (Mines)*
MOIL, Dongri Buzurg Mine, India

1: INTRODUCTION

Mining, which has its very base the destruction of the parts of the environment, and mineral processing, which deals with the recovering with the small parts of mined material that is of use to man and discharging rest as waste, can not help but have an impact on the environment.

The environmental effects associated with mining operations involve a variety of adverse consequences. Mining may cause environmental damage through a number of routes, such as:

- land clearance, erosion of spoil tips,
- hydrological effects,
- impact on ecosystem,
- disruption of natural and human transport system, and through human health and safety.

In the recent years MOIL has demonstrated that the preservation and restoration of the environment during and after mining is possible. It could be seen from the sustainable growth in all the fronts of MOIL as a whole and Dongri Buzurg mine in particular. Introduction of Environmental Management Planning and mechanization of operations at Dongri Buzurg Mine has enabled the company to become more eco-friendly, profitable and improve the quality of life of its employees.

1.1 : DONGRI BUZURG MINE

Manganese Ore (India) Limited, a Public Sector Undertaking and a leading producer of Manganese Ore, in India operates ten mines. The Company is engaged in mining, beneficiation and processing of Manganese Ore. Dongri Buzurg Mine produces Dioxide ore, mainly used in dry batteries and Chemical industries. Ferro and Blast Furnace grade ore is also produced for steel industries. The total production of Manganese ore from the mine is around 0.13 million tones/annum, out of which about 25,000 tonnes is commercial grade dioxide ore.

The mine also operate a High Intensity Magnetic Separation (HIMS)Plant for up gradation of dioxide ore. An Electrolytic Manganese Dioxide Plant, based on indigenous technology for production of Electrolytic Manganese Dioxide (EMD), a vital raw material for dry battery manufacturing is also being operated at this mine. In view of various types of activities at this mine, the environmental management programme forms an integral part of environmental management at the mine.

1.2: MINING AND PROCESSING ACTIVITIES

Different types of ore deposit greatly differs in occurrence and composition of minerals constituting inert, toxic and non-toxic constituents which affects natural life. The course of processing such materials to prevent environmental damage may consist of controlling pollutants produced during various operation. Various activities of mining is as below in short.

- Mining Activities:

The total lease hold area of mine is 170.729 hectors, out of which the mining area is 82 hectors. At Dongri Buzurg mine, manganese ore consists secondary oxide minerals such as Psilomelane, Cryptomelane and primary silicate minerals in the form of Braunite and also Jacobsite. The gangue consists Quartz, iron Minerals and Apatite. The waste rock is mainly mica schist which is lighter and very friable in nature. The central portion of a mine, ranging about a Km. strike length which is almost 1/3 of the property is containing comparatively thicker ore body of high grade manganese ore. This area is worked with the aid of heavy earth moving equipment's keeping a bench height of 7.5 mtrs. in the waste rock and 6 mtrs. in the ore body. As such the central and thicker portion of the ore body is worked by mechanized Open cast method.

The main mining activities with reference to environmental aspect are as follows (i) Drilling (ii) Blasting (iii) excavation (iv) Transportation (v) Dumping & Stacking (vi) Beneficiation and ì (vii) Processing.

2: ENVIRONMENTAL MANAGEMENT PLAN
Environmental Management Plan of Dongri Buzurg Mine has prepared with due consideration of the following four steps for development of eco-friendly atmosphere at the mine:
1. Mine Planning
2. Environmental Management Overview Strategy (EMOS)
3. Plan for Operations
4. Environmental Audit

2.1: Eco-friendly Technologies
Many of the areas where improvements are feasible have already been tried and improvement in production and productivity has occurred. Most of the mechanization programmes were aimed at replacement of manual tools into mechanized ones, such as;
* Providing machines for handling man and materials at mine site
* Value added products at mine sites, like Electrolytic Manganese Di-oxide plant at Dongri Buzurg mine
* Creating more eco-friendly environment by afforestation
* Reclamation & Revegetatation of mine spoil dumps through Integrated Biotechnological Approach (IBA)

3: Protection of Environmental Impacts
Opencast mining drastically disturbs the physical, biological and socio-economical features of the area. The mines which are located in forests threatens the ecological balance of the region by disturbing flora and fauna to a great extent. The hazards with potential deleterious effects on the surrounding environment emanating from the mining industry are air, water and noise pollution. The visual intrusion, because of the activities like deforestation, waste dumping, mining, soil erosion, polluted air and dirty water, has also been a part of negative environmental impact in opencast mining area. In certain specific areas, blasting operations have significant impact on the environment and working condition. The main topics of the concern are contamination and disposal of unused explosives, malfunction of explosives borehole, blast induced vibrations and air over pressures, fly rock hazards, post blast explosion and other energetic reactions.

To protect the negative environmental impacts associated with these activities, MOIL has conducting control blasting in its all opencast mines. Improvement in blast designs has been seen by introduction of non-electric detonating systems i.e. EXCEL, in fly rocks, ground vibrations and noise pollution. MOIL has engaged National Environmental Engineering Research Institute (NEERI) for Rejuvenation of Mine Spoil Dumps at Dongri Buzurg mine. These activities has reduced the pollution in the following major area:
* Air pollution due to wind blown dust
* Surface as well as ground water pollution.
* Land degradation due to changes in land use.
* Visual intrusion

3.1: WASTE DUMP MANAGEMENT
Area of about 200 Mtrs. long and 100 Mtrs. wide is ear marked for dumping waste rock. It is systematically filled with muck keeping a bench height of 15 Mtrs. and width of about 30 mtrs. once the height of the bench reaches the pre-decided configuration, the top bench is earmarked for plantation in the subsequent year and as such the reclamation, and stabilization of dump with afforestation closely follows the active dump area.

Though the height of the waste dump is kept 15 Mtrs. for the purpose of benching but it is formed in three separate layers each of 5 Mtrs. The waste rock consists mainly of Muscovite, micacious schist which are very weak and slippery therefore the 5 Mtrs. thick slice is first consolidated by leveling and rolling the surface. A road roller cum compactor is provided for this purpose. Once the first slice is complete then the second slice of 5 Mtrs. is taken subsequently. For this purpose every bench of the dump are made approachable from the main dump road once the planned height and width is acquired in the dump. It is leveled, stabilized by means of roller and also by growing bushes and heameta grass for the plantation in the subsequent year. Data of the dump in Hectare in is in Table 1.

Table 1 : Dump Area

Total area accepted by Dump	Area established	Percentage
32 Hectare	28 Hectare	87.5

3.1.1: TOP SOIL MANAGEMENT

The Dongri Buzurg Mine is located in a hilly terrain. The mine is known locally as "SAGWAN VAN" (Teak Forest), indicating teak as the main plant grown in the area. This spice is normally growing in rocky and sandy soil, therefore it is very little useful for mass plantation. Historically there was no cultivation in the mine lease hold area showing the fertility of the land, however, as small patch on the foot hill have been found to be cultivable. This land is being purchased by the company to acquire area for dumping and also for obtaining fertile soil for plantation and other social forestry purposes.

The company has purchased around 35 acres of revenue land in the year 1992-93 and purchase of another about 60 acres during the year 1997-98 is in hand. These lands are situated about a kilometer away from footwall and hanging wall of the ore body. The intermediate area is covered by quartile and lateritic rocks and as such these area have been earmarked for spoil dumps.

The top soil is normally recovered from cultivable land in the surrounding of the mine from the acquired area by CK - 90 excavator. Excavation and transportation and also for subgrade mineral handling. A Dozer D-65 is also attached to this fleet for collection of top soil, spreading of top soil, surface road making and leveling in the dump etc.

The details of top soil generated, stored & utilized is given below in Table 2:

Table 2: Top Soil Utilization

Year	Quantity of Top Soil generated	Quantity of Top Soil Stored	Quantity of Top soil Utilized	Percentage of Utilization
1994-95	85000	25000	60500	71%
1995-96	75650	10500	90150	119%
1996-97	156000	20000	96100	91%
1997-98	86950	25375	81575	94%
1998-99	89750	26500	83750	93%

3.1.2: RECLAMATION AND REHABILITATION

The Mine Spoil dumps are located mostly in barren lands comprising of quartzitic & lateritic pebbles are revenue lands purchased from the local farmer. The fertile top soil from revenue land are removed and used in mixture for application in integrated Biotechnological approach to fill the pits. A layer of about one feet is also laid over the spoil dumps for increasing vegetation. Amendment of mine spoils with sugar mill waste (Presumed) at the rate of about 100 tonnes per Hect. increases the moisture holding capacity and helps root proliferation. As such the land is not only reclaimed but the fertility is increased manifold. The land area covered by spoil dumps is around 32 Hectares out of which 27 Hectares has already been covered with plants and vegetation and the balance area is under active dump included in the future plant for afforestation.

The plants surviving in afforested area is about 1800 to 2000 Nos. per hectares having better health and growing capacity. Therefore there is a high level satisfaction for rehabilitation of plants. The fruit growing trees are also being planted to attract the fauna and as such reclamation and rehabilitation programme for the mine is closely following the mine activities not only to maintain the present level of Flora and fauna but to increase the same mine field with socio-economic benefits.

- **Reclamation and Revegetation of Mine Spoil Dumps through Integrated Biotechnological Approach (IBA):**

Rejuvenation of degraded mine land productivity and fertility, massive afforestation with ecologically and economically important plant species and ecorestoration being a major thrust area in Dongri Buzurg Mine, appropriate scientific approach to restore the productivity of degraded land, mine spoil dumps and greening the abamendent lands has been launched in collaboration with National Environmental Engineering and Research Institute (NEERI).

Mining of minerals particularly by opencast method adversely affect the environment resulting in

281

degradation of land on large scale. The overburden dumps created during the process of opencast mining are devoid of good structure, texture, horizonation, useful nutrients, organic matter and microbial activities which operates various ecological cycles. Poor physical conditions do not encourage plant establishment on such lands resulted in severe erosion problems. The restoration of productivity and fertility of such lands is a major environmental concern and a systematic and scientific approach is needed for revegetation of such lands by planting economically, ecologically important and stress tolerant plant species. Considering the nature and extent of problems and concern for conserving the environment, MOIL took a lead for massive afforestation at the mines of the company with special programme on reclamation of mined areas and spoil dumps supported by exhaustive research and development in collaboration with National Environmental Engineering Research Institute (NEERI). The scientific studies were initiated in 1987. An approach plan for rejuvenation of land productivity and reclamation of spoil dumps was prepared based on physico-chemical-microbiological properties of mine land and mine spoil dump and need of technological intervention involving Integrated Biotechnology to achieve stable ecosystem restoration.

The Integrated Biotechnological Approach envisages the use of industrial wastes like pressmud from sugar mill industry as an organic amendment for the spoil and use of nitrogen fixing bacteria like *Rhizobium* and *Azotobacter* which can tolerate high manganese concentration and VFM fungi. The use of biofertilizers reduces the environmental risk of using chemical fertilizers. Plantation up-to date is given below in table 3.

Table 3: Sapling Planted & Existing Plants as on 31-3-2000

Tress planted during 98-99	5000
Tress planted during 99-00	4000
Total number of tress available as on 31-3-2000	61572

4. OUTCOME

Implementation of Environmental Management Plan & IBA at Dongri Buzurg Mine improvement had been seen in the following areas:
- Environmental risk: no environmental risk as on date
- Eco-friendly work culture
- Compliance with statutory requirements
- Standard operating procedure

Improvement in dealing with;
- employees
- union
- environmental group
- regulating agencies
- community group

Production & Productivity:

Production & productivity has improved considerable it could be seen from the below mentioned table at Dongri Buzurg mine. This is turn, contribute in the profitability of the company directly. Production & Productivity in turn with Out put per man shift in Tonne is given in Table 4.

Table 4: Production & Productivity

Year	1996-97	1997-98	1998-99	1999-2000
Production (T)	111625	122120	133419	158087
OMS	0.70	0.77	0.75	0.91

282

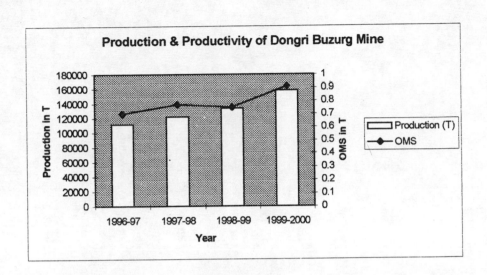

Removal of Copper (II) from Aqueous Solution and Copper Plating Industry Wastewater by a Carbonaceous Sorbent Prepared from Waste Fruit Peel of *Artocarpus heterophyllus*

B. Stephen Inbaraj and N. Sulochana

Department of Chemistry, Regional Engineering College,
Tiruchirappalli 620 015, Tamil Nadu, India

INTRODUCTION

Environmental pollution due to copper arises from industrial and agricultural emissions. It is found in municipal wastes as by-product from the metal finishing and processing industry and agricultural sources such as fertilizers and fungicides. The tolerance limit of copper for discharge into inland surface waters[1] is 3.0 mg L^{-1} and in drinking water[2] is 0.05 mg L^{-1}. Conventional methods for the removal of Cu(II) from wastewater are not cost effective in the Indian context. Adsorption by activated carbon was found to be one of the most effective methods in treating metal-bearing wastes. However, due to the high cost of conventional activated carbons, there is a great upsurge in the development of activated carbon from cheaper and readily available materials. Jack fruit (*Artocarpus heterophyllus*) is one of the most popular fruits in South India and the average annual yield is reported to be 50-100 per tree with each fruit weighing about 23-40 kg. The outer peel (rind), which is mostly fibrous, constitutes about 59 % of the ripe fruit[3].

The present study is undertaken to evaluate the efficiency of a carbonaceous sorbent prepared from jack fruit peel for the removal of Cu(II) from aqueous solution and copper plating industry wastewater. A batch-mode adsorption study involving parameters such as agitation time, initial concentration, carbon dose and pH was systematically carried out.

METHODS

Air-dried jack fruit peel (after removing the carpel fibers) was treated with conc. H$_2$SO$_4$ in a weight ratio of 1:1.8 (peel :acid) to yield a black product which was kept in an air-oven maintained at 160 ± 5°C for 6 h. It was then washed thoroughly with distilled water, until free of excess acid and dried at 105 ± 5°C. The carbon product obtained (JPC) was ground and the portion retained between 89 and 124 μm sieves was used in all the experiments. The characteristics of the carbon such as moisture content, bulk density, ash content, pH, water soluble matter, acid soluble matter, decolourising power, surface area, ion exchange capacity and iron were experimentally carried out and reported in Table 1.

All adsorption experiments were carried out by agitating the carbon with 100 mL Cu (II) solution of desired concentration at pH 5.0 and at room temperature (32 ± 0.5°C) in a mechanical shaker (200 rpm). After the defined time intervals, samples were withdrawn from the shaker, centrifuged and the supernatant solution was analysed for residual Cu(II) concentration using Atomic Absorption Spectrophotometer (GBC 902). Adsorption isotherm study was carried out with Cu (II) solution of different initial concentrations ranging from 10 to 100 mg L^{-1} and agitating with a fixed carbon dose until equilibrium was reached. To study the effect of pH on the adsorption capacity, the initial pH of 20 mg L^{-1} Cu (II)

Table 1 Characteristics of JPC

Parameter	
pH	7.24
Moisture content, %	16.47
Bulk density, g mL^{-1}	0.76
Ash content, %	5.00
Solubility in water, %	1.32
Solubility in 0.25 M HCl, %	9.3
Decolourising power, mg g^{-1}	127.30
Ion exchange capacity, mequiv g^{-1}	1.01
Surface area, m^2 g^{-1}	131.73
Iron, %	0.07

solution was adjusted using small amounts of dilute nitric acid or sodium hydroxide and agitated with a fixed carbon dose. Experiments were repeated at least three times and mean values are reported. Standard deviation and analytical errors were calculated and the maximum error was found to be ± 5%.

RESULTS AND DISCUSSION

Adsorption Kinetics: The kinetics of adsorption of Cu(II) by JPC is shown in fig.1 with smooth and single plots indicating monolayer adsorption of Cu(II) on the carbon. The increase in the rate of Cu(II)

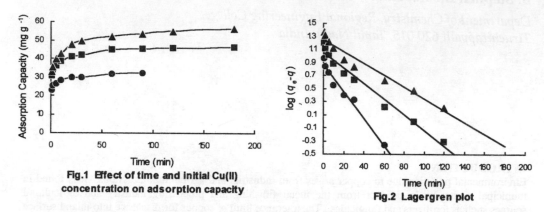

Fig.1 Effect of time and initial Cu(II) concentration on adsorption capacity

Fig.2 Lagergren plot

removal with agitation time may be attributed to the decrease in the diffusion layer thickness surrounding the carbon particles[4]. Figure 1 also shows that the increase in the initial Cu(II) concentration increased the amount of Cu(II) adsorbed. The removal of Cu(II) followed first order rate expression given by Lagergren[5] $(\log(q_e-q)=\log q_e-K_{ad}t/2.303)$. The rate constants calculated from the slopes of the plots (Fig. 2) are 0.0474, 0.0276 and 0.0207 min^{-1} for initial Cu(II) concentrations of 20, 30 and 40 mg L^{-1}, respectively. An examination of the effect of Cu(II) concentration on the rate constant helps to describe the mechanism of removal taking place. Since a direct linear relationship does not exist in the plot of initial cu(II) concentration versus rate constant (Fig.3), it seems likely that pore diffusion limits the over all rate of adsorption[6].

The plot of amount of Cu(II) adsorbed versus square root of time using the relation, $q=K_p t^{1/2}$, depicts that the initial curved portion is attributed to boundary layer diffusion effect, while the subsequent linear portion to intra particle diffusion effect[7] (Fig.4). The prevailing linear portion of the plots indicates that the intraparticle diffusion is the rate-controlling step. The intraparticle diffusion rate constants (K_p) obtained from the slopes of the linear plots are 0.5772, 0.6122 and 0.9601 mg g^{-1} min$^{-1/2}$ for 20, 30 and 40 mg L^{-1}, respectively. Assuming spherical geometry[8] for the carbon, the adsorption rate constant for the process can be correlated to the pore diffusion coefficient in accordance with the expression, $D_p=0.03(r_o^2/t_{1/2})$. The average value of pore diffusion coefficients calculated for different Cu(II) concentrations is 0.6533 x 10^{-11} cm^2 s^{-1}. The pore diffusion coefficient obtained in the range of 10^{-11} to 10^{-13} cm^2 s^{-1}, further shows that pore diffusion might be the rate-limiting step in the Cu(II) adsorption by JPC[9].

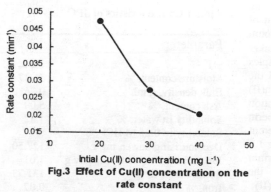

Fig.3 Effect of Cu(II) concentration on the rate constant

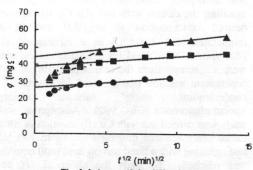

Fig.4 Intraparticle diffusion plots

The resistance to mass transfer from the solution onto the adsorbent surface is inversely proportional to mass transfer coefficient, β_L, which is determined using the following relationship[10].

$$\ln\left[\frac{C_t}{C_o}-\frac{1}{1+mK_L}\right] = \ln\frac{mK_L}{1+mK_L}-\frac{1+mK_L}{mK_L}\beta_L S_s t$$

The β_L values calculated from the slopes of the linear plots of $\ln[C_t/C_o-1/(1+mK_L)]$ versus t (not shown) for different Cu(II) concentrations were found to have the average value of 6.7167×10^{-7} cm s^{-1}. The β_L values obtained reveal that the velocity of Cu(II) transported from the liquid phase onto JPC is rapid enough that the process can be utilised for its removal from wastewaters.

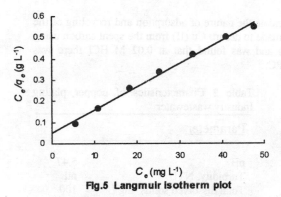

Fig.5 Langmuir isotherm plot

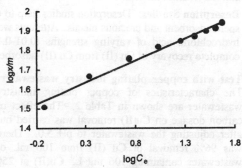

Fig. 6 Freundlich isotherm plot

Adsorption isotherms: The equilibrium data obtained for different initial Cu(II) concentrations fit well with both Langmuir and Freundlich isotherm models. The isotherm data in fig.5 are well described by the linear form of the Langmuir equation[11], $C_e/q_e=1/Q_ob+C_e/Q_o$. The constants, Q_o (capacity parameter) and b (affinity parameter) were found to be 92.59 mg g^{-1} and 0.21 L mg^{-1}, respectively. The essential characteristics of a Langmuir isotherm can be expressed in terms of a dimensionless constant separation factor or equilibrium parameter, R_L, which is defined by $R_L=1/(1+bC_o)$. R_L values obtained (0.0464-0.1958) between 0 and 1 indicate favourable adsorption of Cu(II) on JPC for the concentrations studied[4].

The equilibrium data also conforms to the Freundlich equation (Fig.6) represented as $x/m=K_FC_e^{1/n}$ and its linear form as $\log(x/m)=\log K_F+1/n\log C_e$. The constants, K_F (a measure of adsorption capacity) and n (a measure of adsorption intensity) were found to be 36.97 mg g^{-1} and 4.51, respectively. Values of $1<n<10$ shows favourable adsorption of Cu(II) on JPC[12]. The sorption equation derived ($x/m=36.97C^{0.22}$) can be employed to determine the volume of wastewater that could be treated. The volume of wastewater, containing 5 mg L^{-1} of Cu(II) as initial concentration, that could be treated by 1 g of JPC was found to be 52.7 liters.

Effect of carbon dose: Taken an initial concentration of 40 mg L^{-1} the amount of Cu(II) adsorption increased with the increase in carbon dose and reaches a maximum value (96%) after a particular dose (100 mg 100 mL^{-1}). The increase in Cu(II) adsorption with carbon dose was due to the introduction of more adsorption sites for adsorption (Fig.7)

Effect of pH: The Cu(II) removal by JPC increases with increase in pH and attains 95% at pH 5.0. The pH range of 5 to 9 was effective for the maximum removal of Cu(II) by JPC (Fig.8).

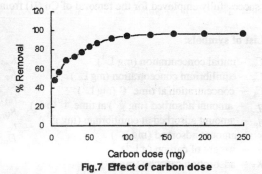

Fig.7 Effect of carbon dose

Below pH 5.0, the increase in the concentration of H$^+$ ions in the reaction mixture competes with Cu^{2+} ions for the adsorption sites, resulting in the reduced uptake of Cu(II). As the pH increases, the concentration of H$^+$ ion decreases, whereas the concentration of Cu^{2+} remains constant and therefore the uptake of Cu(II) can be explained as an H$^+$-Cu^{2+} exchange reaction. This was also confirmed by the decrease in the initial pH observed with the increase in the Cu(II) adsorption, which was due to the release of H$^+$ ions into the reaction mixture.

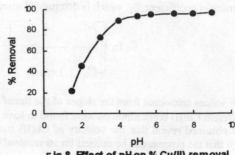
Fig.8 Effect of pH on % Cu(II) removal

Desorption Studies: Desorption studies help to elucidate the nature of adsorption and recycling of the spent adsorbent and precious metals. Attempts were made to desorb Cu (II) from the spent carbon using hydrochloric acid of varying strengths (0.01-0.1M) and was found that at 0.02 M HCl there was complete recovery of Cu (II) from Cu (II)-adsorbed JPC.

Test with copper plating industry wastewater: The characteristics of copper plating industry wastewater are shown in Table 2. The effect of carbon dosage on Cu(II) removal was carried out after adjusting the wastewater to pH 5.0. There was 96% removal of Cu (II) from 100 mL of wastewater containing 106 mg L^{-1} Cu(II) at 250 mg of carbon dosage (Fig.9).

Table 2 Characteristics of copper plating Industry wastewater

Parameter	
pH	5.47
Turbidity, NTU	nil
Total dissolved solids, mg L^{-1}	100
Conductivity, mmho	5.5
Chloride, mg L^{-1}	566
Sulphate, mg L^{-1}	400
Sodium, mg L^{-1}	89
Potassium, mg L^{-1}	9.3
Calcium, mg L^{-1}	13.1
Magnesium, mg L^{-1}	33.2
Copper, mg L^{-1}	106

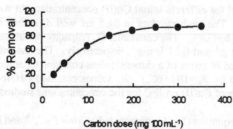
Fig.9 Effect of carbon dose on % Cu(II) removal from copper plating wastewater

CONCLUSION

The foregoing results and discussion reveal that the carbon prepared from waste jack fruit peel can be successfully employed for the removal of Cu (II) from water and wastewater.

List of symbols:

C_o – initial concentration (mg L^{-1})
C_e – equilibrium concentration (mg L^{-1})
C_t – concentration at time 't' (mg L^{-1})
q – amount adsorbed (mg g^{-1}) at time 't' (min)
q_e – amount adsorbed at equilibrium (mg g^{-1})
x – amount adsorbed (mg L^{-1})
m – weight of carbon (g L^{-1})
K_F – adsorption capacity (mg g^{-1})

K_{ad} – adsorption rate constant (min^{-1})
K_p – intraparticle diffusion rate constant(mg g^{-1} ₁
t – time (minutes)
$t_{1/2}$ – time for half change (sec)
R_L – separation factor
r_o – radius of the carbon particle (cm)
D_p – pore diffusion coefficient (cm^2 s^{-1})
K_L – Langmuir constant for kinetic study (L g^{-1}
S – specific surface of the carbon per unit volu

ACKNOWLEDGEMENT

The authors thank Rajiv Gandhi National Drinking Water Mission, Ministry of Rural Development, Government of India for their financial assistance.

REFERENCES

1. Indian Standards Institution, IS: 2490 (Part I), 1982.
2. Indian Standards Institution, IS: 10500, 1991.
3. Chadha, Y.R. ed. The Wealth of India – Raw Materials. Publications and Information Directorate, CSIR, New Delhi, 1985, **Vol.I:A**, pp.448.
4. McKay, G. *J. Chem. Technol. Biotechnol.*, 1982, **32**, 759.
5. Khare,S.K. Panday,K.K., Srivastava,R.M. and Singh,V.N. *J.Chem. Technol. Biotechnol.*, 1987, **38**, 99.
6. Knocke, W.R. and Hemphill, L.H. *Water Resour.*, 1981, **15**, 275.
7. Gupta, G.S. Prasad, G. and Singh, V.N. *J. Indian Assoc. Environ. Mgmnt.*, 1989, **16**, 174.
8. Bhattacharya, A.K. and Venkobachar, C. *J. Envi. Eng. Div.*, 1984, **110**, 110.
9. Singh, A.K., Singh, D.P. and Singh, V.N. *Environmental Technology Letters*, 1988, **9**, 1153.
10. McKay, G., Otterburn, M.S. and Sweeney, A.G. *Water. Res.*, 1981, **15**, 327.
11. Langmuir, I. *J. Am. Chem. Soc.*, 1918, **40**, 1361.
12. McKay, G., Blair, H.S. and Garden, J.R. *J. Appl. Polym. Sci.*, 1982, **27**, 3043.

289

Reclamation of Cr (VI) Rich Water and Wastewater by a Geological Material

Y.C. Sharma[1], N. Sinha[2], J.L. Pandey[3] and D.C. Rupainwar[4]

[1]*Environmental Pollution Control and Research Laboratories, NIFFT, Hatia, Ranchi 834 033, India*
[2]*Student, Department of Environmental Science, Vikram University, Ujjain, India*
[3]*Asstt. Director, NABL, NPL, Dr KS Krishnan Marg, New Delhi 110 012, India*
[4]*Department of Applied Chemistry, IT-BHU, Varanasi 221 005, India*

1.0 INTRODUCTION

Chromium can exist in many valance state like metallic chromium, bivalent chromium, trivalent and hexavalent chromium but only two of the forms trivalent, and hexavalent are important with the environmental view point (1). The hexavalent chromium is regarded approximately 100 times more toxic than the trivalent form (2).

Exposure to hexavalent chromium causes dermatitis, allergic skin reactions and ulceration of intestine. This has been reported to be a potent carcinogen and teratogen(3). The USEPA has suggested a limit of 0.05 ppm of chromium in domestic water supply. Solvent extraction, ion-exchange, reverse osmosis, precipitation and adsorption on activated carbon are the established treatment technologies (4-6) for treatment of Cr (VI) containing water and wastewater. But the above technologies are quite cost intensive and are not suited for developing nations. The present work addresses the reclamation of Cr (VI) rich water and wastewater by a geological, wollastonite to provide an economically viable treatment. The effect of various important parameters on Cr (VI) removal from water has been studied.

2.0 EXPERIMENTAL

Wollastonite is a clay mineral available in plenty in world and in India, Wollastonite reserves are found in Udaipur, Rajasthan. The wollastonite was used as such in the experiments just after crushing it to save the cost of treatment. The crushed sample was sieved to maintain an average particle size to 100 um. The average particle size of the adsorbent was measured by particle size analyzer, model HIAC-320 (Royco Inst. Div., USA) and the surface charge by lazer zee meter, model 500 (Penkem, Inc. NY, USA). The surface area of the powdered adsorbent was determined by a "three point" N_2 gas adsorption method using Quantasorb Surface Area Analyser (Quantachrome Corp., NY, USA) and the porosity by Mercury was determined by following Indian Standard Methods, IS : 1527(8). Batch adsorption experiments were carried out by agitating 1.0g adsorbent with 50 ml aqueous solution of Cr (VI) of desired concentrations and pH in different glass bottles in a thermostat at 100 rpm. After predetermined time interval, the sample bottles were taken out and the reaction mixtures were centrifuged. The progress of adsorption was examined by determining the amount of Cr (VI) in the aliquot by spectrophotometry using spectrophotometer (uv-2100, Shimadzu, Japan). The pH of the solutions was maintained by 0.1M HCI/ NaOH solutions. The ionic strength of the solutions was maintained by 0.01M $NaClO_4$.

3.0 RESULTS AND DISCUSSION

3.1 Characterization of wollastonite: The physiocochemical analysis of wollastonite ws carried out (8)and the results are given in table 1. This table shows that SiO_2 (48.52%) and CaO (48.48%) are the major constituents of wollastonite, all other constituents are present in traces. This reflects that these two species would be responsible for Cr (VI) removal from the aqueous solutions.

3.2 Effect of contact time and concentration on removal of Cr (VI): During the experiments, the removal of Cr(VI) increased up to 100 min. and then it became constant (Table 2). It is also clear from this table that by decreasing the Cr(VI) concentration in solution, the % removal increased from 47.4 to 69.5 at 2.5 pH, 30°C temperature, 0.01M $NaClO_4$ ionic strength by decreasing the Cr(VI) concentration from 2.0 x 10^{-4}M to 0.5×10^{-4}M in aqueous solution. The time of equilibrium, 100 min. for the present system is independent of concentration. The finding of higher removal in low concentration range is of industrial importance(9)

Table1 : Physicochemical analysis of wollastonite

Constituents	% by weight
SiO_2	48.85
CaO	48.48
Al_2O_3	0.52
FeO_3	0.26
Loss on ignition	2.50
Surface area ($cm^2\, g^{-1}$)	1.18
Porosity	0.23
Density ($g\, cm^{-3}$)	2.21
Mean particle diameter (um)	100.0

Table 2: Effect of contact time and concentration on the (%) removal of Cr(VI) by adsorption on wollastonite

% Removal	Time (min)	10	20	30	40	50	60	70	80	90	100
	0.5×10^{-4}M	33.7	45.0	53.4	50.2	62.8	64.7	65.0	66.2	68.0	69.5
	1.0×10^{-4}M	29.0	38.2	45.0	50.6	55.9	59.2	61.0	61.8	62.3	63.0
	2.0×10^{-4}M	15.0	20.4	26.3	30.6	35.0	37.9	40.8	44.3	45.1	47.9

3.3 Kinetic Modelling: The kinetic modelling of the process of removal was carried out using the following model (10):

$$Log (qe-q) = log\, qe - (Kad/2.303) . t \qquad (1)$$

Where qe and q (both mgg^{-1} are the amounts of Cr(VI) removed at equilibrium and at any time respectively, Kad(min^{-1}) is the rate constant of Cr(VI) removal. The values of qe and q were taken from table 1. The value of Kad at 0.5×10^{-4}M Cr(VI) concentration, 30°C, 2.5 pH and 0.01 M $NaClO_4$ ionic strength was found to be 3.0×10^{-2} min^{-1}

3.4 Intraparticle Diffusion Study: The possibility of removal of Cr(VI) from water samples by intraparticle diffusion was tested and intraparticle diffusion was found to play a role in removal. The coefficient of intraparticle diffusion was determined using following equation (11):

$$\bar{D} = (0.03 / t_{1/2})\, r_0^{\,2}$$

Where $\bar{D}$, ($cm^2\, min^{-1/2}$) is the coefficient of intraparticle diffusion, $t_{1/2}$ ($min^{1/2}$), time for half adsorption and r_0(cm) radius of adsorbent particles. The value of D at 30°C was found to be 3.5×10^{-10} $cm^2\, sec^{-1}$ at 0.5×10^{-4} M Cr (VI) concentration.

CONCLUSIONS:
Based on the above studies, the following conclusion may be drawn:
1. Wollastonite shows significant removal of Cr (VI) from aqueous solutions.
2. Higher (%) removals obtained at low initial concentrations
3. The process of removal follows a first order rate kinetics
4. Intraparticle diffusion also plays a role in Cr (VI) removal

A wollastonite is a naturally occurring geological, it incurs no extra cost for treatment and the process can be recommended for treatment of Cr (V) rich wastewaters.

REFERENCES:

1. Kannan K (1995), Fundamentals of Environmental Pollution, First Ed, S Chand and Co. Ltrd., India.
2. Nriagu, J.O. (1988), A silent epidemic on environmental metal poisoning. Environ. Pollut, 50: 139-161.
3. Mukherjee, A.G. (1986), Environmental Pollutionand Health Hazards; Causes and Control, Galgotin Pub. N.d. (India).
4. McKay, G., Otterburn, M.s. and Sweeny, A.G. (1981), Surface mess transfer process during colour removal from effluent using silica, Water Res. 15:327-331
5. Panday, K.K., Prasad, G and singh, V.N. (1984), Removal of Cr(VI) from aqueous solution by adsorption on fly ash and wollastonite, J. Chem. Tech. Biotechnol, 34A:367-374.
6. Sharma, Y.C., Prasad, G and Rupainwar, D.C. (1991), Treatment of Cd(II) rich effluents (kinetic modelling and mass transfer) Int. J. Environ. Anal. Chem. 45:11-18.
7. Sharma, Y.C., Prasad G and Rupainwar, D.C. (1991), Adsorption for removal of Cd(II) from effluents, Int. J. Environ. Studies, 36:315-320
8. Indian Standard Methods of Chemical Analysis of Fire clay and Silica Refractory Materials (1960), IS:1527.
9. Panday K.K. Prasad, G and Singh V.N. (1985) Copper (II) removal from aqueous solution by fly ash, Water Res: 19: 869-873.
10. Srivastava, S.K. Tragi R. Pant, N. and Pal N. (1989), Studies on the removal of some toxic metal ions, Part II (Removal of lead and cadmium by montmorillonite and kaolinite) Environ. Tech. Lett. 10: 275-282.
11. Gupta, G.S.., Prasad G and Singh, V.N. (1988), Removal of color from wastewater by sorption for water reuse, J. Environ. Sci. & Hlth, A23(3): 205-218.

293

Remedial Measures for Acid Producing Minerals

S.D. Prasad[1] and R.K. Mehendiratta[2]

[1]CMD and [2]Dy. Gen. Manager
Mineral Exploration Corpn. Ltd., Nagpur, India

INTRODUCTION

Acid producing minerals can be classified as minerals having tendency to release acid forming ingredients in mining environment and promote the formation of acid. During the course of mining excavation, when such minerals (Pyritic and sulphide ores) are exposed, these make a contact with the environmental air and water resulting in oxidation. The chemical reaction shall be as below:

$$FeS_2 + 7/2\ O_2 + H_2O = Fe^{2+} + 2SO_4^{2-} + 2H^+$$
$$Fe_2 + \tfrac{1}{4}\ O_2 + H^+ = Fe^{3+} + 1/2\ H_2O$$
$$Fe_{3+} + 3H_2O = Fe(OH)_2 + 3H^+$$
$$FeS_2 + 1\ 4\ Fe^{3+}\ H_2O = 1\ 5\ Fe^{2+} + 2SO_4^{2-} + 16\ H^+$$

From the above it may be observed that oxidation of pyrites and other sulphides takes place in steps. In the first instance pyrites is oxidised and ferrous iron is produced. Then ferrous iron is further oxidised into ferric iron. Both the first and second reactions are catalysed by Microbes, Thiobacillus ferroxidans, which increase the overall rate of oxidation by more than a million times. In the third step, ferric iron which is a powerful oxidiser oxidises more pyrites resulting in the production of Ferrous iron. Thus, the process of catalysis becomes a continuous one and ferric iron is produced. Therefore, the acid formation cycle is jacked up spirally and exponentially.

Depending upon the kind of mineral and its tendency to release the acid bearing ingredients, the severity of acid formation is likely to be experienced. Such phenomenon can occur in underground mining as well as in the open cast mining. In open cast mining, to overcome, the water can be channelised through non acidic character rocks and environmental air contact will not be severe due to openness., However, the problem is extremely severe in case of underground mining. In underground mines apart from the seepage the water table is encountered and pumping of water takes a reasonable time. Further, mining activity also involves usage of water mainly in wet drilling and dust suppression. Therefore, the contact of excavation with water cannot be either avoided or instantly separated, thus acid formation occurs. More the delay in handling the water, the concentration of acid increases proportionately. Further, these minerals/rocks remain in contact with mine air and form acidic environment resulting acidic fumes. The severity of acidic fumes depends on the ventilation system of the mine. The velocity of mine air is significant as it dilutes the acidic fumes speedily.

During the course of mine development and production such difficulties are experienced which require special attention so that the workers are facilitated to work in acid free environment. The infrastructural features of mine comprising of mining equipments, service lines and steel structures/track lines etc. also deteriorate very fast due to corrosion in the presence of acidic environment. Although the authors have experienced the workings of acidic Pyrite mines of M/s. Pyrites Phosphates and chemicals Limited, but the contents of the titled subject are not confined to PPCL alone.

The remedial conventional measures during the mining activities shall be as under:

- Speedy handling of mine water by using pumps with non corrosive Material of Construction or lining such as stainless steel or polymer lined .
- Coating the steel structures with anti rust and bitumen i.e. providing a thin barrier to eliminate the mine air and moisture contact.
- Provide adequate ventilation to dilute the acidic environment. This shall depend upon the extent of workings, rate of production and maximum number of persons deployed in any of the shift.
- Make arrangement to supply normal water (pH-7) through service lines for all the operational activities of the mine.
- Wash all the mining equipments after use with normal water (pH-7) being supplied through service lines.
- Maintain good drainage system for diverting total mine water to limestone beds or low lying sump for speedy handling by the pumps. Delivery lines of acidic mine water shall be non corrosive material of construction such as HDPE.
- Protect environment by neutralizing acidic mine water by lime dozing.
- Afforestation of acid resistant trees in the surroundings.
- Storage of shales over limestone boulders to neutralize the acid generated.
- Reduce the contact of air by covering the dumps with lime sludge/clay.
- Plan the mining activities to avoid solid waste dumping at surface. These dumps having low sulphur contents are extremely detrimental to environment. In fact these dumps get exposure to varying weather conditions which promote the acid forming process. Low-grade shales having sulphur contents in the range of 1-10% have been experienced with spontaneous heating with severe emission of smoke. The severity had been excessive during rainy season.

To summarize the above, the significant remedial measures in an acidic mine shall be speedy handling of mine water. Further, transport of acidic water through leakproof non-corrosive pipe lines (HDPE) to an appropriate effluent treatment plant with designated capacity and recycle the water after treatment. These treatment plants shall have automated lime dozing arrangement with agitators, thickeners etc. Further more the plant shall be equipped with pH recorders and alarms for effective monitoring the process of treatment (Refer Fig. No.1). However, such plants have various limitations experienced as under:

- Heavy requirement of high quality precipitated lime. However, this shall depend at the rate of production, extent of workings and pumping arrangement etc. etc. It has been experienced that the Mine with average production in the range of 550 TPD (Pyritic ore) with average make of water about 1600 of LPM required a lime dozing of 3.5 to 4 ton per day, to neutralize total acid mine drainage.
- Frequent maintenance of plant/machinery
- Manpower for operation and maintenance
- Cost intensive
- Complaints from nearby population from time to time of water contamination in wells and damage to agricultural fields.

To overcome the above limitations/constraints MV technologies of OHIO (USA) developed Pro Mac (Professional Mine acid control) product. There is an important saying that "an ounce of prevention is better than a pound of cure". The basic concept of treating the acid mine drainage was changed to preventing the acid formation process. The whole process of acid formation was critically examined and analysed. The acid formation process could be established such as oxidation of sulphides catalysed by the bacteria "Thio bacillus ferrooxidans." Inhibition of these bacteria can mitigate acid mine drainage to the extent of 98% or even more. Thus, the objective became to destruct/arrest bacteria to check the acid mine drainage. Several surfactants are bactericidal to Thio bacillus ferrooxidans. Pro Mac product was developed as bactericidal. This contains active ingredient of Sodium dodecyl benzene sulfonate. The product proved selectively bactericidal to acid producing bacteria. The surfactant action in principle destroys the acid producing bacteria. The nature provided these bacteria with cytoplasmic cell wall (a greasy film); this protects the bacteria from the acid they create. The application of surfactants washes away the cytoplasmic cell wall, so that acid seeps through normally neutral interior of the cell to destroy the bacteria.

Promac product has been developed in powder form for spraying and in pellets form for use in specific conditions. Bactericide treatment can reduce acid production from 3500 to <30 PPM.

The biotechnology can be applied at two different stages of mining activity. One is during operations, such as active stocks piles or waste dumps. The other is in reclamation of acid producing wastes, after mining activity has ceased, to prevent post reclamation acid mine drainage and to allow successful re-vegetation of the land.

The controlled release pellets treatment outlasts the life of the pellets, as revealed during site studies. A single treatment sets up a change in site microbiology to replace acid producing thiobacillus ferro-oxidans bacteria with heterotrophic bacteria and creates a stable site without the need for further treatment. In continuously disturbed operations, such as mineral stockpiles, waste rock dumps, tailing impoundments or coal refuse disposal areas, bactercides are best applied as water based spray. Periodic applications of spray are required because of the disturbances to the site, which exposes new material.

Results obtained with Promac treatment have been found encouraging. Tests were carried out for few weeks and show that the bactericide has begun to control new acid production. The amount of difference between acid from the control columns and acid from bactericide treatment columns would continue to increase if tests were run longer .

The experience of M/s. MVTI of Pro mac product has established the success of bactericide treatment in arresting the acid formation cycle compared to acid control by lime treatment. Water quality monitored at reclamation sites has been shown in Table 1 & 2.
Bacteride treatment improves the vegetation and the same has been shown as site recovery cycle in the Fig. No.2

CONCLUSIONS:
The following conclusions can be drawn from the case studies.
- Bactericides are effective in preventing acid mine drainage in active operations and, in reclamation. The treatment can deliver better results than alkaline addition.
- The effect of bactericides in reclamation is a permanent one that outlasts the life of the controlled release system.
- The economics of using bactericides are quite favourable, compared to other reclamation treatments or perpetual water treatment.

REFERENCE:
Water quality and reclamation management in Mining using bactericides by V. Rastogi (Mining Engineering Journal 1996).

TABLE –1
Water quality from reclamation test plots after two years.

Sl. No.	Sample	pH	Acidity	Sulphate (ppm)	Iron (ppm)	Mn (ppm)	(ppm)	
1.	Lime treated	5.4	3592	12591		2311	30	
2.	Bactericide treated	7.2	27	5680		2		0.4

TABLE –2
Water quality from reclamation site drains

Sl No.	Sample	pH	Acidity	Sulphate (ppm)	Iron (ppm)	Mn (ppm)	Mn (ppm)	Al. (ppm)
1.	Treated (1989)	5.9	19	100		0.2	0.3	0.5
2.	Control (1989)	2.6	844	2.04		104	6.1	387
3.	Control (1994)	3.4	112	9		16	1.6	9.3

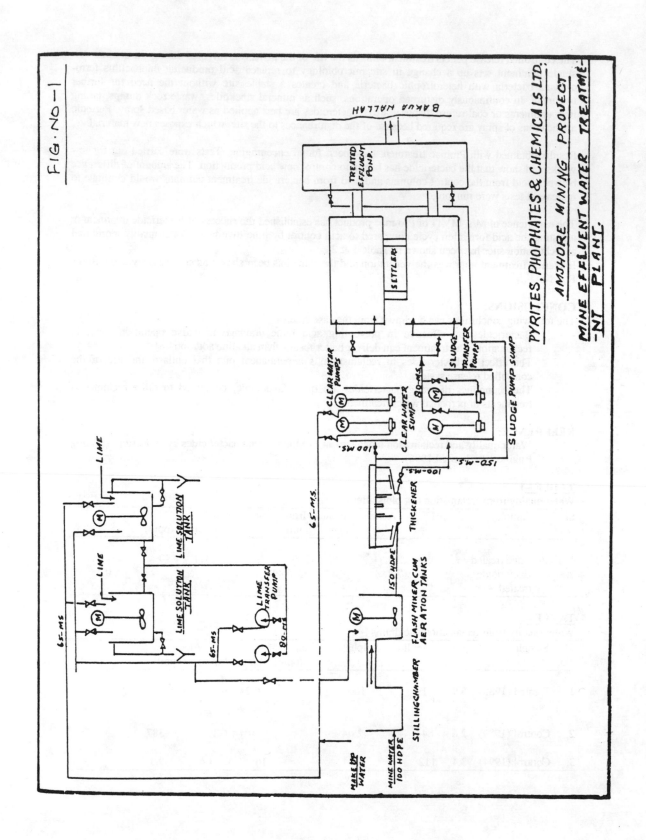

FIG. NO-1

PYRITES, PHOPHATES & CHEMICALS LTD.

AMJHORE MINING PROJECT

MINE EFFLUENT WATER TREATME-
-NT PLANT.

Site Recovery Cycle. The Key to Successful Reclamation with a Single Bactericide Treatment.

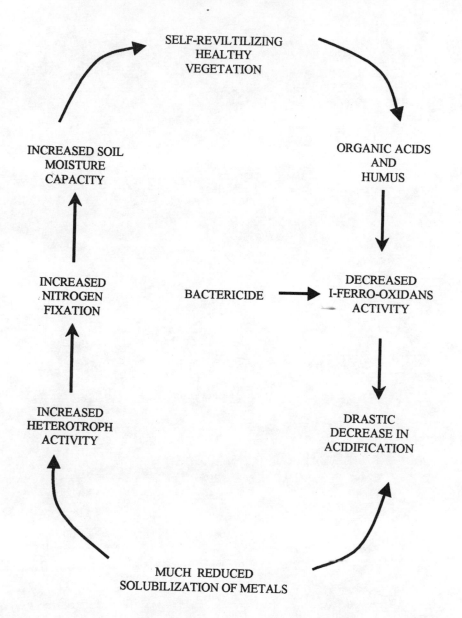

Management of Post Mining Environmental Problems: An Experimental Approach

N.R. Thote[1], A.G. Paithankar[2] and D.P. Singh[3]

[1]*Mining Department, Visvesvaraya Regional College of Engineering, Nagpur, India*
[2]*Retd. Prof. & HOD, Mining, VRCE, Nagpur, India*
[3]*Vice-Chancellor, Lucknow University, Lucknow, India*

INTRODUCTION

Extraction of mineral and fossil fuel becomes the need of present day. This leads in increasing the mining projects and expansion of many new projects. In India, there is drastic increase in opencast production with high degree of mechanization that resulted in large volume of excavation and creation of a large dumpsites. Since the dumpsites are the non-productive in nature hence no attention is given to them especially for its environmental pollution. Dump spoils are the major environmental pollutants like acid mine drainage, contamination of heavy metals in the water and soil etc. As a result of long history of mining, the problem faced by those involved in protecting the environment and in reclamation, rehabilitation or restoration can be very different and very complex.

This paper contains selective work carried out in the Carnon Valley (South West, U.K.) and similar work was done in India as a part of research work under the sponsorship of British Council by our counterpart.

EFFECT OF DUMPSITES ON ENVIRONMENT

Surface mining involves creation of large dumpsites. In case of coal mining, the overburden is normally sandstone or shale and in case of metalliferous mining it is the waste rock that may contain various traces of elements may not be economical hence thrown away. Basically, dumpsite causes environmental problems like land degradation and water pollution (Singh, 1992). Atkinson and Mitchell (1994) opined that oxidation and subsequent leading of residual sulphides present in mining and waste dump is responsible for the generation of acidity and contamination of surface ground water. The Heavy metals ions present in mining waste can be toxic to human, animals and aquatic life (Mandavgane et al., 1994). Al., 1994). It is suggested that removal of heavy ions can be done by ferritization but this is costly and May not be useful for large volume like dumpsites.

Cattle dying due to chronic copper poisoning were reported by Scheepers (1989) in R.S.A. The animals affected originated from farm in the vicinity of copper smelting plant in the Eastern Transvaal Lowveld (R.S.A.). Toxic metal can change the biological structure and produced physiological poisoning by becoming attached to the tissues of aquatic organism and consequent accumulation. Table 1 shows target organs affected by various pollutants.

Table 1. Target organs affected by various pollutants.

Pollutants	Target organs affected
Arsenic, Cadmium, Mercury, Fluoride and Benzene,Lead	Blood
Cadmium, Se, Fluoride	Bones and teeth
Arsenic, Mercury, Lead	Brain
Arsenic, Mercury, Molybdenum, Selenium	Liver
Arsenic, Mercury,Cadmium	Lungs
Arsenic, Mercury,Lead, Chloroform	Kidney

Mining of metallic ore deposit may lead to increase metal concentration of surface as well as ground water. The major sources of metal pollutants are mine drainage and waste seepage through waste dump of overburden(Muthreja and Paithainkar, 1994). Table 2 shows the water quality standards in India hence wherever mine drainage meets source of drinking water, the impurities should be below these values. Table 3 shows typical values in the soil.

Table 2. Water quality standards in India.

PH	6.3-9.2
Total hardness	600 ppm
Turbidity	25 ppm
Chorides	100 ppm
Cynide	0.05 ppm
Fluoride	1.5 ppm
Nitrates(as NO3)	45 ppm
Phenols	2 ppm
Sulphates (as SO4)	400 ppm
Manganese	0.5 ppm
Mercury	0.001 ppm

Iron	1 ppm
Copper	1.5 ppm
Cadmium	0.01 ppm
Selenium	0.01ppm
Chromium	0.05 ppm
Lead	0.1 ppm
Arsenic	0.05 ppm
Zinc	15 ppm
Magnesium	150 ppm

Table 3 Concentration of selected trace elements normally found in soils and plants (after Bowas, 1979.Allaway, 1966)

Elements	Concentration in soil mm/ kg		Concentration in plant (range)
	Range	Typical	
As	0.1-40	6	0.1- 5.0
B	2-200	10	5-30
Be	1-40	6	-
Cd	0.01-7	0.06	0.2-0.8
Cr	5-3000	100	0.2-1
Co	1-14	8	0.05-0.15
Cu	2-100	20	0.1-10
Pb	2-200	10	0.1-10
Mn	100-400	850	1.5-100
Mo	0-2.5	2	1-100
Ni	10-1000	40	1-10

EXPERIMENTAL WORK

Experimental work was carried out to know the water pollution of past mining activities and sediments carried due to natural drainage at Carnon Valley in South West of United Kingdom. The Carnon Valley was chosen for these studies because it was highly polluted due to mining activities. The area is located in Cornwall. This area is full of old mining activities though the mining activities have come up to an end since long time. Over the years ore of tin, copper, lead, zinc, silver, antimony, iron and manganese together with lesser amounts of tungsten, cobalt, nickel, uranium, barite and fluorspar have been won from the mineralized areas associated with the granite intrusions. This area has following features: water stream flowing from west to east and between two small watercourses arise from underground is Nangiles Adit and County Adit. One more underground opening is exiting but water does not flow through it. There is a huge slag dump on north side. Therefore, the entire watercourse was running from copper mine to low land area. pH value and temperature of water was measured as shown in Table 4. Water samples were collected and analyzed for copper concentration as shown in Table 5 by Atomic Absorption Specto- photometer at ERC(Exeter), U.K.

Table 4. pH value and the temperature of the samples from Carnon Valley.

Sample No	pH	Temp. in degrees
W1	5.60	8.9
W2	6.0	9.5
W4	4.46	12.9
W6	5.61	11.2
W7	3.86	14.3
W8	5.04	11
W9	3.80	11.3

Table 5. Water analysis results for the trace of water for the traces of copper (by AAS)

Sample No.	Copper concentration (ppm)	Copper concentration after adding lime (ppm)	% change in copper concentration
W1	0.70	0.67	3
W2	0.075	0.050	33
W3	0.082	0.050	39
W4	1.200	1.175	2
W5	1.250	1.175	6
W5	0.850	0.850	0
W7	3.075	2.950	4
W8	1.375	1.300	5
W9	1.075	0.975	9

Sediments near by the watercourses were collected and analyzed by XRFS PW- 1400 System the laboratory of ERC,Exeter(U.K.) for their concentration but only dominant element traces are shown in Fig.1.

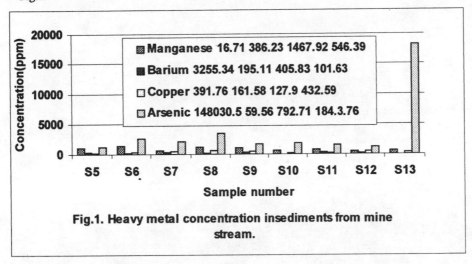

Fig.1. Heavy metal concentration in sediments from mine stream.

CONCLUDING REMARKS

1. Copper concentration increases from stream source to low land it indicates seepage of copper contamination added due to waste dump by percolation. The present copper concentration is far above the threshold values hence water is harmful for flora and fauna as well as for drinking

2. From the pH analysis it is clear that acidity of drainage water increases from source to low land. This shows percolation of acidic metal elements into the stream from dumpsites.

3. Addition of lime in the stream decreases copper concentration this may be due to formation of precipitates of higher density, which settled down.

4. Sediment analysis shows exorbitant high values of poisonous metal concentrations like Arsenic, Copper, Barium and Manganese. Arsenic concentration found was excessive and can cause brain, liver, kidney damage if this water is added to drinking water source. This may be very much dangerous to animals who are gracing in this area, plants and aquatic life.

5. A lesson should be learned from this old mining dumpsites that proper measures should be taken to either remove the toxic metal traces from the dumpsites or chemical treatment should be done with the percolated water from the mine and then leave it to the stream.

ACKNOWLEDGMENTS
Authors are thankful to Principal V.R.C.E. Nagpur for his permission, my collogue S.S.Gupte, M.J.Heath, John Merefield and other staff of Earth Resources Centre, U.K. who worked with me on this project. Thanks are due to British Council for their financial help.

REFERENCES
Gupte, S.S. and Thote, N.R., Stone I. and Roger C.G.D. (1994). Environmental impact of mine dumpsites in cornwall- A case study of Carnon Valley(U.K.), Int. Proc. Environmental Issues in Mineral Industry, New, Delhi, 159pp.

Atkinson A and Mitchell P.B.(1994). The environmental impact of mining and some novel methods of reclamation, Int. Proc. Impact of Mining on Environment, Org. by VRCE Nagpur (India), 305pp.

Singh T.N.(1992). Dimension of surface mining and impact on ecology and land usage. Proc. 4th Nat. Sem. Surface Mining, ISM, Dhanbad, 4.2.1pp.

Muthreja I.L. and Paithankar A.G.(1994). Impact of mining on water regime – with special refernce to heavy metal contamination, Int. Proc. Impact of Mining on Environment., Org. by VRCE Nagpur (India), 117pp.

Scheepers G.J.(1994). Cronic copper toxicity in ruminant's –prophyltic salt lick, Int. Proc. Impact of Mining on Environment, Org. by VRCE Nagpur (India), 113pp.

Mandavgane S.K.and Dara D.M.(1994). Reclamation of heavy metal ions from mine water by ferritization, Int. Proc. Impact of Mining on Environment, Org. by VRCE Nagpur (India), 103pp.

THEME 4 :
Air Pollution

Interpretation of Air Quality Data Using Indicator Kriging Technique

B.S. Sastry[1] and G. Vijay[2]

[1]*Prof & Head, Dept. of Mining Engineering, Indian Institute of Technology, Kharagpur, 721 302, India*
[2]*Graduate Student, Dept. of Computer Science, Iowa State University, Ames, Iowa, 500 11, USA*

INTRODUCTION

The NAAQS (National Ambient Air Quality Standards) as stipulated by CPCB of the Government of India are intended to protect with sensible margin of safety, the people, vegetation, and property of the nation. Air quality data generated during the EMP/EIA studies in the mining areas is commonly interpreted with the help of the NAAQS   (Table 1). The interpretation requires spatial description of the pollution parameters. Based on this it would be possible to predict the level of contaminant, as well as to estimate the probability of exceedence of a cut-off value of an air quality parameter at any given location. This information helps in various decision making exercises relating to pollution control strategies to be taken up by contributors of the pollution as well as different public administration agencies.

The factor that distinguishes the earth science data sets from others is that the data belongs to some location in space. Spatial features of the data set such as the location of extreme values, the overall trend, the degree of continuity, are often of considerable interest in decision making (Isaaca and Srivastava, 1989). The key concept of geostatistics is that of the regionalised variables, which has properties intermediate between a truly random variable and one completely deterministic. Even though a regionalised variable is continuous in space, it is not usually possible to know its value everywhere in the region. Instead the values are known only through samples which are taken at specific locations. The characteristics of the sample in terms of its size, shape, orientation, and spatial arrangement constitute the support of the regionalsied variable. A change of support will result in a change in the characteristics of the regionalised variable. If a data analysis makes use of eight-hour average values, then the results are also to be interpreted in the same context. That is, no conclusions should be drawn , say, about the annual means from this data.

AIR QUALITY DATA

The data being investigated in the present context is obtained from the EIA studies conducted by IIT in four connected mining blocks in Eastern India. A total of 18 air quality monitoring stations spread around the four blocks comprise a spatial extension of 40 km by 60 km approximately are included in this investigation.. Monitoring was done for SPM, SO_2, and NO_x values. Measurements were taken for 12 hours in a day, and for 15 to 30 days in a season. Such assessments were conducted season-wise during summer, monsoon, and winter. The EIA studies were taken up as separate projects, and therefore, the data for all the blocks was not available for the same time frame. The studies covered a period ranging from 1988-1992.

During the present analysis the summarised data was used. Investigation of temporal variation of the pollution data was not done, since the available data was for different time frames. However, for the purpose of modelling the pollution dispersion, the year-wise variability during the period of study was assumed negligible. In other words, the yearly mean level of a parameter such as SPM (as obtained from the 12 hourly samples whose number varied from station to station) is assumed to have remained constant for a given location during the duration of study for the EIA projects. Considering no significant Industrial activity with large scale consequences on air quality were initiated in the region, the assumption may not be improper.

1. Summary Statistics

In Table 2 summarised data is presented for all the stations that formed the study. The annual mean values and the associated standard deviation values are given for the parameters. Bi-variate statistical analysis was performed among the variables. The Parameters SPM and NO_x exhibited fair degree of correlation.

The mean values for stations belonging to Block 1 appear to be significantly lower in terms of SPM. Standard deviations in general varied from 50 to 100 percent of the mean values. Although not presented in the table, the skewness of the distributions is essentially positive. The bi-variate correlation between SPM and NO_x suggests negative trend in block 2. The correlation is good in Blocks 1 and 4, whereas, it is fair in Block 3.

GEOSTATISTICAL MODELLING

1. Semivariogram

Semivariance is a basic measure of geoostatistics which is used to express the rate of change of a regionalised variable along a specific orientation. It is a measure of the degree of spatial dependence between the samples. From the measurements made at scattered sampling points, and from the knowledge of the semivariogram, it is possible to estimate the values at any unsampled location. The estimation technique of Kriging determines an optimal set of weights that are used in the estimation of the variable at the unsampled locations. The semivariogram being a function of distance and orientation, the weights change according to the geographic arrangement of the samples.

The experimental semivariogram is constructed from the coordinates and values of known samples. For a lag of h, in a given orientation, the expected value of the square of paired sample value difference is computed. The experimental semivariogram is half of the above computed expected value. Thus γ (h) is given as,

$$\gamma (h) = 0.5 * E((Z(x)-Z(x+h))^2)$$

The plot of the γ (h) values as a function of h, is known as the experimental semivariogram. Often times the pattern of the semivariogram may not vary with orientation, or the variation may not be established due to sparse data. In such case, an isotropic semivariogram is assumed to define the distribution of values. In the investigations carried out the orientation dependence was not studied. The semivariograms generated thus are essentially isotropic.

Variogram modelling further processes the experimental semivariogram. The exercise is much akin to fitting a probability density function to a histogram obtained from known samples. Several possible variogram models are feasible. The selection of the appropriate model is a very important aspect of geostatistics. The model brings about an overall nature of the variance of the distribution of values within the domain of interest

2. Indicator Transformation

The transformation of the ordinary variables into indicator variables is a practice that helps in determining the probability distributions for sample values. For example an indicator transform based kriging can provide the probability for the SPM value to be in excess of a cut-off in all locations in the region of interest. The 'soft transformation' converts the value of the actual variable into a value that lies within a range. In the present context the cut-off values of 200 $\mu g/m^3$ is taken for SPM corresponding to Residential Area cut-off category of NAAQS. For the cases of NO_x, and SO_2 a cut-off value of 30 $\mu g/m^3$ is taken corresponding to Sensitive Area category.

For a cut-off Z_j, the value of the indicator variable value is taken to be the probability that the actual variable value lies below the cut-off. This probability is determined from the number of observations in a sampling location as follows.

P ($Z<Z_j$) = (Number of observations with measured value below Zj) / (Total number of observations)

The indicator variable thus takes a value between 0 and 1.0. In Table 4, the associated indicator variable values corresponding to different sample locations are shown. The column for the data values 'SPM+NO_x' in the Table is used for the cokriging analysis on the data.

3. Point Kriging

Point kriging is the technique of estimation of an unknown value at a given location. The value at the unknown location is considered to be a weighted mean of the values of the neighbouring know points.

$$U = \sum (a(i) * Z(i))$$

Where, $a(i)$ is the weight attached to sample $Z(i)$. Unbiased condition of estimation requires that the sum of the weights is one. If the true value at this location is U_0 then the amount of error being made in the estimation is $(U - U_0)$. The expected value of the square of this error is known as the estimation variance. That is,

$$\sigma^2_E = E((U-U_0)^2)$$

In kriging this estimation variance is minimised with respect to the weights $a(i)$. The minimisation procedure (Journel and Huijbregts, 1978, and, Isaacs and Srivastava, 1989) yields a set of linear equations in terms of the γ values. The solution of this set of equations known as the kriging equations yields the optimum (BLUE) set of values for $a(i)$. The resultant estimation variance is known as kriging variance, the square-root of this being kriging error. Through kriging, therefore, one can generate not only the best possible estimator, but also quantify the degree of error being made in such estimation. The precision of estimation is also dependent on the accuracy of variogram modelling.

4. Cokriging

Sometimes, one variable may not have been sampled sufficiently to provide estimates of acceptable precision. The precision of this estimate may then be improved by considering the spatial correlations between the primary variable and other better sampled secondary variables. The estimates of the primary variables may improve if good cross-correlations exist between the primary and the secondary variables (Carr, et.al. 1985). The cokriged estimator is given as a linear combination of primary and secondary data, where the variable V represents the secondary variable. Thus,

$$U = \sum (a(i) * Z(i)) + \sum (b(i) * V(i))$$

Unbiased conditions can be such that, $\sum (a(i)) = 1$, and $\sum (b(I)) = 0$. From this basis the rest of the estimation process follows that of kriging.

ANALYSIS OF RESULTS

The air quality data is processed using the geostatistics software, GEOEAS 1.2.1 of the USEPA, and the GSLIB routines of the Stanford University. Kriging and indicator kriging are performed with GEOEAS, whereas, the cokriging exercise is done using the GSLIB routines.

The semivariogram models fitted to the experimenrtal data are illustrated in Figure 2 for NO_x, and in Figure 3 for PNO_x. The models fitted have a nugget of 0, and range of 6.5 min. The model parameters for all the variables are given as below.

PARAMETER	TYPE	SILL	NUGGET	RANGE
SPM	Gaussian	12000	0	6.5
NO_x	Gaussian	135	0	6.5
SPM⏐NO_x	Gaussian	16000	0	6.5
PSPM	Gaussian	0.07	0	6.5
PNO_x	Spherical	0.035	0	6.5

Kriging estimation on indicator transforms, or cokriging estimation is performed at 100 unknown locations falling on a regular grid in the region of interest. The probability contour plots for PSPM, and PNO_x are illustrated in Figures 4 and 5 respectively. The kriged estimates for SPM and NO_x are computed and represented as contour plots in Figures 6 and 7 respectively. Likewise, the Figures 8 and 9 give the cokriged estimates for the respective parameters. The mean estimation error is computed for SPM and Nox for the eases of kriging and cokriging considering the 100 locations. The values are.

Mean Estimation Error	Kriging	Cokriging
SPM	38.107	12.23
Nox	4.046	1.29

The availability of good correlation between SPOM and NO_x can be assumed to be the reason for the improved estimation of values from cokriging.

CONCLUSIONS

Geostatistics provides a scientifically firm basis to analyse spartial variables. The present exercise shows an approach of the use the technique for air quality data modeling and prediction. Based on the summarised annual mean parameters of SPM, Nox, and SO_2 measured at eighteen stations, variogram modelling, indicator transformation, kriging, and cokriging are performed. The suitability of the method to generate probability contours based on a desired cut-off is illustrated. Results of kriging and cokriging are compared to bring out the improved estimation from cokriging. No site specific inferences are drawn from the findings. The reasons being, the limited knowledge of the polluting sources in the domain, the sparseness of the data points, and the grouping of the data ignoring possible temporal variability.

Acknowledgement

The authors are grateful to Prof S D Barve, and Prof N Mukherjee (Retd) of the Department of Mining Engineering at the IIT Kharagpur for providing the baseline data used in the study.

REFERENCES

Carr, J.R., Myers, D.E. and Glass, C.E. (1985) . Cokriging – A Computer Program. *Computers & Geosciences*, Vol. 11, No.2, pp.111-127.

Isaaks, E.H. and Srivastava, R.M. (1989). *Applied Geostatistics* First Ed., Oxford University Press, New York, 551pp.

Journel, A.G. and Huijbregts, Ch.J. (1978). *Mining Geostatistics*. First Ed., Academic Press, London, 600 pp.

Table 1 : National Ambient Air Quality Standards

Pollutant	Averaging Period	Industrial Area	Residential / Rural Area	Sensitive Area
SO_2	Annual	80	60	15
	24 Hours	120	80	30
NO_x	Annual	80	60	15
	24 Hours	120	80	30
SPM	Annual	360	140	70
	24 Hours	500	200	100
RPM	Annual	120	60	50
	24 Hours	150	100	75
Pb	Annual	1.0	0.75	0.5
	24 Hours	1.5	1.0	0.75

Note : Ambient concentration measured in micro grams per cubic metre

Table 2 : Results of Uni-variate Statistical Analysis

Station	No of samples	SPM Mean	SPM SD	NO_x Mean	NO_x SD	SO_2 Mean	SO_2 SD
1	72	29.355	15.965	4.543	2.176	Na	Na
2	72	55.003	28.693	11.518	8.748	Na	Na
3	72	36.21	16.859	8.182	5.991	Na	Na
4	72	33.704	14.629	8.539	5.320	Na	Na
5	72	37.278	14.176	6.569	3.865	Na	Na
6	72	18.557	9.43	5.082	2.812	Na	Na
7	48	225.829	212.608	5.835	5.003	25.7149	18.718
8	64	267.606	191.761	8.668	6.327	19.481	13.373
9	64	312.85	227.172	4.828	3.831	23.559	23.778
10	64	209.446	129.821	8.053	9.192	26.317	19.979
11	64	266.844	169.516	35.872	26.248	27.083	27.846
12	64	298.75	174.092	29.95	15.674	21.253	17.782
13	64	238.5	150.246	25.852	12.908	18.40	15.244
14	64	195.609	139.778	23.57	16.46	17.148	15.615
15	64	278.578	157.544	42.08	24.351	29.598	28.195
16	32	99.844	87.007	19.109	14.247	9.994	10.959
17	32	180.594	130.705	23.987	11.398	17.438	13.374
18	32	127.5	70.185	23.106	11.367	16.6	14.939

Note : Ambient concentration measured in micro grams per cubic metre

Table 3 : Bivariate Analysis – Correlation Between SPM and NO$_x$

Station	Region	Latitude (min)	Longitude (min)	Coef of Correlation
1	Block 1	63.89	47.88	0.712
2	Block 1	61.24	47.43	0.725
3	Block 1	65.32	48.65	0.747
4	Block 1	62.22	47.16	0.634
5	Block 1	62.59	48.56	-0.253
6	Block 2	63.02	46.81	-0.446
7	Block 2	61.67	42.48	-0.307
8	Block 2	57.59	43.17	0.177
9	Block 2	59.84	41.95	-0.324
10	Block 2	59.70	44.00	-0.173
11	Block 3	54.46	42.05	0.519
12	Block 3	55.73	41.51	0.461
13	Block 3	50.81	39.32	0.544
14	Block 3	52.50	39.13	0.704
15	Block 3	56.83	41.51	0.682
16	Block 4	63.40	38.76	0.781
17	Block 4	65.24	39.20	0.729
18	Block 4	62.20	39.35	0.677

Table 4 : Annual Mean Data and the Associated Probabilities

Station	Latitude (min)	Longitude (min)	SPM	NO$_x$	SPM+ NO$_x$	PSPM	PNO$_x$	PSO$_2$
1	63.89	47.88	29.355	4.543	33.898	1	1	NA
2	61.24	47.43	55.003	11.518	66.521	1	0.9722	NA
3	65.32	48.65	35.210	8.1825	43.392	1	1	NA
4	62.22	47.16	33.704	8.539	42.243	1	1	NA
5	62.59	48.56	37.278	6.569	43.847	1	1	NA
6	63.02	46.81	18.550	5.080	23.630	1	1	NA
7	61.67	42.48	225.82	5.835	231.66	0.5833	1	0.6667
8	57.59	43.17	267.61	8.668	276.27	0.4844	1	0.8
9	59.84	41.95	312.85	4.820	317.67	0.4219	1	0.7647
10	59.70	44.00	209.44	8.053	217.50	0.625	0.9615	0.6562
11	54.46	42.05	266.34	35.872	302.22	0.4375	0.5781	0.75
12	55.73	41.51	298.75	29.950	328.70	0.3281	0.6875	0.8438
13	50.81	39.32	238.50	25.852	264.35	0.4844	0.7500	0.8571
14	52.50	39.13	195.61	23.590	219.20	0.5469	0.8125	0.8281
15	56.83	41.51	278.38	42.080	320.45	0.4531	0.3594	0.6719
16	63.40	38.76	99.840	19.100	118.94	0.9062	0.875	0.6375
17	65.24	39.20	180.59	23.980	204.57	0.6875	0.8125	0.9062
18	62.20	39.35	127.50	23.106	150.61	0.8125	0.875	0.9062

Note :   PSPM : Probability for SPM to be below 200 $\mu g/m^3$
PNO$_x$: Probability for NO$_x$ to be below 30 $\mu g/m^3$

PSO$_2$: Probability for SO$_2$ to be below 30 $\mu g/m^3$

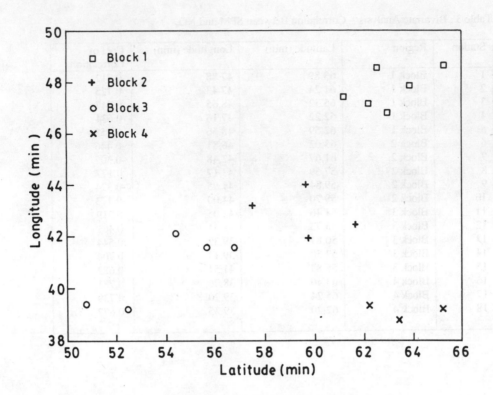

Fig 1. Distribution of Air Quality Stations

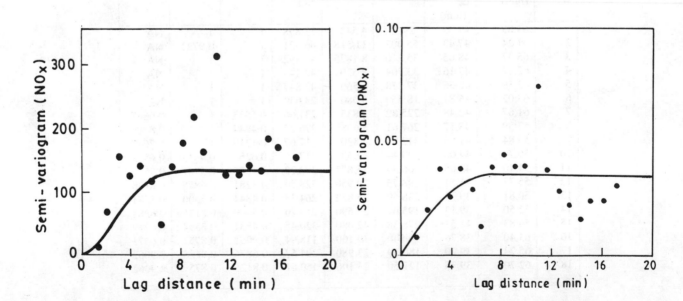

Fig 2. Isotropic Semivariogram for NO_x

Fig 3. Isotropic Semivariogram for PNO_x
(Prob for $NO_x < 30$ μg/m³)

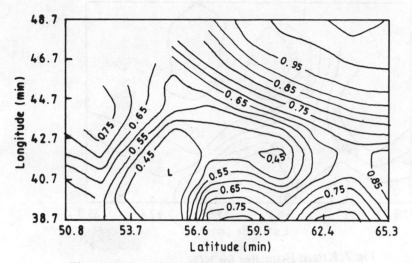

Fig 4. Probability Contours for SPM < 200 µg/m³

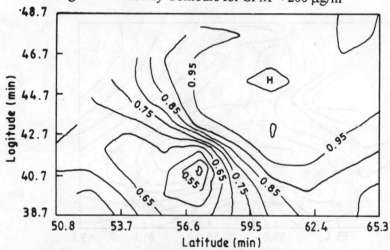

Fig 5. Probability Contours for NOx < 30 µg/m³

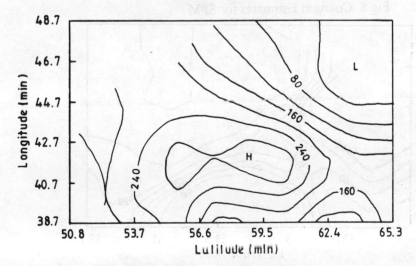

Fig 6. Kriged Estimates for SPM

313

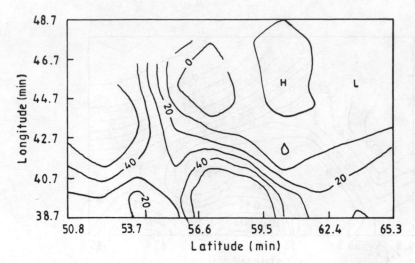

Fig 7. Kriged Estimates for NOx

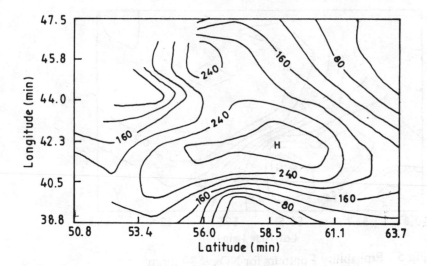

Fig 8. Cokriged Estimates for SPM

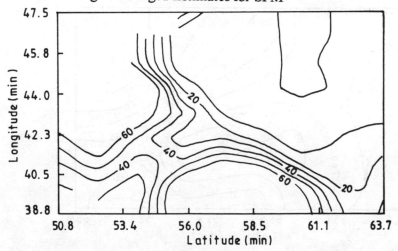

Fig 9. Cokriged Estimates for NOx

314

A Trial Application of Chemical to Suppress Mine-Environmental Dust-Techno Economic Evaluation

Virendra Singh[1] and A.R.P. Reddy[2]

[1]*Professor and Head of the Department of Mining Engg., University College of Engineering (K.U.)., India*
[2]*Additional Manager, 5 Incline Group of Mines, S.C.C.L, Kothagude, India*

INTRODUCTION

Dust is an undesirable by product of mining operations. Coal dust larger than 74 microns in size may cause explosion whereas finer than 5 micron causes pneumoconiosis. The problems I has been there since the very beginning of mining operations but the introduction of mechanised mining systems have aggravated the menace. In opencast mines a considerable amount of particulate matter is released to the mine atmosphere. While the dust generated by drilling can be contained locally, the one generated and located along the haulroads create menace peculiar to the vehicular movement or the direction and intensity of wind. Cloud of dust along the haulroads, particularly during dry months, is a typical sight. Continuous and increased frequency of traffic may result in the presence of dust clouds an everlasting affair.

1.1 DUST EMMISION STANDARDS IN MINES.

Different countries have prescribed different standards of permissible concentration of dust in the mines.
(A). IN USA, THE COAL MINES AND SAFETY ACT PRESCRIBES
a) Maximum dust 2 mg /m 3 of air at working place
b) If dust contains free silica >5% the above mentioned limit is lowered by (10 /x) mg/m3 where x stands for x% of free silica in dust.

(B). IN UK THE COALFACE IS CLASSIFIED AS
a)   Approved, if the mean concentration of dust is < 7mg/m3 of air.
b)   Not approved, if the mean concentration of dust is > 7 mg/m3 of air.

(C). IN THE EARTWHILE USSR
1.   Permissible maximum amount of respirable dust =10 mg/ m3 for free silica < 10%
2.   Permissible maximum amount of respirable dust =2mg /m3 of air if free silica is > 10%

(D). IN INDIA
As per D.G.MS circular No.16 of 1975, the permissible respirable dust concentration limit:
a)   3 mg/m3 of air for dust containing less than 5% free silica.
b)   (15/x) mg/m3 where, x is free silica in the dust when it is more than 5%.

1.2 GENERATION OF DUST:

Genesis of dust (siliceous and carbonaceous) is the resultant of various operations performed to handle the overburden and coal produced viz., drilling, blasting, loading, transportation, crushing and retrieval for dispatches or backfiring, Whereas most of the functions have local effect, the operations of blasting and transportation have wide spread effects and hence, the major areas of concern. Between the two the dust generated can also be controlled at source by selecting proper blasting pattern and explosive.

Haul roads are required for the movement of dumpers, and other vehicles. They are of two types 1) temporary haulroads at each bench.2) permanent or main haulroads leading to crusher and dumping yards.

The width of main haulroad is 25 m with drains of 2m-width and 1 m depth (x section 2m x1m) on to its either side. The gradual gradient being 1 in 15, 1in 20. The temporary haulroads are made on the floor of the benches, which are soil, sandstone, clay, coal depending upon whether it is overburden or coal bench.

1.3 DUST PROBLEMS ON HAULROADS
Dust problem on over burden bench haulroad:
O.B benches 16m high and 60 m wide are formed using 10 m3 shovels and 85 t dumpers. The sequence of formation benches is as given below

Bench I: soil and Morrum Bench II: Grey sandstone Bench III: Grey sand stone and clay
Bench IV: coal bands shale, clay.

The dust problems, as has been said above, are due, mainly to blasting and loading operations as also due to the vehicular traffic on the benches. Dust originates mainly from loose soil, sandstone and shaly strata. The dust concentration recorded averaged around 4.5 mg/m3.

Dust problem on coal benches:
Coal is being extracted from 60-m high slices with 40.0m wide benches using 4.5 3.3 cum. Hydraulic shovels in conjunction with 50 t rear dumpers. The air borne dust survey conducted on coal benches revealed 3.5-mg/m3 airborne dust.

Dust generation on haulroads:
Major component of dust pollution created along haulroads is, due to the movement of vehicles. It depends upon the type and frequency of vehicles, weight of the moving machines and their speeds. The heavier, faster and more frequent machines contribute the larger component of dust. The type and the number of such machines at the site and at the time of experiment was as follows.
Dumpers 85 T (43); 50 T (9): 35 T (32); water tankers (5); dozzers (12); motor graders (4); and cranes (10).
The airborne dust survey conducted on haulroads indicated 6 mg/m3 of dust.

2.0 SOLUTIONS EARLIER TRIED
2.1 WITH WATER SPRAYING:
The simplest method of preventing the almost from being airborne in by wetting it. But it has been found that water spraying alone isn't sufficient in (I) wetting the entire dust and hence requires very large quantity of water; (ii) water evaporates quickly and hence needs very frequent spraying particularly during dry season.

2.2 CONSOLIDATING ROADWAY DUST:
The two commonly applied methods of consolidating the haulroad dust, besides wetting, are:
I) Calcium chloride method, requiring large amount of continuous water sprinkling
ii) Salt crust process: Leading to corrosion of the vehicles

Neither of these two methods found favour with mining engineers. The search therefore, has been continuing for more effective chemicals, and methodology of treatment. Water spraying alone, as has been found, has some limitations. To keep dust along haulroad of 10-km length at the experimental mine under control required 20 trips in an 8-hrs shift. Given that a water tanker of 28,000-litre capacity could cover a distance of 23-25 km; it required 4 tanker / sh. to keep the entire length of haulroad wet. Depending upon weather conditions, a tanker is required to make 5/ 7 trips shift. Since the location is having tropical climate the relative humidity during summer comes down to 35% from otherwise average of 55% .

This further, puts considerable demand on the water spraying system and entails to stoppage of work either for the shortage of water for spraying or for inadequate suppression of dust i.e. up to the permissible limit. This leads to higher cost as well as loss of production.

2.3 NEW TECHNOLOGY OF DUST SUPPRESSION :
M/s Syntron Industries offered to supply chemical 'DUSTRON PC' compound formulated by them to be sprinkled mixed with water in the water tankers already being used by the mines.
Mechanism of functioning of the Dustron PC compound
It is claimed that the compound functioned on the following mechanism.
a)   It generates interfacial tension between the particles of dust, which facilitates the agglomeration of micron size particles.

316

b) It lowers the surface tension of water thereby forming greater water spray dispersion for more efficient dust suppressing area coverage.

c) It prevents the formation of dust film restricted to water surface alone and allows dust to penetrate into water droplets thereby permitting or enabling less moisture to do more efficient job.

d) Wets the dust particles, which are water repellant by converting them from hydrophobic to hydrophilic.

e) Agglomerate the finer particles, and /

f) Reduces the water surface tension from 72 dynes/sq.in. to 28 dynes/sq.in.

3.0 FIELD INVESTIGATION:
3.1 PLANNING OF FIELD INVESTIGATION:

The entire length of haulroad 10.9 Km was divided into 4 categories as mentioned elsewhere. On each haulroad the following tests were conducted under varied situation.

a. Measurement of Impact Strength of the haulroad surface.

b. Measurement of Moisture content of the haulroad-surface-dust,

c. Measurement of Airborne Dust along the haulroads.

d. Measurement of the fineness of the haulroad-surface-dust.

The tests were conducted during the period from 17-3-96 to 30-4-96 because of the limited quantity (175-kg) of compound supplied by the manufacturer for trial purposes. All the tests, were conducted during the first shift i.e. from 7.00 am to 3.00 PM. The available 28-Kl tankers were used for water as well as for compound mixed water spray. The sample collection or field tests were conducted at time intervals of 2 hrs, 4 hrs, and 6 hrs after the spraying. 1 kg, 2 kg, 3 kg and 4 kg of compound was added to tanker load of 28,000 litre of water in order to find out the best mix. Compound mixed water was sprayed during the first shift only. Normal water to be sprayed in the subsequent shifts.

3.2   METHODOLOGY OF THE FIELD TESTS & INSTRUMENTATION

a   Impact Strength Test:

To assess the effect of the added compound on the strength of the haulroad surface, a Simple device, to cause an object to drop from a given height to impinge on the haulroad surface, was fabricated. The size of depression (dia and depth) was measured. The relative changes in the sizes of depressions found under different conditions of application represent the change in the impact strength of the haulroad surface.

b   Moisture Measurement Test:

Were conducted to assess the influence on moisture retention capacity of the haulroad surface due to the application of the chemical compound. Samples collected were analysed in laboratory for moisture content (%) in a hot air oven.

c   Penetration Test;

To find out the action of the chemical compound on the binding of loose dust layers which in turn adds to the toughness of the haulroad surface. It has been found that 85 T dumper, the heaviest earth moving machine, when loaded, exerts a pressure of about 9 kg/cm^2. For conducting penetration test device was improvised to exert a pressure of 15 kg/cm^2 when inserted on the surface. The reduction in the depth of penetration indicates improvement in toughness of the surface.

d   Airborne dust Measurement:

JCB/MRE Gravimetric Dust Sampler was used to estimate the concentration of airborne dust.

e   Sieve Tests:

Was conducted to estimate the effect of spraying on the binding of loose dust particles i.e. to assess the coagulation effect of the chemical added. It has been found that the dust finer than (-) 212 mesh sieve size gets airborne due to vehicular traffic leading to air pollution.

Samples were collected at some predecided locations along each haulroad by brushing and packing in polythene bags. They were reduced to the laboratory size of 100 gms by coning and quartering before they were analysed.

4.0 EXPERIMENTAL SITE:

The experiment of dust suppression along haulroads was carried out at an o. c coalmine producing about 2.5 million tones of coal for an overburden equivalent of 101 lakh cum. The salient features of the experiment site at the time of experiment were as given below:

Thickness of the seam	: 6.10 m to 31.19 m
Average seam gradient	: 1 in 6.5
Area of excavation	: 500 hectares (approx.)
Max, depth of the quarry	: 155 m

The mining scheme includes :
Removal of topsoil by scrapping and main overburden by drilling and blasting. Loading and transport by shovel and dumper (10m3 and 85 t dumpers combination)
Mining of coal by drilling and blasting Shovel and dumper (4.5/3.3 m3 and 50 t dumpers combination)

4.1 SUB DIVISION OF ROADS:
The entire of 10-km haulroad was divided into 4 sections covering the entire mine
a) Yamuna road : It covers the O.B bench 1 and OB bench road 2 , 4.5 km long from the face to dump yard no.1 i.e. eastern side tg gradient is 1 in 20.
b) Tungabhadra road(T<B> road) : It covers ob bench 3 and 4 It is 2.5 km long from the bench faces to the refilling dump yard . It slopes 1 in 16,.
c) Coal bench road (CB road) : It covers coal extraction bench. It is 0.8 km long at 1 in 20 gradient.
d) Coal haul road (CH road) : It extends from the edge of coal bench upto 800 tph feeder breakers.
The floor of the haulroad has soling of 100 to 150-mm gravel on It was the best-maintained road since the max.100 -150 mm gravel . It was the best maintained road since the max. no of HEMM ply on that. It was 1.5 km long sloping 1 in 20.

4.2 PROCEDURE OF APPLLCAITION
The following procedure to be followed:
a) The compound to be well mixed with water in the desired proportions
b) The quantity of water required to be planned @ 3.8 to 4,4 sq. of road surface per litre of water.
c) The Dustron mixed water to be sprayed around 10.00 am and then with plain water the rest of the day To begin with it has to be followed for 3 consecutive days before which the desired results are achieved. Thereafter the required number of water tanker trips can be reduced.
d) The properties of surface tension and capillarity of the mix to be tested in the laboratory before it is applied in the field.

5.0 RESULTS: Are as given below:
5.1 IMPROVEMENT IN THE IMPACT STRENGTH OF ROAD:
The results of observation on changes in the impact-strength of haulroad surface, due to application of the compound, are as given below:

TABLE.1   IMPACT STRENGTH (Kg/Cm2)

LOCATION CONDITION OF APPLICATION	YAMUNA	T.B.LOAD	COAL HAUL	COAL BENCH
1. Without water apary	25.25	25.44	25.38	25.36
2. With Water spray	31.3	32.32	30.46	29.62
3.With compound:				
i. PC 1 Kg	38.22	37.94	37.86	38.34
ii. PC 2 Kg	45.62	44.32	43.19	44.63
iii. PC 3 Kg	54.90	55.92	54.68	54.76
iv. PC 4 Kg	64.50	65.60	64.50	64.54

It is thus found that 100% improvement in the strength is possible when the compound is 4 K g/ 28000.00 litter tanker as compared to the strength achieved with water spray alone

IMPROVEMENT IN MOISTURE RETENTION:

The moisture content of the surface roads in a shift of 8 hours at different situation is as follows:

TABLE.2

(I)	Without water spray	1.67% to 3.60%
(II)	With water spraying	1.89% to 6.36%
(III)	With 1 Kg of compound	2.24% to 8.48%
(IV)	with 2 Kg of compound	4.29% to 11.4%
(V)	With 3 Kg of compound	5.69% to 14.39%
(VI)	With 4 Kg compound	8.56% to 15.93%

The increase in the moisture content of soil along haulroads ranges from 57% - 65% as compared to that with plain water spraying. The increased moisture retention indicates better dust suppression and consequently reduced frequency of wetting with plain water.

5.2 IMPROVED HARDNESS OF HAULROAD SURFACE:

Penetration studies made reveal improvement as indicated by the results of observation given below:

TABLE.3

Location	Plain water Spraying	With 1 Kg Compound	With 2 Kg Compound	with 3 Kg Compound	with 4 Kg Compound
i. Yamuna Road	3.00	2.80	2.40	1.70	1.00
ii. T.S. Road	2.70	2.50	2.25	1.60	0.90
iii. Coal Haul Road	2.60	2.50	2.00	1.30	0.75
iv. Coal Bench Road	3.00	2.30	2.18	1.40	0.700

Thus there is 3 to 4 fold reduction in penetration under conditions of application of chemical as compared to the condition with plain water spraying alone. This reductions increases with increase in the content of the chemical from 1 kg to 4 kg

5.3 REDUCTION IN FINES IN AIRBORNE DUST CONCENTRATION:

The results of sieve size analysis of the airborne dust concentration sampled at selected locations, vis-à-vis different quantity of the compound added, gives the values of -212 μ size as given below:

TABLE.4   dust mg/m3 of air

Location	With plain water spray	Chemical compound mixed water spray			
		1 Kg	2Kg	3Kg	4 Kg
Yamuna Road	9.43	7.41	5.52	4.53	3.00
T.B.Road	11.67	8.41	6.41	4.59	3.10
C.H. Road	11.87	9.59	6.78	5.41	3.15
C.B.Road	13.54	10.69	7.59	5.12	3.20

6.0 TECHNO ECONOMIC EVALUATION:

The economic analysis of the technological, advantages accruing out of chemical compound – added–water spray gives the following plus points:

1. Reduction in water spray tanker trips from 10 to 6 per shift. It results in the reduced amount of water handling i.e., in spraying as well as subsequent pumping.
2. Reduction in machine maintenance
3. Less frequent operation of motor grader due to better compaction of the haulroad surface.
4. Increased output for same investment, This is because of increased availability and occupancy of haulroad by production related activities / operations.
5. Improvement in the health conditions due to the resultant clean and Eco-friendly environment.
6. Reduction in water consumption by approx. 40%.

7.0   CONCLUSION & ACKNOWLEDGEMENT:

The unambiguous inference of the study is that such type of treatment results in improved health of haulroad. This , in turn, yields better technical performance of equipment operating , on it. The reduced generation of dust yields more Eco friendly environment. Improved economy is a natural outcome. The savings are substantial.

It is recommended that the compound may be used by more mines so that much more reliable information is generated. Also, it may result in the development of much more effective compounds as well as much more convenient and cost saving techniques of application.

The Authors acknowledge with gratefulness the cooperation and facility extended by the mines authorities of M/s. S.C.Co. Ltd. for conducting trials. Acknowledgements are also due to the authorities of Kakatiya University for permitting the presentation of the paper. Thanks are due also to Sri.G.Venkateshwar Rao, for transferring the script in computer floppy. We also thank M/s.Syntrone Industries pvt.Limted for the free supply of the Chemical for trial. The main contents of the paper have been extracted from the dissertation submitted by Sri ARP Reddy for his M.Tech degree of Osmania University.

REFERENCES:

1. Literature supplied by M/S. Syntrone Industries Pvt. Limited.
2. Reddy, ARP. 'A Trial Application of dust Suppression Chemical on Haulroads of OC-II. Techno Economic Evaluation of Chemical.
3. DGMS Circular No.16 of 1975.

Physico-chemical Characterisation of Different Size Fractions of Flyash—A Case Study

A. Sarkar[1] , Ruma Rano[1], N.C. Karmakar[2], S. Bhattacharya[3] and I.N. Sinha[4]

[1]Department of Applied Chemistry, [2]Department of Mining Engineering, [3]Department of Fuel & Mineral Engineering [4]Centre for Mine Environment
Indian School of Mines, Dhanbad 826 004, India

Abstract

Flyash collected from a local thermal power plant was fractionated into five different size fractions. Physico-chemical properties of each of the fractions were determined. The results reveal that the properties are largely dependent upon the size range of the particles. Density as well as surface area increases with decrease in size. The carbon content increases with increase in particle size. Morphological characterization leads to the conclusion that the particles are highly porous, the coarser fractions are macroporous whereas the finer ones, microporous. Cenospheres are present in the smaller fractions. Spherules are quite abundant particularly in smaller fractions. Physico-chemical characterization of the various fractions of flyash will be of help in the design process of its proper utilization. The paper gives some useful indication towards this aspect.

Introduction

Flyash, a thermal power station waste, has attracted a lot of attention over the past two or three decades. With increasing demand for power, a large number of thermal power stations have been set up in the recent past. The average yearly generation of fly ash in India alone is to the tune of 100 million tonnes. Naturally, this poses a great threat to the environment. The answer to such an imposing threat lies in its effective utilization.

Flyash is being utilized for a number of distinct purposes. Its main utilization is in land filling and road making where it can be used in bulk quantity. A large amount of flyash is utilized in the preparation of flyash brick and mortar. Apart from this, flyash is utilized as soil nutrients, in the making of ceramic tiles, as cement additives etc. However for all these end uses, the characteristic properties of flyash, namely physico-chemical and morphological ones, are of prime importance. For example, the pozzolanic property is the most important factor in its use as brick making material. Strength, surface area and density are to be considered as very useful properties in the preparation of ceramic tiles. Hence, it is very important to have a thorough investigation on the physico-chemical and morphological properties of flyash which is the subject matter of the present paper.

Experimental

The experiments consist of three parts, namely, sample collection, sieve sizing, chemical analysis and instrumental characterization.

Flyash was collected from the chimney stack of a local thermal power plant operated on the technique of fluidised bed combustion. It was properly mixed following the standard coning and quartering procedure. Representative sample was then dry sieved using sieves of mesh size 150, 250, 300 and 350 respectively. Weight percentage distribution, density, particle size and surface area of each fraction was determined following the standard procedures. Each fraction was chemically analyzed for silica, alumina, iron and carbon contents. FTIR spectra for each fraction were obtained using KBr-pellet method of sample preparation. For morphological study, Scanning Electron Microphotographs of each of the fractions were obtained. For a few samples, Transmission Electron Microphotographs were also obtained in order to have a detailed morphological characterization. XRD analyses were done for all the fractions.

Table 1 : Various Physico Chemical Properties of Flyash Fractions

	A	B	C	D	E
Wt. percentage distribution (%)	9.19	12.25	34.00	9.04	35.50
Density (gm/cc)	1.56	1.79	1.84	1.92	1.98
Surface area (m^2/g)	0.060	0.084	0.198	0.217	0.413
Carbon content (%)	36.57	29.98	8.32	6.35	2.57
SiO$_2$ %	36.1	50.4	50.5	50.6	50.9
Al$_2$O$_3$ %	8.04	9.70	12.84	12.90	11.58
Fe %	0.56	0.50	0.84	0.45	0.53
Morphology class	D	D	-	-	-
Aliphatic CH$_2$	Moderately strong band	Moderately strong band	Absent	Absent	Absent
Spherules	-	-	Sparse	A few	Present
Surface crystals	-	-	-	-	Visible on Some Spheres
Cenospheres	-	-	-	-	Present
Pleurospheres	-	-	-	-	Present

A = Fraction retained by m.s.150 (i.e. +150 mesh)
B = Fraction retained by m.s.250 (i.e. −150 +250 mesh)
C = Fraction retained by m.s.300 (i.e. −250 +300 mesh)
D = Fraction retained by m.s.350 (i.e. −300 +350 mesh)
E = Fraction passed through m.s.350 (i.e. −350 mesh)

Results and Discussions

Table 1 represents the data related to physico-chemical properties for the various fractions of flyash, namely, +150, (-150 +250), (-250 +300), (300 +350) and −350 mesh size. Weight percentage distribution reveal that abundance of the finest (-350 mesh) fraction is maximum, followed by the size fraction (-250 +300 mesh). These two fractions account for 70% of the total weight. Other fractions constitute the rest, with the coarsest particles (+150 mesh) accounting for only 9%. A correlation between weight percentage distribution and carbon content of each fraction leads to the inference that the Fluidized Bed Combustion Process is not very effective. The flyash contains high proportion of unburnt carbon. The percentages of carbon content in different fractions are 36.57, 29.98, 8.32, 6.35 and 2.57 respectively. Combining the abundance of each fraction with the carbon content data, it is found that the first two coarser fractions contribute to the maximum extent (61.89%) towards the carbon content of the flyash. The first three contributes 86.15% and the last two a mere 13.49 %. This clearly shows that though the abundance of first two fractions are much less, their carbon contents are so high that the overall carbon content of the flyash increases. FTIR spectroscopic data also reveal that the first two fractions have essentially high content of aliphatic carbon (CH$_2$ asym. = 2920 cm^{-1}). A plot of average density versus carbon content shows that there is a direct correlation between the density and carbon content. Surface area per unit weight of the particles decrease on increasing the size.

322

Morphological examination using Scanning Electron Microscope (SEM) of the particles reflects interesting features (Figs. 1-3). Particles are, in general, of irregular shape possessing a large number of pores. However, as size decreases the particles assume more regular shape. Spherical particles become prominent in the finest size fraction. Cenospheres and pleurospheres are present in the finest particles. Magnetite spheres are abundant in the magnetic portion of the finest fraction. Transmission Electron Microscopy (TEM) reveals that spheres have thick outer shells (Fig 4). In some cases, concentric shells constitute the spheres (Fig.5).

SEM does not show the presence of surface crystals on the spherules, whereas TEM clearly reveals that on some spheres surface crystals have formed.

Formation of mullite crystals on the spheres has been observed at least in one case as evident from TEM study.

Another interesting feature revealed from SEM study at higher magnification is that these particles, particularly the coarser ones, not only contain pores but also a large number of deep cracks or crevices (Fig. 3). Particles of smaller dimension do not contain much of this type of cracks/crevices. This indicates that the mechanism of formation of larger particles is perhaps different from that of the smaller ones.

Significance of the Study

The present investigations are very significant from the view point of proper utilization of flyash.

Carbon content of flyash is a direct measure of unburnt carbon as supported by FTIR study. Higher carbon content increases the amount of ash carried off by the flue gas. Since FTIR data reveals that the carbon present in the coarser particles is essentially aliphatic in nature, it can be concluded that partially burnt hydrocarbons are most likely to be spewed into the atmosphere by coarser particles. Strength of bricks, mortars prepared using flyash will be dependent upon the carbon content. Higher carbon content leads to lowering of strength of the building materials. Thus, a recommendation follows that only the finest fraction of flyash is to be used in the preparation of bricks etc.

Morphogenesis study revealed that the coarser particles are of irregular shape containing large number of macropores, voids etc. It is recommended that coarser particles should not be used in the preparation of building materials. The voids are intense points of weakness, and flyash brick and mortar prepared using such particles are likely to be weak.

One more reason for suggesting the use of only the finest fraction in the preparation of brick and mortar is that only this fraction contains cenopheres and pleurospheres which are known to be very hard and chemically resistant.

Conclusion

From the physico-chemical and morphological studies the following conclusions may be drawn.

- Large particles (+150 mesh) have a quite different surface characteristic from the rest. Also, they differ in physical properties like density and surface area significantly from the rest of the particles and the same trend is followed in case of morphogenesis, carbon content, aliphatic carbon content etc.

- The particles in the finest fraction all together constitute a different class. These particles are rounded, regular and contain less of macropores.

- Cenospheres and pleurospheres are present mostly in finest fraction.
- Carbon content of coarsest fraction is maximum while the finest fraction contain least amount of the carbon.
- Only the finest fraction of flyash is sustainable for the preparation of brick and mortar.

323

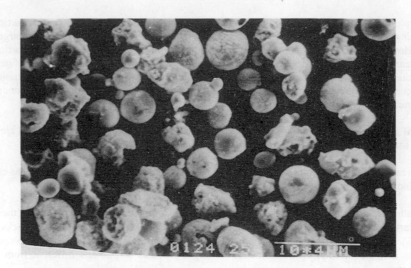

Fig. 1   SEM of finest fraction of flyash (-350 mesh)

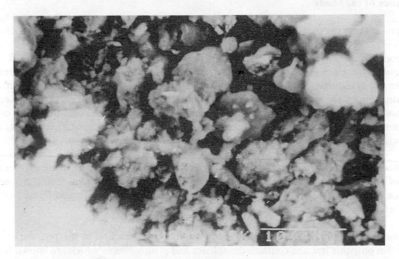

Fig. 2   SEM of coarser fraction of flyash (-150 +250 mesh)

Fig. 3   SEM of coarsest fraction of flyash (+150 mesh)

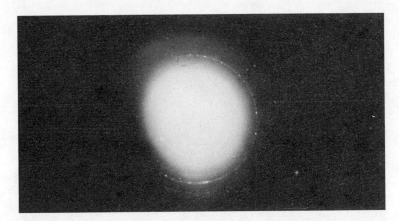

Fig. 4   TEM of a spherule

Fig. 5   TEM of a spherule with concentric shells

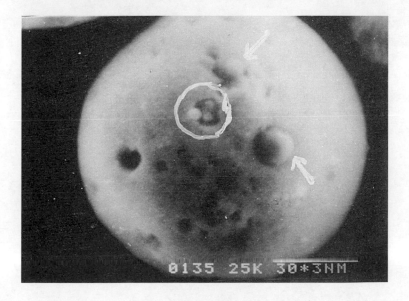

Fig. 6   Cenospheres popping out from a plerosphere

THEME 5 :
Policies of Government on Environment

Mineral Development and Environment: Policy Issues in India

A.N. Bose

Controller General, Indian Bureau of Mines, Civil Lines, Nagpur 440 001, India

ABSTRACT:

Natural resources are the vital components for sustaining the life support system on the earth. Due to increasing pressure of population and the consumeristic life styles, these vital resources have been dwindling over the year. The scenario is not different in the mineral sector too. Quite often, mineral rich areas are located in the areas which are ecologically fragile and full of biological diversity. Therefore, it is necessary to take a balanced view so that mineral development and conservation of environment keep pace with each other.

Our National Mineral Policy also emphasised to minimise adverse effects of mineral development on the forest, environment and ecology through appropriate abatement measures. Therefore, managing the interface of mining activities and environment is a formidable task and has to be given careful attention so that while economic development is not hampered, the environment is also protected. The present paper highlights the policy issues related to mineral development vis-à-vis protection of environment in Indian mines.

INTRODUCTION

The process of economic and social development of a country always involves industrialisation which, in turn, is linked with core sector growth. The development of infrastructure and core sector is directly linked with increased production of minerals. However, mining activity, in spite of best practices adopted, invariably involve a conflict with nature specially land. This is more so in open cast mining technology which is adopted for quick and economic extraction with higher percentage of recovery. Apart from leaving the scars on the earth's surface, there are other problems associated with mining activity such as air and water pollution, noise, vibration, subsidence, effect on flora and fauna, loss of top soil, resettlement and rehabilitation of land oustees. The severity of these adverse environmental impacts can be reduced best by incorporating environment concerns in the production planning process itself, rather than adding environmental measures onto a process already finalised on the basis of techno-economics. In fact, the integrated planning is an ongoing process throughout the project life.

SCENARIO OF INDIAN MINING INDUSTRY

The Indian mining sector has registered a rapid growth since independence. During the last five decades, the production of all the principal minerals except gold and mica increased enormously. Presently as many as, 86 minerals comprising of 4 fuel, 10 metallic, 50 non-metallic and 22 minor minerals are being exploited aggregating about 500 million tonnes of production per annum. During the year 1999-2000, there were about 3120 working mines in the country including about 550 for coal and lignite. The total value of mineral production during the same period was about US $ 9.54 billion (excluding atomic minerals). Over the past decade, the Indian mineral industry has been maintaining an average growth rate of about 6% per annum

In 1991, India initiated a wide ranging programme of economic reforms. In tune with this, a new era was ushered in when the new National Mineral Policy was pronounced in March, 1993. The policy envisages foreign technology and equity participation in exploration and exploitation of scarce and high value minerals. The new policy also emphasised to minimise adverse effects of mineral development on the forest, environment and ecology through appropriate protective measures.

MINING AND ENVIRONMENT POLICY
General

The concern of protection of environment in Indian scene first appeared in Article 48 A in the directive principles of Indian Constitution amended in 1976 (42nd amendment) which reads as "the State shall endeavour to protect and improve the environment and to safeguard the forests and wild life of the country." Consequently, item 17 A – Forests and 17 B – Protection of Wild Animals and Birds were transferred from State list to concurrent list of Seventh Schedule. It was also mentioned in Article 51(g) in fundamental duties of citizen under Constitution to protect environment. The first official recognition of conservation and protection of water resources came through the enactment of the Water (Prevention and Control of Pollution) Act, 1974 and Rules and Water Cess Act of 1976 and Rules made thereunder. In early eighties a full fledged Department of Environment, under the Ministry of Environment and Forests came into existence in the country.

Environmental Policy in Mining

Keeping in view of the adverse impacts of mining operations on the environment, Government of India have initiated legislative and administrative measures to achieve sustainable development of mineral extraction. The Ministry of Environment and Forests, serves as the nodal Ministry in the administrative structure of the Central Government for planning, promotion and coordination of environmental and forestry conservation
programmes. Indian Bureau of Mines, an organisation under the Union Ministry of Mines, has also shouldered with the responsibility of administration of certain Acts and rules relating to environmental protection in mining areas. The programmes of environmental assessment, protection and conservation are supported by comprehensive regulatory framework. The major Acts are :

1. Wild Life Protection Act, 1972 amended in 1983, 1986 and 1991.
2. The Water (Prevention and Control of Pollution) Act, 1974 amended in 1988.
3. The Air (Prevention and Control of Pollution) Act, 1981 amended in 1988.
4. Forest (Conservation) Act, 1980 amended in 1988.
5. Environment (Protection) Act, 1986.

Besides, specific provisions for environmental safeguards have been incorporated in the Mines and Minerals (Regulation and Development) Act, 1957, and rules made thereunder.

In order to ensure the continued and sustained progress of mining industry, Government of India has brought out the National conservation Strategy and Policy statement on Environment and Development in June, 1992. The policy statement clearly lays down the guidelines under which a nine point Action Plan has been proposed to prevent and to mitigate environmental repercussions in mining and quarrying operations :

- Mined area rehabilitation and implementation of the environmental management plans concurrently with the on-going mining operations to ensure adequate ecological restoration of the affected areas;
- Rehabilitation of the abandoned mined areas in a phased manner so that scarce land resource can be brought back under productive use;
- Laying down of requisite stipulations for mining leases regarding tenure, size, shape and disposition with reference to geological boundaries, and other mining conditions to ensure systematic extraction of minerals alongwith environmental conservation;
- Emphasis on production of value added finished products from mining so as to reduce indiscriminate extraction;

330

- Upgradation and beneficiation of minerals at the source, to the extent possible in order to ensure utilisation of low grade mineral resources and to reduce the cost of transportation, processing and utilisation;
- Environmentally safe disposal of the by-products of mining;
- Restriction on mining and quarrying activities in sensitive areas such as hill slopes, areas of natural springs and areas rich in biological diversity;
- Discouraging selective mining of high grade ore recovery of associated lower grade ore during mining;
- Environmental impact assessment prior to selection of sites for mining and quarrying activities.

National Mineral Policy, 1993 and Environment

In line with the new Industrial Policy announced by the Government of India in 1991, National Mineral Policy was formulated in 1993, which emphasised the need to minimise adverse effects of mineral development on the forest, environment and ecology through appropriate protective measures. It further states that mining operations shall not ordinarily be taken up in identified ecologically fragile and biologically rich areas. Strip mining in forest areas should as far as possible be avoided and it should be permitted only when accompanied with comprehensive time bound reclamation programme. As far as possible, reclamation and afforestation will be processed concurrently with mineral extraction. Efforts would be made to convert old unused mining sites into forests and other appropriate forms of land use.

NOTIFICATIONS UNDER EP ACT FOR MINING

Under Section 3(2)(v) of the Environment (Protection) Act, 1986 (EP Act) the Central Government have the power to take such measures for protecting and improving the quality of environment by restriction of areas in which industries, operations etc., shall not be carried out or shall be carried out subject to certain conditions. The rule 5 of the Environment (Protection) Rules, 1986 made under the EP Act have laid down the factors to be taken into consideration while regulating such activities, and procedures to be followed for such final notification in the official gazette.

Accordingly, the Government of India have issued the notifications with respect to certain ecologically sensitive/fragile areas from time to time as follows :

(i)   **Restriction on mining on Doon Valley in Uttar Pradesh**

Dehradoon valley which is popularly known as Doon Valley being a ecologically sensitive area, number of public litigations against carrying out mining in this Valley and setting up cement plants in the area were filed in early eighties and the cases went upto the Apex Court, which directed to stop future limestone mining in this area. Therefore, a notification restricting mining activity in Doon Valley had been issued.

(ii)   **Restriction on mining in Aravali Hill range in Gurgaon District in Haryana and Alwar district of Rajasthan**

All new mining operations including renewals of mining leases, existing mining leases in sanctuaries/National Park and areas covered under project Tiger and unauthorised mining in this hill range have been prohibited without proper permission. Any person desiring to undertake mining activity in the said area shall have to submit an application in prescribed form alongwith EIS and EMP.

(iii)   Restriction of mining in wild life, sanctuaries, national parks, near to national monuments, areas of cultural heritage also in ecologically fragile areas rich in bio-logical diversity, gene pool etc.

(iv)   **Restriction of mining in Coastal Regulation Zone (CRZ)**
India has a coast line of over 7,500 km and to regulate the mining operations along the coastal zone with safeguarding the interest of environment, there is a statutory provision that the coastal stretches of seas, bays, estuaries, creeks, rivers and backwaters which are influenced by

tidal action (in the landward side) upto 500 meters from the High Tide Line (HTL) and the land between the Low Tide Line (LTL) and the HTL as Coastal Regulation Zone and imposes the mining of sand, rocks and other sub-strata materials, except those rare minerals not available outside the CRZ areas. Dredging and under water blasting in and around the formations are also not permitted.

(v)   **Environment Impact Assessment of Development Projects**

The Ministry of Environment and Forests is assigned the responsibility for appraisal of projects (including mining projects) with regard to their environmental implications. The environmental appraisal of mining projects commenced in the year 1982 through an Environmental Appraisal Committee for mining project and which is now called Expert Committee (Mining). This Committee consists of experts from various fields including Controller General of Indian Bureau of Mines. Based on environmental impact assessment and issues arising thereto, decisions are taken by the competent authorities in respect of the projects including selection of sites. With respect to the mining projects, the Government of India issued the notifications which inter-alia provides the following :

i)   Site clearance for prospecting and exploration of major minerals in areas above 500 hectares.
ii)   Site clearance for mining projects involving major minerals with lease in excess of 5 hectares in area.
iii)   Environmental clearance for mining projects involving major minerals with leases in excess of 5 hectares in area.

In addition to these, public hearing has become mandatory for clearance of any mining project. However, no prior site clearance including public hearing is needed for test drilling on a scale not exceeding one bore hole per hundred square kilometers.

IMPLEMENTATION OF MMDR ACT

The Mines and Minerals (Regulation and Development) Act, 1957 was amended in 1986 to include provision for environmental protection and rehabilitation of flora. A separate sub-section 5(2)(b) was introduced for submission of mining plan duly approved by the Central Government. The procedure of submission of mining plan has been elaborated in Rule 22 of Mineral Concession Rules (MCR), 1960, in case of fresh grant or renewal of mining lease and under Rule 11 of Mineral Conservation and Development Rules (MCDR), 1988 in case of existing mines. Both MCR and MCDR have been promulgated under this Act.

More recently, the Mines and Minerals (Regulation and Development) Act has been amended in December, 1999 to make it more compatible with the present day scenario of economic liberalisation and renamed as Mines and Minerals (Development & Regulation) [MMDR] Act thereby giving more thrust on development rather than regulation.

As per MMDR Act, mining operations are allowed to be carried out only in accordance with the mining plan duly approved and enforced by Indian Bureau of Mines in case of non-coal mines. A Mining Plan includes an Environmental Impact Assessment and Environmental Management Plan which, interalia, identifies the likely environmental impact to be caused due to mining activities and suggest for its mitigative measures.

Apart from mining plan, the mining lease holder is also required to submit Scheme of Mining for every 5 years which review the implementation of mining plan and proposal for next five years.

ENFORCEMENT OF FC ACT

Forest (Conservation) [FC] Act, 1980 deals with the protection and conservation of forest. For the purpose of FC Act, mining leases can be divided into two parts.

(a) renewal of leases which were granted prior to 25 October, 1980 (i.e. date of commencement of FC Act); and
(b) grant of fresh mining leases for which diversion of forest land would be required.

In case of renewal, somewhat lenient view is taken in the light of the fact that mining operations to be continued in the already broken up area. Diversion of forest land is being considered for mining purpose against the compensatory afforestation over the non-forest land to be earmarked by the State Government which has to be transferred and mutated in favour of the Forest Department. If the lease area is more than 20 Ha., an Environment Management Plan is also needed. In such cases the proposals have to be cleared both by the Expert Committee (Mining) as well as Forest Committee.

STRATEGY AHEAD

Environment management is a process of natural resources optimisation for short and long term human welfare. Ensuring environmental sustainability of the development process through social mobilisation and participation of people at all levels is need of the hour. Exhaustible or vulnerable natural resources should be priced appropriately in order to prevent over exploitation and technologies which conserve the use of natural resources. In this connection, an important objective of ninth plan is to ensure environmental sustainability of the development process.

For mining industry, the following strategy deserves to be adopted.

(a) Mining and forest policies should be harmonised to facilitate optimisation of natural resources – forest, biodiversity, minerals, water and land.
(b) Detailed maps should be prepared clearly demarcating inviolate areas for total protection. This inviolate area, to be identified by a multi disciplinary expert group, may inter-alia cover :

- Mangroves
- Areas rich in biodiversity and genetic stock
- Wildlife habitats including migratory routes and corridors.

(c) R&D promotion for ex-situ conservation of biodiversity.
(d) Development of new facilities, or strengthening of the existing ones, for ex-situ conservation of flora and fauna.
(e) Develop data base, through case studies, on impact of existing mining operations in selected ecologically sensitive and protected areas.
(f) Demarcation of abandoned mines with time-bound Action Plans for their rehabilitation.
(g) Catalogue best practices in mining operations alongwith national and international bench-marks and formulate Action Plans for their adoption in a time-bound manner.
(h) Detailed strategies and rehabilitation methodologies for biological reclamation of mining areas in different geo-climatic regions of the country.

CONCLUSIONS

Mining is an important sector which plays a vital role in the economy of our country. However, often mining industry is projected as a threat to the environment. But it is possible to achieve a balance between the protection of environment and mineral development. The Government is very much concerned about the preservation of environment. The policies declared so far are eco-friendly. Yet, we have to concentrate our efforts for its implementation in the true sense and spirit.

The views expressed in this paper are author's own and not necessary of the organisation to which he belongs.

-oo0oo-

The Role of Governments in Environmental Management in Nigeria

Osa Sunday Omogiate

*Managing Director/Chief Executive, Earth Resources Improvement Agency (NIG) Limited
149 Uselu Lagos Road, P.O. Box 10195, Benin City, Nigeria*

INTRODUCTION: The structure of Nigerian Federal system into three tiers of government (Federal, States and Local Government Councils) between 1954 and 1976, have further expanded the scope of Nigeria environmental management with each level of government establishing commissions or agencies/departments to formulate, implement and evaluate environmental protection programmes and feedbacks. The broad responsibilities of Federal government in environmental management in Nigeria as enshrined in the national policy on environment is to coordinate environmental protection and natural resources conservation for sustainable development. The primary mandate of all governments therefore, is to achieve environmental objectives as stipulated in section 20 of 1999 Constitution of the Federal Republic of Nigeria, which is essentially to protect and improve water, air, land, forest and wildlife in Nigeria.

The Federal Ministry of Environment (FME) also initiated an environmental Agenda called "Environmental Renewal and Developmental Initiative" (ERDI) with its primary objectives of taking full inventory of our natural resources, asses the level of environmental damage and design and implement restoration and rejuvenation measures and to evolve and implement additional measures to halt further degradation of our environment. The Federal, State and Local governments are responsible for drainage construction, road rehabilitation and water resources management. Although, the local councils are usually responsible for providing solid waste disposal and water services, it completely lacks, the technical and financial resources to render these services effectively. Since the council have a vital role to play in urban and rural environmental management strategy, other tier of government should assist to develop the capacity to handle the task through funding, training and re-orientation or through international development organisations.

There are fundamental elements of environmental management which are crucial to the attainment of the goals of environmental standards. These basic elements are mainly managerial functions which are not mutually exclusive. These managerial principles are:
1. Leadership
2. Planning
3. Policies and options
4. Capacity building
5. Organisation and coordination
6. Conceptual framework
7. General control and feedback or public comments.

Environmental management at federal, states and local councils are not only functioning at strategic managerial level but equally performed at tactical and operational levels in order to protect the natural resources for mutually beneficial agreement.

Structurally, instability of government since 1960, when Nigerian attained her independent created set-back in environmental management with adhoc approaches through states and local councils in managing environmental sanitation (disposal of solid waste, clearing of weeds etc) nation-wide which was not sustainable to general environmental protection and control and hence the development of sound environmental standard has remained obscure and elusive. Other problems confronting environmental management are:

a. Low capacity of professional (technical know how and skilled manpower
b. oil spillage
c. urban solid waste
d. striking a balance between government policy and industrial growth
e. infancy stage of environmental science
f. potholes on our road
g. lack of proper coordination of researches into environmental programmes and hence in haphazard form.
h. penalty for defaulters of environmental laws are unarguable too small compared to offences committed.

All tiers of government are now taking bold steps in establishing a sound environmental management which is now an essential requirement for achieving sustainable economic development and attracting lending from foreign financial organisations such as world bank, Asia bank, Africa development bank, bilateral and multilateral organisations. In the same vain, local and international donor agencies have become imperative (financial contribution and training expertise) while individual corporate organisation like Earth Resources Improvement Agency Ltd, Benin or other non-governmental organisation's participation in environmental management has become imminent. For instance, Lagos state government in Nigeria (1997) set up a pilot scheme which allows private sector participation in waste management. This scheme is known as Private Sector Participation (PSP) programme where domestic waste collection and transportation are being carried out by individual operators that have the where-withal to do the job. This Job that was originally done by local government councils but are now operated by private sector. The fee is determined by Lagos state government and paid by residents.

The Federal Ministry of Solid Mineral Resources recently reviewed the mineral act of 1963, which provides for restoration of excavated land of solid mineral sites into (1) fish pond (2) tourist sites (3) fertile agricultural land.

Development of Environmental Management Commissions and Agencies/ Departments.

The federal Government policy on environmental management dated back to 1958 (pre-Independence Nigeria) when the colonial ruler set up.

1. Sir Henry Willink's Commission: the Willink report led to the establishment of "The Niger Delta Development Authority (NDDA)" Some of the major functions of NDDA include the tackling of ecological problems of area which occupies about 70,000sqkm of southern Nigeria, Environmental flooding, erosion, pollution and degradation in coastal environment created "peculiar difficult terrain for the area, thus made development a near impossible task" and hence relative high cost of infrastructure development.

Secondly, the NDDA failed because the authority lacks empathy, courage, and political will to develop the environmental needs of the region due to lack of funds.

2. National forestry Act (1958) was enacted by Federal Government to guide amongst others (i) the national policy on conservation which stipulates that at least 20 percent of the country land area should be put under permanent vegetation cover (ii) forestry law forbids setting fire or allowing fire to spread to any forest growth unless such forest growth is being or has been felled for farming purposes. The law also stipulates imprisonment of six months for offender.

3. Department of petroleum resources (DPR): The oil and gas pollution control unit of the Department of petroleum resources in the ministry of petroleum resources acted as regulatory body and guided the oil/gas industries before federal Environmental Protection Agency (FEPA) was created in December 30, 1988. As a Government Department, it lacks, the strong political will to enforce Laws and Regulations.

4. Urban Development and Environment: after the Stockholm, Germany Conference on the environment in 1972, Federal Government established a division of urban development and environment within Federal Ministry of Economic Development in 1975.

5. Federal Department of Water Resources: Was established under Federal Ministry of Agriculture in 1975 to carry out the following mandate:

(a) to quantify the water resources by maintaining an official machinery of data collection and analysis.

(b) to prescribe quantity standard for sewage treatment and effluent disposal, thus controlling environmental and water pollution.

(c) to control the exploitation of the resources by issuing licenses to the River Basin Development Authority and Water Board.

(d) to prescribe quality standard for water supplies for various purposes and liaise with Ministry of health to ensure that these standard are kept.

6. State Water Board: Were established to:
    a. supply water to consumers which include public industries, Agricultural establishments and Amusements Park etc. Also to issue sub-licenses to consumers for direct abstraction where it is unable to meet the demand of the customers.

7. Bendel state forestry law (1976) as applied to Edo State of Nigeria is to function as:
    (a) the protection and management of forest estate and wildlife resources of the state.
    (b) Advice government in formulating policies on management of the forest and wildlife resources.
    (c) The protection and control of Exploitation of economic timber trees in forest reserve and areas outside the forest reserve.
    (d) Enforcing the provisions of the forestry and wildlife laws and advising industries on forest products.

8. Forestry Law of Bendel State (1976), defines the role and functions of local government as:
    (a) Establishment of wood lots to protect watersheds and river courses.
    (b) Protection of forest and farm trees in arable land against fire and illegal felling of trees.
    (c) Protection of wildlife against poaching.
    (d) Other forestry and wildlife functions which the state government may delegates.

9. River Basin Development Authorities (1980).
    Came into existence with the following functions.
    (a) To undertake comprehensive development of underground water resources for multipurpose use.
    (b) To undertake schemes for the control of flood and erosion and for water shed management.
    (c) To control pollution in rivers and lakes etc.

10. Directorate of Food, Roads and Rural Infrastructure (DIFFRI)-1986.
    With the objective of constructing feeder roads, water supply facilities to enhance the agricultural productivities of rural dwellers.

11. Federal Environmental Protection Agency (FEPA) –1988.
    FEPA'S mandate includes:
    (a) Advice federal government on national environmental policy formulation and priorities, setting standard and regulation, compliance monitoring and enforcement.
    (b) Overall responsibility for the protection and development of the environment, biodiversity conservation and sustainable development of Nigeria's natural resources.
    (c) Advice the president on the utilisation of the Ecological fund which is 2% of national revenue for the protection of the environment.

12. Environmental Impact Assessment (EIA)-1992.
    EIA was promulgated into law as stipulated by law number 86 of 1992 amongst others:
    (a) To facilitate inform decision making by providing clear and well structured dispassionate analysis of the effects and consequences of proposed projects.
    (b) To assist in the selection of alternative including the selection of the best practicable or most suitable once.
    (c) Serve as an adapter, organisational learning process in which the lessons of experience are integrated into policy, institutional and project design and ensure effective consultation with locally affected people.

13. Other agencies include oil mineral producing and development commission (OMPADEC) established by decree number 23 of 1992 to administer and disburse the special 3% fund from federation account. Solely for the rehabilitation and development of oil mineral producing areas (nine states), with respect to tackling of ecological problems that have arisen from exploration and exploitation, pollution, oil spillage and general environmental degradation.

14. Petroleum Trust Fund (1994)PTF: Amongst its mandate are –
    (a) To rehabilitate roads that have been degraded.
    (b) Helps to rehabilitate national water scheme throughout Nigeria.
    (c) Dredging of rivers e.g. River Niger from Escravos to Boro (Niger State) by adhering to environmental impact assessment.

15. The National Water Resources Institute Kaduna.

16. State Agricultural Development Projects etc.

17. The Federal Ministry of Environment (May) 1999.
    On May 29, 1999, a new ministry of Environment was created by the presidency to bring together all the activities within government machinery that relates to environment and sustainable development. This is an effort to give environmental issues the top priority attention in its development programmes. Relevant Department units of some Federal ministries were transferred to compliment the function and scope of activities of the newly created ministry in order to address the enormous environmental problems in Nigeria.

The relevant department units affected in the transfer exercise were as follows:

a. Environmental Health and sanitation Unit of the ministry of Health.
b. Oil and Gas pollution control unit of the department of petroleum Resources of the Ministry of Petroleum Resources
c. Forestry Department (including wildlife, forestry monitoring, Evaluation and coordinating unit (FORMECU) of the Ministry of Agriculture.
d. Costal Erosion Unit, Environmental Assessment Division, Sanitation Unit of the Ministry of Works and Housing.
e. Soil erosion and flood control department of the Ministry of Water Resources.

The new mandate of the new Ministry, if properly implemented will protect and improve water, air, land, forest and wildlife in Nigeria as enshrined in section 20 of 1999 Constitution of the Federal Republic of Nigeria.

The following framework for the structure of the Ministry has been put in place in order to implement policy strategies.

Services Departments.
- Administration and finance department
- Planning, research and statistic department.

Technical Department.
- Environmental conservation department
- Environmental assessment department
- Erosion, flood and coastal zone management department
- Forestry department
- Pollution control and environmental health department
- Drought and desertification amelioration department

NEW ROLE IN ENVIRONMENTAL MANAGEMENT.

The new ministry of Environmental (June 1999) has proffered the following solution to the environmental problems in Nigeria.

1. Eliminating Gas Flaring and Spillage: In 1969, Nigeria government introduced legislation, the petroleum (Drilling and Production Regulation of 1969) which mandated oil/gas companies to use associated gas from their operation within five years. This was largely ignored by oil companies by 1974, in 1979, the associated gas re-injection Act was enacted as a possible means of forcing oil companies to comply with time limit of April, 1980 but both government and law were ignored. The 1997 constitutional conference recommended 2010 as possible end to gas flaring and oil spillage but in October 1999, the new Federal Ministry of Environment (regulatory body of environment) sent invitation to 53 oil and gas companies operating in Niger delta and adjoining basins to summit. "An Inventory of their post-imparted sites as well as its environment management plans that will help it achieve zero pollution and eliminating gas flaring within shortest possible time (2003-2008). This is so because about 70% of daily production of Three(3) billion standard cubic feet of gas is flared.

2. Niger Delta Clean Up Fund: Another environmental management technique by Federal Ministry of Environment is Niger Delta clean up fund. Although a company like South Atlantic Petroleum has indicated positive contribution inspite of its commitment to (a) taxes (b) royalties (c) joint industry projects aimed at environmental conservation and preservation.

3. Provision of geographical information system (GIS). The Federal Ministry of Environment has initiated a plan of action for both industrial and nature – induced (e.g. erosion and flooding) environmental pollution and degradation with a view to generating data for the preparation of a plan of action and fulfilling the objectives of the planned for clean up of the Niger Delta.

a. First step towards effective management of flood and erosion menace is detailed inventory of all areas impacted by the menace and extent of impact.
b. Government intends to spend 70.5 million naira to inventorise major erosion and flood impacted sited in oil producing states.
c. Develop a region – wide emergency preparedness plan.
d. Prepare an action plan for the remediation of past impacted area
e. Government is also inviting experts from other countries that may like to provide more information about geographical information system.

4. Social responsibility of government towards oil producing communities environmental management (Now and Future): Inspite of the already communities based public relations strategies as evident from the president and Federal Ministry of Environment, other measures being adopted or being proactively involve are:

338

a. Monitor strictly the operation of oil/gas companies to comply with health and safety standard.
b. Consider effects of oil exploration/drilling and gas flaring on public health, site of value, plants, animals and general environmental impact assessment on mining impact on host community.
c. Making oil companies to fulfill their promises to host communities in order to eliminate youth restiveness and wanton destruction of pipelines criss-crossing the Niger Delta or seizure of wells.

5. Checking diversion of Ecological Fund:
The Federal government recently initiated plans to put in place appropriate measure to ensure that funds released for ecological problems are not diverted to any other conceivable use. The threats to our normal life in recent times include environmental degradation in Niger Delta, gully erosion in the east, coastal erosion of land bordering the Atlantic oceans, the restless drought and desertification in the Northern parts of Nigeria and industrial pollution in cities. This funds ensures that National Policy on environment are pursued with particular emphasis on environmental protection, conservation of natural resources and poverty alleviation. On May 29, 2000, the Federal Government from 2% Ecological funds agreed to spend ₦1.28 billion naira on flood and erosion control.

6. Afforestion programme: A memorandum submitted to federal government revealed that Edo and Imo States occupied the first two positions in the destruction of forest reserves in the country. Between 1996 and 1997, Edo state had depleted over 96 percent of its forest reserve while Imo, 97 percent which constitute great danger to our ecosystem.
An old value system which involve afforestation is needed by state governments through ecological funds to restore the Nigerian environment and the ecosystem.

7. Incessant Coastal Erosion on are bordering the Atlantic ocean. In 1999, the Federal Government of Nigeria constituted an interministerial committee with about 6.5 billon Naira to solve the incessant Atlantic Ocean over flowing its bank. This is to enable urgent solution to Victoria Island in Lagos (a high class residential area) from being submerged by Atlantic Ocean and other menace of ocean current ravaging Lagos Bar beach.

8. Environmental Impact Accessment: All Improvement projects at Federal, State and local government levels are now to be subjected to environmental impact analysis right from their early budgetary states under a new National Policy already being implemented. This will ensure the integration of environmental concerns into national policy plan and economic evaluation of environmental costs and benefits. In project appraisal to planning and budgetary purposes.

9. Desert encroachment: Poses a serious environmental problems, to daily life of the people in especially sahelian region of Nigeria. Some states in the North have lost 25 percent of their agricultural land; hence people now migrate from rural areas to urban cities.
A legislation is being planned by house of representatives to combat the problem of desertification in the country, known as "National Arid Land Development Authority Bill" as well as Katsina state Government has introduced and implementing various measures to tackle the scourge of desertification at its own sector of Sahara desert.

10. Water Hyacinth: The shipping industry is being seriously threatened by the proliferation of water hyacinth on our water ways which government intends to solve, by clearing it from water ways.

11. Sanitary inspectors: Federal Ministry of Environment intends to introduce relevant laws to discourage environment polluters. Sanitary inspectors are to visit domains to ensure conformity with minimum hygienic standard.

12. Waste to wealth programme: Federal Government plans to transform waste generated across the country into wealth. Experts assessment of disposal problem show that wastes generated in the country contains four main materials: (a) Plastic (b) Metal (c) textile and (d) Papers. Nine cities have been designed as test ground for a pilot land waste management programme by Federal Government. The cities are Aba, Lagos, Abuja, Ibadan, Port-harcourt, Kano, Kaduna, Abeokuta and Onitsha, under the pilot scheme, waste will be separated into four materials and recycled, those that could not be recycled would be burnt and residue taken to a land fills with special linings which ensure that residue does not go under to pollute the water.

13. Gully Erosion: There are states, which have special problems of gully erosion. States like Imo, Borno, Gombe, Kogi etc. and various parts of Niger Delta are worse. The 2% ecological funds of federal revenue are meant to address this issue.

14. Ecotourism: Wetland and national park have almost ceased to be in existence in Nigeria. New Ministries of Environment and tourism is set to boost the industry.

15. New curriculum on Education: New curriculum into the educational system so that pupils and students ate taught environmental habits. This will latter be extended to tertiary institutions.

16. Articulated Policies: The agenda of environmental protection relies on existing policies that are being fine turned by Federal Ministry of Environment to suit the need of the time.

339

17. Funding the trans-border of Niger and Nigeria by United Nations Environmental programme: The United Nation Environmental Programme (UNEP) has concerted to making funds available for the development of a national report toward an integrated management of land and water in the shared catchments of the trans boundary area between Nigeria and Niger republic. The project, which is, coming under a large programme, focuses on trans-boundary management of land degradation and shared water resources along Niger and Nigeria border. This scheme is to gulp about 20 million U.S dollars.

18. Complete compliance attainment: The Federal Ministry of environment has issued directives to all industries and facilities generating waste managers to take necessary step to attain full compliance between year 2004-2006 with United Nation environmental standard.

19. Environmental sanitation Edict for waste management in Lagos state: Lagos State Government is to enact an edict that would punish offenders and empower sanitary inspectors to monitor compliance in waste management in the state.
    a. Boost and make the activities of private sector participation (PSP) in waste management in the state effective and efficient
    b. Pilot projects to be reviewed periodically.
    c. Those who register with private sector participation scheme (PSP) to receive support from the government.

20. Smoke and noise pollution control: Smoke and noise environment has brought about pollution and degradation of air, land (forest), seas and river, not much attention has been given to these two vital areas.

CONCLUSION: Government should concentrate in setting up the necessary infrastructure, legislations and awareness as well as encourage more non-governmental organizations and community based concerned groups/agencies to work with organization/institutions which generates these pollutants and government regulating bodies. It is hoped that the concept and result of environmentally sustainable development would be achieved.

REFERENCES (BOOKS).

1. Ogunsanya Mobolaji, Osaghae E. etal (1992) citizenship Education for Junior Secondary Schools Book I, Page 12 and 13 published by Macmillan Nigeria publishers.
2. Nwadian Mon (1998) Educational Management for sub-sahara Africa. Published by Nigerian Society for Educational Planning (NSEP) in collaboration with Moose amalgamates.

REPORTS.

1. Hassan Adamu (2000) Nigeria Minister of Environment, Protecting our environmental and natural rsources. The Guardian, Wednesday, February 16, 2000.
2. Newswatch (1999): Waste to Wealth of December, 14, 1999.
3. Omogiate Sun Osa (1998): Geological parameters for environmental protection of the coastal environment of the African continent (Nigeria: a case study), A report submitted to (pasicom-maputo) July, 1998.
4. United Nations Environmental Programme to fund Nigeria-Niger border scheme (1999): Guardian of December, 20, 1999 Page 39 and 43.
5. Government of Lagos plans urban renewal project (1999), Guardian of December 20, 1999.
6. Environmental Ministry should coordinate all environmental agency by Nigeria Conservation foundation (2000) This day Monday January 17. page 40 vol. 3, No. 1730.
7. 302 Billion to combat desert (2000), Guardian Tuesday, Feb. 8, 2000.
8. Izeze Ifeanyi: Agenda 21, seven years beyond RIO: Guardian Monday June. 28, 1999.

340

THEME 6 :
General on Environment, Population Control etc.

Impact of Human Population on Resources and Ecosystem

Y.V. Rao, Harsha Vardhan and M. Aruna

Department of Mining Engineering, Karnataka Regional Engineering College, Surathkal 574 157, India

1 INTRODUCTION

At one time, a human was just another consumer somewhere in the food chain. Humans fell prey to predators and died as a result of disease and accident just like other animals. The tools available to exploit their surroundings were primitive, so these people did not have a long term effect on their surroundings. Minerals and energy resources were only minimally exploited. Aside from occasional use of surface seams of coal or surface oil seeps, most of the energy needs of early humans were filled by biomass. As human population grew, and as their tools and systems of use became more advanced, the impact that a single human could have on his or her surroundings increased tremendously. The use of fire was one of the first events that marked the capability of humans to change ecosystems. As technology advanced, wood was needed for fuel and building materials, land was cleared for farming, streams were dammed to provide water power, and various mineral resources were exploited to provide energy and build machines. These modifications allowed larger human populations to survive, but always at the expense of previously existing ecosystems. Today with over 6 billion people on the earth, nearly all of the surface of the earth has been affected in some way by human activity.

There are about 115 million people in Bangladesh, where the annual population growth rate is 2.5% . This means that the annual increase in the population of the country is 2.87 million. Recently more than 100,000 people died in Bangladesh when storms pushed ocean water onshore, flooding the low lying coastal lands that make up most of the nation. This great human tragedy is made even more tragic by the fact that, given the normal increase in the population in Bangladesh, those 100,000 people were replaced in just 2 weeks.

Bangladesh is one of the poorest nations in the world, and this poverty affects human survival. The average number of calories of food available per person is only 85% of that required for good health. Less than half the population has access to safe drinking water, and less than a fifth has access to adequate modern sanitation. Average life expectancy is about 50 years. With adequate resources for each individual and a rapid growth rate, Bangladesh struggles to maintain even its existing poor standard of living. For Bangladesh, it is difficult to talk about solving major environmental problems, conserving biological diversity or optimizing production of fisheries and vegetation when people barely have sufficient resources to survive and the growth of the human population erases any advances.

Let us assume that in 1994, the world population numbers approximately 5.5 billion people and the annual growth rate is approximately 1.8%. At this rate , 93 million people are added to the earths population in a single year - a number equal to the 1990 population of Mexico. About 90% of this growth takes place in developing countries, such as Bangladesh, India, countries in Africa, Asia, and Central and South America. In these regions, the average population growth rate in 1990 was 1.9%, where as in developed nations the growth rate was less than 1% and in some cases much lower.

At mid year 1998, world population stood at 5,926,000,000, according to estimates prepared by the Population Reference Bureau. This total represented an increase of 86 million over the previous year, firmly establishing that world population would reach the six billion mark during 1999. Given that the fifth billion was achieved as recently as 1987, global population was on track to add this next billion during the shortest time in the history. The annual rate of increase declined to about 1.41% from about 1.47% in 1997, once again the result of birth rate declines in some less developed countries. The 1998 rate of increase if maintained, would double world population in 49 years. Approximately 137 million babies were born world wide in 1998, 2 million fewer than in 1997. Just over 90% of the births in 1998 occurred in less developed countries. About 53 million people died in 1998; 78% of those deaths were in less developed countries. The continuing youthfulness of the less developed countries ensured that their populations would continue growing for many decades.

Life expectancy at birth was 64 years for males and 68 for females in 1998, the same as in the previous year. In the more developed countries the same figures were 71 and 79 and in the less developed countries, 62 and 65. The 1998 world infant mortality rate stood at 58 infant deaths per 1000 live births, a slight decrease from 59 in 1997. The lowest infant mortality rates were in western and northern Europe, at 5 and 6 infant deaths per 1000 live births, respectively. Finland reported the lowest rate of 3.5.

In 1998 the population of less developed countries grew at 1.73% per year, 1.99% for less developed countries outside China. These rates were slightly lower than in 1997, in part owing to a decline in the growth rate in India. The population of less developed countries constituted 80% of the world total. Of the 86,000,000 people added annually to the world population, 98% were in the less developed countries. In the less developed countries women averaged 3.3 children each down from 3.4 in 1997. In less developed countries excluding China, however women averaged 3.9 children each.

Africas population in 1998 totaled 763 million, 20 million more than in 1997. The continents annual growth rate was 2.5%, by far the worlds highest and sufficient to double the population in only 27 years. In 1998 life expectancy at birth in Africa was the lowest at 50 years for males and 53 for females. Infant mortality was the worlds highest at 91 infant deaths per 1000 live births.

Asias population rate declined from 1.6% in 1997 to 1.5% in 1998, largely owing to a decline in the growth rate in India.Figure 1 and 2 shows the Indias demographic transition in the 20th century and Indias population growth rate in the 20th century respectively. Life expectancy in Asia in 1998 stood at about 64 for males and 67 for females.

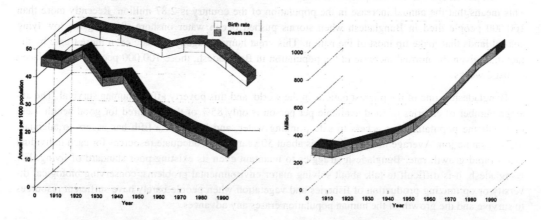

Fig.1 Indias demographic transition in the 20th century   Fig.2 Indias population growth rate in the 20th century

Woman in Asia averaged 2.8 children in 1998, but the average was 3.3 in countries outside China. In China women averaged only 1.8 children, a result of the national population program. In India women averaged 3.4 children, down slightly from 1997.

The population of more developed countries was 1,178,000,000 in 1998. The growth rate during the year was an extremely low 0.1%. Much of that growth rate was in the U.S. In Europe in 1998 there were more deaths than births, as was also the case in 1997. The population of no fewer than 13 European countries experienced this natural disease in 1998, among them Germany, Italy, and Russia. The Czech Republic, Italy, Latvia and Russia shared the worlds lowest fertility in 1998, averaging only 1.2 children each. Life expectancy at birth in Europe was 69 for males and 77 for females.

3 POPULATION AND TECHNOLOGY

The current danger to the environment has two main factors: the number of people and the impact of each person on the environment. When there were few people on the Earth and the technology was limited, human impact was local. In that situation, the overuse of a local resource had few or no large or long lasting effects. The fundamental problem now is that there are so many people and our technologies are so powerful that our effects are no longer local and unimportant. Our old habits, however, do not change quickly.

As biologist and educator Paul Ehrlich has pointed out, the total impact of the human population on the environment can be expressed through a simple relationship: The total environmental effect is a product of the impact per individual times the total number of individuals. An increase in either the individual impact each of us has on the environment or the total number of people results in an increase in the total human effect on the environment. The combination of rapid increases in both population and technology has geometrically increased our effect on the environment.

Technology not only increases the use of resources, it causes us to affect the environment in many new ways, compared with hunters and gatherers or with people who farmed with simple wooden and stone tools. For example, before the invention of chlorofluorocarbons to be used as propellants in spray cans and coolants in refrigerators and air conditioners, we were not causing depletion of the ozone layer in the upper atmosphere. Similarly, before we started driving automobiles there was much less demand of steel and little demand of oil- and much less air pollution.

An important consequence of the "population times technology" equation is that the addition of each new individual to the population of an industrialized nation leads to a greater effect on the environment than does the addition of each new individual in the population of a poor, undeveloped nation. Today human population trends vary greatly among countries. We tend to criticize and worry about the undeveloped nations whose populations continue to grow at a rapid pace, but we need to be aware that smaller industrialized nations, even though their populations may not be increasing so rapidly, have the greater per capita effect owing to their higher standard of living and powerful technology. Nations that have large populations as well as high technology, such as the United States, Japan, and the combined European community, have a huge effect on the environment.

Understanding the causes of increase and learning how to calculate rates of increase are two steps for understanding the increase in population. People who study human population have defined a number of terms and concepts that help us to achieve this understanding. These include

- Age structure
- The demographic transition
- Total fertility
- Relationships between the human population and the environment
- Factors that increase the death rate, including why death rates increase with crowding and why death rates will increase if we do not decrease birth rates
- How a higher rates of living correlates with decreased birth rates, death rates and growth rates.

The study of these and the related concepts is called human demography. Demography means the study of populations. Demographers count a population and attempt to project changes into the future. Growth of human population can be viewed in four major periods

- An early period of hunters and gatherers, when total human population was probably less than a few million.

345

- A second period beginning with the rise of agriculture, which allowed a much greater density of people and the first major increase in the human population.
- The industrial revolution, with improvements in health care and the supply of food leading to a rapid increase in the human population.
- The present situation where the rate of population growth has slowed in a wealthy, industrialized nations, but population continues to increase rapidly in poorer and less developed nations.

Pollution is usually defined as something that people produce in large enough quantities that it interferes with our health or well being. Two primary factors that contribute to the damage done by pollution are the size of the human population and the development of technology that invents new form of pollution. When the human population was small and people lived in a primitive manner, the wastes produced were biological and so dilute that they often did not constitute a pollution problem. People used what was naturally available and did not manufacture many products. Humans like any other animals fit into the natural ecosystems. Their wastes products was biodegradable materials. Pollution began when human population became so concentrated that their waste materials could not be broken down as fast as they were produced. As the population increased, people began to congregate and establish cities. The release of large amounts of smoke and other forms of waste into the air caused an unhealthy condition because the pollutants were released faster than they could be absorbed by the atmosphere.

In general we rely on science and technology to improve the quality of life. However technological progress often offers a short term solution to a specific problem but in the process can create an additional pollution problem. The development of steam engine allowed for machines to replace human labor, but increased the amount of smoke and other pollutants in the air. The modern chemical industry has produced many extremely valuable synthetic materials but has also produced toxic pollutants.

4 MINERAL RESOURCES

Not only are the mineral resources unevenly distributed, but those that are easiest to use and are the least costly to extract have been exploited. Therefore if we continue to use mineral resources, they will be harder to find and more costly to develop. As with energy, North America is one of the primary consumers of the worlds mineral resources. Reasonable estimates are that North America consumes over 30% of the minerals produced in the world each year, which is a disproportionate share, given that the combined population of the united states and Canada is about 5.2% of the worlds population.

Costs are always associated with the exploitation of any natural resource. These costs fall into three different categories. First, the economic costs are those monetary costs necessary to exploit the resource. Money is needed to lease or buy land, build equipment, pay for labor and buy the energy necessary to run the equipment.

A second category is the energy cost of exploiting the resource. It takes energy to extract, concentrate, and transport mineral materials to manufacturing sites. Since energy costs money, energy costs are automatically converted to economic costs.

A third way to look at costs is in terms of environmental effects. Air pollution, water pollution, animal extinction, and loss of scenic quality are all environmental costs of resource exploitation. Environmental costs are often deferred costs. They may not even be recognized as costs at first but become unimportant after several years. Environmental costs are also often represented by lost opportunities or lost values because the resource could not be used for another purpose. Environmental costs are also being converted to economic costs as more strict controls on the pollution of the environment are enacted and enforced. It takes money to clean up polluted water and air, or to reclaim land that has been removed from biological production by mining.

These three categories of costs (economic, energy and environmental) are associated with the several steps that lead from the mineral resource in its undisturbed state to the manufacture of a finished products. These steps are exploration, mining, refining, transportation and manufacturing. In addition some areas, such as national parks and reserves, have been off-limits to mineral exploration. As current reserves of mineral resources are used up, pressures will build to explore in these protected areas, resulting in increased environmental costs. Once a mineral resource has been located and the decision made to exploit it, the resource must be taken from the earth. Mining involves large expenditure of money to pay for labor and the construction of machine and equipment's. Energy must be purchased for operation. In addition to the economic and energy costs are significant environmental costs. Mining affects the environment in several ways. All mining operations involve the separation of the valuable mineral from the surrounding rock. The surrounding rock must be disposed of in some way. These pieces of rock are usually piled on the surface of the earth, where they are known as mini tailings and present an eyesore. It is also difficult to get vegetation to grow on these deposits.

Many types of mining operations require vast quantities of water for the extraction process. The quality of this water is degraded, so it is unsuitable for drinking, irrigation or recreation. Since mining disturbs the natural vegetation in an area, water may carry soil particles into streams and cause erosion and siltation. Some mining operations such as strip mining, rearrange the top layers of the soil, which lessens or eliminates its productivity for a long period of time. Strip mining has disturbed approximately 75,000 square kilometers of U.S. land, an area equivalent to the state of Maine. After mining, extracting materials from ores also involves environmental costs in the form of air and water pollution. These environmental costs are being converted to economic costs as regulations on industry require less environmental damage than had been previously tolerated.

Many areas of the world, such as tropical rain forests, currently support ecosystems that are little affected by humans but are under threat because they are capable of supporting agriculture or other human uses. The primary factor that will determine the survival of these natural areas is population pressure. As the human population grows, it will need more food and more space. This ultimately leads to the destruction of natural ecosystems and their replacement by human modified ecosystems. Wilderness Act of 1964, defined wilderness as "an area where the earth and its community of life are untrampled by man, where man himself is a visitor who does not remain". Recently there has been intense pressure from within the U.S. federal government to allow oil exploration activities to be conducted in areas currently designated as wilderness, such as the Arctic National Wildlife Refuge. As population increases and resources become more scarce, this pressure will become greater throughout the world.

5 FUTURE POPULATION TRENDS

The world bank, an international organization that makes loans and provides technical assistance to developing countries, has made a series of projections based on current birth rates and death rates and assumptions about how these will change. Their critical assumptions are
- Mortality will fall everywhere and level off when female life expectancy reaches 82 years.
- Fertility will reach replacement levels everywhere between 2005 and 2060.
- There will be no major world wide catastrophe.

Even assuming a rapid achievement of replacement fertility, this approach projects an equilibrium worldwide population of 10.1 to 12.5 billion. Developed countries would only increase from 1.3 billion today to 1.9 billion, but developing countries would increase from 4.1 billion to 9.6 billion. Bangladesh would reach 257 million; Nigeria, 453 million; India, 1.86 billion. In these projections, the developing countries contribute 95% of the increase.

6 HOW CAN WE STOP POPULATION GROWTH

The simplest and one of the most effective means of slowing population growth is to delay the age of first childbearing by women. As more women enter the work force and as education level and standards of living increases, this delay tends to occur naturally. Social pressures that lead to deferred marriages and childbearing can be very effective.

Countries with high growth rates have early marriages. In south Asia and Africa south of the Sahara, about 50% of the women marry between the ages of 15 and 19. In Bangladesh women marry at age 16 on the average, whereas in Srilanka the average age of marriage is 25. The world bank estimates that if Bangladesh adopted Srilanka's marriage pattern, families could average 2.2 fewer children. Increases in the marriage age could account for 40% to 50% of the drop in fertility required to achieve zero population growth for many countries. Age at marriage has increased in some countries, especially in Asia. For example, in Korea the average marriage went from 17 in 1925 to 24 in 1975. China passed laws fixing minimum marriage ages, first at 18 for women and 20 for men in 1950, then at 20 for women and 22 for men in 1980. Between 1972 and 1985, China's birth rate dropped from 32 to 18 per thousand people, and the average fertility rate went from 5.7 to 2.1 children. China's leaders have the goal of reaching zero population growth in the year 2000, at a level of 1.2 billion people.

Birth Control In Developing Countries

Another simple means of decreasing birth rates is breast-feeding, which can delay resumption of ovulation. This is used consciously as a birth control method by a women in a number of countries; in the mid 1970's the practice of breast feeding provided more protection against conception in developing countries than did family planning programs, according to the world bank.

Much emphasis is placed on the need for family planning. Traditional methods range from abstinence to induction of sterility with natural agents. Modern methods include the pill, which prevents ovulation through control of hormone levels; surgical techniques for permanent sterility; and mechanical devices. Although now safe medically in most of the cases, abortion is one of the most controversial methods from a morale perspective. Ironically, it is also one of the most important birth control methods in terms of its effects on birth rates.

National Programmes To Reduce Birth Rates

Reduction in birth rates requires a change in attitude, a knowledge of the means to control birth, and the ability to afford these means. Change in attitude can arise simply with an increase in the standard of living. In many countries, however it has been necessary to provide formal family planning programmes to explain the problems arising from rapid population growth and the benefits to individuals of reduced population growth and to provide information about birth control methods, as well as to provide access to these methods. The choice of population control methods is an issue that involves social, moral, and religious beliefs, which vary from country to country. It is difficult to generalize about what approach to use throughout the world.

In 1974, the World Population Conference in Bucharest, Romania, was attended by representatives of 136 countries. They approved a world population plan that recognized that individuals have the right to decide freely the number and spacing of their children and to have access to information telling how to achieve their goals.

The first country to adopt an official population policy was India, whose program was initiated in 1952. But few developing countries had official family planning programmes before 1965. From 1965 to 1975 there was widespread introduction of such plans, and by 1976 only Burma, North Korea, and Peru did not provide some support for family planning. Although most countries now have some kind of program, the effectiveness varies greatly. International aid for family planning reached $260 million by 1977.

A wid evariety of approaches have been used, from simply providing more information, to promoting and providing some of the means of birth control, to offering rewards and extending penalities. Penalties are usually in the form of taxes. For example, Ghana, Malaysia, Pakistan, Singapore and the Philippines have used a combination of methods including limits on several benefits such a tax allowances for children, maternity benefits, and tax deductions after one or two children. Tanzania has taken another approach, restricting paid maternity leave for women to a frequency of once in three years. Singapore does not take family size into account in allocating government built housing, so larger families are more crowded. As an example of the use of rewards, Singapore increased the priority of school admission to children from smaller families.

China has one of the oldest and most efective family planning programs. In 1978 China adopted an official policy to reduce the countries human population growth from 1.2% in that year to zero by the year 2000. An emphasis was placed on single children families. The government uses education, a network of family planning that provides information and means of birth control, and a system of rewards and penalities. Women are given paid leave for abortions and for sterilization operations. Although there are benefits to families with a single child, including financial subsidies in some areas, in some parts of China families that have a second child most return the bonuses received for the first. Other rewards and penalities vary from province to province.

Other countries including Bangladesh, India, and Srilanka have paid people who have voluntarily been steralized. In Sri lanka, this practice has applied only to families with two children, and a voluntary statement of consent must be signed.

7 CONCLUSIONS

Population growth is a fundamental determinant of rising demand for natural resources. In country like India, further population growth increases pressure on the resources required for subsistence. Land for producing crops and grazing animals, forests required for fuel wood and other products and clean water for drinking immediately come to mind. As India's large population continues to grow, there can be little doubt that the sheer number of people will further strain the supply of resources, requiring careful management to keep stocks and flows in balance with demand. In many places in India, where mining activity is intense, the temperature in those places are high because of large deforestation operations which are carried out because of mining.

Apart from the effect of population on the ecosystem, most of the problems being faced by the developing countries like employment, education, poverty, crime and others have the root cause of population. Therefore, the government in these countries should give top priority to the population problem and take appropriate steps to bring down the population growth to a minimum.

REFERENCES

1. Kent, M.M. and Crews, K.A. (1990). World Population: Fundamentals of Growth. Washington, D.C.: Population Reference Bureau.
2. Britannica book of the year, 1999. Encyclopaedia Britannica, Inc.

Environmental and Ecosystem Related Controls and Consequent Inordinate Delay in River Valley Projects in India

R.N. Padhi[1], S.K. Sarangi[2], I.B. Chhibber[3] and M.N. Rahangdale[4]

[1]*Former Dy. D.G. G.S.I., Chairman, CREST (N.G.O. Nagpur Chapter)*
C/24 Manav Sewa Nagar, Seminary Hills, Nagpur 440 006, India
[2]*Managaing Director, Geomin Consultants (Pvt.) Ltd., President, CREST (All India)*
301, Kharvela Nagar, Bhubaneswar 751 001, India
[3]*Former Director G.S.I., Executive Member, CREST (Nagpur Chapter)*
178, Museum Road, Civil Lines, Nagpur 440 001, India
[4]*Geologist, Geomin Consultants (Pvt.) Ltd. Member, CREST (Nagpur Chapter)*
Plot No. 23, Shree Nagar, Nagpur 22, India

INTRODUCTION: -

India is a large country and has a one billion population. Major river valley projects located in the Himalayas are snowfed and comprise Ganga, Jamuna, Sindhu, Bramhaputra basins and their tributaries. In Central and South India major river valley projects rainfed are Narmada, Tapti, Godavari, Krishna, Kaveri and Mahanadi and their major tributaries. These river valley systems have been partially developed by constructing multipurpose dams, but there is scope for further development. India has a vast potential for hydel power generation, to the tunc of 84,000 M.W. at 60% load factor, out of this less than 25% has been utilized. It is seen that USA, Russia and Scandivian countries have exploited the hydro potential upto 95% to 98%. The Government of India has drawn up a hydroelectric development policy. Therefore, large dam at suitable sites, run of the river schemes as well as development of small, mini and macro hydel project (0.5 M.W—15 M.W.) are planned for execution by the Central, State Governments and Private Enterprise and Corporate houses in collaboration with Foreign companies. The major dams constructed between 1950 and 1980 in the country did not involve the environmentalist and their activities. This lead to their fast development, utilization and management of irrigation potential, huge agricultural production, fisheries and water supply for big industries. This gave rise to "Green revolution", "White revolution" and industrial development like major steel and fertilizer plants etc. These completed projects had not created any serious environmental problems. Prior to it India was importing wheat under PL 480 from USA as well as other countries like Russia which has been stopped after India became self sufficient in foodgrains. By a suitable balance between developmental activity and environmental protection major multipurpose projects Tehri, and Sardar Sarovar, can be completed fast. At present these are opposed by the so-called environmentalists. No objection was raised against any project by them between 1950-1980, when major river valley dam projects were constructed and completed. This lobby, appears now to be supported by vested interest of some parties, to hamper the progress and the development of the country.

MAJOR AND LARGE DAMS: -

During 1950-1980 major river valley projects were completed and the benefits shared by the people of India. Some of the major dams have described below in order to understand their impact for the faster, reliable and developments in irrigation, Hydro Power and agriculture and fishery, water supply and forest protection as well as in the improvement of quality of the flora and fauna.

1. BHAKRA DAM: - It is 226 m high and 518 m long straight gravity dam across Satlej. The power generation is 1204 M. W. It has storage capacity of 4621 m cum. Forest area submerged 5746 ha. The irrigation potential, which benefited states of Panjab, Harayana & Rajasthan, is 23,72,000 ha. Gross area submerged is 16,600 ha. Construction was started in 1954 completed in 1964 at costs of Rs.1031.77 million for irrigation, Rs. 724.23 million for left bank power house and Rs. 593.20 million for right bank power house.

2.   IDDUKKI H.E. PROJECT: - It is 168.91 m high and 365.58 m long. First arch dam constructed in India, it ranks among 40 high dams in the world. It has two straight gravity 135.3 m high and 650.9 m long dams and Saddle dam 100 m high and 385.06 m long. Power generation 780 MW with a head of 669.48 m. Construction commenced 1963 and completed 1974.

Srisailam Multipurpose project :- Across Krishna River, its height is 143.3 m and length is 512.1 m Power generation 770 M.W. (7 units of 110 M.W.) work started in 1963 & was completed in 1974.

Nagarjuna Sagar Multipurpose project :- Height is 125 m Masonary, length is 1450 m and earthern portion 3414.6 m. Gross storage 11,558.7 m cum. Date of start 1956 and date of completion 1964, power generation installed capacity of 110 M. W. Irrigation Benefits 0.83 m ha. Cost of the project was Rs. 1645 million.

HIRAKUND DAM :- Height 59 m and it is the longest (26 km) earthern dam with dykes in India. The gross submerged area is 72700 ha. Area benefited for Irrigation is 2,44,430 ha. Forest area submerged is small. It is an irrigation, power and flood control project Installed capacity is 270 M. W. The total River channel at the dam site is 5 km of Mahanadi River.

Other major projects constructed and completed during above period are viz. DVC Multipurpose in Bengal & Bihar, Matatila Dam Multipurpose projects. Ramganga Multipurpose UP, Koyna Hydel project, Pench Multipurpose project, Maharashtra, Ukai and Kadana in Gujrat, Ghandhisagar Multipurpose dam, Tawa Multipurpose dam, Bargi   Multipurpose dam Madhaya Pradesh.

There are more than 5000 - 6000 medium and minor projects completed in the country for the irrigation and water supply purposes, and agricultural and proper land use management.

The projects completed brought to the country an over all progress in irrigation, hydropower development and agriculture production and industrial plants have made India a fast developing country.

Hydroelectric power projects are almost pollution free. The hydel projects promote better environments for the people, aquatic life, flora and fauna for its growth and development of the area. These projects are eco-friendly and no adverse environmental effects noticed. Some of the run-of-river schemes suggest that hydel power plays an important role to boost the economic conditions of the people living in the area by providing electricity to the villages , towns and cities. Some of the hydel projects for power generation are :- (1) Garhwal Rishikesh - Chilla 140   M. W. (2) Baira suil Hydel 180 M. W. (3) Chemera I hydel 540 mw have since been completed. (4) Napth Jhakri Hydel - 1500 mw (5) Chemera II Hydel 500 M.W. are in advance stage of construction. There are many hydel projects, which generate power in almost all the major states of India. There is a possibility of large dams for development and production of hydel power in the country. It needs the involvement of the different state and CEA and power ministry to execute these projects for the benefit of country and thereby have surplus power generation for the industrial development, energising water pumps for agriculture and for the welfare of poor people living below the poverty line.

Large Dams Versus Environmentalists :- There was a strong lobby after 1980 onwards to oppose, hamper the working and thus delay the major projects viz. Tehri Multipurpose and Sardar Sarovar project, Vishnu Prayag H. E. Project,   Silent valley projects, Bhopal patnam and Inchampalli on Indravati and Godavari in Central India. Silent valley H. E. Project was called off in Nov. 1983. It was a collosal loss of power generation and revenue loss for the people of Kerala where there is a severe shortage of power. The submerged area was only one tenth of the total valley area. It would have generated 120 M.W. Vishnuprayag is run of river scheme, located a few km. downstream of Badrinath temple, across Alakananda River and Diversion barrage across Puspavati with a head of 943 m. The power house is underground on the other side of Joshimath, Power generation is of the order 480 M. W. Being underground project it is not likely to affect the environment of the area.

Narmada valley development of water resources in Madhaya Pradesh envisages the construction of 29 major dams out of which 19 irrigation, 7 hydel and 3 multipurpose, 450 medium and over 5000 minor schemes are in different stages of planning and construction.

Resettlement of Displaced Persons of Major Projects :- There is no rehabilitation policy worth the name. Maharashtra is the only state in India which has legislation to recognise the rights of people displaced by irrigation, power and other public utility projects to get land in the area benefited by the projects. The Maharashtra resettlement of project   displaced persons Act 1976. It was a direct outcome of two decades of sustained agitation by the people affected by building of dams. The people are not against the construction of dams but they oppose the discrimination in the distribution of the benefits.

In all the major, medium dams the main problem is resettlement of the displaced persons. It should neither be considered as an obstacle to nor an extra financial burden on the project. There is need of the involvement of the people likely to be displaced and to provide facilities for their suitable rehabilitation. The resettlement is a very delicate issue the project authorities, Government Department should handle it with care with proper support to the uprooted people to supplement their income. The total cost of the multipurpose dams can be recovered in a few years.

Concluding Remarks

1. River systems are complex and each one may have its own characteristics, which influence the stability in an ecosystem. Large dams in the country are must for irrigation, Hydel power generation, high yield of agriculture production, Industrial development viz. Steel, Textile, Sugar, Mining, etc. It is said "Electricity is the back bone of the world economy. Barometer of economic prosperity growth is consumption of power."

2. A transparent policy for the displaced persons should be evolved for all states, to expedite the construction work of large dams by specific time frame. Rules can be changed and modified for creating a healthily atmosphere.

3. Interlinking the Central and South Indian rivers, should be studied and implementation after smoothening obstructions faced at different stages. This is for getting more water from Inter-basin transfer, and co-ordination of the state involved to expedite these projects.

4. Conducting environmental impact studies within set time limits are important for guidance and implementation and for not delaying the projects indefinitely. Geo environmentalists should be accountable in the execution of the major river development projects.

5. Water-shed management studies will augment water, agriculture forest growth and its development along with its environmental management in order to improve the life of poor people, which constitute 80% of Indian population. Studies should be carried out as done by C. G. W. B. for Maharashtra state for artificial recharge. Similarly the areas for artificial recharge and water harvesting methods should be identified in the other states, so that water, a scare commodity, can be saved and used in the service of the people in the new Millennium.

References :-

1) Centre for Science & environment :- The state of India's environment 1984 - 85. The second citizen's report.

2) C.G.W. B. - 2000 Master Plan for Artificial recharge to ground water in Maharashtra.

3) Journal of Engineering Geology Proceeding of ISEG symposium on Environmental Management in Relation to water Resources Development and major constructions (27-29 act 1987, Lucknow) Sept. 1987.

4) G. S. I. Miscellaneous Publication No. 29 (Part 1) 1975.

5) Geotechnical Features of Major Dams in INDIA ISEG Publication (1982).

Environmental Impacts of the Steel Industry and Their Solutions

B.V. Petkar

Deputy General Manager (Retd.), Mishra Dhatu Nigam, Hyderabad. 88, Vidya Vihar, Ranapratap Nagar, Nagpur 440 022, India

INTRODUCTION

The Indus valley civilisation in the third millenium BC reveals that the people inhabiting those areas had a knowledge of metallurgy proved by the presence of copper tools and weapons with tin and arsenic as alloying elements. The Iron pillar near Kutub Minar at Delhi, of a much later period also bears testimony to the developments in the field of Iron and Steel in India. During the 19th century considerable progress took place in Europe and U.S.A. in the development of steel making technology and they became the leading producers of steel in the world. In comparison practically no development took place in India, and it had to wait till the dawn of the second decade of the 20th century to see the establishment of the first integrated iron and steel plant.

STEEL INDUSTRY IN INDIA - AN OVERVIEW DURING THE LAST NINE DECADES

One of the parameters for measuring the degree of development of any country is the per capita consumption of steel in it. It is therefore worthwhile casting a glance over the years since the first integrated iron and steel plant was established. The Tata Iron and Steel Co. Ltd. at Jamshedpur - the first integrated steel plant in India - commenced production in the year 1912 and had an initial capacity of 0.1 million tons of steel. At the time of independence the country was producing about a million tons of steel per annum with a per capita consumption of 3kg. During the four decades, after the commencement of steel production in the country, the growth of steel industry was at a snail's pace. However after independence, during the second five year plan, with the establishment of three new integrated steel plants at Rourkela, Bhilai and Durgapur and doubling the capacity of the Tata Iron and Steel Co. Ltd. at Jamshedpur steel production got a big boost. Since then, till now, new units were established in the various parts of the country and in 1998, India consumed about 23 million tonnes of steel with a per capita consumption of 26 kg. The present capacity to produce has been variously estimated at 25 - 30 million tonnes.

AWARENESS ABOUT ENVIRONMENTAL IMPACT.

In the early years, throughout the world, very little attention was paid to the impact of environmental pollution in general and due to the steel industry in particular. However during the last 30 years increasing global attention has been focussed on the environmental pollution accentuated by large scale industrialisation. The steel industry also had its share in causing environmental degradation. In India also, as the pace of industrialisation accelerated and steel production increased, it became increasingly necessary to save environment. Though the problems in India are not as serious as other highly industrialised countries, it has taken up the challenge to fight pollution alongwith other nations of the world.

STEEL MAKING ROUTES – THEIR IMPACT ON POLLUTION

The major portion of steel in India is produced in the large integrated iron and steel plants. During recent years a number of units producing steel through direct reduction of iron ore followed by melting the sponge in the Electric Arc Furnace, have been established in the various parts of the country. Their adverse impact on environment however is much less compared to large integrated plants due to much lower tonnage produced, and the type of process employed by them. The environmental aspects of the integrated steel plants only are therefore considered in this paper.

VARIOUS UNITS IN INTEGRATED PLANTS AND THEIR CONTRIBUTION TO POLLUTION

An integrated steel plant under Indian conditions normally has the following facilities.

1. Coke ovens to produce metallurgical coke for smelting the iron bearing raw materials in the blast furnace.

2. Sinter plant for sintering the iron ore.

3. Blast Furnace for producing hot metal.

4. Steel making units – L.D. Converters or Open hearth furnaces.

5. Rolling Mills to shape the steel in various required shapes.

The following table gives in general environmental pollution caused by the above facilities.

Unit		Environmental Pollutants.
1. Coke ovens and By – Product plant.	:	Emission of gaseous oxides of Sulphur and Nitrogen. Coal particle emission from oven doors, lids, ascension pipe caps. Liquid effluents containing toxic pollutants like ammonia, phenol and cyanide.
2. Sinter Plant	:	Emission of gaseous oxides of Sulphur and Nitrogen. Dust emission.
3. Blast Furnace and Gas Cleaning Plant	:	Solid waste materials Slag. Flue dust from dust catchers, Gas Cleaning Plant sludge Gaseous fumes during tapping of the hot metal.
4. Steel Making Open hearth process	:	High particle emission from furnaces Emission of oxides of Sulphur, Carbon in flue gases, Open hearth slag.
B.O.F. Converter process	:	Emissions of gases and particles from the vessels. B.O.F. slag.
5. Rolling Mills	:	Emission of oxides of Sulphur and Carbon in flue gases. Fumes generated in the work place by heating / burning of lubricants etc. Splattering of mill scale during the process of rolling.

Apart from the pollution caused due to various pollutants given above, water pollution in steel plants is not to be forgotten. The high temperature metallurgical process operations in steel production require large amount of water for cooling, gas cleaning etc. resulting in polluting the water used for the purpose.

The operation of a steel plant, thus gives rise to water and air pollution and generates lot of solid waste material. On an average the water consumption rates in Indian steel plants vary from $10 - 30$ m^3 per tonne of saleable steel while the solid waste generated is approximately $1000 - 1500$ kg for every tonne of crude steel. Strange as it may sound, approximately 2.5 tonnes of carbon dioxide is released to the atmosphere for every tonne of steel produced apart from the oxides of Sulphur and Nitrogen. The total impact of these conditions on the neighborhood of steel plants and on the health of the people inhabiting those areas can be well imagined, if determined efforts are not made to improve the environmental conditions.

STRATEGY FOR POLLUTION CONTROL

At the outset it must be clearly understood that inspite of best efforts pollution cannot be eliminated altogether. In such a situation the next best alternative is to minimise and control it within such limits that its effect is within the accepted norms of friendly environment. To achieve this objective, appropriate strategies have to be developed and followed. The first step in this direction is to analyse the causes leading to pollution and then devise suitable means which could control it. The next step is the proper utilisation of materials polluting the environment generated in the steel industry, to the extent possible, thereby preventing their uncontrolled spread in the neighborhood which can cause serious environmental degradation.

The main raw materials used in the operation of an integrated steel plant are iron ore, coal and limestone / dolomite. All these materials contain impurities in the as mined condition and also have varying physical characteristics. It is axiomatic that lower the impurities in the raw materials lesser would be the waste products resulting from the operation. The reduction in the consumption of raw materials would also lower the generation of the total quantity of waste products. The physical condition of the raw materials also has a great impact on the overall pollution primarily on account of suspended particle matter, the friable and powdery form leading to higher suspended particle matter compared to the lumpy or solid state.

From the foregoing it would follow that the strategy for pollution control would involve :

(i)   Selection and use of proper quality of raw materials with regard to chemistry and physical condition. In case proper quality is not available in the as mined condition, these will have to be beneficiated to bring them to the required quality.

(ii)   Use of modern technology to reduce the quantity of pollutants resulting from operations. It is worth mentioning here that technologies which generate low pollution also maximise productivity and improve quality.

(iii)   Use of pollution control equipment.

(iv)   Utilisation of the waste products and by - products to the extent possible to prevent their spread in the neighborhood and thus minimise their harmful influence.

QUALITY CONTROL OF RAW MATERIALS FOR POLLUTION CONTROL

Iron Ore:

An iron ore quarry will neither have iron ore of uniform chemical composition nor of uniform physical characteristics throughout the extent of the deposit. Some portions will have richer grades compared to others. Likewise some areas may have hard lumpy mass of ore while other areas may have soft, friable, powdery type. From the point of view of conservation of resources and avoiding generation of large quantity of waste in mines, it would be unwise to utilise only the best quality with respect to chemistry and physical characteristics, leaving the poorer qualities as a mining waste. The proper strategy to be followed would be to make a judicious blend of the various fractions to arrive at an optimum quality in respect of chemistry as well as the physical condition. In certain cases, methods of beneficiation like washing, magnetic separation for magnetic ores may have to be adopted to enrich them. Further, sintering / pelletising may have to be done to make the fines, as a suitable feed for the blast furnaces. The proper preconditioning of the iron ore results in the smoother blast furnace operation resulting in better productivity and lower pollution.

Coal:

The blast furnace productivity and the hot metal quality depend to a very large extent on the quality of coke used as a raw material. Lower the coke ash, lower would be the slag volume and better would be the blast furnace productivity. Unfortunately in India, the reserves of coking coal with less than 16 – 17% ash in coal are now almost on the point of exhaustion and consequently coals having higher percentages of ash have to be employed for coke making, after washing the run of mine coal to bring down their ash levels to less that 17%. The middlings and high ash rejects can be utilised for power generation by combustion in specially designed boilers. Of late, coals containing low ash are being imported from Australia and China.

Limestone / Dolomite:

Limestone and dolomite are mainly used as fluxing materials. These should necessarily contain low insolubles to have good 'available base', to reduce the slag volume. The fines can be used for sintering. The requirement of limestone / dolomite is much less compared to iron ore and coal.

MODERN TECHNOLOGY AND POLLUTION CONTROL

Till the eighties the Indian steel industry was characterised by high pollutant emissions in the environment. However in the nineties the situation changed in most of the integrated steel plants with cleaner technologies and employment of the state of art pollution control equipment. It need not be emphasized again, that the use of technology maximising productivity and improving quality, generates lower pollutants.

Some broad aspects of technology upgradation in respect of the major polluters in the integrated steel plants are given below:

Coke Ovens:

Coke ovens are considered as one of the major polluters in the integrated steel plants. Newer technologies like stamp charged coke oven batteries and Non recovery/ Heat recovery type of ovens reduce pollution to a considerable extent. While the former are being used in India, Germany. France, Poland and Romania the latter are in operation in U.S.A. only in one or two plants, on account of economic and some technological constraints.

As of today, the conventional top charged ovens are the major producers of coke for the steel plants. Improvements like self sealing doors, door cleaning machines, use of coke transfer cars with waste gas scrubbing facility, dry quenching of coke also reduce pollution considerably.

The effluents resulting from coke ovens and by - product plant contain ammonical liquor, phenol, cyanide etc. The ammonical liquor can be converted into useful fertiliser - Ammonium Sulphate and the residual liquid effluents can be treated in a biological oxidation plant which reduces the toxic pollutants to a very large extent

Sinter Plant:

Sinter which in the past was primarily intended to utilise the iron ore fines for rendering them suitable for blast furnace feed has now replaced the raw iron ore as feed material universally. Apart from utilising the iron ore, limestone and dolomite fines, a part of the B.O.F. converter slag is also used for making self-fluxed sinter, considered as an eminently suitable raw material for smooth blast furnace operation. Most of the plants these days have a raw material yard, where blending of materials is carried out to produce high quality sinter. This has not only the advantage of utilising the iron ore, limestone or dolomite fines, as also a part of B.O.F. converter slag, which would otherwise have been waste materials, but also brings about consistency in quality of raw materials, ultimately resulting in lesser environmental pollution.

Blast Furnace:

The blast furnaces continue to be the major producing units of liquid iron. Some of the important developments which increase productivity and decrease coke rate leading to decrease in pollution are given below:

(i)   Use of movable throat armour for uniform burden distribution.

(ii)   Use of belt conveyor for charging the furnace.

(iii) Use of humidified blast.

(iv)   Use of oxygen enriched blast

(v)   Hydrocarbon injection in the furnace.

The decrease in pollution due to above is on account of increasing the productivity, decreasing slag generation and lessening of the water usage.

Steel Making:

The basic open hearth process of steel making which reigned supreme in forties and fifties has now been almost completely replaced by the B.O.F. converters. Apart from other metallurgical advantages of B.O.F. process, it has the particular advantage of reducing the particulate emission compared to basic open hearth steel making. In the Indian context the B.O.F. steel making has a very important role to play in view of the shortage of reusable scrap in the country.

Pollution Control Equipment in Iron & Steel Industry

Examples of some pollution control equipment used in iron and steel industry are given below:

1.	Coke Ovens	: Bag filters in coal / coke transport. Self sealing doors, water sealed A.P. caps Biological effluent treatment plant for effluents.
2.	Sinter Plant	: Electrostatic precipitators and bag filters.
3.	Blast Furnace	: Bag filters in stock house & gas cleaning plants Electrostatic precipitators in gas cleaning plants.
4.	Steel Melting Shops	: Venturi scrubbers for process emissions Bag filters and electrostatic precipitators for fume and dust extraction.
5.	Rolling Mills	: Filters for water filtration and sludge removal. Oil skimmers for oil recovery. Ventilation systems.

Disposal and use of waste products:

By adopting modern technology and the use of pollution control equipment, significant reduction in atmospheric pollution due to gaseous emission is achieved.

The water consumption in the steel plants can be brought down by recycling the treated reclaimed water. This greatly reduces the quantity of highly polluted waste water needing disposal, thereby greatly reducing the aqueous pollution.

The problem of solid wastes viz. blast furnace, open hearth and B.O.F. slags as also the gas cleaning plant flue dust and sludge, limestone and dolomite fines has received considerable attention over the past many years. Blast furnace slag is being increasingly used for cement manufacture after granulation, apart from uses as rail road ballast, in construction industry and slag wool manufacture. The flue dust, gas cleaning plant sludge and mill scale can be used in sinter plants.

while the open hearth slag finds use in road making, preparation of stock yards, parade grounds etc. The B.O.F. slag is employed for recycling at blast furnace and sinter plant to replace lump limestone and for soil conditioning. Limestone and dolomite fines are recycled at sinter plants. In Indian conditions the recycling and reuse rates of solid waste resulting in steel plants varies from 30-45% only, compared to 90-95% in steel plants in developed countries.

The scenario of the solid waste generation and utilisation in some of the leading steel making countries vis-à-vis India is given in the following table.

Waste type	Country							
	GERMANY		JAPAN		U.S.A.		INDIA	
	Gene-ration	Utilisa-tion %	Gene-ration	Utilisa-tion %	Gene-ration	Utilisa-tion %	Gene-ration	Utilisa-tion %
BF slag kg/thm	255	100	284-342	92-100	175-290	100	340-421	30-50
BF dust sludge, kg/thm	1.5	100	9-27	100	17-40	20-40	28	0
L.D. slag kg/tcs	103	96	96-155	90-100	70-170	60-100	200	25
L.D. sludge, kg/tcs	18	100	15-55	100	15-40	30	15	0
Mill scale, kg/tss	17.5	100	35	100	6-20	100	22	100
Mill sludge, kg/tss	3.5	100	2-10	100	1-10	0	12	10

Note : thm – tonne of hot metal; tcs – tonne of crude steel; tss – tonne of saleable steel
Source : IISI

The Indian industries have since realised that the proper solid waste management can lead to improvement in profitability and environmental protection and steps are being taken in that direction.

With the measures outlined above effective pollution control can be brought about while operating the steel plants, maintaining environmental friendly conditions in the vicinity.

CONCLUSION:

Even though many new materials have been developed to replace steel for its various applications, it will continue to dominate the world scene in the foreseeable future. The adverse environmental impact of the steel industry can be greatly controlled by minimising the generation of the waste products and their proper disposal and management. With increasing attention being paid to environmental aspects, it is hoped that the integrated steel plants in future would no more be considered as a threat to environment.

REFERENCES:

Books

1. Vats M.S. The Cultural Heritage of India Volume I, The Ramakrishna Mission Institute of Culture, Calcutta. Publisher 118 pp.

2.   The Making Shaping and Treating of Steel (1985) 10th edition United States Steel
        publisher edited by William T. Lankford Jr., Norman L. Samways. Robert F. Craven. Harold
        E. Mc Gannon (Editor Emeritus)
        Dry quenching of coke 169, 170 pp.
        Stamp Charging 187, 188 pp.
        Biological oxidation 349 pp.
        Fluxes in Iron and Steel Making   325, 326 pp.
        Slags for Non metallurgical uses 334 pp.
        The manufacture of pig iron in the blast furnace 554 – 556 pp.

Journals

1.   S.M.R. Prasad , S.S. Gupta, R.P. Sharma and A.S. Dhillon (1997)
        Environment friendly production of iron and steel – the case of Tata Steel.
        Tata Search 1997 113 – 117

2.   G.S. Basu, P.K. Sarkar, R.P. Sharma, A. Ahmad and A.S. Dhillon (1997).
3.   Recycling and reuse of solid waste at Tata Steel.
        Tata Search 1997 118 – 120

4.   Amit Chatterjee and Kamal Simlai (1997).
        Present coke scenario and technology selection for future coke making in India.
        Tata Tech 28 41 – 43

5.   B.D. Pandey, U.S. Yadav and U.K. Jha (1998)
        Innovative blast furnace iron making technology global vis – a – vis Tata Steel's Scenario
        Tata Tech 29 17 – 20.

Reports, Bulletins and other irregular publications.

1.   I.I.M. Metal News Volume 2 No.5 October, 1999: 1-3
        Dr. J.J. Irani (1999) Luncheon Address by Dr. J.J. Irani, M.D. Tata Steel at the the
        International Conference on " Steel Survival Strategies – XIV " at New York on 22 June 1999.

2.   I.I.M. Metal News Volume 3 No.2 April, 2000: 40
        H.R. Murthy A.G.M. (Environment), Bhilai Steel Plant: The talk on Innovative Waste
        Management in the Bhilai Steel Plant on 11th January, 2000.

Regenerating a Clean and Green Environment in Dimensional Limestone Mining: A Case Study

S.C. Agarwal

Vice-President (Mines), Associated Stone Industries, (Kotah) Ltd. Ramganjmandi, India

INTRODUCTION

Extractive industries like mining, are viewed as ecology pariahs that spoils nature and feed over consumption of finite resource. Indeed, the result of past practice, often insensitive to the environment contributes to mining poor public image in much of the developing countries. With the global awareness about the environmental degradation, there is increasing pressure on the mining industry to restore the ravage done by the open cast mining. The so-called "green society" will play a major role in the next decade. Recycling, conservation & preservation will be a way of life.

KOTAH STONE

The state of the Rajasthan in India is the biggest producer of dimensional stone e.g. marble, granite, slates, sandstone and limestone widely known as Kotah stone. Kotah stone is an excellent flooring stone. It is naturally riven flooring stone. Mining for Kotah Stone has started some 70 years back with a production of few thousand square feet & multiplied to present level of production is 250 million Sq.ft. Mining of Kotah Stone over the years has been all manual & selective. Only those layers, which could be splitted upto a maximum of 3"thickness and could be cut to size by chisel and hammer, have been mined. The rest have been thrown as waste.

ENVIRONMENTAL PROBLEMS

The mineral recovery has never been more than 25% The huge quantity of waste has generated which either has been back filled in exhausted area or dumped and stock piled elsewhere. Man made mountain could be seen all around the mining belt. This symbolizes the reckless mining practice degrading aesthetic values of the area and creating many environmental problems. The land and precious soil has been depleted very fast creating severe scarcity of drinking water. There has been general increase in ambient temperature. Solid waste has been the biggest polluter.

With the global awareness of environmental protection and financial constraints due to increasing cost of waste handling it became very essential to upgrade the quarrying technology so as to improve mineral recovery, reduce generation of waste and to reclaim the degraded land and rehabilitate the waste dumps by dense afforestation to regenerate a green environment.

RESTORATION OF ENVIRONMENT

To restore a clean & green environment post mining, several steps have been taken. An environmental policy has been drafted. It has included 7 commandments, for rigorous implementations & regeneration of clean & green environment.

PRINCIPAL COMMANDMENT

The quarrying technology should be upgraded which is sustainable to environment, society and future generation.

Procedures Adopted

A concept was evolved by the Author of cutting/sizing the Kotah Stone layers insitu at quarry floor prior to separating/splitting along cleavage plane. For this purpose portable electric driven machine was designed fitted with circular diamond tipped steel blade in varying dia of 24"/30"/36". Machine with 36'dia blades proved to be most productive, cutting upto 12'' depth in 2 passes. The layers were cut to in the required size of 2' x 2' to as much as 10' x 6' in – situ.

The concept proved to be revolutionary as it helped in improving recovery in reduction of waste & proved to be highly productive, cost effective, much safer as it reduced undue human fatigue. Water used as cooling & flushing agent has created a very cool & comfortable working environment.

SECOND COMMANDMENT

The mine waste shall not be disposed off on any fresh land either within lease boundaries or outside. This will be filled back in the voids created after final extraction of Kotah Stone.

PRACTICE & PROCEDURES ADOPTED

Opening out land for open pit extraction of Kotah Stone can not be avoided. It has been made mandatory that all waste shall be filled back in excavated area only. The amount of Solid waste has grown improportionately to the amount of Kotah Stone extracted & used and has exceeded the sink capacity. Instead of dumping such extra waste on fresh land, the dumps height has been raised as much as 20-25 mtr above G.L. to accommodate the waste. However with the introduction of upgraded mining technology since 1993, the waste generation has reduced appreciably. It is expected the future waste dumps will be only a few mtr above G.L.

To ensure slope stability benches of 8-10 mtr have been formed. After reaching the final dump heights, the dumps are levelled to a width of 30-50 mtr and are kept ready for capping with soil over long benches 500-600 mtr

THIRD COMMANDMENT

The black topsoil from the advancing face of quarry is to be preserved & conserved.

Practice & Procedures Adopted:

After the dumps have been levelled, we first excavate subsoil from the advancing face & spread over the top of the bench to a height of about 0.5 mtr. This subsoil when levelled fills back the voids and minimise water percolation. It also provides smooth surface for the vehicles to run while bringing soil.

It is then fresh top soil is excavated and transferred directly on the dumps above subsoil. The soil is first spread on the slopes covering it fully and thereon the top of dumps, where we maintain a soil cover of about 1 – 1.5 mtr. using dozer/spreader. Neither the soil nor sub-soil is stock piled & instead it is transferred directly to waste dumps after excavation. This avoids any bacterial degradation or loss between stock piling & transfer.

FOURTH COMMANDMENT

The land which has been excavated open for extraction of Kotah Stone shall be restored/reclaimed either for original use or for any social use.

Practice & Procedures adopted.

Because of improportionate generation of waste stock piled in form of man made mountain., the land cannot be restored to original level and original use. Instead the dump are being rehabilitated with dense afforestation on the slopes as well on the dump top.

1. **Along the dump slopes :** To avoid Soil erosion from the slopes, afforestation along slopes had been planned. For the purpose contour benches at 1M interval & 30/0.5M. wide had been prepared. Plants of Species like, Subabul, Juliflora, Babool , Jetrofa & Anwal had been planted. These species being very hardy requiring less of water & giving wide green outlook, had been preferred. Growth of these species along the slopes has been fairly good. However no weeding or hoeing or watering was possible on the slopes. The toe of each bench has been protected with a stone wall, which additionally added to the beauty of the bench.

2. **On the dump top :** The species of plantation in this area were not selected biologically but ornamental trees have been grown to give aesthetic beauty. However shadow trees of Neem, Shisham, Fruit Plants, Bud, Pipal, Creepers, Gulmohar, Eucalyptus, have dominated. Spiral road with ascending steps and creepers on both sides adds to the beauty of area. Some 6000 plants have grown in full trees over the years besides other growing plants on an area known as Shiva Udyan.

3. The Shiva Udyan extends over a stretch of over 25090 ft and has been tastefully developed into a beautiful picnic spot. Two worship temples have been constructed to attract general mass to the cause of environment along with religion. Rest places, cooking space, eating places etc. have also been constructed in dry masonry with hand dressed Kotah Stone waste pieces.

An adjoining part of a quarry has been temporarily abandoned and used as storage pit for water. Booster pumps are used to supply the water through a network of pipeline.

Mass Plantation
In continuation of the area, yet another 15 hector of waste dumps have been rehabilitated with dense plantation. Rows of different species have been planted leaving approach roads only. These included:

a) Shisham   b) Neem   c) Juliflora   d) Eucalyptus   e) Su-babul

d) Cassia   g) Bamboo   h) Ratanjot (Jetrop) I) Gulmohar   j) Amaltas

Watering & maintenance
The plants grown in this area mainly depended on rain water and atmospheric moisture. However an attempt has been made to carry water by tanker during first year. It is observed that Sheesham plants have recorded the best growth, followed by Eucalyptus & Juliflura. A survival of 67% has been recorded.

About 75,000 plants have been grown in this reclaimed portion of waste dumps. Mining & reclamation of the mine site is an ongoing process, which goes on simultaneously.

FIFTH COMMANDMENT
Steps shall be taken to ensure effective Management of Virgin land in lease area to permit ground water in filtration during rainy season and develop a green belt, till it is required for extraction of Kotah Stone.

Practice & Procedure Adopted.
Mining lease for extraction of mineral / Kotah Stone is normally granted over a large land area to last 20 to 100 years. Entire lease area, is never opened out at any one time. As such a large area, to a fair estimate 2000 hectors of good agriculture land, has been abandoned and is now void of any vegetation. This has reduced the water percolation during rainy season and major part of rainwater runs off as surface water. Vegetation accelerates infiltration and retention of water due to greater porosity. Decision has therefore been taken to raise a green belt by dense plantation on such barren land, within lease boundaries, leaving area required in next 10 years for mining operation. The land has been densely afforested with permitted species. These include Prosopis, Juliflora, Israile Babool, Eucalyptus, Su-babool & Ardu. This is to avoid any legal restrictions on felling the trees under Indian Forest Act.

Good growth with over 73% survival has been recorded. Mine water has been diverted to nurse these plants. A deep trench has been excavated all around to avoid cattle grazing. A dense fence of Su-babool has come up further along these safety trenches. Such afforestation has retarded the speed of surface run off and expected to increase the water infiltration during rainy season.

SIXTH COMMANDMENT
Water is life. It should not be wasted. Various techniques should be adopted for effective management of rain water in quarries & for restoration of water regime.

Practice & Procedures Adopted
Extensive mining over seven decades has created severe scarcity of drinking water. It is a sorry state of affairs that inspite of *34" average rainfall, water has become a scarce commodity specially in months from January to next rainfall each year. Excavation in contiguous to a depth of over 30m below ground level has disturbed the water regime.

Indiscriminate dewatering of deep quarries soon after rains, leads to syphoning of ground water from adjoining well & soil resulting in drying of wells. Fast depletion of soil along with degradation of soil by waste along with degradation of soil dumping has hampered water percolation, causing scarcity of ground water. It has become obligatory to conserve Soil & Water. Techniques adopted and procedure followed included the followings :

Conservation of soil : Soil & sub-soil helps in water in – filtration in ground. It is felt much essential to

365

minimise the rate of depletion of soil. For this it is made mandatory not to dump waste on fresh land, reduce rate of land excavation by upgrading quarrying technology and reduce/retard surface run off. Also all soil & subsoil excavated from the advancing mine face should be utilized for afforesting waste dumps.

Conservation of mine water : To conserve the good rain water it has been decided to retard rate of dewatering equivalent to rate of mining progress instead of enmass bottoming the quarry. The process of dewatering has been spread over 9 months instead of 3 – 4 months earlier.

This has helped in reduced syphoning of water from wells & fractures, making available mine water for drinking & industrial use even in peak summer days and making use of water for vegetation & forestation in mining area.

SEVENTH COMMANDMENT
Each & every one associated with Kotah Stone Mining & processing directly or indirectly must be made aware of the necessity of environmental protection. It should be underlined that polluter has to pay.

Practice & Procedures Adopted
Environment has been included in the Curriculum of Initial training & Refresher training compulsory for each worker. He is briefly told as how environment is getting effected at his working place & surrounding, how it can be taken care of & what is his role in creating a better environment.

CONCLUSION
The code of practices & procedures adopted in regenerating clean & green environment demonstrate that, contrary to previous belief, economic growth can go hand-in-hand with environmental responsibilities that works. Self imposed regulation more than Government imposed rules have to be observed. Without retarding utilisation of natural resources to meet social need environmental issues have to be resolved through re-engineering the process and without deferring commitment towards environmental protections.

Natural Stone Sector leaders are now putting greater efforts into restoration to meet the increasing demand & restrictions placed by society, N.G.O. groups and Environmental legislation instead of fighting and defending the industry. The urgency of conserving natural resources & management of waste can not be overemphasized to regenerate a clean & green environment.

ACKNOWLEDGEMENT
The author wishes to convey gratitude to the Management of A.S.I.(K) Ltd. For giving consent to present this paper. The views expressed are of author and not necessarily of the organisation he is serving.

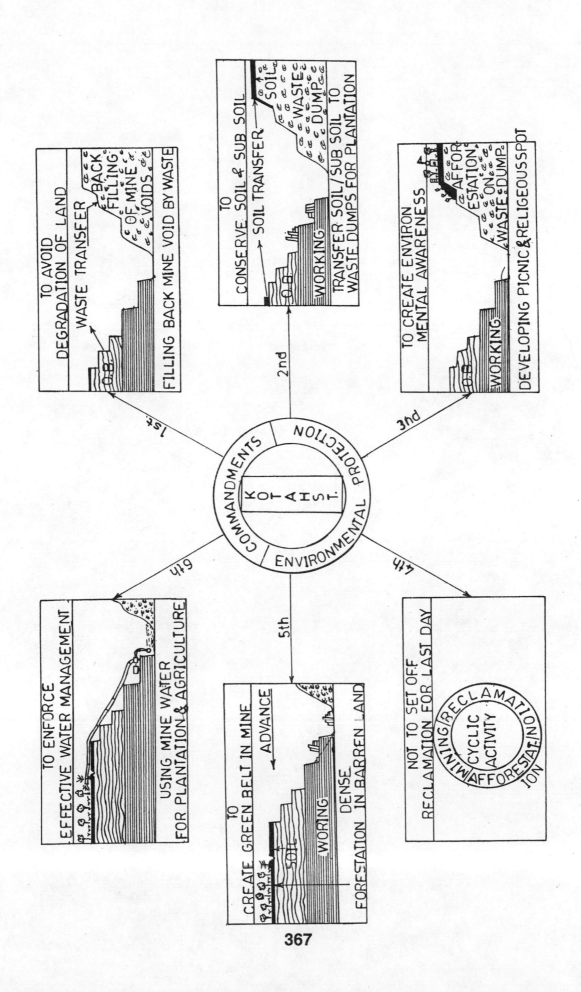

367

Dynamic Model for Environment Restoration Measure Cost in Lignite Mines

Rajesh Arora and Sunil Nakra

Surat Lignite Power Project, Gujarat Industries Power Company Limited, Nani-Naroli, Surat 394 110, Gujarat, India

INTRODUCTION

In recent years environment concerns have seriously threatened the very existence of mining industry. Environmental concerns have assumed a great deal of significance at the same time ecological awareness and sensitivity has thrown up challenges that has spawned a new commitment.

This paper outlines the influence of mining activities on the environment, including the socio-economic aspects of the region as well as to find out a reasonable cost of mitigation to ensure that the damage caused to the environment and society is contained within acceptable limits. In this regard a dynamic economic model has been developed, which describes the exploitation of lignite and its interaction with rest of the economy including the environmental consequences. This model has been validated for environment protection in a lignite mining area in Western India.

COST – THE MAIN ISSUE IN ENVIRONMENT PROTECTION

Estimates of the costs as a proportion of total production costs of course vary from negligible to crippling. Those who are trying to demonstrate that the costs are low point to examples of mines, where investments attributable for environmental reasons have led to higher productivity and lower costs. While those who aim to show that the costs threaten the industry's existence quote the full costs of investments, including elements that have little if anything to do with environmental requirements. It is easy to find examples supporting either version of the facts, partly because it is very difficult to identify the exact portion of the investment that is related to environmental objectives and partly because circumstances differ from one operation to another depending on geology, climate and other factors.

DYNAMIC MODEL FOR ENVIRONMENT PROTECTION IN LIGNITE MINES

In the light of what has been acknowledged in foregoing an attempt has been made to build a dynamic model to address the issue rationally. The dynamic economic model is in principle the most formal since the forecast is based on an explicit mathematical model. The model states in detail and in quantitative terms the way in which the various aspects of the economy are interrelated.

The dynamic economic model provides the forecaster with a record of the prediction with a clear statement of the assumptions concerning exogenous variables and the model solution.

The key issues involved in development of a Dynamic Model for Environment Protection in Lignite Mines are as follows:

- Is there any relation between stripping ratio, depth, and cost of production and cost of environment protection measures?
- If relation exist, what is the relation between these variables?

To determine the financial justification of environment protection, Dynamic model for environmental measures is developed which is shown in Exhibit 1.0. The model is based on input-output flow diagram. The design parameters considered are:

- Cost parameters (Fixed and Variable)
- Environmental parameters (Noise level, Air quality, water quality, etc)

- Socio-Economic factors (Health, Population-Migration, Employment, Literacy etc.)

Based on these factors and the policy of the local, state and central governments, a model is conceived, initially a schematic line diagram is made on which a suitable mathematical model is built.

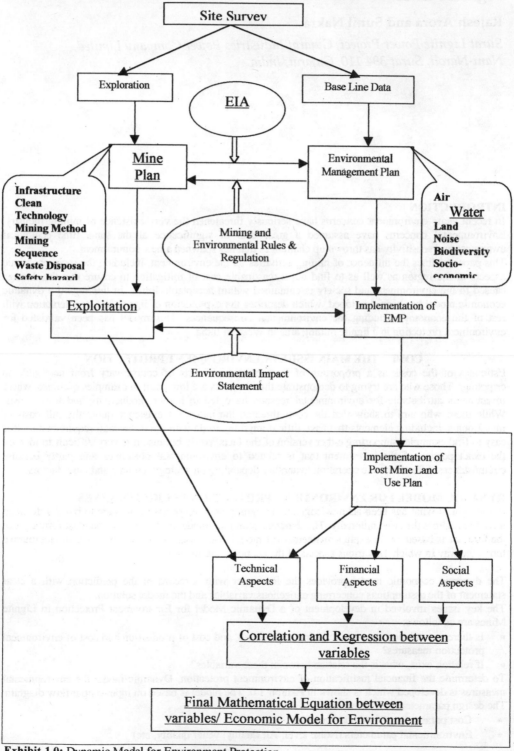

Exhibit 1.0: Dynamic Model for Environment Protection

To frame a dynamic model for environment protection measure during implementation of EMP, correlation and regression analysis has been carried out.

CORRELATION ANALYSIS

Correlation analysis attempts to determine the 'degree of relationship' between variables. In this analysis variables are cost of environment protection measures, stripping ratio, depth, amount of excavation and cost of production.

Regression Analysis

The objective of regression analysis is to study the nature of relationship between the variables so that we may be able to predict the values of one on the basis of another. The closer the relationship between the variables, the greater the confidence that may be placed in the estimates. In regression analysis one variable is taken as dependent while other as independent, thus making it possible to study the cause and effect relationship.

The following relationship is derived from correlation and regression analysis:
(1) Relationship between Cost of Environment Protection Measures v/s Stripping ratio
(2) Relationship between Cost of Environment Protection Measure v/s Cost of Production at 40, 80 and 110 meter depths.
Out of these relationships Dynamic Model for Environment Protection Measure in Lignite Mines in Western India is derived.

Table 1.0 Depth in meter v/s Cost of production in Rs./Cum & Rs./Tonne

Depth	Cost of Production	
	Rs/Cum.	Rs./Tonne
40	43.47	41
80	49.47	48
110	58.47	57

Tables 2.0 Stripping ratio (Cum/Tonne) v/s Depth in meter v/s Cost of production (Rs./Tonne) of lignite

Stripping ratio	Depth		
	40	80	110
12	562.64	641.64	758.64
11	519.17	592.17	700.17
10	475.7	542.7	641.7
9	432.23	493.23	583.23
8	388.76	443.76	524.76
7	345.29	394.29	466.29
6	301.82	344.82	407.82
5	258.35	295.35	349.35
4	214.88	245.88	290.88
3	171.41	196.41	232.41
2	127.94	146.94	173.94
1	84.47	97.47	115.47

Table 3.0 Stripping ratio v/s Environment Protection Cost in Rs./Annum

Environment Cost in Rs./Annum	Stripping Ratio				
	12	8	6	4	2
Green Belt	19.2	19.2	19.2	19.2	19.2
Community Farming	4.42	4.42	4.42	4.42	4.42
Dump Plantation	2.84	1.96	1.5	1.0	0.65
Monitoring	2.3	2.3	2.3	2.3	2.3
Desilting	18.67	18.67	18.67	18.67	18.67
Dust Suppression System	28.4	19.7	15.3	10.9	6.5
	75.83	66.25	61.39	56.49	51.74

371

Assumption

1. Annual Lignite production is same for all the stripping ratio and depths.
2. Lease area or mine property size and shape remains unchanged for all the stripping ratio and depths.
3. Stripping ratio changes with the size of lignite deposit, seams thickness, strike, dip, geology etc.
4. Environment Protection Measure Cost is function of Stripping ratio, depth and cost of production for lignite mines in India. In the calculation of cost of production at various stripping ratios and at various depths, the depth factor is taken into account.

Exhibit A.0 Correlation and Regression of Environment Protection Measure in Rs. per Tonne of Lignite and Stripping Ratio(X) in cum/Tonne

No	X	dx (x-7)	dx^2	Y	dy(y-4.904)	dy^2	dx dy
1.	12	+5	25	5.833	0.929	0.8630	4.315
2.	10	+3	9	5.460	0.556	0.3090	0.927
3.	8	+1	1	5.096	0.192	0.0368	0.192
4.	6	-1	1	4.711	-0.193	0.0372	0.193
5.	4·	-3	9	4.345	-0.559	0.3124	1.677
6	2	-5	25	3.980	-0.924	0.8537	4.620
	Σx=42	Σdx=0	Σdx^2=70	Σy=29.425	Σdy=0	Σdy^2=2.41217	$\Sigma dxdy$=11.924

$$Y-\bar{Y} = \frac{\gamma\sigma_Y}{\sigma_X}(X-\bar{X})$$

OR,

$$\frac{\gamma\sigma_Y}{\sigma_X} = \frac{\Sigma dxdy - \frac{\Sigma dx \times \Sigma dy}{N}}{\Sigma dx^2 - \frac{(\Sigma dx)^2}{N}} = \frac{11.924}{70} = 0.17034$$

From Exhibit A.0 put the respective values

Y - 4.904 = 0.17034 (X - 7)

OR, $\boxed{Y = 0.17034X + 3.7116}$ ------------------------------------ (I)

Where Y = Environment Protection Measure in Rs. per Tonne of Lignite
and X = Stripping Ratio in cum/tonne

Coefficient of Correlation

$$\gamma = \frac{\Sigma dxdy}{(\Sigma dx^2 \times \Sigma dy^2)^{1/2}}$$

From Exhibit A.0 put the respective values

$\boxed{\gamma = +0.917}$

The coefficient of correlation measures the degree of relationship between two sets of variables.

In this study $\gamma = +0.917$, it reveals the positive relationship between the variables, cost of Environment protection Measures and Stripping Ratio.

Exhibit B.0 Correlation and Regression of cost of Environment Protection Measure in Rs. per Tonne of Lignite and Production Cost (X) in Rs/Tonne at 40 mts depth

S No	x	dx (x-345.29)	dx^2	Y	dy (y-4.904)	dy^2	dx dy
1.	562.64	217.35	47241.02	5.833	0.929	0.8630	201.48
2.	475.70	130.41	17006.76	5.460	0.556	0.3090	72.24
3.	388.76	43.47	1889.64	5.096	0.192	0.0368	8.25
4.	301.82	-43.47	1889.64	4.711	-0.193	0.0372	7.99
5.	214.8?	-130.41	17006.76	4.345	-0.559	0.3124	73.16
6	127.94	-217.35	47241.02	3.980	-0.924	0.8537	201.26
	Σx=2071.74	Σdx=0	Σdx^2=132274.84	Σy=29.425	Σdy=0	Σdy^2=2.41217	$\Sigma dxdy$=564.38

372

$$Y-Y = \frac{\gamma\sigma_y}{\sigma_X}(X-X)$$

OR, $\frac{\gamma\sigma_y}{\sigma_X} = \frac{\Sigma dxdy - \frac{\Sigma dx \, X \, dy}{N}}{\Sigma dx^2 - \frac{(\Sigma dx)^2}{N}} = \frac{564.38}{132774.84} = 0.00425$

From Exhibit B .0 put the respective values

$Y- 4.904 = 0.00425 (X - 345.29)$

Or, $\boxed{Y = 0.00425X +3.4385}$ ------------------------------------(II)

Where

Y = Cost of Environment Protection Measure in Rs. per Tonne of Lignite

and   X = Cost of Production at 40 meter depth in Rs./Tonne

Coefficient of Correlation

$\gamma = \frac{\Sigma dxdy}{(\Sigma dx^2 \, x \, \Sigma dy^2)^{1/2}}$

From Exhibit B .0 put the respective values

$\boxed{\gamma = +0.99}$

The coefficient of correlation measures the degree of relationship between two sets of variables.

In this study $\gamma = +0.99$, it reveals the perfect positive relationship between the variables, Environment protection Measure and Cost of Production at 40-meter depth.

From equation (I) and (II)

$\boxed{Y = 0.17034X + 3.7116}$ ----------------------- (I)

Here X=X1=Stripping ratio

$\boxed{Y = 0.00425X +3.4385}$ ------------------- (II)

Here X=X2 = Cost of Production

and Y= Cost of Environment Protection Measure in Rs. per Tonne of Lignite

$\boxed{Y_{40} = 0.08517X1 +0.002125 \, X2 + 3.575}$ ----------------------- (III)

Where X1 = Stripping ratio for 40 meter depth

and   X2 = Cost of Production for 40 meter depth

Table 4.0 Cost of Environment Protection at 40-meter depth & various Stripping Ratio

$$Y_{40} = 0.08517X1 +0.002125 \, X2 + 3.575$$

S. No.	Stripping Ratio in cum/Tonne	Cost of Production	Environment Protection Measure in Rs./Tonne of lignite
1.	12	562.64	5.79
2.	10	475.70	5.43
3.	8	388.78	5.08
4.	6	301.82	4.72
5.	4	214.88	4.37
6.	2	127.94	4.01

Exhibit C.0 Correlation and Regression of cost of Environment Protection Measure in Rs. per Tonne of Lignite and Production Cost(X) in Rs/Tonne at 80 mts depth

S No	X	dx (x-394.29)	dx^2	Y	dy (y-4.904)	dy^2	dx dy
1.	641.64	247.35	61182.02	5.833	0.929	0.8630	229.29
2.	542.70	148.41	22025.52	5.460	0.556	0.3090	82.219
3.	443.76	49.47	2447.28	5.096	0.192	0.0368	9.399
4.	344.82	-49.47	2447.28	4.711	-0.193	0.0372	9.102
5.	245.88	-148.41	22025.52	4.345	-0.559	0.3124	832.58
6	146.94	-247.35	61182.02	3.980	-0.924	0.8537	229.046

$\Sigma x=2365.74$ $\Sigma dx=0$ $\Sigma dx^2=171309.64$ $\Sigma y=29.425$ $\Sigma dy=0$ $\Sigma dy^2=2.41217$
$\Sigma dxdy=642.314$

$$Y-Y = \frac{\gamma \sigma_y}{\sigma_X} (X-X)$$

OR, $\dfrac{\gamma \sigma_y}{\sigma_X} = \dfrac{\Sigma dxdy - \dfrac{\Sigma dx \, X \, dy}{N}}{\Sigma dx^2 - \dfrac{(\Sigma dx)^2}{N}} = \dfrac{641.314}{171309.64} = 0.0037494$

From Exhibit C.0 put the respective values
$Y- 4.904 = 0.00374 (X – 394.29)$

OR, $\boxed{Y = 0.00374X + 3.4313}$ ---------------------------------------(IV)
Where,

Y = Cost of Environment Protection Measures in Rs. per Tonne of Lignite
and   X = Cost of Production in Rs./Tonne at 80 meter depth.
Coefficient of Correlation

$\gamma = \dfrac{\Sigma dxdy}{(\Sigma dx^2 \, x \, \Sigma dy^2)^{1/2}}$

From Exhibit C.0 put the respective values
$\boxed{\gamma = +0.99}$

The coefficient of correlation measures the degree of relationship between two sets of variables.

In this study $\gamma = +1.0$, it reveals the perfect positive relationship between the variables, Cost of Environment protection Measure in Rs. per Tonne of Lignite and Cost of Production in Rs./Tonne at 80 meter depth.

From equation (I) and (IV)

$\boxed{Y = 0.17034X + 3.7116}$ -- (I)

Here X=X1=Stripping ratio

$\boxed{Y =0.00374X + 3.4313}$--------------------------------- (IV)

Here X=X2 = Cost of Production
and Y= Cost of Environment Protection Measure in Rs. per Tonne of Lignite

$\boxed{Y_{80} = 0.08517X1 + 0.00187X2 + 3.57145}$ ------------------------------- (V)

Where X1 = Stripping ratio for 80 meter depth
and   X2 = Cost of Production for 80 meter depth

Table 5.0 Cost of Environment Protection at 80-meter depth & various Stripping Ratios

$$Y_{80} = 0.08517X_1 + 0.00187X_2 + 3.57145$$

S. No.	Stripping Ratio in cum/Tonne	Cost of Production in Rs./Tonne of Lignite	Environment Protection Measure in Rs./Tonne of lignite
1.	12	641.64	5.79
2.	10	542.70	5.43
3.	8	443.76	5.08
4.	6	344.82	4.72
5.	4	245.88	4.37
6.	2	146.94	4.01

Exhibit D.0 Correlation and Regression of cost of Environment Protection Measure in Rs. per Tonne of Lignite and Production Cost(X) in Rs/Tonne at 110 mts depth

S No	X	dx (x-466.3)	dx^2	Y	dy (y-4.904)	dy^2	dx dy
1.	758.64	292.35	85468.52	5.833	0.929	0.8630	271.008
2.	641.70	175.41	30768.66	5.460	0.556	0.3090	97.177
3.	524.76	58.47	3418.74	5.096	0.192	0.0368	11.093
4.	407.82	-58.47	3418.74	4.711	-0.193	0.0372	10.758
5.	290.88	-175.41	30768.66	4.345	-0.559	0.3124	98.405
6	173.94	-292.35	85468.52	3.980	-0.924	0.8537	270.716

$\Sigma x = 2797.74$ $\Sigma dx = 0$ $\Sigma dx^2 = 239311.84$ $\Sigma y = 29.425$ $\Sigma dy = 0$ $\Sigma dy^2 = 2.41217$ $\Sigma dxdy = 759.157$

$$Y - \overline{Y} = \frac{\gamma \sigma_y}{\sigma_x} (X - \overline{X})$$

OR,

$$\frac{\gamma \sigma_y}{\sigma_x} = \frac{\Sigma dxdy - \frac{\Sigma dx \times \Sigma dy}{N}}{\Sigma dx^2 - \frac{(\Sigma dx)^2}{N}} = \frac{759.157}{239311.84} = 0.0031722$$

From Exhibit D.0 put the respective values

Y- 4.904 = 0.0031722 (X – 466.29)

Or, $\boxed{Y = 0.00317X + 3.4278}$ ---------------------------------------(VI)

Where,

Y = Cost of Environment Protection Measures in Rs. per Tonne of Lignite

and   X = Cost of Production in Rs./Tonne at 110 meter depth.

Coefficient of Correlation

$$\gamma = \frac{\Sigma xy}{(\Sigma x^2 \times \Sigma y^2)^{1/2}}$$

From Exhibit D.0 put the respective values

$\boxed{\gamma = +1.0}$

The coefficient of correlation measures the degree of relationship between two sets of variables.

In this study $\gamma = +1.0$, it reveals the perfect positive relationship between the variables, Environment protection Measure in Rs. per Tonne of Lignite and Cost of Production in Rs./Tonne at 110 meter depth.

From equation (I) and (VI)

$\boxed{Y = 0.17034X + 3.7116}$ -------------------------------------- (I)

Here X=X1=Stripping ratio

$\boxed{Y = 0.00317X + 3.4278}$ -------------------------------- (VI)

375

Here X=X2 = Cost of Production
and Y= Cost of Environment Protection Measure in Rs. per Tonne of Lignite

$$Y_{110} = 0.08517X1 + 0.001585X2 + 3.5697$$ -------------------------------- (VII)

Where X1 = Stripping ratio for 110 meter depth
and   X2 = Cost of Production for 110 meter depth

Table 6.0 Cost of Environment Protection at 110-meter depth & various Stripping Ratios
$$Y_{110} = 0.08517X1 + 0.001585X2 + 3.5697$$

S. No.	Stripping Ratio in cum/Tonne	Cost of Production in Rs./Tonne of Lignite	Environment Protection Measure in Rs./Tonne of lignite
1.	12	758.64	5.79
2.	10	641.70	5.43
3.	8	524.78	5.08
4.	6	407.82	4.72
5.	4	290.88	4.37
6.	2	173.94	4.01

CONCLUSION

- The principles of sustainable mine development in Lignite Mines in Western India is based on scientific mine planning are followed for the protection of environment giving due consideration to the socio-economical aspects.

- The implemented dynamic model for environmental protection measure in Lignite mines in Western India derived as follow:

$$Y_{40} = 0.08517X1 + 0.002125 X2 + 3.575$$

Where
Y_{40} = Cost of Environment Protection Measure in Rs. per Tonne of Lignite,
X1 = Stripping ratio upto 40 meter depth
X2 = Cost of production of lignite upto 40 meter depth

$$Y_{80} = 0.08517X1 + 0.00187X2 + 3.57145$$
Y_{80} = Cost of Environment Protection Measure in Rs. per Tonne of Lignite,
X1 = Stripping ratio upto 80 meter depth
X2 = Cost of production of lignite upto 80 meter depth

$$Y_{110} = 0.08517X1 + 0.001585X2 + 3.5697$$
Y_{110} = Cost of Environment Protection Measure in Rs. per Tonne of Lignite,
X1 = Stripping ratio upto 110 meter depth
X2 = Cost of production of lignite upto 110 meter depth

- The efficacy of Dynamic model for Environment Protection Measure is realized in Lignite Mines in Western India, where it forms the basis for future course of action that shows optimizing return at the minimum expense of inputs. The model based on various design parameters is benchmark for budgeting in Environment Protection Measure.
- The significance of Dynamic Model for Environment Protection cannot be over-emphasized. Irrespective of uncertainties that may exist or arises model is worthwhile and advantageous to all future management action. Above all this model is introduce in the management and planning methods that integrate environmental impacts in the project planning process from the outset of a project and that do not see environmental mitigation as something to be added as an afterthought once plan has been executed.
- As per the recommendation of Environment monitoring agency, the annual recurring environmental control cost for adoption of various control measures will be 1.5 to 2.0 % of production cost of lignite. This is controversial as we look into the Table 4.0, 5.0 and 6.0 that the cost of production increases as depth increases but environment protection measure cost remains unchanged.

LIMITATION
1. Since mining is site specific, it may not be applicable to other mining site.
2. Whole life of the mining project has to be taken into account.
3. The model should built on appropriate scenario based on various simulation models with the help of known computer model.
4. Sufficient data should be acquired so that correct relationship can be established.

REFERENCE
1. Dhar, B.B. (1999) Mining Environment Scenario Beyond 2001(1999); Proceeding of International Conference on Mining Challenges.
2. Dhar, B.B. (1996): Environmental Management System for Closure and best practice in Indian Mining Industry: Sixth Workshop of the Mining and Environmental Research Network 4-9 August, Harare.
3. Dube, A.K. (1999): Life Cycle Assessment Analysis for Indian Mining Industry _ A Clean Environment Approach: International Conference on Mining Challenges of the 21st Century during 25-27 November 1999 at Hotel Taj Palace, Delhi.
4. Ghosh, R. & A. Rani (19990 A step toward Eco-friendly coal mining in India, In ; proceeding of the International Symposium on Clean Coal Initiatives, N. Delhi, pp 587-592
5. Hendnickson, C., Horvath, A. Joshi, S & Lave, L. 1998. Economic Input-Output models for Environmental Life-Cycle Assessment, Environmental Science & technology, April 1, p p 184A.
6. Editorial – Global economy and the Indian minerals sector Journal of Mines, Metals & Fuels, June-July 1998, p213.

Population Explosion Leading to Serious Water Scarcity

R.S. Varshney

Secretary General Honoraire, International Commision on Irrigation and Drainage Formerly Engineer-In-Chief & Head Irrigation Deptt., U.P., India

ABSTRACT:

God has been kind to India in bestowing sufficient water supply. He gave us enough water resources, sufficient to have decent quality of life. Though this precious resource was concentrated only in four monsoon months in a year and too unevenly distributed in the vast span of the country, nevertheless by proper storages, proper distribution and careful management, there was sufficient to satisfy drinking, irrigation and other development needs.

However, because of explosive population growth and haphazard and ill planned urbanization and industrialization, this seemingly sufficient but 'limited' water resource is having a rapid fall in the per capita availability of water. From a fairly comfortable level of water availability per person (3450 m3 in 1951) in the early stages of development, a significant steep slide has occurred: 1076 m3 in the year 2000, which would reduce to about 700 m3 by 2050. Water availability in the year 2010 would touch the critical limit of 1000 m3 per person, which will start hampering health, economic development and human well being.

On the other hand there are increasing evidences of rise in pollution of both ground and surface waters. The irony is that an effort to transfer some water to less fortunate regions or to store water behind dams in reservoirs is opposed by vested interests in the name of environment.

Solutions are suggested which can lead to control of the situation to some extent. But if population increases in the same way as at present, it is doubtful if any remedy can cure the malady. Perhaps God will administer the remedy.

1.0 WATER AVAILABILITY

Of all the water on the earth (about 1400 x 1015 m3), 97 percent is contained in oceans and seas. Of the remaining 3 percent, only 0.03 percent is available in rivers to be used for mankind. The distribution of run-off (in cm per year) is variable throughout the world (R.S. Varshney Engineering Hydrology 1986). North and South America are more favoured with water (31.5 and 45 cm respectively) compared to other continents of the world. Most of the arid areas lie in Asia (17 cm), Africa (20.5 cm), Australia (7.6 cm). Though Asia's water availability is quite low, India is very favourably situated with 1683 x 109m3 run-off of nearly 369 million ha, yielding an overall annual run-off of nearly 45.7 cm; about as much as the richest continent of South America. However, the water supply of India is not well distributed in space and time: concentrated in only 4 monsoon months in a year and having extreme variation in different regions. Although water is a renewable resource, it is also a finite one. The term 'mean supplies per person' is only a broad indicator of overall security. This indicator drops as population grows. As compared to 1974, the world's per capita supplies have come down by 33 percent, as 2 billion more people have been added to the population (world population 1804 1 billion; 1974 4 billion; 1999 6 billion; 2028-8 billion). If water availability becomes less, the health of the people and development activities suffer. According to standard definition, for water availability from 1000 m3/capita/year to 1700 m3/capita/year, water supply is a primary constraint to life.

According to the latest estimate by CWC (MOWR GOI Jan. 1999), the annual mean flow (water resources) has been assessed as 1953 billion m3 and utilizable water resources from conventional schemes is 690 billion m3, which is about 35% of the total surface water resources. The gross groundwater resource, as per assessment of CGWB, is 432 billion m3 whereas utilizable ground water resource is 396 billion m3 . Thus the assessed gross available and utilizable water resources of the country are 2385 and 1086 billion m3 respectively.

2.0 POPULATION PROJECTIONS
CWC (January 1999) have projected rural and urban population at different time horizons, on the basis of different forecasts made by experts. The projections (average of extremes given by CWC) for different time horizons are:

Time horizon	2000 A.D.	2010 A.D.	2025 A.D.	2050 A.D.
Population million	1009	1168	1363	1611

Based on 1991 population, the available and utilizable water resources per capita per year were 2830 m3 and 1288 m3, respectively. These figures will reduce to 2042 m3 and 930 m3 respectively in 2010. This shows that alarm bells will start ringing in the next decade.

These national level statistics mask large disparities between basin-to-basin and region-to-region. The utilizable water resources per capita per year vary between 3020 m3 in Narmada basin and about 180 m3 in Sabarmati basin and basins of inland drainage (basis 1991 population). Out of the twenty basins, four basins only had more than 1700 m3 water per capita per year, while nine basins had utilizable water resource between 1000 to 1700 m3, five basins had between 500 to 1000 and two basins had less than 500 m3 in the year 1991.

The above figures show that with the present trend of population growth, the water availability in the year 2010 will start hampering health, economic development and human well being. According to the standard definition of scarcity, five river basins are already facing water scarcity, while two basins are critically water scarce.

3.0 SOME STUDIES MADE BY OTHER AGENCIES
Almost 90 % of the growth in population in this world in this century will be in the water deficient regions in the developing countries.

3.1 IWMI STUDY
IWMI Sri Lanka have made a study of water scarcity (ICID- News Update March 1999) for 118 countries over the 1990-2025 period. According to them, 17 countries in the Middle East, South Africa, and dry regions of the Indian sub continent and north China may not have enough water to maintain 1990 levels of per capita food production from irrigated agriculture and meet industry, household and environmental needs. Twenty-four countries, mainly in the sub Saharan Africa are extremely water scarce. These countries will have to more than double their efforts to extract water to meet their requirements. The remaining countries of the world especially in North America and Europe will have less pressure on water supplies.

3.2 WORLDWATCH INSTITUTE REPORT (TOI 23.06.2000)
The report points out that by 2050, India will add 519 million people and China 211 million. Pakistan is projected to add nearly 200 million, while Egypt, Iran and Mexico are slated to increase their population by more than 50 percent.

The report estimates that 70 percent of water consumed globally including diversion from rivers and underground water, is used for irrigation, 20 percent goes to industry and 10 percent for residential purposes. This is why India uses 1000 t of water to produce 1t of wheat worth around $200. But the same amount of water can be used to raise industrial output to the value of $10,000.

Affluence in the country will generate additional demand for water. People will consume more foodstuffs, which means more grain and therefore more water.

As a result of water scarcity and population growth, North Africa and the Middle East have become the world's fastest growing grain import markets. It is said future wars will be fought over water. But given the difficulty of winning a water war, the competition for water seems more likely to take place in the world grain market. The countries that will win this war will be those that are financially strongest, not those that are militarily strongest. The report warns that unless the governments in water short countries act quickly to stabilize population and to raise water productivity, their water shortages may soon become food shortages.

4.0 PROBLEMS AND SOLUTIONS
4.1 EMERGING PROBLEMS
The following are the two main problems:
1: GROWING POPULATION AND WATER SCARCITY
The reasons for population growth in this country are well known to everyone. They are religious and political. The result is that the water availability will go on reducing till the population stabilizes. Experts feel that it should be around 2050 A.D. Hence the per capita availability, in terms of annual average utilizable water resources of India, which was 3450 m3 in the year 1991 is 1076 m3 in 2000 and shall reduce to danger level of 674 m3 in the year 2050. This means that the water availability in the year 2010 will start hampering human well-being and development.

Apart from availability wide inequity exists in the availability at various levels and regions.

2: WATER POLLUTION
There are increasing evidences of pollution of both ground and surface water. Massive increase in pollution of fresh surface water resource is expected due to higher fresh water withdrawn for urban domestic supplies between 2000 A.D. and 2050 A.D. After 2030 A.D. urban population will become more than the rural population in India.

The pollution, which normally is expected to increase in proportion to the increase in population, has increased many a times more because of haphazard and ill planned urbanization. All the recharge ponds are leveled for making colonies and urban infrastructure. Communication lines are laid after cutting drainage lines, thus creating environmental hazards and flood plains are encroached impunitively. Urbanization and that too ill planned adds to additional water demand.

4.2 SUGGESTIONS
Due to pressure from the population even in the world's largest river basin of Ganga/Brahmputra/Meghna dwindling water supplies and deteriorating water quality will threaten the lives of millions of Indians and Bangladeshis (World Water March/April 2000). The problem can be eased only with the co-operation of the people. Purely seminaring will not help, because such meetings have low participation of those who have worked in the field and thus they cannot fully visualize the problems. However the following methods if adopted with people's participation will certainly ease the situation.

1: CONSERVATION AND AUGMENTATION OF WATER RESOURCES
Utilizable quantum of water in space and time can be increased through increase in storage capacity to hold monsoon water. The total storage built up in various river basins through major and medium projects up to 1975 was about 174 billion m3. With the ongoing projects this can go up to 415 billion m3. But when it can materialize is difficult to say. There are various persons who are bent upon stopping construction of reservoirs. They will certainly help the environment, if they teach people not to pollute the water. Had they gone to Gujarat during this year drought, they would not think to oppose Sardar Sarovar and Tehri. Their genuine demands should be fulfilled, but by stopping projects they are doing the greatest harm to the country

2: INTER-BASIN TRANSFER OF WATER
A plea is made that interbasin transfer of water can solve the water scarcity to a great extent. Unluckily the political atmosphere in our country has become such that surplus states would not like to part with the water in their area and those who do not have that water will argue to have as much share as possible from the inter state river. If per chance agreement is reached between two or more governments, the next government first tries to undo the agreement. The problem can be sorted out if people on different reaches are taken into confidence and their views are heard and analyzed.

The Central Government for political reasons makes matters worse by declaring bifurcations and trifurcations of states. The states in upper reaches do not then wish to give even the legitimate demand of lower reach states. It is certain that with the creation of Uttrakhand UP will face music in the future.

The problem of the interbasin transfer of water can be solved by first taking up inter sub-basin links

(Varshney Jan- 1998). After creating goodwill and showing practical results of conservation and augmentation of water resources, people can be persuaded to agree on sharing of water.

3: REDUCING GAP BETWEEN DEMAND AND SUPPLY OF WATER

Irrigation utilizes maximum water. Water demand can certainly reduce if people are made to understand that giving more water to fields does not mean more grains. The problem is that our agencies that are supposed to advice the farmers by practical demonstration do not do much generally. There are a few WALMIs like that of Aurangabad, who have done good work and with their constant touch with the farmers have succeeded in making farmers to use less water to get optimum produce. Our farmers have generally small holdings and are poor. To expect them to switch over to drip and sprinkler irrigation is too much. Therefore a good management of present day farming's method can help in reducing water demand and thereby conserve water.

4. STOPPAGE OF WASTAGE OF WATER

It may look paradoxical to think that water can be wasted by people in our country and even in such areas where water is scarce. But this is a fact that people in posh colonies and affluent zones get more water and they waste also. Unless people in all areas get equal amount of water, whether enough or less, this practice will not stop. There are some states (like Delhi), which cry for more and more water (surely for their ever growing population) and force U.P. and Haryana to give more water even after cutting their irrigation needs, do not fulfil their obligation of treating the effluent, when out-letting the effluent in Yamuna. Only two percent of Yamuna passes in through the Delhi territory but the pollution contribution is 71 percent by Delhi in Yamuna. Actually people on higher pedestal should set an example of not taking more water than what their county men are normally getting.

5.0 RECAPITULATION

The water availability for persons in the country is steeply declining because of population explosion. Already certain areas in the country are facing scarcity of water and the situation will worsen in days to come. It is necessary that population growth should stabilize. Though government will not take serious action because of vote bank politics, however strong discentives, like debarring persons to stand for election, denial of jobs in the government and parastatal bodies, denial of subsidized items etc can force people to have population control.

Storage capacity for water has to be increased many a times.

People's cooperation is necessary for conservation and augmentation measures.

If we do not take serious steps to control population growth, the already strained and inadequate water supply systems will crumble, resulting in misery and epidemics. Is this going to be the answer for population control?

382

Are We Heading for Water Wars in the Near Future Because of Population Explosion?

Anand Paithankar

Retd. Prof. & Head of Mining, V.R.C.E., 12, Saraswati Vihar, Trimuti Nagar, Nagpur 22, India

1.0   INTRODUCTION:

It is well known that the human population on our planet is continuously increasing. According to one estimate if the same trend continues the present population level standing at 6 billions will rise to 10 billions by 2050 which will be an alarming situation. Most of the increase is in developing or underdeveloped countries. Most of the developed countries have kept their population under control, but the advantages of this plus point have been neutralized by indiscriminate industrialization.

Countries like India are in the worst situation because of population explosion and also constantly increasing industrial activities. Both consume enormous quantity of water. According to BBC environment correspondent Alex Kirly the world has no more water than it had 2000 years ago when the population was less than of 3% present figure which comes to 180 millions only.

The net result of present situation is severe water shortage throughout the world and water conflicts.

2.0   SOURCES OF WATER:

2.1   **Rain:** We are fortunate that the nature has provided us this wonderful source of fresh water. It not only provides life-sustaining water by feeding our numerous rivers and streams but gladdens the heart of all human beings. Ask any Indian how he feels when the first rain drop of Mansoon after a long period of skin burning summer falls on his skin. Poets of the past and present have written innumerable poems giving expression to their delight and poets in future will continue to write on this evergreen subject.

2.2   **Ice Caps and Glaciers**: We have on some of our mountain tops ice caps and glaciers, which is a frozen source of water. In summer these caps and glaciers melt and feed fresh water in to our rivers. For example in summer, in some Indian rivers like Ganga and Brahmaputra water level rises.

2.3   **Ground Water**: Some part of the subsurface strata is fissured and porous and is a reservoir of water. When the rain comes these natural subsurface reservoirs are charged.

2.4 **Lakes**: Fresh water lake, natural and artificial (created by dams) accumulate large quantity of rainwater. There are many such lakes scattered all over the world. Some of the biggest natural lakes are in Canada.

2.5 **Sea Water**: Conversion of sea water in to drinking water is uneconomical for several reasons and therefore not-discussed in this article.

3.0 **Population Explosion and Its Impact on Water**: As the population grows and the planet warms water tables around the world are falling. According to one estimate the water table is falling at an average rate of 3 meters per annum. Six years back the water level in the hottest month i.e. May was at a depth of just 2 meters below ground level in a well at the author's residence. This year the level has gone down to 10 m, the average rate of fall is 1.33 m per year during the last 6 years in Nagpur. In coastal region of Sourashtra in India the author observed that water table has gone down 40 m below ground level and sea water is mixing with ground water.

According to a BBC report the ground water in the Thames vally in U.K.has fell to the lowest level in a country. Meanwhile domestic consumption has doubled over the last 30 years.

According to one estimate (1) the Earth's annual water deficit now stands at 160 billion cubic meters.

4.0 **Water Marketing**: Because of scarcity of clean water sale of water has become a lucrative business. In Indian homes even today whenever a guest arrives a glass of water
is first offered because that is considered as the cheapest drink. But because of increasing
fear of water pollution water is treated to remove pollutants and water has become costlier. In hotels and restaurants in India a bottle of mineral water is sold at about Rs. 15/-. In Spain this author noted that a bottle of water.costs more or less same as a bottle of beer. Oil-rich but water-poor Saudi Arabia purchases half of its water abroad. Israel imports 87% of its water and Jordan imports 91%. Some 31 countries – mostly in the Middle East and Africa – are now listed as 'water-stressed'. In another 25 years, 48 countries with more than one third of the world's population will suffer from water starvation.

5.0 **Water Conflicts**: Because Canada's lakes and rivers hold approximately one-quarter of the Earth's fresh water, global entrepreneurs are vying to ship billions of litres of Canadian water to Customers in California, Mexico, Japan and the Middle East. Several US companies already have laid claim to Canada' water under the North American Free Trade Agreement (NAFTA) and they have threatened legal action through the World Trade Organisation if their plans are blocked (References: Gar Smith, The Hitavada May 2000) In India a long queue at public water taps is observed in every summer and sometimes people also quarrel. A long time political battle is being fought by two states in India, Tamilnadu and Karnataka over sharing of Water of River Kavery. There is also a dispute between India and Bangladesh over sharing of water of the river Gangas. In many European countries river water is a subject of conflict because water pullulated by one-country affects quantity of water consumed by the countries on the downstream side of the river.

6.0 **Water Pollution**: All human activities cause water pollution, right from kitchen and toilets to mega-industries. According to Joseph Jenkins, the author of human cure handbook the human societies can be divided into two categories: "Those who shit in their drinking water and those who don't". U.S. domestic water consumption according to Eco forum, the magazine of the Nairobi based U.N. Environmental programme, standard flush toilets use 200 t of fresh water to flush each tone of human waste.

All industries cause some from of water pollution. In mining industry different operations contribute to the pollution of water. The mining of metallic deposits and coal deposits containing pyrites (through acid generation) may lead to increased contamination of surface as well as ground water. The major sources of metal pollutants are mine drainage, water seepage through waste dumps, tailing impoundment, surface run off and ground water interception with excavation.

Agriculture because of use of chemical fertilizers and insecticides is causing water pollution. Chemical industries in several industrial areas have polluted the water sources. The toxicity of metal depends on the capacity of a metal to affect adversely any biological activity. Toxic metal can get in to food chains through aquatic organisms.

7.0   **Water Reclamation**:   In view of the increasing water pollution water reclamation has become imperative. This is done right from home at mini level to large cities at mega level. Varity of water fillers are now available in the market, which remove sediments and also effectively deal with micro- organisms. The age-old method of boiling is still practised in many Indian homes even today. Water pollution is not due to population explosion alone, but is due to industrial explosion also otherwise why should western countries who have controlled their population experience water crisis?

SOLUTIONS TO THE PROBLEM:

1. Redesigning rain catchment areas for water harvesting.
2. Harvesting of rain water flowing off roof tops.
3. Avoiding wasteful use of water at every level.
4. Education to control "Population".
5. Education staring form school level to train the young minds to live in harmony with the nature.

CONCLUSION: - If people learn to follow the laws of nature the water crisis can be easily managed

Stability Assessment of Baira Siul—Pong Transmission Line, Himachal Pradesh, India

Y. Pandey and R. Dharmaraju

Central Building Research Institute, Roorkee 247 667, India

INTRODUCTION

Some of the important hydroelectric power projects are located in hilly areas due to obvious logistic reasons. Huge reservoirs are constructed to create high heads of water for running the electricity generating turbines and the power is taken out with the help of long transmission lines, as load centers are usually away from the hills.

Chamera Hydroelectric Project is one of the prestige projects of National Hydroelectric Power Corporation (NHPC). It is located in the Himalayan hills of Chamba district of Himachal Pradesh. The power generated at the project site is taken out through the Chamera-Moga transmission line, owned by Power Grid Corporation of India Ltd. (PGCL). Substantial portion of the line (126 transmission towers) passes through the hills of Himachal Pradesh. In the hilly portions, the transmission towers were generally located on the ridges or on the hill slopes. These hills commonly have steep slopes and are susceptible to rapid erosion, sliding, sinking, slumping, distressing, subsiding and rock failure etc. The area receives normal to heavy rainfall during winter and monsoon seasons. The area of study falls on high seismic belt in the seismic zone V of the seismic zoning map of India.

After the Chamera project reservoir was impounded and power generation got started, it was observed that some of the transmission towers started showing distress. Unhealthy transmission towers posed serious safety and financial problems to the owner and the end users. Some of the towers were relocated, at some points a particular stretch of the line was shifted to another hill and the strategically important power supply was disrupted causing loss of crores of rupees.

At this point of time, PGCL approached Central Building Research Institute (CBRI) to study the geological stability of hilly tower location and classify the towers into highly unstable, unstable and stable categories.

A semi-quantitative methodology developed by the Institute has been followed for the study to assess the stability of concerned slopes. Normal geotechnical study of soil and rock samples collected from the site was also undertaken. Some of the unstable towers were instrumented to study their distressing behavior during one year, and suitable remedial measures were suggested to improve the functional stability of the individual transmission towers.

GEOLOGY OF THE AREA

Geologically, the area under study belongs to the rock formations of Precambrian to Recent deposits. The main rock types exposed in this region are quartzites, phyllites, schists, carbonaceous phyllites, granites, granite gneisses, sandstones, shales, siltstones, limestones, conglomerates, fluvioglacial deposits & river terraces. The rocks have been subjected to tectonic disturbances and resulted into highly folded and faulted lithounits. The terrain is characterized by steep to high angled slopes. The area along the tower alignment is traversed by four major thrusts, viz, Jutogh Thrust, Shali Thrust, Murree Thrust and the Main Boundary Thrust.

METHODOLOGY

Assessment of slope stability of hilly terrains requires thorough understanding of various geo-environmental factors, which contribute to the instability of slopes. In view of this a quantitative macro-zonation approach based on the basic factors controlling the stability of the slope has been developed to delineate the hazardous or vulnerable areas of slope instability. Several parameters have been considered for evaluation of slope stability around tower locations. The parameters taken into account are slope morphometry, lithology, orientation of discontinuities, presence of major geological structures, soil thickness/overburden, extent of weathering, vegetation density, hydrological conditions and presence of landslides.

Parameters Considered for the Stability Assessment

The parameters considered for the assessment of geological stability of tower locations along the transmission line are described below:

The slope morphometry of the area consisting of slope angle and its direction and $>45^0$ was considered to be more susceptible to slope failure. The average slope angle had been divided into five categories $<15^0$, 15^0-25^0, 25^0-35^0, 35^0-45^0, $>45^0$.

The structural orientation of discontinuities like beddings, foliations and joints in relation to slope have been identified as dip slope, where discontinuity plane dips outwardly in the same direction of slope; oblique slope where discontinuity plane dips inwardly or obliquely to the slope; and massive slope where no discontinuity plane exists. Out of these three categories, the dip slope may be considered as the most unfavorable condition for slope instability problems.

The thickness of soil cover or overburden material plays an important role in causing instability of slopes as it can slide (surfacial erosion) or subside on the firm rock strata. In the field, it had been estimated and grouped into three categories i.e. <1 m, 1-3 m & >3 m. Out of these categories, the overburden having thickness >3 m was considered as the most susceptible condition for slope instability.

Lithology includes the type of rocks or nature of soil present on the slopes. The nature of rock can be categorized as very strong, strong, weak or very weak rocks, whereas the slope mass material consisting of soil entirely can be classified as very soft, soft to very stiff soil.

The slope is effected by the seepage zones and surface drainages created due to the presence of major nalas on and around the slope. Surface drainage may be considered to be very critical parameter for the stability of the slope. For the present study surface drainage has been categorized as low, moderate and high.

The extent of weathering influences the strength of the rocks, in general, and promotes the instability of the slopes. A rock may be strong if it is unweathered. Hence, the degree of weathering of slope mass material is important for assessment of the stability of slope. The extent of weathering has been categorized into high, moderate and low.

The growth of vegetation always increases the slope stability by holding the soil and reducing the erosive action. Barren slopes are generally more prone to failure as compared to forested slope. In the present methodology, it has been categorized into four types i.e. thick forest, moderate forest, sparse forest and barren slope.

The presence of major thrusts and faults in the hilly terrain make the planes weak, producing highly deformed, shattered and fractured rocks. The slopes located within this zone are generally vulnerable to landslides and characterized by sheared and fractured rocks. For assessment of susceptibility to slope failure, this parameter has been categorized into three as major structure, minor structure and absence of any weak planes.

Terrain morphology is an important factor, which deals with the nature of slope. The slope may be concave, convex or straight. In most of the cases it has been observed that convex slopes are more vulnerable to instability than other type of slopes.

Rainfall is one of the most important factor for slope failure. To initiate the slope movement, there is always an active factor, which actually triggers failure. It is well known that rainfall acts as a triggering factor in most cases of landslides/slope failures. Slope failures usually begin with intense rainfall within a short span of time which brings the entire soil cover to saturation. As the piezometric head in the stratum increases, the frictional force holding the soil on slope decreases. This, combined with decreased cohesion due to water inflow, reduces the shear resistance significantly to create slope failure.

The presence of existing landslides on a slope is the clear indication of the instability of the slope. The slide area may be active or dormant (old). From slope stability point of view it has been considered as active slide, old slide and absence of slide.

Numerical Rating Assignment

All the above-mentioned parameters had been numerically quantified by assigning numerical ratings to each. For these, each parameter was assigned certain numeric value (weightage) in order to their share of contribution towards slope failure. The unfavorable condition of the parameters had been assigned lower rating than the favorable condition, e.g., gentle slope had been assigned a weightage 10 while very steep slope has been assigned weightage 0. Similarly for orientation of discontinuities, massive slope having no discontinuity plane has been assigned weightage 10 and dip slope 0. Similarly the ratings have been assigned to each other category of the parameters. The individual weightage given on the basis of field investigation to each category of the parameter existing on a slope are finally added to obtain the total weightage (score) for a particular slope. This represents the score for stability assessment. The lower the score more will be the chances of failure. The score are divided into five classes of slope stability, i.e., highly stable, stable, moderately stable, unstable and highly unstable.

Table 1 : Weightage of Parameters

Sl.No.	Parameters	Categories	Weightage
1.	Rock/Soil type	Very strong rock	07
		Strong rock	05
		Weak rock	03
		Soft soil	01
		Very soft soil	<1
2.	Slope morphometry	Very gentle slope	10
		Gentle slope	07
		Moderate slope	04
		Steep slope	02
		Very steep slope	00
3.	Slope aspect relative to discontinuities	Massive slope	10
		Oblique slope	05
		Dip slope	00
4.	Overburden	< 1 m	10
		1-3 m	05
		> 3 m	00
5.	Surface drainage	Low	10
		Moderate	05
		High	00
6.	Weathering	Low	10
		Moderate	05
		High	00
7.	Vegetation Density	Thickly dense	10
		Moderately dense	06
		Sparsely dense	03
8.	Landslide existence	Absence of slide	10
		Old slide	05
		Active slide	00
9.	Geological structure	Absence of structure	10
		Minor structure	05
		Major structure	00
10.	Terrain morphology	Straight slope	10
		Concave slope	03
		Convex slope	00

SLOPE STABILITY ASSESSMENT AND ZONATION

The field data collected for different parameters were analyzed to determine the stability conditions of the slopes. To assess the slope stability conditions, scores for each slope were calculated by adding the weightages assigned to the categories of each parameter present on that slope. Thus calculated scores were further classified according to the slope stability assessment rating class as shown in Table 2. The classification and the relative stability in each slope have been worked out, a typical section around Chamera dam shown Fig. 1.

Table 2 : Description of Slope Stability Assessment Rating (SSAR) Classes

Class	V	IV	III	II	I
SSAR	0-22	22-44	44-66	66-88	88-110
Description	Very bad	Bad	Normal	Good	Very good
Stability	Highly Unstable	Unstable	Moderately Stable	Stable	Highly Stable

GEOTECHINCAL INVESTIGATIONS

At the time of the inspection of tower sites, soil and rock samples were collected from four most critical tower locations namely 53, 54, 63 and 83. These samples have been analyzed in the laboratory for grain size distribution, Atterberg's limits and shear strength parameters. The presence of fine fractions of soil particles was considered to be the real cause for the reduction in shear strength in undrained condition, thus creating sliding and sinking of the ground in the vicinity of tower locations. This confirmed the findings of the geological investigations too.

INSTRUMENTATION AND MONITROING

During the course of survey, few highly unstable towers were selected for instrumentation and monitoring and the Electronic Distance Meter (EDM) was the main instrument employed for the purpose along with normal survey tools. EDM is an infrared-based optical instrument mounted on a Theodolite, with an accuracy of 5 mm +5ppm.

Transmission Tower no. 52 was one of the model towers selected for instrumentation study. This tower was located on moderately steep convex slope with sparse vegetation. The rocks present were schists with talc content showing slippery nature. The slope material was mainly rock pieces and weathered schist giving rise to pulverized clay material of about 1 m thickness. Unfavorable bedding was creating planar failure on the slope. Major nalas on both the sides of Siul nala at the toe of the slope continuously saturated the slope material at the base. The existing landslide along the nala was also affecting the slope. Indication of subsidence was seen in the area and cracks in the retaining wall was visible. The tower location was found to be quite close to Jutogh Thrust.

For the purpose of study, six markers fabricated from mild steel pipes and plates (1.5 m long) were installed around the tower. Observations were taken between December 1996 and June 1997, and the results are shown in Table 3.

Table 3 : EDM Data Analysis of tower no. 52 for the period Dec.'96 to Jun.'97

Sl. No.	Marker Reference No.	Lateral Movement in mm.	Vertical Movement in mm.	Spatial Displacement in mm.
1.	1	17	24	29
2.	2L1	49	25	55
3.	3L1	42	23	47
4.	4W1	44	08	45
5.	5W2	25	26	36
6.	6W3	29	30	42

The lateral movement had been found to vary from 17 mm to 49 mm where as vertical movement varied from 8 mm to 30 mm. The trend of movement data had clearly indicated that the tower location was unsuitable as two legs of the tower were showing considerable downward movement.

On the basis of the instrumented observations of tower no. 52, it was concluded that the slope, in general, was not in good condition due to unfavorable discontinuities and presence of nalas on both

390

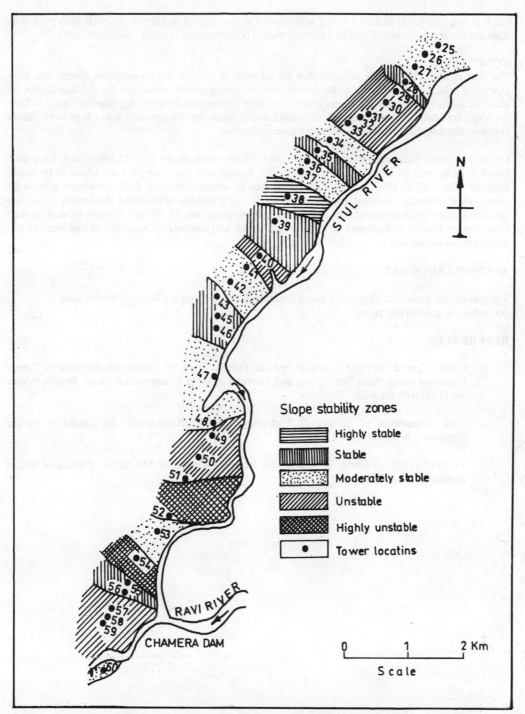

Fig. 1 : A typical section around Chamera dam

sides. It was suggested that a) a breast wall should be constructed to control the slide area present in between tower nos.51 and 52 and b) a toe wall should be constructed to check the toe erosion.

CONCLUSION

The present study has clearly revealed that the location of most of the transmission towers had been arbitrarily selected without giving due consideration to topographical conditions for the foundation. At some locations no adequate slopes and drains have been provided to drain off the rainwater quickly from the tower foundation. At various towers puddle marks under the transmission tower legs were clearly visible indicating the accumulation of water during monsoon.

Out of 126 tower locations surveyed in Baira Siul - Pong transmission line, 25 tower sites have been found to be located on unsafe zones – three towers located near the reservoir were found to be highly unstable, where as 22 other towers were categorized as unstable. Among these the slopes around 10 towers need immediate suitable control measures to strengthen the stability of the towers. The most vulnerable zone had been found near the reservoir where tower nos. 52, 53 and 56 were located. In fact, these towers needed realignment in view of the safety and commercial viability of the strategically located transmission line.

ACKNOWLEDGEMENT

The authors are grateful to Director, Central Building Research Institute, Roorkee for his kind permission to publish this paper.

REFERENCES

1. CBRI. Central Building Research Institute (1998). Study of Geological Stability of Tower Locations along Baira Siul – Pong and Chamera – Moga Transmission Lines. Project Report No.G (S) 009. Roorkee.

2. DST. Department of Science & Technology (1994). Methodology for Landslide Hazard Zonation – A Report. New Delhi.

3. IS 1893 (1975). Criteria for Earthquake Resistant Design of Structures. Bureau of Indian Standards. New Delhi.

Geoethical Backgrounds for Geoenvironmental Reclamation

Václav Nemec[1] and Lidmila Nemcová[2]

[1]*Consultant, University of Economics, Prague, Czech Republic*
[2]*University of Economics, Prague, Czech Republic*

INTRODUCTION

Geoethics as a new discipline covers ethical problems and dilemmas which are to be solved in the fields of geology, mining activities and energy resources. Ethical standards in any activity dealing with the use of mineral resources should be studied and developed by geoethics. The scope of this article is to present a short review of the main existing and potential problems dealing with geoethics and to focus the attention on specific problems connected with the geoenvironmental impact of mining activities, land reclamation etc. Some technical instruments which may be applied in the geoenvironmental reclamation are discussed.

GEOETHICS - A NEW DISCIPLINE FOR SUSTAINABLE USE OF MINERAL RESOURCES

Since 1991 the authors have tried to establish geoethics as a new discipline in the family of earth sciences. The reason was **to create an instrument to help in decision making whenever some ethical dilemmas occur in problems connected with the use of mineral resources.** The idea has been accepted and various specialized meetings on both national and international bases have been already organized including regular sessions organized as part of the Mining P□íbram Symposia in the Czech Republic since 1992 (in five sessions until 1999 altogether 131 contributions from authors from 15 countries have been published) and also symposia at the International Geological Congresses in Beijing (1996) and in Rio de Janeiro (2000).

Theoretical aspects of geoethics include definition of geoethics, philosophical background and interrelations especially with other disciplines of earth sciences as well as of applied ethics. Geoethical codes, geoethical accounting, specific social and educational problems, some historical as well as futurological reflections, etc. are to be considered as **practical aspects of geoethics**. Both theoretical and practical aspects **are to be cultivated in an equilibrium.** A good co-operation with business ethics and with ecological (environmental) ethics is needed. In several areas geoethics can be also considered in connection with professional ethics, especially for people working in earth sciences, mining industry etc.

The main reason for establishing and developing geoethics is **the non-renewability of mineral resources.** Their exploitation brings many problems. Sometimes an immediate impact on the environment or even substantial changes of the landscape occur. But even a full respect to all needs of ecological ethics during the mining process does not guarantee that real geoethical interests and needs will be also taken into consideration. In many situations these interests and needs concern not only the local area of the deposit but a far larger region if not the whole planet.

Without a full respect to the principles of geo- and eco-ethics the exploitation of mineral resources can be very dangerous. Such a danger usually should be considered also as a problem of **responsibility**

towards future generations for whom a clean environment and also a sufficient potential of mineral resources should be conserved.

Mineral resources are needed to ensure the further existence of the mankind on our planet. Unfortunately they are not distributed evenly. Their differentiation in the space has caused various tensions. In the history - as presented by A.S.Trembecki - the rather stabilized periods have been alternated by demographic catastrophes. Geoethical rules for an internationally harmonized use of mineral resources should protect the mankind against such catastrophes.

In highly developed countries sometimes a tendency exists to shift heavy geoethical problems to less developed countries (e.g. when the mining activities are transferred from their own territories into the less developed regions and countries). This phenomenon can be presented as **a lack of global solidarity.**

Various geoethical dilemmas appear as **conflicts among individual and public interests.** Many of these problems should be taken into consideration and solved also in a global context. A highly responsible approach to decision making in any problem concerning strategical sustainable use of mineral resources and reserves is needed. It should be prepared and cultivated on any level. Therefore geoethical principles are to be studied not only by specialists in earth sciences but also by managers, economists as well as by representatives of public administration, politicians and statesmen. Geoethical principles are to be obviously taught in any education system. Of course for future specialists in earth and environment sciences a more detailed knowledge of these principles is to be provided not only in special courses of geoethics but also by introducing geoethical aspects into individual disciplines of earth and environment sciences.

Principles of the human partnership, solidarity and sustainability should be applied for solving geoethical problems. It is necessary to create instruments for monitoring and controlling both local and transnational operations with mineral reserves which might result in havy dangers for the life of future generations.

When creating and applying geoethical codes more general principles are to be combined and interrelated with detailed rules taking into consideration **specific local and regional problems**. In accordance with some Russian authors (M.A.Komarov, G.S.Gold) the system of using mineral resources should correspond to the specific situation of these resources in the given country, with the historical development and cultural level of the country, with the mentality of local population, with the social and political situation, with ecological restrictions, with the situation at the world market of raw materials, etc.

When examining moral respect of the man for the nature from the point of view of ecological ethics then in case of mining activities the minimization or abolishment of any harm to the environment is in the focus of attention. Problems connected with mineral reserves themselves and with their sustainable use are not taken as priority. This can be easily understood if we take into mind that the irreversibility point of the actual growing ecological crisis is being estimated in a time horizon of about 20 years - i.e. in case of continuing with the present system of deteriorating the environment after 20 years an ecological catastroph will become unavoidable (in another further time horizon). The time horizons for exhausting mineral resources depend on individual kinds of these resources and also on the eventual success of finding some new materials to replace them - in each case they seem to be far more distant. Nevertheless the limited mineral potential of our planet represents the same danger for the sustainability of life as the approaching ecological catastroph.

Many examples of **plundering ore deposits** are known from the past as well as from actual practice when only pure profit criteria are applied for making decisions concerning the mining strategy on individual deposits. Such an approach brings a danger that in some cases a time horizon for irreversibility of making impossible the complete exploitation of the given deposit will be far less distant and that in the future the unexploited part of the deposit will become also unexploitable. The lifetime of individual deposits can be terribly minimized and from this point of view a time horizon for exhausting mineral reserves in fact can become less distant than it seems to be nowadays. A direct liaison of geoethics and sustainable use of mineral resources is evident.

CURRENT GEOETHICAL PROBLEMS IN THE MINING INDUSTRY

Because each deposit of mineral reserves is unique, numerous facts are to be taken into consideration when the exploitation process is being prepared and controlled.

From the geoethical point of view following steps are to be ensured **before starting any mining activity** or as a priority in the course of the exploitation process:

- the complete evaluation of disposable resources

- finding the optimal scenario (time model) how to exploit the deposit with special regards to non-renewability of mineral resources as well as to principles of both sustainability and geoenvironmental reclamation (as described in the further text)

- all needed social aspects as well as revitalization problems are to be taken into consideration in any scenario of possible exploitation processes

- the externalities should be taken into consideration as well because possible changes in the state policy can change the resulting decisions in the future (alternative solutions are to be prepared in advance).

In the course of any mining activity:

- the mining activities should respect all current standards of business ethics

- the problems of exporting the exploited mineral resources should be solved with the full respect to the long-term strategy for the optimal and sustainable use of these resources (short-term advantages are to be excluded or at least minimized in any decision making, shifting of serious geoethical problems to other countries in the benefit of the local population should be completely eliminated)

- the existing systems of ecological and/or social audit should be complemented by a "geoethical and/or sustainability audit" emphasizing an appropriate way of long-term using of non renewable resources

- the geoenvironmental reclamation should be a consistent part of the appropriate ecological and geoethical audit.

The complete realization of all steps as presented above is supposed **before starting any decision making process leading to stop (temporarily or definitely) any further mining activity.** From the geoethical point of view not only managers of mining entreprises but also respective state and communal authorities as well as the population are to be convinced that everything has been done either to conclude the mining process by the complete exploitation of the whole deposit or - in case of a temporary stop - to achieve such a degree of exploitation where any discontinuity of mining activities will not cause any irreversible impossibility to continue with the mining activities in the future.

Special attention should be paid in advance to social problems occurring in connection with closing mining activities as well as to problems of revitalization of the exploited space.

When the appropriate state authorities declare their approval with closing the mining activity special attention is to be paid also to the problem where the documentation of the deposit and of the mining activities will be ensured for a possible and useful research in the future. The questions of how to make all historical, technical and geological exclusivities accessible to the public, how to conserve the old mining traditions etc. are to be discussed. Also needed financial means are to be ensured for technical or natural monuments or musea.

The actual process of globalization brings new problems for geoethics as well. **The respect for and protection of the rights of future generations anywhere in the world for needed mineral resources should be taken into consideration.**

Decision making process at any level should be accompanied by **the personal and corporate responsibility for any possible consequence of any decision.** Max Weber ideas are valid until nowadays. Various recently developed theories (e.g. corporate citizenship, stakeholder theory etc.) can be successfully applied in the practical life including geoethics.

GEOENVIRONMENTAL RECLAMATION

Many specific problems of geoenvironmental reclamation - connected with mining activities - **can be solved by means of the appropriate time models of ore deposits.** The 3-D space of a deposit has to be submitted to the conditions of exploitation. The time model consists of **simulating the mining activity with the full respect to geological, technical, economic as well as geoethical and geoenvironmental conditions.** In practice it is only the space model of the deposit with fairly detailed and reliable data which permits us to create the time model of the same deposit. The accessible parts have to be found in it where the required qualitative parameters (e.g. mean grade of metal, mixture of components for final production etc.) can be reached by the mining process in defined time periods. The original methodology was described or commented in previous works of the author (V.N☐mec, 1988, 1990). The advantage of this methodology in various practical applications has been proved especially thanks to the possibilities of combining automatic and interactive solutions of individual modelling steps.

The mathematical method of the selection of appropriate accessible exploitable parts of the deposit is based on the profit function (theory of games). It is quite possible to include into the system of profit functions **conditions for minimizing the land use** in any period of exploitation which will make it possible to realize also **an early revitalization process of the deposit environment.** Similar results can be achieved by **considering specific geometric constraints for the development of the open pit** in the given exploitation periods. Various other environmental and ecological criteria can be considered and respected as well.

CONCLUSION

Moral and specific geoethical aspects are to be incorporated into any decision making when any needs and real possibilities of a sustainable use of mineral resources are to be considered. Future needs of the mankind must be respected anywhere. A high responsibility to future generations is to be cultivated not only among earth scientists and mining engineers when any scientific or technical problem connected with the use of non renewable resources is to be solved but also among managers, politicians and statesmen, especially when long-term decisions for mineral economics and policy are to be made.

Many further aspects (not explicitly explained in this article) are covered by geoethics. **Social problems** (security of labour or unemployment problems in the mining industry etc.), **cultural problems** (including a needed increase of respect to the nature as well as to the history - e.g. old mining traditions, conservation of archives, technical monuments etc.) and also **specific educational problems** are of first-rate importance. **The ethical background is to be obviously cultivated** in any education system not only for specialists in earth and environment sciences but on a corresponding level also for representatives of political life and for broad civic circles.

Challenges and risks resulting from the actual practice of operating with mineral resources generate geoethical dilemmas for the next century and millenium. They have to be labeled satisfactorily clearly. Individuals and institutions must be ready to solve them efficiently. The principles of the human partnership, solidarity and sustainability should be used when geoethical problems are to be solved. It is necessary to create instruments for monitoring and controlling both local and transnational manoeuvres which might result in havy dangers for the life of future generations.

Geoethical problems in fact did exist much sooner than the new science - geoethics - devoted to solve them and our own work is not at all the first application of ethics in the earth sciences. Nevertheless the present situation on our planet needs to find new ways of thinking and new ways of planning and controlling mining activities all over the world when national borders are being suppressed and new visions of globalization exist. A considerable improvement of ethical climate on our planet is needed and geoethics should take an active part in this process.

REFERENCES

G. S. GOLD (1998): Ethical aspects for substantiating criteria for an effective use of mineral resources - The Mining P□íbram Symposium, section Geoethics, paper G 6

M. A. KOMAROV, G. S. GOLD (1999): Geoethics and geological education. - The Mining Peíbram Symposium, section Geoethics, paper G 7

L. NEMCOVÁ (1999): Challenges to the future development of geoethics. - The Mining Peíbram Symposium, section Geoethics, paper G 10

V. NEMEC (1988): Geomathematical Models of Ore Deposits for Exploitation Purposes. - Sci. de la terre, Sér. Inf., nancy, 27, pp. 121 - 131

V. NEMEC (1990): New Experiences with Time Models of Mineral Deposits. - APCOM, Tech. Univ. Berlin, pp. 127 - 135

V. NEMEC (1998): Actual Problems of Developing Geoethics. - The Mining Peíbram Symposium, section Geoethics, paper G1

V. NEMEC (1999): To the Recent Development of Geoethics. - The Mining Peibram Symposium, section Geoethics, paper G 11

V. NEMEC, L. NEMCOVÁ (1998): Geoetika a trvale udreitelný rozvoj. (Geoethics and the sustainable development). - Charles University, Prague

A. S. TREMBECKI (1998): Geoethical conditions of mining activities. - The Mining Peíbram Symposium, section Geoethics, paper G 10

Ground Vibrations Excited by Pile Driving—A Case Study

A.J. Prakash, P.K. Singh and R.B. Singh

Central Mining Research Institute, Dhanbad 826 001, India

INTRODUCTION

The term 'geo-environment' refers to the upper most part of the lithosphere which is affected by the human activities. Human activities influence the geological, physical, chemical and bio-chemical processes taking place in rocks and soils (Aswathanarayana[1]). Ground vibrations which may be generated due to construction blasting or during pile driving is a very prevalent geo-environmental problem as these vibrations may cause damage to residential structures in their vicinity.

Ground vibration induced by pile driving, their propagation and frequency characteristics were studied at a site where vertical pile driving were carried out for casting concrete piles in-situ. Observations were taken during driving of five different end bearing piles at different locations of the same site.

GROUND VIBRATION

Ground vibrations excited by pile driving are of impulsive character and these are transient phenomena like those induced by blasting in mining and construction activities. Recent developments in construction projects such as industrial area, construction of multistory apartments nearby existing residential buildings are coming up in different parts of the country where pile foundations are needed. Thus understanding of the geo-environmental problem arising from such construction is essential.

From Israel, a case has been reported by Alpan and Maidev[2]. Nearby residents had complained about excessive vibrations when driving was carried out with adjacent pile driver giving out one blow per second, the rated power out put being 2KW. Vibrations were recorded in buildings about 15 m away, at the nearest point to the pile driver. A preliminary survey to observe the cracks in the building was made before test driving commenced. The particle velocity measured inside the building did not exceed about 15 mm/s but accelerations reached to 0.7g, no noticeable damage appeared to the adjacent building.

The problem of investigating ground vibrations excited by pile driving and their propagation had been discussed in relatively few publications (Baba & Toriumi[3], Wiss[4], Dvorak[5], Frydenlund[6], Ceisielski et al[7, 8], Rosset et al[9]). These contributions concern investigations of parameters involved in ground vibrations. Knowledge of these parameters is important during pile driving in the immediate vicinity of the buildings which will enable to optimise the height of the hammer drop (Ciesielski et al[10]). Dynamic characteristics of ground by studying ground vibrations due to pile driving at some sites were carried out in India (Prakash[11]).

Li et al[12, 13] monitored ground vibrations of hydraulic filled land induced by construction activities i.e. piling, sheet piling and compaction sheet piling were monitored by instruments. The empirical formula for attenuation of vibration on ground surface for each kind of construction activity was developed based on the monitoring results. Piling causes wave to propagate in soil media. These waves are distributed as body waves and surface waves.

For waveform generated by impact type pile driving of end bearing or friction piles, Attewel and Farmer[14] postulated a qualitative model. Their model assumes that the major portion of the impact energy is conveyed to the pile top where the energy is transmitted to the surrounding soil medium in the form of strain energy propagated as the elastic compression and shear waves.

EXISTING VIBRATION STANDARDS

Structural damage due to blast vibration is a prevalent geo-environmental problem. As peak particle velocity (PPV) is the most pertinent parameter for structural damage, PPV of ground vibration generated due to pile driving is apprehended to cause disturbance and / or damage to structures in its vicinity. Various standards in this regard have been put forward globally by different researchers working on this line.

Some investigators included frequency also along with PPV in the damage criteria. Australian Standards (Ca – 23-1967) and Swedish Standard (After Persson et al, 1980) included amplitude in the damage criteria. Some most common standards are given in Tables 1- 3.

Table – 1: After DGMS (India) Standard (Technical Circular No. 7 of 1997)

Type of structure	Dominant excitation frequency, Hz		
	< 8 Hz	8- 25 Hz	> 25 Hz
(A) Buildings / Structures not belonging to the owner			
1. Domestic houses(Kuchcha, bricks & cements)	5	10	15
2. Industrial Buildings	10	20	25
3. Objects of historical importance and sensitive structures	2	5	10
(B) Buildings belonging to owner with limited span of life			
1. Domestic houses /structures	10	15	25
2. Industrial buildings	15	25	50

Table-2: After Australian Standards (Ca-23-1967) (Just and Chitombo, 1987)

Type of structures	Maximum value
Historical building, monuments and buildings of special value	0.2 mm displacement for frequency less than 15 Hz
Houses and low rise residential buildings, commercial buildings not included below	19 mm/s resultant ppv for frequency greater than 15 Hz.
Commercial buildings and industrial buildings or structures of reinforced concrete or steel construction	0.2 mm maximum displacement corresponds to 12.5 mm/s at 10 Hz and 6.25 mm/s at 5 Hz

LOCALE AND MONITORING METHOD

Ground vibration due to pile driving was studied in Haldia at the site of Fertilizer Plant. Experimental program was carried out at the pile driving site where alluvial ground was containing layers of soft soils, moist clay, fine and medium grained sand, its average density was 2.8 g/cc (t / m^3).

Vertical pile driving was carried out for casting concrete pile in-situ. Observations were taken during driving of piles at five different locations of the same site.

Table-3: After Sweden Standard (After Persson et al, 1980)

Type of structures	Limiting vibration parameters		
	Amplitude (mm)	Velocity (mm/s)	Acceleration (mm/s^2)
Concrete bunker steel re-inforced	-	200	-
High rise apartment block- modern concrete or steel frame design	0.4	100	-
Underground rock cavern roof hard rock, span 15 – 18 m	-	70 - 100	-
Normal block of flats-bricks or equivalent walls	-	70	-
Light concrete building	-	35	-
Swedish National Museums – Building structure	-	25	-
Swedish National Museum-Sensitive exhibits	-	-	5
Computer center	0.1	-	2.5
Circuit breaker control room	-	-	0.5 – 2

Pile Driving System

Cylindrical steel piles used were of 0.400 - 0.450 m diameter. These were driven by impact of a vertically falling hammer. Weight of each hammer was 3.5 t and its fall were between 1 and 1.5 m. Piles were driven using two different types of driving equipment (i) drop hammer and (ii) single acting hammer.

Drop hammer

The drop hammer consists of a single piece of metal. It was being lifted by means of a cable operated by a winch and was released from a fixed height by some automatic device. This system was comparatively slow in operation.

Single acting hammer

In single acting hammer, the hammer was raised to a fixed height by means of steam energy. The driving energy on the pile was provided by the weight of dropping hammer only.

Experimental Details

Before starting the experiment on pile driving, pile was marked at an interval of one meter from bottom up to the pile top. At each blow of the hammer the pile depth was observed and consequent ground vibrations were monitored. Observations were taken till the whole pile was completely driven or till the bottom end of the pile reached the firm stratum of the ground.

Sensor was mounted at different distances from the pile center and ground vibrations were monitored for each blow of hammer on the pile top. A few observations were also taken with varying height of the hammer drop and a sensor mounted at a certain distance from the pile center.

RESULTS

A brief description of five different test piles is given in Table-4. Length of the test piles having 0.400 - 0.450 m diameter, varied from 25.2 To 32.0 m. Hammer each of weight 3.5 ton was falling vertically from 1.0 – 1.5 m height.

Table 4: Brief description of five different test piles

Pile No.	Site	Weight of hammer (t)	Height of hammer drop (m)	Diameter of pile (m)	Length of pile (m)
1	Ammonia plant rectsol	3.5	1.0	0.400	32.0
2	Ammonia plant rectisol	3.5	1.0	0.400	30.0
3	N.K.P. Prilling tower	3.5	1.5	0.450	25.2
4	N.K.P. Prilling Tower	3.5	1.5	0.450	26.5
5	Ammonia plant carbon recover	3.5	1.0	0.450	29.6

Peak particle velocity (PPV), amplitude and frequencies of ground vibration were monitored at distances ranging from 6.0 to 75.0 m. Vibrations were recorded for a series of blow on each pile. PPV were 8.07 – 0.725, 14.5 – 1.95, 15.4 – 0.56, 4.21 – 1.02, 8.35 – 1.34 m x 10^{-3}/ s, amplitudes were 43.88 – 3.65, 65.0 –

9.5, 25.32 – 6.5, 36.25 – 5.12, 40.6 – 8.78 m x 10⁻⁶. Frequencies of ground vibrations at different piles were 26 – 38 Hz, 28 – 37 Hz, 11 – 31 Hz, 17 – 34 Hz, and 17 – 33 Hz respectively. Magnitudes of ground vibration and its frequencies monitored during driving of five different piles are given in Table-5.

Table –5: Ground vibrations monitored during driving of five different piles

Test Pile No.	Sensor distance from pile center (m)	Amplitude X 10⁻⁶ (m)	PPV X 10⁻³ (m /s)	Frequency (Hz)
1	15.0 – 55.0	43.88 – 3.65	8.07 – 0.725	26-38
2	8.0 – 50.0	65.0 - 9.5	14.5- 1.95	28- 37
3	25.0 – 75.0	25.32 – 6.5	15.4- 0.56	11- 31
4	6.0 – 21.0	36.25 - 5.12	4.21 – 1.02	17- 34
5	11.0 - 30.0	40.6 - 8.78	8.35 – 1.34	17- 33

Damping Co-efficient

Ground vibration records corresponding to each blow of hammer were analysed and thus damping coefficients (β) were determined. Considerable change in damping co-efficient with distance for the same pile was observed (Fig.1).

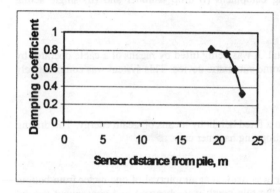 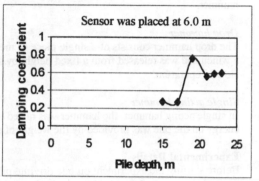

Fig. 1: Variation of damping coefficient at a site.

Empirical Equation

As per standards, usually PPV is considered as the most reliable parameter for assessing damage to a structure, some standards suggest for amplitudes too (Table-2-3).

Recorded data of a test pile were analysed using software ' Linreg' for linear regression and an empirical equation has been established. Examination of these data indicates that the attenuation of the observed amplitudes may be approximated by the following formula:

$$A = 80.42 \, D^{-0.9145}$$

Where A = Amplitude of vibration, (x 10⁻⁶ m) micron
       D = Distance from pile, m

Table - 6: Validity of an empirical equation A = 80.42 D⁻⁰·⁹¹⁴⁵

Distance from pile, m	Displacement (x 10⁻⁶ m) Observed	Computed	Difference	Percentage
12.0	8.0	8.28	0.28	3.5
15.0	6.7	6.758	0.058	0.86
20.0	5.0	5.19	0.19	3.8
25.0	4.2	4.235	0.035	0.83
30.0	3.6	3.58	- 0.02	0.55

Validity of the empirical equation derived for the site using the aforesaid software 'Linerag' may be justified from the Table-6 as only 3.5 to 0.55 percent deviation were found between observed and computed data. Here amplitudes are taken into considerations. Similar equation may be derived by using PPV of ground vibration. The empirical equation is very similar to that of Li, Chung and Chang[13].

400

Effect of Height of Hammer Drop

In drop hammer pile driving system the height of drop is normally 1.0 m. For experimental purpose, distance of sensor from pile was kept constant and fall of hammer was varied keeping at 0.5, 1.0 and 1.5 m and consequent ground vibrations were monitored. Observations were repeated for a particular pile keeping the sensor at 6.0 and 9.0 m from piles. Plot shows (Fig. 2A & 2B) that ground vibrations were increased with height of hammer drop but no significant change in frequencies was noticed.

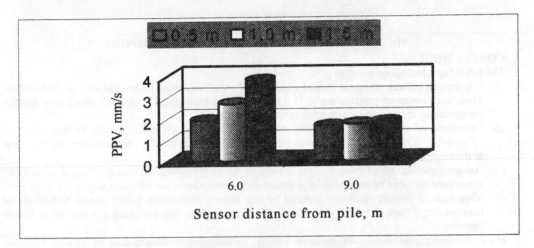

Fig. 2A: Effect of height of hammer dropped on ground vibration

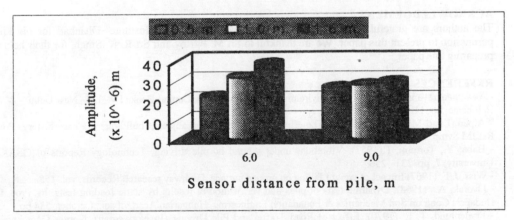

Fig. 2B: Effect of height of hammer dropped on ground vibration

Ciesielski et al[7] conducted studies to estimate the influence of hammer drop on the vibrations induced during piles driving. After analysis they found a distinct change in the value of amplitude with increase in height of hammer drop, but did not find any influence on the frequency of ground vibrations. Their findings are found to agree with the observations presented in the paper under discussion.

Effect of Pile Driving System

Drop hammer and single acting hammer were deployed at the same site for pile driving. Keeping sensors from pile at 11.0, 15.0 and 30.0 m, the observed ground vibrations are plotted in the bar chart (Fig. 3). Single acting hammer was observed to generate higher magnitude of ground vibration than drop hammer. No significant change in frequency was observed. Thus ground vibrations were found to vary with the piling system. This agrees with the findings reported by Li[12, 13].

401

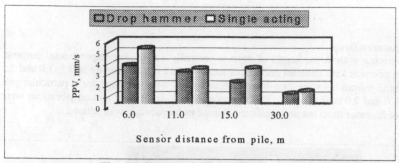

Fig. 3:Influence of pile driving systems on ground vibration

CONCLUSIONS

The following conclusions are drawn:

(1) Pile driving induces transient ground vibrations of significant magnitudes and various frequencies, These were observed propagating to 75.0 m and beyond from piling. These vibrations may damage the structures in its vicinity, either industrial or residential.

(2) Magnitude of ground vibrations may be estimated to a considerable extent by an empirical relationship. Constants associated with the equation depend on the site characteristics but the trend of the relationship remains unaltered.

(3) Damping coefficients of ground vibration induced during piling at a site change. Thus, it is desirable to compute its value by analysing large data at different distances for the same site.

(4) Magnitude of ground vibrations induced by pile driving depends on piling system and height of hammer drop. Thus, it can be controlled by controlling these two variables at a site for a limited distance.

(5) Considering environmentally cautious society, geo-environmental problem of ground vibration generated during pile driving is of paramount importance. The presented study reflects a guideline in this area for the pile operators whenever it is carried out in built up area.

ACKNOWLEDGEMENT

The authors are grateful to the Director, Central Mining Research Institute, Dhanbad for his kind permission to present this paper. We are thankful to Sri M. P. Roy and Sri R. K. Singh, for their help in preparing this paper.

REFERENCES

- Aswathanarayana, U. (1995). Geo-environment An Introduction, Capital Books Pvt. Ltd., New Delhi – A. A. Balkhema, pp. 1-10.
- Alpan, I and Meidev, T. (1963). The effect of pile driving on adjacent buildings = a case history. Proc. RILIM Symp, Budapest, 2, 171 pp.
- Baba, Y., Toriumi, I (1957). Vibrations under ground by pile driving, Technology Reports of OSAKA University, 7, pp. 231– 256.
- Wiss, J. F. (1967) Impact of weight falling on ground surface. Highway research Records, vol. 155.
- Dvorak, A. (1968). Dynamic tests of piles and verification of results by static loading tests. In: Proc. III Budapest Conf. on Soil Mechanics & Foundation Engineering, Hungarian Acad. of sci., October, 534 pp.
- Frydenlund, T. E. (1970). Effect of Road Traffic and Pile Driving (In Norwegian). Norges Geotekniske Institutes Stipend, 1969 –1979, Veglab
- Ciesielski, R., Maciag, E., Stypula, K. (1977 a). Propagation of vibrations of soil caused by pile driving. Czechoslovak Conf.: Dynamics of Engineering Structures, Smolenice – Bratislava, pp. 249–256.
- Ciesielski, R. Maciag, E., Stypula, K. (1977 b). Results of Initial Research on reductions of vibrations induced by pile driving in the soil – building system. Polish Conf.: Experimental investigations on Structures (In Polish. Technical University of Cracow, Cracow – Janoice), pp 53-66.
- Roesset, J. M., et al (1994). Impact of weight falling into the ground. J. of Geotechnical engineering, vol. 120, no. 8, August. pp. 1394-1412.
Ciesielski, R., Maciag, E., Stypula, K. (1980). Ground vibrations induced by pile driving some results of experimental investigations. Int. Symp. on Soil under Cyclic and Transient loading/Swansea/ 7-11, January/ 1980, pp. 757–762.
- Prakash, A. J. (1983). Studies of Dynamic properties of soil and rock. Ph. D. Thesis, Indian School of mines, Dhanbad, 322 pp.
- Li, J. C., Chung , Yu-Tung, Chang & Hsing – Chung (1990). Ground motion associated with piling and soil improvement construction. Tenth South east Asian geotechnical Conf., 16- 20 April, Taipei, pp. 425– 430.
- Li, J. C. (1984). Pile driving induced vibration on ground surface. Journal of Civil and Hydraulic Engineering, Vol.10, no.4: 45 – 59 (In Chinese).
- Attewel, P. D. and farmer, I. W. (1973). Attenuation of ground vibration from pile driving. Ground Engineering, Vol. 6, No. 4: 26–29.

Identifying Suitable Plants for Reclamation of Degraded Lands in and Around Opencast Mine: A Case Study

A.K. Agarwal[1], R.T. Jadhav[2], A.G. Paithankar[3] and P.K. Jha[4]

[1]*Assistant Professor , Mining Department, RKNEC, Nagpur, India*
[2]*Professor and Head, Chemistry Department, RKNEC, Nagpur, India*
[3]*Retd. Professor and Head, Mining Department, VRCE, Nagpur, India*
[4]*Professor and Head, Mining Department, RKNEC, Nagpur, India*

INTRODUCTION:

Land degradation and water pollution due to mining activity is a common and major problem confronted by the nations all over the World. Vast areas of denuded land and polluted water bodies are seen at many places in the vicinity of mining area. In order to minimize adverse impacts of mining activity a study has been carried out at a large open cast mine located in Vidarbha region in central India.

The investigations covered study of premining conditions and critical examinations of the current method adopted for Land Reclamation. The study has helped in providing guide lines for post mining land use which includes identification of suitable plant species for reclamation of the degraded land.

The fieldwork essentially involved several visits to mine site and collection of basic data relating to premining conditions and present practice of reclamation and collection of soil and water samples at several critical locations. The laboratory work was carried out to study the physical and chemical properties of soil and water. Four saplings of different plant species, which commonly grow in and around the area, were planted in pots in the soil brought from the field and without any amendments of chemical or Bio fertilizers were closely monitored for their growth for three months. The project work helped in identifying the most suitable plant species that can be grown on the reclaimed area of the mine.

IMPACT OF OPEN CAST MINING ON LAND:

Land degradation due to open cast mining is one of the primary reasons for inviting environmental problems. Large-scale open cast mining results in complete alteration of the surface topography, leaving an area of more or less parallel ridges or of peak mounds of material which overlays the excavated deposit. Spreading of waste and overburden material from mines over a large area out side the pit limit makes the land unusable and affects natural equilibrium.

Surface mining destroys vast amount of cropland/ Forest due to excavation, over burden dumping causing large-scale removal and burial of soil and subsoil beds. Agricultural land may also have to be acquired for development and construction of plants, service station, workshops, offices, roads, residential complexes etc. In India, it is estimated that for every million tonne of coal produced 4.0 ha of land is degraded directly and another 4.0 ha indirectly.

CASE STUDY:

A large open cast coal mine situated in the Vidarbha region of central India was selected to investigate the dimension of land degradation and identification of different plans that can grow in the area after mining is over with minimum investment was undertaken Before opening the deposit, the majority of lease out area was barren with Babul trees and Shrubs and only small portion was agriculture land.

ANALYSIS OF WATER SAMPLE:

Water samples were collected from three different places from the mine site. These three sites are: River Water, Sump Water and Face water. The water samples collected from these sites are named as Sample 1, Sample 2 and Sample 3. These samples were analyzed in laboratory for: Temperature, pH, Suspended Solids, Total Dissolved Solids, Chloride, Alkalinity and Hardness.

403

TABLE – I
REPORT OF WATER ANALYSIS

PARAMETER	SAMPLE I	SAMPLE II	SAMPLE III
Temperature(^{0}C)	21	22	21
PH	6.8	7.3	7.6
Suspended Solids(mg/lit)	17	180	165
Total Dissolved Solids (mg/lit)	546	744	338
Chlorides (mg/lit)	3048.74	24999.6	6097.4
Alkalinity (mg/lit) CO_3^{--}	14.5	14.58	21.87
HCO_3^{--}	153.09	127.575	87.98
Hardness(mg/lit) $CaCO_3$	363.4	413.88	302.838
Mg^{++}	191.8	131.23	111.04

ANALYSIS OF SOIL SAMPLE:

Although a large number of elements are required for healthy plant growth , only three are usually lacking in wastes to such an extent that corrective action is required. These are Nitrogen (N) , Phosphorus (P), and Potassium (K), without which no plant growth will occur. The soil samples were collected from three different sites from the mine and these were analysed for N,Pand K. The results are presented in table II.

TABLE – II
REPORT OF SOIL ANALYSIS

SR. NO.	PARAMETER	CONCENTRATION
1.	N Kg/ha	280
2.	P Kg/ha	32.5
3.	K Kg/ha	12

PLANT GROWTH MONITORING:

Soils samples were collected from overburden dumpsite of the mine and four different species of plants were planted in the soil samples collected, in the laboratory condition. These four species namely Gulmohar, Badam, Neelgiri (yellow) and Neelgiri (Blue), were planted in the soil collected without using any fertilizer. The plant growth was monitored continuously for three months and the results are indicated in table no. III.

TABLE – III
REPORT OF PLANT GROWTH MONITORING

INITIAL HEIGHT ON	GULMOHAR	BADAM	NEELGIRI (YELLOW)	NEELGIRI (BLUE)
26/10/99	76 cm	60 cm	45 cm	65 cm
3/10/99	No change	No change	No change	No change
10/10/99	No change	No change	No change	No change
17/10/99	No change	No change	No change	No change
24/10/99	77 cm	61 cm	No change	66 cm
31/10/99	77 cm	61 cm	46 cm	67 cm
7/11/99	77 cm	62 cm	46 cm	67 cm
14/11/99	77 cm	62 cm	49 cm	67 cm
21/11/99	78 cm	63 cm	49 cm	67 cm
28/11/99	78 cm	63 cm	50 cm	68 cm
5/12/99	78 cm	64 cm	51 cm	69 cm
12/12/99	79 cm	65 cm	52 cm	70 cm

404

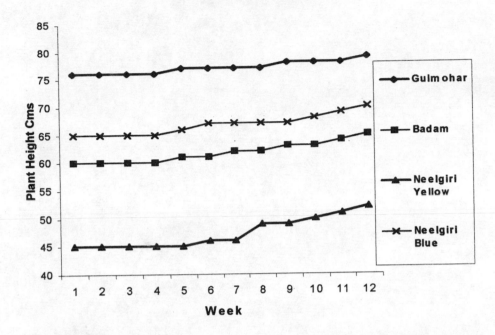

CONCLUSION:

The area of Wardha Valley coal fields is blessed with thick black cotton soil and tropical climate, where the rate of growth of trees is generally very good and reclamation by afforestation is recommended as one of the best method. In the particular case under discussion, the same observation has been noted. The plants which are selected for experimentation were locally grown trees. The soil also rich in sodium, phosphorus, potassium (NPK) and other nutrients. As a result all the plants have shown good growth.

Biogas Generation a Technique of Protecting Environment from Biomass

V.S. Deshpandey[1], P.G. Babrekar[2] and R.D. Askhedkar[3]

[1]*Head (Industrial Engg.) Priyadarshani College of Engineering,*
Hingna Road, Nagpur, India
[2]*Retd. Head, (Agricultural Chemistry & Soil Science), P.K.V. Akola, India*
[3]*Prof. (Ind. Engg.) & Dean Academics, P.C.E. & A., Nagpur 19, India*

1.   INTRODUCTION

Agriculture is the most important occupation in rural India and about 80% of rural population are engaged in agriculture. Dairy is another popular business, which goes hand in hand with agriculture. In rural areas proper latrines and sewage system is not available or open drainage system are available. Environment is polluted due to human excreta. Similarly animal dung also pollutes the environment. Presently dried dung cakes (1) is used as a household fuel by rural people for which small children and women spend a lot of time for collecting cow dung at the cost of education and other important activities.

Present energy sources are petroleum based, the lifetime of which is finite. The fast depletion of this important energy source creates the need for concentrated efforts to search alternative renewable source such as Solar, tidal and wind energy, small-scale hydroelectric generation and finally biogas generation from organic wastes. Biogas generation from animal/human excreta will clean the rural environment and simultaneously provide cheap and clean fuel for cooking and lighting in rural areas.

Our country has played a dominant role in development of anaerobic digestion of nightsoil, cattle dung and residual slurry as quality manure.

This paper reviews the various techniques of biogas generation and the technical and R & D issues for improving the waste utilization for generation of Biogas at optimum cost.

2.0 BIOGAS GENERATION

2.1   Technology

The term, BIOMASS refers to all organic matter except fossil fuels and its non food category, comprises the complete range of terrestrial and aquatic plants including algae, agricultural and forestry surpluses and all organic wastes from agriculture, fisheries, forestry and human communities. Biomass conversion energy recovery systems are of two types – (1) Thermo-chemical which include combustion, pyrolysis and gasification and (2) Biological such as the biogas generation through anaerobic digestion of human, animal and agricultural wastes, and ethanol production from straws, wood and other materials rich in ligno-cellulose complexes. Among these biological conversion processes, anaerobic conversion into biogas along with stabilized spent slurry of high manorial value, has been studied extensively and adopted in many developing countries, largely rural and facing shortages of crude petroleum. Cost per unit of the alternative energy and fabrication and erection of biogas digesters locally, was found consistent with reasonable economy. Major raw materials–farm, human and animal habitation wastes are abound in rural areas and rural populace, especially vulnerable to the price and supply fluctuations of conventional fuels, benefit directly or indirectly from biogas plants (2 and 3).

The Indian Agriculture Research Institute, New Delhi established that cattle dung and cellulose farm wastes can be used in producing biogas and manure and developed a pilot plant utilising cattle dung as feed on indigenous needs and resource. Two conflicting aspects of providing fuels to villages and at the same time supplying quality manure to farms are reconciled through this research. Schematic block diagram for biogasification of organic wastes is given in Figure 1. More than 2 million gobargas plants on an individual home basis based on this technology are established in rural areas by Khadi and village industries association, Mumbai.

2.2   Digesters
The digesters for biological conversion of organic wastes in to biogas and stabalised spent slurry are of two types.

1. Batch digesters : The feed slurry in water after seeding (inoculculating) is allowed to ferment completely before the system is recharge, it is characterized low capital cost, low efficiency and high labour requirement. The gas production, initially slow while establishing the microbial population, rises to a peak and then declines as the feed stock becomes exhausted. The retention time (R.T.) in the case of batch digesters is the time in days from the start to the end of digestion and varies from 60 to 90 days.
2. Continuous or Semi-continuous digesters : These are with input – output facilities. The digesters can be set for periodic addition and removal. Volume of feed slurry equals volume of spent slurry. The volume of feed slurry relates to retention time (RT) of substrate in the digester, RT should not be too short or too long.   The category is characterized by high capital cost, high efficiency and reduced labour requirement. Initial period of acclimatization (habituation with new environment) is long – nearly 45 to 60 days for each type of feed and the production is more uniform than batch digester.

Fig.2 represents the simplest biodigester design currently available in rural area. These digesters are semicontinous type and suitable for digesting cattle dung. And human excreta.

Anaerobic digestion of organic wastes produces biogas with of carbon dioxide (CO_2) 30-40 percent and methane (CH_4) 50-60 percent with lesser amount of hydrogen (H_2), nitrogen (N_2) and hydrogen sulphide (H_2S). Its calorific value is 5500 kcal/m^3.

3.   OPTIMIZATION OF PRODUCTION OF BIOGAS
The objective of optimization of biogas generation is to maximize the generated volume of biogas (methane) in shortest possible time (3 and 4). The volume of biogas generated is a function of 1) substrate 2) Carbon : Nitrogen ratio 3) pH of fermenting slurry in digester 4) Organic loading rate (OLR) 5) Concentration of solids in the feed. These factors are briefly discussed below.

Substrates : Organic materials (substrates) must have a high carbon content in the form of celluloses, hemi-celluloses, sgars, lipids and proteins. The substrate should also be low in lignin, which is non-biodegradable and hinders the microbial breakdown of the celluloses and hemi-celluloses.

Carbon-Nitrogen Ratio : The optimal C/N mass ratio is 30:1 A lower C: N ratio usually indicates the loss of nitrogen during digestion.

Temperature : Optimum anaerobic digestion takes place at 30°C in mesophilic range (25-40°C). Biodigestion essentially stops at 10°C.

pH range for fermenting slurry : The optimal pH range of fermenting slurry is from 7.0 to 7.2 account of but the pH may vary between 6.6 and 7.6. The pH usually drops due to excessive acid production that may be difficult to correct. Careful addition of lime to raise the reaction to near neutrality may restore the system to normal.

Agitation : Some daily mixing is desirable to keep heavier material distributed throughout the system. The amount of agitation required is highly dependent on the type of feedstock used.

Concentration of Solids : Terrestrial plant materials and animal habitation wastes generally require dilution to reduce the solid concentrations to 7-9 percent. A minimum amount of water should be added to aquatic vegetation since the initial solid content is generally below 7percent of fresh weight.

Organic loading rates(OLR) : Optimum organic loading rate depends on digester design, type of substrate and temperature. Adjustment of OLR is difficult and requires prior bench/laboratory scale studies followed by trial and adoption on field scale.

4. R AND D ISSUES IN BIOGAS PRODUCTION.

Important technical, R and D issues for improving prospects of biogas production are :

I.   Feed stock evaluation : It is aimed at better understanding of the process, improving biodegradability and predicting the performance under anaerobic digestion conditions and that too, without conducting long term batch tests, such as the effect of molecular structure on fermentation and rapid judgement based on substrate composition, effect of either physical size reduction or pretreating with acid or alkali on quality or additives commensurable with available resources.

II.   Reducing the construction costs of biogas plants : The most expensive parts of plant are digester and gasholder. Investigations relating, use of less expensive materials other than concrete, steel sheet and galvanized iron pipes and improvement of current designs, alongwith using solar energy/heat exchangers to maintain constant temperature for most effective fermentation throughout the year appear important. Minimizing corrosion, increasing stability, preventing contamination of drinking water and easy maintenance are special considerations.

III.   Using Digested slurry : The material needs to be evaluated to predict its value fully as manure/soil amendment through long term field trials and associated analyses of soils and crops at various locations in the country. There is greater need to find out the most effective method of treating and using the effluent on account of its dependence on the nature of feed stocks and local weather conditions.

5. CONCLUSION

A good number of plants fed with cattle dung in the country have been reported either not working satisfactorily or operating at lower capacity on account of its paucity created due to its traditional use as dried cakes. Pollution creating night soil (human faecal waste and urine) constitute another suitable substrate. The source is not fully utilized in our country and wasted through unhygienic decay because of varying degrees of psychological and religious barriers. Inhibitions against using night soil for biogas digesters are disappearing gradually as evident from connecting household latrines to gobar gas plants in rural areas and community installations working purely on night soil in Maharashtra & Gujrat.

1.   If the plant is adjusted/set properly the expected Rate of gas production per kg of fresh cattle dung will vary from $0.09m^3$ in summer months with mean temperatures of $30^\circ C$ to $0.04m^3$ in winter with mean temp. of $15^\circ C$. The biogas yields per kg. Dry matter of plant residues varies from 0.10 to 0.30 m^3 and it can safely be estimated at 200 to 300 m^3 biogas and 0.5m.t. biomanure (slurry)per m.t. of dry matter.

2.   There has been growing interest in developing ecologically, economically and socially sustainable, Integrated farming systems (IFS) by integrating plant – animal framing along with utilization of waste products after carefully considering soil conditions, agro-climatic features, water availability and marketing opportunities to fix the choice of crops, farm animals and aquaculture. Since biogas production by anaerobic digestion is a component in such IFS, it can provide pollution control (Fig.3) by utilizing sludge as quality manure after further decay in pits and effluents for growing algae in ponds and in various types of aquaculture. Water from fishponds can be used for fertigation of vegetables and field crops (3 and 5).

REFERENCE :

Joshi, K.G.   (1982)   an approach to social forestry: Planning for a village or a group of villages. Seminar on Social Forestry. February 20-22, 1982, Nagpur.

NAS   (1977) Methane generation from human, animal and agricultural wastes. National Academy of Sciences Washington. DC (USA)

NAS   (1981)   Supplement to energy for rural development : Renewable resources and alternative technologies for developing countries, National academy of Sciences, Washington, DC, (USA)

Chittenden, AE, Head, SW. And   Breag. G (1980) Anaerobic digesters for small scale vegetable processing plans. Report No.G.139. Tropical Products Institute, London

Swaminathan. M.S.   (1999) Integrated natural resources management   Key to sustainable advances in agricultural productivity. Lecture delivered at the 4th Agricultural Congress Jaipur, National Academy of Agricultural Science on Feb. 22, 1999, Science and Culture 65 (5-6) : 115-123.

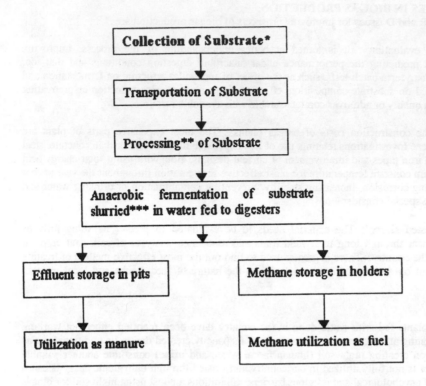

*Substrates or feed stocks used are human and animal habitation wastes, shredded leaves/forest litter, crop residues, weeds and unpalatable grasses and by-products from agro-based industries.

** Processing include size reduction, mixing, slurring with water and feeding to digester with proper C:N ratio.

***Limit for TS Conc. Of digester contents is 10% and operating volume should be 60% of total volume of digester.

FIGURE : 1
SCHEMATIC BLOCK DIAGRAM FOR BIOGASIFICATION OF ORGANIC WASTES

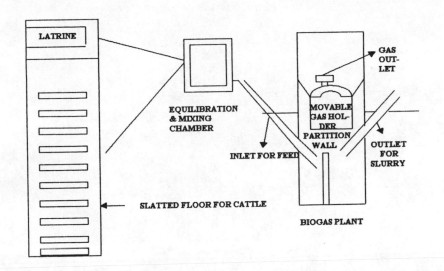

FIGURE 2
BIOGAS PLANT FOR THE GENERATION OF METHANE FROM NIGHT-SOIL AND

CATTLE-SHED WASTES.

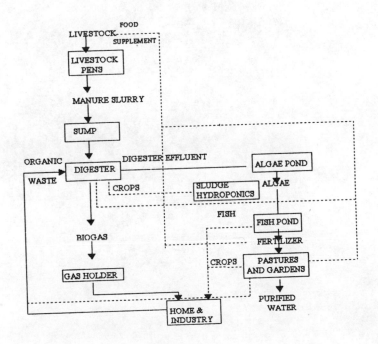

FIGURE 3
INTEGRATED FARMING SYSTEM

411

Physical, Chemical and Biological Processes of Geoenvironmental Reclamation at Gevra Open Cast Project (SECL)—A Case Study

S.K. Puri

Additional G.M., Gevra Project, SECL, India

1.   INTRODUCTION

Gevra Mine of S.E.C.L. has the distinction of single largest coal producing mine in India. During 99-2000 it produced over 18 million tonnes and removed 12 million cubic meters of O.B. Gevra open cast mine was started in the year 1981. The present extent of mine is 5 kms in strike & 2 kms in dip direction. Over burden removal is done with 10 cubic meter shovels in conjunction with 120 Te and 85 Te Dumpers 80 % of OB is back filed in the quarry and 20% is dumped in external dumps at present. Major coal production is done with front end loader 1.9 cubic meter and tipping truck of 10 to 15 Te capacity. Existing Gevra mine has a plan to expand from existing sanction capacity 12 MTPA to 25 MTPA with deployment of 42 m^3 shovels and 240 tonnes Dumpers.

SALIENT FEATURES OF MINES

1. Manpower 1340 (March'99)
2. Leasehold area 2023 Hectares
3. Active mining Area 987 Hectares
4. Main consumers of coal NTPC, MPEB, GEB, Cement Plants and others.

ENVIRONMENT MANAGEMENT CONSIDERATIONS

In 1981 Govt. of India decided that no mining project would be sanctioned by Govt. of India unless it was examined from environmental angle by MOEF. To aid this environmental appraisal committee for Mining projects was set up in 1981 under department of environment MOEF. Over the years there have been spate of instructions and guide lines.

Further in 1987, the world commission on Environment and Development issued its report entitled over common future. The report was discussed in U.N. General assembly and was adopted .

IT PROVIDES SUPPORT FOR THE VIEW THAT

i   Environment management plan should be integrated into the economic planning.
ii Public participation and involvement of community is required.
iii In good economic growth, poverty alleviation and environmental management are in many cases mutually consistent objectives.

MAJOR ENVIRONMENTAL IMPACTS ASSOCIATED WITH MINING ARE

1. Degradation of land, 2. Water pollution, 3. Air pollution, 4. Noise pollution inclusive of Air Blast and Ground vibration, 5. Population affected by Mining, 6. Diversion of Infrastructure built over Mining area & Diversion of nature old resources like rivers, Nullahs etc under which coal/mineral deposits may be existing, 7. Health hazard to workers.

Since 1986 Coal Mining Industry came to be identified as major contributing factors to environmental degradation particularly land degradation. Mine fires, due to large scale subsidence in Raniganj & Jharia Coal fields & total disregard to land reclamation by erstwhile mine owners have left ugly scars on the landscape.

NEEDS FOR LAND RECLAMATION

Less than 30% of total land of globe of 148950×10^3 sq. km is available for human use. Users of land are increasing at the rate of 2 %, so per capita availability of land is decreasing. In India 16 % of global population is accommodated in 2-3 % of global land. So per capital availability of land is 0.3 Ha. Again 5 - 7 % of Indian land mass is desert and quit a large part chronically desert prone. More than 40 % of land in India is seriously affected by erosion of one kind or the other.

The Indian rivers carry to the sea annually about 1500 M Te of eroded materials. The rate of erosion is higher where man has indiscriminately used the land and natural balance has been disturbed due to reshaping , modifying & defacing of natural topography by mining and other activities. This accelerated rate of erosion in the country contributed loss of valuable land resource at the rate of 5 - 7 M Ha per year. Siltation in the reservoirs (used for power generation) caused a reduction in their life to the extent of 1/4 or $1/5^{th}$ of the original life span.

The land is damaged because of excavation, spoil heaps, land erosion etc. Due to mining the top soil which may be dumped near by a place would be degraded, besides there is loss of flora & fauna. Top soil takes hundred years for formation as existed previously. Reclamation must take in to account not only forestry but also agricultural land formation. OB and top soil should be dumped & stocked separately. Land slides, soil erosion can be prevented in the hill top deposits provided sufficient amount of trees are planted around the periphery of mine. The water course or nallah need to be diverted from the active mine zone.

THE CUMMULATIVE EFFECT OF MINING ON ENVIRONMENT

Barren lands generated by cutting of greenery's, lowering of water table etc. are the problems, which once started have cumulative continuous cycle effect.

It is well established that open cast mining damages greenery as well as water table. These two natural resources are capable to protect all aspects of environment hence also the land. Forests acts as climatic stabilizers, inhibit flooding and sedimentation and can produce fuel and fodder.

Cooling capacity of one grown tree equals that of five average air conditioners. Tree roots provide extra shear strength to the soils to add in resisting gravity movement and surface erosion. This influence may continue 3 to 5 years after cutting. The forest canopy acts to dampen rain drop impacts, the tree ground litter further reduces erosion potential. The decaying vegetal matter enriches the nutrients in the top soil horizons and increase permeability to allow water percolation. The direct run off is minimised and sheet flooding prevented.

ADVERSE EFFECTS OF DEFORESTATION ARE

(1) Increased erosion and soil loss (2) Increased down slope sedimentation and its accompanying consequences such as more frequent flooding and silting of reservoirs (3) Upset of hydrologic regime and lowering of water table (4) Loss of soil nutrients (5) Destruction of wild life habitats (6) Reduction in aesthetic qualities. Deforestation has adverse effects on the climate, lowers the minimum temp. 1 deg. to 7 deg. F in all seasons and to increase the maximum temperatures by 1 to 6 deg. F in warm seasons and 1 to 3 deg. F in cold seasons. Humidity decreases by 2 to 25 % and rainfall reductions of 1 to 10%.

The reclamation work includes back filling, regrading, rehabilitation restoration and regeneration. It would be wise to pre plan the land to be used after mining operation either for the existing pattern or for alternative land use. Generally it could be desirable to restore the land reclamation in such a manner that reclaimed land can be used in a better way. To fulfill the objectives of reclamation, rule and regulation laid down should be strictly followed after considering the post mining land use. And its feasibility, provisions of necessary public facilities, economy, acceptance standards for adequate land stability, drainage vegetative coverage, hazards to public health & stability, adverse effects in flora & fauna. It is very important that top soil is saved for later use in vegetation etc.

The void created by extraction of mineral and rock should be back filled and graded approximately to its premining contour and quality condition or if these can be improved so much the better. The amount of void created by the extraction of mineral and rock depends upon the depth of extraction, the final slope of the pit, the slope of the back filled spoil dump, ultimate width of the bottom most benches etc.

The main criteria to select reclamation machineries are governed by the following factors :-
(1) High simplicity, high mobility, capable of doing both production & reclamation work, low cost, low maintenance, operating cost, longer life easy to operate & maintain.
(2) Suitability in all operating condition with high degree of safety.
(3) They must be suitable for specific mining, geological & topographical condition.

CASE STUDY AT GEVRA PROJECT
CASE STUDY OF RECLAMATION AT GEVRA PROJECT

Gevra open cast project (R12 Mty) is an existing mine. It is proposed to increase the production capacity of this project from 12 Mty to 25 Mtr to meet the increasing demand of power grade coal from Korba Coalfields and SECL. The life of the project will be 25 years. The maximum Quarry depth will be 218 m. The production program proposed is given below.

Year	Coal Production	O.B. Removal
1st year	16.8	11.35
2nd year	16.8	11.35
3rd year	16.8	10.79
4th year	20.0	18.04
5th year	20.0	18.04
2005-2006 6th year	25.0	26.44
2005-2006 7th year	25.0	30.24

GEVRA PROJECT PRESENT LAND STATUS (AS ON 1.3.99)

1.   Total acquired land for Gevra Project   = 2946 Ha

2.   Land given to Dipka Project   = 192 Ha
    Land given to Laxman Project   = 730 Ha
                      Sub Total   = 922 Ha

3.   Total land available with Gevra Project   = 2946 - 922
                                            = 2024 Ha ------------------ [A]

3a.   Land used for infrastructures/office buildings   = 89.83 Ha

3b.   Land used for Colonies   = 133.41 Ha

3c.   Land occupied by ext. OB dumps   = 210.75 Ha

3d.   Total Excavated Area   = 770.00 Ha
                     Sub Total   = 1203.99 Ha---------------- [B]

4.   Vacant land + MGR + Junadih Siding   = 820.00 Ha
         [[A] - [B]]

5.   The total excavated area   = 770 Ha.
                  a)Internal Dump Area   = 130 Ha.
                  b) Area Decoaled = 125 Ha.

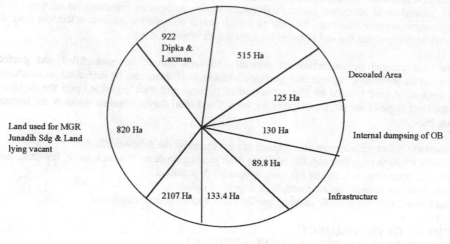

Pie chart labels:
- 922 Dipka & Laxman
- 515 Ha — Decoaled Area
- 125 Ha
- 130 Ha — Internal dumpsing of OB
- 89.8 Ha — Infrastructure
- 133.4 Ha — Colony
- 2107 Ha
- 820 Ha — Land used for MGR Junadih Sdg & Land lying vacant

SL.No	Particulars	Existing Area in use (Ha)	Additional land Requirement in (Ha)	Total in (Ha)
1	Land Acquisition	2024.00	732.00	2756.00
			Laxman addition 730.00 for 25 MTY Project	-------------- 3486.00 --------------
2	Infrastructure	89.83	30.00	119.83
3	Colony	133.41	12.00	145.41
4	Land occupied by Ext. O.B. dumps	210.75	300.00	510.75
			(A) Sub Total	775.99
5	Area for Safety zone		277.00	277.00
6	Area for Rehabilitation		30.00	30.00
			(B) Sub Total	307.00
			A + B	1082.99
7	Vacant land, MGR, Siding	820.00	(C)	820.00
	Quarry Area	=Acquired land	- (A+B+C) = 3486 -	1903.00

			=	1583.00
8	Internal OB dump land occupation	130.00	950.00	1080.00
9	Void Created would be (By 25 MTY Project for 35 year life)			503.00

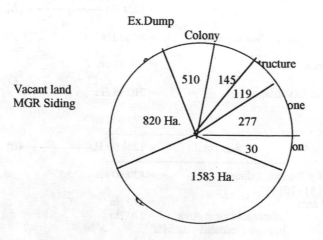

Pie chart labels:
- Ex.Dump
- Colony
- Structure
- 510
- 145
- 119
- one
- Vacant land MGR Siding
- 277
- 820 Ha.
- 30
- on
- 1583 Ha.

PLANTATION (GEVRA PROJECT)

Plane Area plantation   = 356 Ha

External OB Dump plantation Area = 51 Ha

Internal OB Dump plantation Area = 27 Ha   } 78 Ha.

                Total   = 434 Ha

PROPOSED LAND AVAILABILITY FOR PLANTATION AND OTHER AGRICULTURAL USE

From,
External OB dumps   = 510 Ha
Internal OB dumps   = 1080 Ha

                       1590 Ha

Area already planted   =   78 Ha

                       1512 Ha

STATUS OF FINAL LAND USE

1.	Total reclaimed land available	=	1080 Ha
2.	Water body	=	503 Ha
3.	Rehabilitation	=	30 Ha
4.	Thickly afforested Safety zone	=	277 Ha
5.	Thickly afforested Ext. OB dumps (Stabilised for any use Safely)	=	510 Ha
6.	Colonies	=	145 Ha
7.	Infrastructure, office buildings and Roads	=	120 Ha

                       2665 Ha

8.	Vacant land, MGR, Siding etc		820 Ha

                Total   = 3485 Ha

The Total reclaimed as well as slopes have been nicely vegetated by plantation of green grass like Dinanath, styto, flower trees, shrubs, fruit trees, food growing plants. A tourist spot named Amrakanan has been developed by planting fruit trees on 1.6 Ha of land on a total of 16 Ha of Internal dump on the west side of quarry.

Post mining land use is given separately. Reclamation plans for various stages of mine development have been drawn and are given in the plate no. Mining development is being done as per Post Mining Land use plan developed during the planning stage.

PROPOSED LAND USED FOR 25 MTY PROJECT
POST MINING LAND

1.	Land already acquired	=	2946 Ha
2.	Additional land to be acquired	=	732 Ha

            Sub Total(A)   = 3678 Ha

3.	Present land used for infrastructure	=	89.33 Ha
	Present additional land for infrastructure	=	30.00 Ha

            Sub Total(2+3)   = 119.33 Ha

| 4. | Present land used for colony | = 133.41 Ha |
| 5. | Additional land for colony | = 12.00 Ha |

| | | 145.41 Ha |

6.	Present land occupied by Ext. OB dumps	= 210.75 Ha
7.	Additional land occupied by Ext. OB dumps	
	Final stage of Mining activity	= 300.00 Ha
	(35 years at 25 MTY)	
	90 MM^3 OB for Ext dumps	
	Approx 15 MM^3 required 50 Ha	

| | Sub Total(B) | = 775.99 Ha |

| 8. | Present land occupied by Int. OB dump | = 130.00 Ha |
| | Proposed additional land occupied by Int. OB dump | = 1500.00 Ha |

| | Sub Total(C) | = 1630.00 Ha |

9. Vacant unreclaimed land would

$$A - (B + C)$$
$$3678 - (775 + 1630)$$
$$= 1273 \text{ Ha}$$

This includes safety zone Area & Area allocated for rehabilitation & land given to Laxman & Dipka 922 Ha.

| Area for safety zone | = 277.00 Ha |
| Area for Rehabilitation | = 30.00 Ha |

| | 307.00 Ha |

THE HIGHEST & LOWEST RL OF INTERNAL DUMPS & EXTERNAL 1 TO 6 ARE GIVEN IN THE FOLLOWING TABLE

Internal Dump no. 1		Internal Dump no. 2		Int.Dump no. 3		Int. Dump no. 4	
Premining	Present	Premining	Present	Preming	Present	Preming	Present
RL	RL	RL	RL	RL	RL	RL	RL
304 m	320m	312 m	340m	318m	346 m	299m	322m

Internal dump	Location	Area	Premining RL
5	South of J2 K2		
	Tail end	31800 sqm.	310m

Internal Dump	PMRL	Present RL av	Area
6	312	300 m	85200 sqm.

LAND RECLAMATION

The project involves 1550 Ha of land for Quarry area. The total volume of excavation within the boundaries is 1546.00 M cub insitu. This includes 835 Mt of coal and 1036 M cum of OB. It is estimated that about 800 cubic meters of OB will be accommodated in the back filling. The last working cut & haul road of quarry having void volume of 746 m cum will remain open. This may be used for working dip side property at a later date.

Presently back filling is being done in the quarry & small quantity OB is being dumped in the external dumps. After 4 to 5 years all OB would be accommodated in the internal dumps. The OB dumps will be terraced at the height of 30 m. The top bench would be graded & leveled to facilitate drainage of water away from the quarry.

BIOLOGICAL RECLAMATION

Simultaneously the top surface of the backfilled area is being given a cover of top soil for plantation .Tech & biological reclamation of the back filled area is given in the following table :

Year Till 1999	Plantation on Back filled Dumps	Area	No of plantations done
		21 Ha	50, 150
	Plantation on Ext. OB Dumps	79 Ha	226, 800 plants
	Total dump area plantation	100 Ha	276, 800 plants
			48, 000 grass plants
	Plain area plantation	348 Ha	2,497,250 plants
	Proposed 2000 - 2001		1,96,500 plants

The work of plantation has been awarded to MP V.V.N and they are paid on the basis of survival rate for 3 years.

CONCLUSION:-

There is appreciable improvement in climate and environment due to judicious measures taken against Land degradation at Gevra Project. All the ingradients of environment namely Air, Water, Noise, land, Fauna & Floura are taken care by proper implementation of a well planned reclamation programme. In the U.S for agricultural land , the original productivity of the premining agricultural land use is to be restored by Soil conditioning & use of chemicals. Our endeavor should be to return the land resource to the society in a better form than the one we received for mining purposes. The mining has a finite life , but the land use by the society would continue indefinitely . Whatever improvement we can do would be for the benefit of original inhabitants & migrants to that society.

ACKNOWLEDGEMENT:-

The author is thankful to the mine management for providing facilities, information and valuable suggestions.

BIBLIOGRAPHY:-
1.   Draft Project Report for 25 MTY of Gevra Project Exp.
2.   Various records of Survey and Envt. Dept. Of Gevra Project.

Controlled Surface Blasting in an Operating Refinery Environmental Concern and Control Measures

Nagesh V. Hanagodu

Sr. Manager (Civil), Mangalore Refinery & Petrochemicals Ltd., Kuthethoor P.O. Mangalore 574 149, India

1.   INTRODUCTION

Mangalore Refinery & Petrochemicals Ltd., (MRPL) is located about 20 km. north east of Mangalore in South Karnataka and is spread over an area of 1650 acres of land. The terrain is hilly and undulating with levels varying from +6M. MSL to +80M MSL. The 1st Phase of MRPL's 3.6 MMTPA refinery was completed in mid 1996 and the 2nd Phase for 6.0 MMTPA was commenced in December 1996 and completed in July 1999.

The problem discussed below pertains only to the development of Phase 2 crude tank farms (the tank farm consisted of five 65M dia 20 M high crude tanks). The crude tank area identified for Phase 2 was hilly with level upto + 60M and the whole area had been levelled to +10M to accommodate the crude tanks. The strata was varying widely with sheet rocks of granite overall with red murrum and interspersed with boulders of all types, shapes and sizes varying from +1M to +8M with depth of boulders up to 3M

During the execution of Phase 1 site grading work, all operation requiring surface blasting were carried out as per normal practice as there were no other facilities present at that point of time around the area of blasting. It was not possible to use this existing blasting practices in Phase 2 as the areas for crude tanks of Phase 2 were surrounded by existing Phase 1 plant structures, and any accidental fly rock or excessive vibrations could lead to severe damage and fire of existing facilities.

2.   BLASTING PROCEDURE FOLLOWED AT MRPL FOR PHAE 2 CRUDE FARM AREA

It was decided to adopt the most suitable controlled blasting technique with the help of blasting experts and MRPL engaged the services of National Institute of Rock Mechanics to assess the condition under which blasting operations were to be carried out. Initially trial blasts were conducted in the area to be site graded. The initial trial blasts were conducted using jackhammer holes and cartridge explosives with ohm meter, firing cable and exploder short delay detonators with muffing arrangements and parameters considered were layout of hole, depth of hole and charge, special gelatine (SG) having a diameter of 25 mm and weight of 125 grams per stick were used. The holes were initiated with short delay electric detonators and detonators were tested for continuity and resistance. The blast area were covered by wire mesh and sand bags to control fly rocks and ground vibration was also monitored at two locations and the fly rocks was contained to less than 5M by these trials.

Subsequent to the trial blasts, the blast design for sheet rock was finalised as follows. Hole diameter 34-38 mm (jackhammer only) larger diameter 100 mm was not recommended because of the risk of fly rock.

- Burden 1.0 M
- Spacing 1.2 M
- Hole depth 0.5 M to 1.5 M depending on site condition
- Number of holes per blast 40 max.
- Explosive per hole 1 stick per hole of depth greater than 1M and 1/2 stick for hole depth less than 1 M.
- Explosives to be used special gelatine from ICI or Kelvex 220 from KLL.
- Stemming material clay stick
- Initiation short delay detonators
- Number of blasts per day – no restriction – only at fixed times
- Muffling arrangement with wire mesh and sand bags
- Ammonium nitrate with fuel oil was no recommended because it does not perform properly in small diameter holes and quantity of explosive per hole is difficult to control and it cannot be used in rainy season.

3. CONTROL OF FLY ROCK WAS ACHIEVED AS FOLLOWS
- By selecting suitable hole diameter, burden, spacing and depth of holes
- By optimizing charge per hole
- By correct connection of delays
- By covering of the blast (muffling)
- By creating free face, and avoiding crater blasts

4. CONTROL OF GROUND VIBRATIONS WAS ACHIEVED AS FOLLOWS
- By using delay detonators
- By restricting number of holes
- By selecting suitable blast design
- By creating free faces and avoiding crater blasts

In an irregular blasting pattern as was being adopted here, it was essential to create more numbers of working areas and deploy large number of drilling and blasting crew to complete the job at the earliest.

5. BLAST DESIGN GUIDELINES FOR BOULDER BLASTING
Method of blasting – POP shooting only
Number of holes – 1 or 2 depending on the size of boulders
Hole depth – about 2/3 thickness of boulders
Charge per hole –1/3 stick of special gelatine half stick of special gelatine is permitted only after proper judgement by the supervisor
Initiation device –ordinary electric detonators
Muffling procedure – one sand bag on each hole before blasting to control fly rock
Large size embedded boulders were treated as sheet rock and all design aspects applicable to sheet rock blasting were followed.

6. ON SITE TRAINING FOR DRILLING & BLASTING PERSONNEL
The following topics were covered for on site training to carry out safe blasting.
- The fundamentals of blast design, selection of hole diameter, burden, spacing, depth, stemming with correct marking of holes depending on site conditions were explained.
- The difference between controlled blasting versus conventional blasting.
- The types of explosives available and the selection criteria of explosives.
- Influence of geology on rock blasting.
- Correct priming procedures for small diameter holes with concept of delay blasting.
- Practical ways of controlling ground vibrations and fly rock with general precautions and safety measures in blasting.

7. FIELD OBSERVATIONS & CONCLUSION
- If the soil around the boulders was cleared, the breakage was better compound to the boulders embedded in the soil.
- If the boulders required two or more holes the drillers usually drilled the holes giving more importance to their convenience to stand on the boulders rather than the placement of holes to minimise fly rock.
- During heavy rains the blast holes are filled with water and stemming with clay sticks in the

water holes are tight resulting in fly rock. Sand as a stemming material worked very well.

- Jackhammer could drill 25-30 holes in boulders whereas 40-45 holes are drilled in sheet rock.
- For two boulders separated by a seam of clay, the hole position should be adjusted to the centre of the boulders.
- If two boulders are lying one above the other with thick clay seam at the middle they should be blasted separately as the hole depth is restricted by the thickness of the upper boulders. This is very important for the front row so as to avoid the escape of gas energy through the clay seam, which may result in excessive fly rock.
- The faces should be thoroughly inspected before drilling. This has to be done to prevent the gas energy from escaping due to inadequate burden, which will result in fly rock. Any undercut or cavities are to be noted and the position of holes is to be adjusted accordingly.
- As per blast design, the hole depth should not be less than 0.5M when the depth was between 30 CM and 40 CM the charge was blown out.
- It is very essential to survey the are and finalise the level before taking the final lift of the blast. No hole should be left having thickness less than 1M, as this will give rise to excessive fly rock while final levelling.
- For some of the blast ground vibration were monitored with seismo graphs and vibration meter at certain location and the reading were found to be less than 1 mm/sec.
- Boulders blasted on top of hillocks, the noise generated was higher due to echo from existing tanks while valley blasting with same charge and number of holes, and the noise level was low.
- Free face was developed opposite to the pipe racks/structures to ensure safety of the structures.
- With the design procedure followed, there is no risk of ground vibration and fly rock throw was restricted.
- As a precautionary measure, a zone of 200 M from blast site should be treated as danger zone for men and materials and should be properly guarded at the time of blasting.

Phytoremediation of Mine Spoil Dumps using Integrated Biotechnological Approach

Asha Juwarkar, Kirti Dubey, Rahul Khobragade and Milind Nimje

National Environmental Engineering Research Institute, Nehru Marg, Nagpur 440 020, India

Introduction

The environmental pollution due to heavy metals is not a recent phenomenon but its magnitude has highly increased during last decades. Metalliferous mine and smelting industries and other industrial emissions, waste disposal, metal contaminated sludges, pig slurry, etc. are all the important sources of metal dispersion and enrichment in the environment. It has been observed that around mining area the dumps are exposed to erosion, thus facilitating the disposal of metals in soil and water through leaching.

Mining is one of the major contributor for the creation of wasteland in the country. In India, around 0.8 million hectares of land is under mining affecting the local environment drastically. Mining specially opencast, drastically destroys the land ecosystem, structure, and microbial community. Several microbial processes such as nitrogen and carbon cycling, humification and soil aggregation are practically non-functional and hence do not support biomass development. It is also responsible for contamination of water bodies, stream, groundwater and adjoining lands leading to severe environmental problems.

Most metals are phytotoxic when present in soil in excess even if they are essential plant micronutrients. The toxicity of metals in soil can be inferred from the reduction in species diversity. The tolerance may be suggested by the survival of plants on the soil in which they are growing while other plants perish under similar conditions of soil (Macnair and Baker, 1994). In view of ever increasing problem of heavy metal accumulation in the soils studies are required on the growth responses of plants grown in metal contaminated of soil/spoil and metal uptake by different parts of plants is urgently warranted as a means of phytoremediation of contaminated soil/spoil. The study of these parameters will help in understanding the effect of vegetation on leaching of heavy metals in ground water. Phytoremediation is a new technology employed for removing excessive toxic metals from contaminated sites. It is believed that phytoremediation with the use of specific plants is a relatively unexpensive operation. The method of phytoremediation is simple, safe and dependable.

The present paper describes the study on growth performance and metal uptake by plants in pot culture studies using iron, copper and zinc mine spoil separately amended with soil, farm yard manure and biofertilizer strains. This study focus on to assess the effect of vegetation on leaching of heavy metals from the heaps of iron, copper and zinc mine spoil dumps into ground water.

Methodology

Collection of Iron, Copper and Zinc Mine Spoil for Phytoremediation Study

Mine spoil samples from following three different metal mine sites were collected:

- Iron ore open cast mine under Sesa Goa Ltd., Kodli, Goa
- Malanjkhand open cast copper mine under Hindustan Copper Ltd., Malanjkhand, Madhya Pradesh and
- Zawar underground zinc mine under Hindustan Zinc Ltd., Udaipur, Rajasthan

Pot Culture Studies

Pot culture studies were conducted to ascertain the growth response of selected plant species and their metal uptake ability. Growth performance of plants was analysed in terms of height and metal uptake analysis was done by using Inductively Coupled Plasma Spectroscopic Method (APHA, AWWA, WEF, 1998). The growth performance was studied after a regular interval of 3 months and metal uptake study was done after every six months in triplicate. Two plant species were selected for each metal mine spoil. Plants selected for pot culture studies for different types of spoil are as follows :

Iron mine spoil	: *Pongamia pinnata* (karanj) and *Euginea jambulina* (jamun)
Copper mine spoil	: *Azadirachta indica* (teak) and *Dendrocalamus strictus* (bamboo)
Zinc mine spoil	: *Pongamia pinnata* (karanj) and *Dendrocalamus strictus* (bamboo)

Following treatments were screened under pot culture study :

T1	: Spoil as such
T2	: Soil as such
T3	: 1 part soil + 4 part spoil
T4	: 1 part soil + 4 part spoil + 50 t/ha FYM
T5	: 1 part soil + 4 part spoil + 50 t/ha FYM + biofertilizer

For each treatment three replications were maintained.

Results

Effect of different Blends of Iron, Copper and Zinc Mine Spoil, Soil, FYM and Biofertilizers on Plant Growth

Effect of different blends was studied on the growth performance of plants after 3,6,9 and 12 months of plantation in pot cultures in terms of height and health of the plant.

Plant Growth on Iron Mine Spoil :

Pongamia pinnata and *Eugenia jambulina* showed good response in spoil amended with FYM and inoculated with biofertilizers as compared to other treatment (**Fig. 1**). The plants showed very poor response in spoil alone. Addition of FYM as ameliorative material along with top soil boosted the plant growth. However, inoculation with biofertilizers further enhanced the plant growth. The biofertilizers enabled the plants to develop stress tolerance and healthy root system.

Plant Growth on Copper Mine Spoil

To study the growth response of plant on copper mine spoil *Dendrocalamus strictus* and *Azadirachta indica* were selected. *Dendrocalamus strictus* showed comparatively good response to amendments than *Azadirachta indica* (**Fig. 2**). The plants showed good response in spoil amended with FYM and inoculated with biofertilizers as compared to other treatments. However, the growth of plants in copper mine spoil is slower as compared to the growth performance in iron and zinc mine spoil.

Plant Growth on Zinc Mine Spoil

The response of plants in pot culture studies of zinc mine spoil, Udaipur is shown in **Fig. 3**. The study revealed that the spoil amended with FYM and biofertilizer at the rate of 50 t/ha increased the growth of *Pongamia pinnata* and *Dendrocalamus strictus* from 28 to 75 cms and 40 to 120 cms respectively. Very poor response was observed in T1 treatment representing spoil only due to lack of nutrients and substrates availability for rhizospheric microorganisms.

The increase in height is due to establishment of a microbial population in the rhizosphere of the plants and gradual initiation, and stabilisation of biogeochemical cycles. An improvement in the nutrient status of spoil amended with organic matter and inoculated with biofertilizers is observed in pot culture studies.

The garden soil also shows the presence of heavy metals but the uptake of metals by plants in soil is arrested due to its non-availability. The garden soil being rich in microbial population and organic matter render the metals non-availability to the plants.

Metal Uptake Studies

The metal uptake studies of metal mine sites were done to assess the effect of vegetation on the leaching of different metals in groundwater. The impact of soil amendment with organic matter and the site specific biofertilizers is being studied in pot culture study.

Metal Uptake by different Plants (Pot Culture Studies - Goa)

After 6 months of Plantation

The plants selected for iron mine spoil are *Eugenia jambulina* and *Pongamia pinnata*. Chromium is not detected in any of the plant species, because of the trace availability of chromium in spoil (**Table 1**). It has been reported that chromium uptake is linearly proportional to the amount of Cr in soil (Jackson *et al.*, 1999). Kodli mine being an iron mine, the spoil shows large amount of Fe. The addition of biofertilizers and organic amendment has decreased the uptake of zinc in both the plant species. Similarly, the uptake of Cu, Mn and Cd are observed to be reduced due to organic amendment.

After 12 months of Plantation

After 12 months of plantation Cr, Pb, Cd and Ni did not show translocation to the shoots and hence were not detected. However, uptake of Zn, Mn, Fe, Cu decreased after addition of site specific biofertilizers in T5 treatment. The amendment with organic matter reduced the uptake of metals (**Table 1**).

Metal Uptake by different Plants (Pot Culture Studies - Malanjkhand)

After 6 months of Plantation

The plants selected for studies on metal uptake were *Dendrocalamus strictus* and *Azadirachta indica*. After 6 months of plantation, trace metals of Cr, has been detected whereas Pb, Cd and Ni were not translocated to the shoots. After amending the spoil with organic matter, there was comparative decrease in the metal uptake. On treatment with site specific biofertilizers (T5) there was further decrease in the metal uptake in plants (**Table 2**).

Metal Uptake by different Plants (Pot Culture Studies - Udaipur)
After 6 months of Plantation

The plants selected for pot culture studies for zinc mine spoil, Udaipur were *Pongamia pinnata* and *Dendrocalamus strictus*. Chromium was not detected in both these plants while *Pongamia pinnata* showed no uptake of Cd and Ni also (**Table 3**). But *Dendrocalamus strictus* showed trace amount of Ni and Cd. The uptake of Zn by both these plant species was comparatively more than that of the spoil itself. This may be due to presence of zinc in available form in remarkable amounts in the farm yard manure which is used as amendments. This zinc in addition to that of spoil is translocated into the plants and increases the accumulation of metal in leaf. The other metals like lead, iron and copper shows reduced uptake. This is due to the reaction of the developing microflora and biofertilizers which alters the acidic pH. It has been reported that the acidity of soil increases the availability and uptake of metals (Nicolson *et al.*, 1997). The above metals are reported to be leached out in acidic pH but the organic amendment result in higher population density of bacteria in the rhizosphere which protects the plants against heavy metal toxicity by precipitating heavy metals from the proximity of plant roots. The organic matter also shows the presence of trace available heavy metals and the plant species shows uptake of metals which is within the toxicity range to the plant species and hence the plants are growing luxuriantly in pot culture studies.

Table 1 : Metal Uptake by different Plants (Pot Culture Studies - Goa)

Plants	Treatments	Metal concentration, ppm							
		Cr	Zn	Pb	Cd	Ni	Mn	Fe	Cu
After 6 months									
Eugenia	T1	ND	0.69	ND	0.30	ND	4.03	105.00	0.19
Jambulina	T2	ND	0.53	ND	0.28	ND	1.01	24.90	0.09
(Jamun)	T3	ND	0.67	ND	0.29	ND	2.83	50.00	0.15
	T4	ND	0.65	ND	0.29	ND	3.22	52.10	0.14
	T5	ND	0.52	ND	0.28	ND	1.66	22.60	0.13
Pongamia	T1	ND	0.72	ND	0.32	ND	2.98	104.40	0.21
pinnata	T2	ND	0.46	ND	0.29	ND	1.31	16.60	0.16
(Karanj)	T3	ND	0.98	ND	0.30	ND	3.34	38.70	0.11
	T4	ND	0.65	ND	0.30	ND	2.26	24.40	0.18
	T5	ND	0.64	ND	0.29	ND	1.15	21.10	0.15
After 12 months									
Eugenia	T1	ND	0.39	ND	ND	ND	2.56	45.30	0.12
jambulina	T2	ND	0.48	ND	ND	ND	1.75	28.90	0.14
(Jamun)	T3	ND	0.40	ND	ND	ND	2.19	25.80	0.13
	T4	ND	0.49	ND	ND	ND	3.44	52.00	0.15
	T5	ND	0.38	ND	ND	ND	1.70	22.30	0.13
Pongamia	T1	ND	0.46	ND	ND	ND	2.34	9.86	0.13
pinnata	T2	ND	0.48	ND	ND	ND	0.46	4.84	0.13
(Karanj)	T3	ND	2.12	ND	ND	ND	2.26	4.49	0.11
	T4	ND	0.60	ND	ND	ND	1.49	10.50	0.16
	T5	ND	0.51	ND	ND	ND	1.10	9.10	0.14

T1 - Spoil as such; T2 - Soil as such; T3 - Soil + Spoil; T4 - Soil + Spoil + 50 T/ha FYM; T5 - Soil + Spoil + 50 T/ha FYM + Biofertilizer

Table 2:Metal Uptake by different Plants(Pot Culture Studies-Malanjkhand)after 6 months of Plantation

Plants	Treatments	Metal concentration, ppm							
		Cr	Zn	Pb	Cd	Ni	Mn	Fe	Cu
Dendrocalamus	T1	0.03	0.54	ND	ND	ND	1.13	6.28	0.15
Strictus	T2	0.02	0.63	ND	ND	0.01	0.82	4.71	0.01
(Bamboo)	T3	0.01	0.42	ND	ND	ND	0.88	5.98	0.15
	T4	0.01	0.42	ND	ND	0.01	1.06	3.54	0.15
	T5	0.00	0.40	ND	ND	ND	0.82	2.98	0.14
Azadirachta	T1	ND	0.30	ND	ND	ND	0.52	5.28	0.15
indica	T2	ND	0.41	ND	ND	ND	0.63	1.97	0.23
(Neem)	T3	ND	0.44	ND	ND	ND	0.58	2.63	0.12
	T4	ND	0.40	ND	ND	ND	0.54	2.64	0.17
	T5	ND	0.30	ND	ND	ND	0.50	2.50	0.13

ND - Not detected.T1 - Spoil as such; T2 - Soil as such; T3 - Soil + Spoil; T4 - Soil + Spoil + 50 T/ha FYM; T5 - Soil + Spoil + 50 T/ha FYM + Biofertilizer

After 12 months of Plantation

After 12 months of plantation trace amounts of Cr, Pb and Cd was detected where as Ni was not detected in any of the treatments in *Pongamia pinnata*. However, none of the above metals were detected in *Dendrocalamus strictus*. Uptake of Zn was more in T5 treatment as compared to T2 treatment (spoil as such) showing similar trend of uptake as that of 6 months. Other metals like Fe, Cu, Mn showed decreased uptake after treatment with FYM and biofertilizers.

Conclusion

Growth performance and metal uptake studies by selected plant species have shown that plants of different types can be used for phytoremediation of metal mine spoil dumps and such type of approach will be helpful in bioremediation and recovery of contaminated soils into productive lands. Vegetation establishment on metal mine spoil dumps ensures no surface erosion by water and wind, reduction of leaching and lessening of amounts of toxic elements released into ground water thereby protecting the nearby environment.

References

APHA, AWWA, WEF. 1998. Metals by Inductively Coupled Plasma (ICP) Method. In. Standard Methods for Examination of Water and Wastewater, pp. 3-37

Jackson, B.P., Miller, W.P., Schumann, A.W. and Sumner, M.E. 1999. *J. Env. Quality.*, **280**, pp. 639

Macnair Mark R. and Alan J.M. Baker. 1994. Metal tolerant plants : An evolutionary perspective. In Margaret E. Farago (ed.)., *Plants and the Chemical Elements.*

Nicholson, F.A., Chamber, B.J. and Shepherd, M. 1997. In Extended Abstracts. *Fourth International Conference on Biogeochemistry of Trace Elements.* Clark Kerr Campus, University of California, California, pp. 499

Table 3 : Metal Uptake by different Plants (Pot Culture Studies - Udaipur)

Plants	Treatments	Metal concentration, ppm							
		Cr	Zn	Pb	Cd	Ni	Mn	Fe	Cu
After 6 months									
Pongamia	T1	ND	2.98	1.06	ND	ND	1.72	12.70	0.11
Pinnata	T2	ND	0.25	0.44	ND	ND	0.31	4.36	0.04
(Karanj)	T3	ND	1.50	0.56	ND	ND	1.46	7.82	0.04
	T4	ND	2.06	0.42	ND	ND	0.64	6.15	0.25
	T5	ND	2.68	0.36	ND	ND	0.66	4.44	0.11
Dendrocalamus	T1	ND	1.31	1.59	0.32	0.07	0.48	5.74	0.11
strictus	T2	ND	0.74	0.94	0.32	0.09	0.63	12.40	0.11
(Bamboo)	T3	ND	1.93	1.46	0.33	0.07	0.51	7.68	0.12
	T4	ND	1.40	1.45	0.34	0.08	0.59	7.06	0.13
	T5	ND	3.22	1.34	0.30	0.09	0.83	12.70	0.10
After 12 months									
Pongamia	T1	ND	2.87	0.70	ND	ND	1.50	10.10	0.12
pinnata	T2	ND	0.37	ND	ND	ND	0.93	4.82	0.10
(Karanj)	T3	0.02	4.60	ND	ND	ND	1.46	13.30	0.12
	T4	0.01	0.55	ND	ND	ND	0.64	9.49	0.13
	T5	0.01	0.42	0.21	0.01	ND	0.61	5.38	0.06
Dendrocalamus	T1	ND	3.14	ND	ND	ND	0.84	3.49	0.11
strictus	T2	ND	0.45	ND	ND	ND	0.40	4.51	0.11
(Bamboo)	T3	ND	2.05	ND	ND	ND	0.59	3.42	0.13
	T4	ND	1.08	ND	ND	ND	0.51	2.77	0.10
	T5	ND	3.15	ND	ND	ND	0.40	1.91	0.08

ND - Not detected, T1 - Spoil as such; T2 - Soil as such; T3 - Soil + Spoil; T4 - Soil + Spoil + 50 T/ha FYM; T5 - Soil + Spoil + 50 T/ha FYM + Biofertilizer

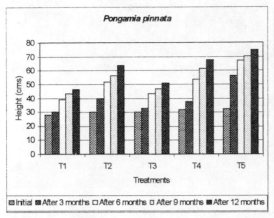

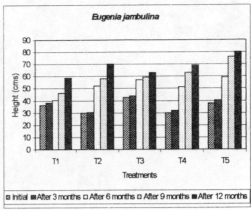

T1 - Spoil as such; T2 - Soil as such; T3 - 1 part Soil + 4 part spoil;
T4 -1 part Soil + 4 part spoi + 50 T/ha FYM; T5 -1 part Soil + 4 part spoi + 50 T/ha FYM + Biofertilizer

Fig. 1 : Growth Performance of Selected Plant Species Planted under Pot Culture with Iron Mine Spoil, Goa

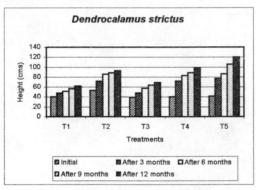

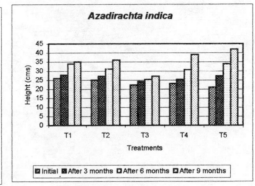

T1 - Spoil as such; T2 - Soil as such; T3 - 1 part Soil + 4 part spoil;
T4 -1 part Soil + 4 part spoi + 50 T/ha FYM; T5 -1 part Soil + 4 part spoi + 50 T/ha FYM + Biofertilizer

Fig. 2 : Growth Performance of Selected Plant Species Planted under Pot Culture with Copper

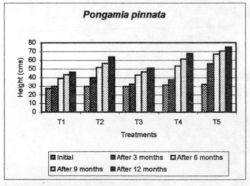

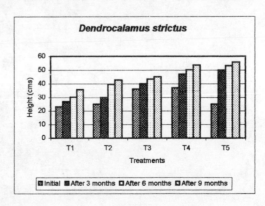

Mine Spoil, Malanjkhand

T1 - Spoil as such; T2 - Soil as such; T3 - 1 part Soil + 4 part spoil;
T4 -1 part Soil + 4 part spoi + 50 T/ha FYM; T5 -1 part Soil + 4 part spoi + 50 T/ha FYM + Biofertilizer

Fig. 3 · Growth Performance of Selected Plant Species Planted under Pot Culture with Zinc Mine Spoil, Udaipur

Environmental Impacts of Mining Operations: Some Practical Solutions

P.B. Rastogi

Ministry of Environment & Forests, Regional Office (Central), Lucknow, U.P., India

INTRODUCTION

Minerals constitute the backbone of the industrial economy of the State as well as whole Country but mining activities involving mineral extraction and its processing lead to wide range of environmental impacts on the air, water, land and socio-economic conditions of the people. The magnitude and significance of the environmental impacts of the mining and beneficiation activities depend mainly upon the way the mineral being mined, the scale of operation involved which is linked with the total production, location and nature of deposits, method of mining etc. However, mining needs effective management of all the surrounding natural resources along with the exploration of mineral.

STATUTORY REQUIREMENTS FOR MINING PROJECTS

EIA Notification issued by the Ministry of Environment & Forests, Govt. of India in January, 1994 (as amended in May, 1994) makes environmental clearance statutory for all the mining projects (major minerals) with leases more than 5 ha (MOEF, 1994, Singh *et al*, 1994). Environmental clearance is also necessary for all the mining projects located in ecologically sensitive/fragile areas e.g. Doon Valley in Uttrakhand, U.P. and Aravalli Range in Rajasthan etc. irrespective of the cost and area involved (MOEF, 1989 and 1992). Besides, all the mining projects located in/near forests; wildlife sanctuaries, national parks, wetlands, mangroves, biosphere reserves, hill and mountain areas etc. also need environmental clearance. Comprehensive guidelines are published for the control of environmental pollution caused by the whole gamut of mining operation and for the reclamation of mined areas.

In the present paper, an effort is made to highlight likely environmental impacts of mining and its practical solutions so that they are effectively used by the entrepreneurs leading to mitigation and significant reduction of anticipated impacts.

IMPACTS OF MINING OPERATIONS AND IT CONTROL

1. Air Pollution and its Control

Air Pollution in mines in mainly due to fugitive emissions of suspended particulate matter such as silica, fluoride, asbestos, metallic mineral fumes and gaseous emissions like SO_x, NO_x, CO, HC, CH_4 etc. In coalfields, soft coke making is a major source of air pollution. Besides, mining operations like drilling, blasting, movement of the heavy earth machinery on haul roads; collection, transportation and handling of minerals, screening; sizing and segregation units are the major source of such emissions. Underground mine fire is also a major source of air pollution in some coalfields.

Ambient air levels for coal mines have been notified and efforts should be made to control emissions of flume gases from beehive or regular coke oven plants and processing plants within acceptable limit. Air pollution in open cast mining can be drastically reduced by simple spraying of water on the haul roads used for transportation of minerals. Provision of hoods at transfer points, vulcanising of conveyer belt joints, underbelt cleaning and installation of dust suppression and/or extraction system on the conveyor belts also reduces air pollution drastically. Air borne dust during transportation of mineral can be easily controlled by using covered trucks, railway wagons etc. and

431

spraying water on the top. Similarly, air pollution can be controlled in coal/ore handling plants, crushing and screening units by providing dust suppression and extraction facilities. Regular maintenance of vehicles and machineries, tarring of roads from mine to ore collection site also reduces air pollution at site. Road side plantation acts as an effective barrier. Development of green belt all along the periphery of the mine lease area, ore stock pile and waste rock dump also reduces air emissions.

Direct exposure of air emissions to the construction workers at site can also be avoided by using mouth masks, creating awareness among the workers about the benefits of using it, briefing hazards of air pollutant and strict implementation of safety rules. It is needless to mention here that high levels of suspended and respirable particulate matter increases respiratory diseases such as chronic bronchitis and asthma whereas gaseous emissions contributed towards global warming besides causing health hazards to the exposed population.

2. Water Pollution and its Control

The major sources of water pollution in the mining operations are - Spent water from handling plants, dust extraction and dust suppression system ; liquid effluent from preparation and beneficiation plants; leachates/wash offs from waste/ tailing dumps and mine water pumped out from the drainage system.

Acid mine water is produced when mine (particularly pyrites) interacts with the water table, aquifer, perched water body or when surface water finds its way into a mine where sulphides are present in the ore or country rock. It has low pH, high levels of sulphates, iron and total dissolved solids which depletes oxygen level in the water and increases toxicity due to presence of heavy metals. Quantity of acidic mine water can be reduced by simply modifying mining methods viz. Sealing of mine or part after closure, surface reclamation, water diversion and control of ground water flow system by well fields. Acid mine drainage can also be controlled by deep well injection, subsurface dams and grout curtains. Recycle and reuse of waste water should also be ensured. To achieve this, drain should be constructed all along the road to collect the water. A garland drain around the waste rock dump and ore stock pile should also be developed. Sump with pumps should also be installed to trap runoff water in monsoon to prevent any spoil outside mine premises i.e. near shaft, ventilation fan, stock pile, boundary wall, block plantation etc. The waste water so collected may be used to irrigate plantation on over burden dumps, reclamation of the mined site, green belt development around the mine site and also for spraying on the haul roads after proper treatment. Standard waste water treatment methods for neutralisation and removal of dissolved solids can also be used like lime or lime stone treatment accompanied by aeration or oxidation process to convert ferrous iron into ferric iron, neutralisation with soda ash, caustic soda and anhydrous ammonia; reverse osmosis; ion exchange; electro-dialysis evaporation; ozone oxidation; desulphating; sulphide ion removal; microbiological control and permanent iron removal.

3. Surface/Ground Water Regime and its Management

Stream courses may be affected due to solid wastes generated due to mining operations which may cause over spilling and create artificial barriers in surface water bodies which should be checked at source itself. Surface water streams traversing the proposed area, if necessary to be diverted, should be diverted in such a way that it should not affect the regime and downstream users. Due consideration should be given to the catchment and command area involved, hydrological aspects, rehabilitation of flora and fauna, change in the drainage pattern etc. Water for mining or colony requirement should be obtained from a different source since its withdrawal from the same source in upstream may reduce the flow and may adversely affect the pollution absorption/dilution capacity of the stream.

Tapping of ground water specially in underground mining operations leads to lowering of the ground water table and depletion of aquifers which, in turn, affects wells, tanks, hydrodynamic conditions of rivers, underground recharge basins, aquifers (due to leakage of fines through leaching and percolation) and hydrochemical alternations (due to chemical pollution from mining).

Local geological, geomorphological and hydrogeological studies should be carried out in great detail before any activity is initiated to minimise adverse effects on the water regimes. Recommended control measures to be adopted include construction of settling ponds, sealing of shafts against aquifer zones, filling up of mined/stopped zones, controlled subsidence and location of tailings/waste disposal sites away from the water bodies.

4. Solid Waste Management

The main solid wastes generated in the mining operations include over burden, mine waste and tailing disposal. Segregation of the stones in the coal handling plant and the coal breeze also contribute to the solid waste generation. Its proper management is necessary.

Proper disposal site located on a secure and impervious base (solid rock if possible) should be identified at the planning stage itself for disposal of over burden and mine wastes. The site should be away from natural water streams and shallow aquifers and should ensure minimum leaching effects. Sanitary landfills are recommended for wastes with high concentration of sulphides or other reactive metals. Over burden mine wastes should be properly utilized by back filling and its proper reclamation, restoration and rehabilitation of the terrain and should not affect drainage and water regimes. Low-lying areas can also be utilized for over burden and waste dumping which can be profitably utilized after proper levelling and providing soil cover. Mine wastes may also be used in road construction. Over burden dumps of proper heights must be properly graded, terraced with contour drainage. Proper angle of the slopes and stabilization should be ensured so that over burden dumps can be revegetated. Top soil and over burden should be stored separately. Top soil should be retained for subsequent rehabilitation of the land since it contains majority of seeds, soil micro-organisms and more readily recycled plant nutrients. Extensive plantation around the waste dump and ore stock pile should be carried out which helps in suppression of dust generated due to winds. A green belt should be developed along the periphery, waste rock dump, ore stock pile, ventilation fan, roads and on block plantation and in selected open areas.

Regarding tailing disposal, effort should be made to reuse and recycle maximum mine waste water. Instead of tailings disposal into surface waste, land disposal should be adopted but these areas should not be connected with any drainage system of the area. Tailings may also be used for filling in the mine working. Tailing can also be disposed off by building a tailing dam converting it into dry land. Tailing dams should be constructed at the site having lowest permeability, deep ground water table, away from shallow aquifers, streams or rivers. No seepage and leakage should be allowed. Tailing dams should be designed in such a way that it provides storm water run off. Tailings can also be discharged in deep water after the quality of effluents is within the limits of aquatic life/planktons. No tailings should be disposed of in forests, sanctuaries and reservoir catchments. Toxic metals/chemicals should be disposed off in impervious tailing dam only. Peripheral plantation is necessary to avoid air borne tailing. All the tailing areas should be revegetated before being abandoned.

5.   Land Degradation and its Management

Mining is accompanied by extensive road activity, creation of new settlements, destruction of forest areas and introduction of new activities, which causes irrepairable damage to the soil cover resulting in loss of land fertility. Soil conservation and afforestation measures can be intermeshed with engineering solution to provide long term stability. Preparation of land use plan is necessary to achieve minimisation of adverse environmental impacts and effective and economic rehabilitation of the closed mine site to a productive afteruse. General survey must include present land use pattern, human settlements in the area, characterization of local ecosystem, climate of the area, geological, geomorphological, hydrogeological studies along with soil and water analysis.

An action plan for minimizing adverse environmental impacts due to proposed mining may be prepared including all the three phases of mining i.e. pre-operational, operational and post-operational.

During pre-operational phase, vegetational barriers and contours in the hills should be raised to prevent soil erosion and arrest mine wash. Check dams should be constructed and stabilized by vegetation. Banks of the streams should also be vegetated to prevent discharge of sediments into the streams.

During operational phase, vegetational barriers should be constructed along the periphery of mining areas, which will help in top soil preservation, lessening of adverse visual impacts, noise baffling and dust suppression etc. Tree cutting and felling should be restricted to barest minimum and planned well in advance. During the course of surface mining, top soil gets totally disturbed due to inherent nature of mining. The flora and fauna are totally affected. Top soil gets buried in the deeper horizons where as deep seated rocks are brought and laid on the top surface of the mine spoils. Thus, top soil characteristics of mine spoil are totally changed. It is deficient in NPK, humus and organic matter, soil organisms like bacteria, fungi, algae, protozoa, arthropods and other soil microbes as well as burrowing animals. The first task should be to stabilise the soil by planting grasses and leguminous species to provide soil binding, soil cover and soil enrichment etc. Main purpose of biological reclamation is to restore the mined out area to at least equivalent to if not better to that of premining stage. Plantation should be undertaken on waste rock dumps, near shafts, compressor and ventilation paths, along both sides of the roads, periphery of the acquired boundary.

During post-operational phase, mine areas should be properly levelled, reclaimed and rehabilitated for useful purposes e.g. agriculture, pasturage, pisciculture, recreation, setting up of wild life habitats/sanctuaries etc.

6. Noise, Vibration and Subsidence and its Control

While open cast mining operations cause noise pollution to the workers at site, in the neighbouring communities and damage to the nearby structures, underground mining poses serious occupational health problem for the workers. Main sources of noise pollution are blasting, drilling, ventilation fans, instruments used for underground mining, heavy earth moving machinery, dumpers, crushers, material handling and cleaning equipments etc.

Govt. of India (GOI) has notified ambient noise levels and should be strictly followed by keeping noise levels within limit. Noise can be abated by choosing right machinery and equipments. Noise can also be prevented at source by proper maintenance of compressors, ventilation fans and mine shaft by oiling and greasing and installing noise insulating enclosures. Rubber line should be provided to ore bunkers, skips etc. Noise transmission at compressors and ventilation fans should be intercepted by planting bushy trees around the source of noise generation. Residential areas and townships should be planned away from the mining areas. Boundary walls and green belts between the township and mining site should be developed which work as an effective acoustic barrier.

Ground vibrations are caused by blasting operations, rock bursts and bumps. Blasting operations damage buildings and regular vibration cause annoyance. Ground movement due to vibration can be controlled by avoiding overcharging, use of delays and improve blasting technology. Tectonism (geological formations) and seismicity (terrain stability) should be kept minimum. Buildings likely to be affected may be protected by trenching.

Subsidence causes extensive degradation of natural environment and damage to man made structures. It affects ground water regime, surface drainage pattern and usable lands. Highways, buildings, bridges, water and gas mains may be sheared, broken or twisted due to strain and slope changes. Subsidence damage can be alleviated by adopting precautionary measures on surface to protect installation or modifying mining methods to minimize deformation of the surface. Mining site should be away from the geographical discontinuities/weaknesses (faults/thrusts/fractures etc.). Structures should be designed with due consideration of subsidence (rigidity, flexibility). Controlled subsidence should be adopted. Adequate support should be provided in the underground mining to avoid subsidence. Worked-out areas should be filled with sand or other suitable packing materials.

7. Rehabilitation and Resettlement Problem

Generally, mining activities are confined to remote areas where population is very sparse. However, this small population should be properly rehabilitated and resettled. Rehabilitation plan must be prepared. Local traditional life styles of tribals should not be disturbed. Local population should get the maximum benefits. Land for land on pro-rata basis and compensation should be provided. Grants should be provided for construction of house and cattle sheds etc. Common infra-structural facilities in the township like school, panchayat ghar, well, post office, drinking water, sanitary system, park, tree plantation, space for grazing the cattles, recreational facilities should also be provided. Total replacement cost for any asset destroyed should be provided.

8. Health

Provision and use of dust respirators to all the workers working in dusty atmosphere should be made mandatory. Large water bodies created due to mining result in the introduction of water-borne diseases e.g. Malaria, Filaria, Schistosomiasis which leads to disaster rather than anticipated prosperity in the nearby mine area. Regular medical check up of the workers for any air or water borne disease should be carried out to detect any outbreak of disease and take necessary remedial measures immediately.

9. Socio-economic

The impacts of mining projects on the socio-economic plays an important role in the over all impact analysis of the project. It involves many factor such as aesthetics and local sentiments and requires a careful approach. Mining operations generate employment directly as well as indirectly. Local villagers may be employed in mining work. Some may be involved in plantation at site. Recruitment of people from the study area should be carried out by giving preference to weaker section. Literacy rate should be enhanced due to better economic conditions during mining operations, which will lead to better social status and thereby improved life styles.

10. Sites of cultural, historical, religious and scenic importance

Mining in such areas should be undertaken only after taking sufficient safeguards to protects these sites since these sites can not be restored once perished.

434

DISCUSSION

It is practically very often observed at the time of site inspection that mined area is left behind with scars of unusable land due to excavation and haphazard dumping of solid wastes. No proper reclamation of the mined area is carried out. No effort is made to reduce or minimise air, water, noise pollution at site when mining activity is in progress resulting into serious health problems in the workers directly exposed to the pollutants. It is all due to callous attitude of the project proponents. One of the reason may be that funds spent on environmental management do not add to their profitability and such investments do not yield any direct benefits.

Environmental management in the mining area deserves immediate attention of the entrepneurers, research scientists, policy planners and others involved in the mining activities. The remedial measures suggested in the EMP should be strictly implemented *in toto* and *pari passu*. An Environmental Management Cell (EMC) involving all the concerned persons from the site itself may be constituted and made responsible for implementation of environmental control measures suggested in EMP. Appropriate technology should be adopted which help in controlling pollution problems. Effective management and proper utilization of all the mine products should be the main motto during mining operations. This can easily be achieved by making serious efforts and finding its own solutions through intersectoral interaction and coordination. Technological innovations developed under research and development programmes in various technical institutes and universities should be tried and implemented to improve the surrounding environment and resolve the local environmental problems locally. Finally, abandoned mine should not be left as such but efforts should be made to utilise it for suitable purposes, for example, play grounds, residential complexes, industrial developments etc.

ACKNOWLEDGEMENTS

The author wish to express deep sense of attitude to Shri M.B. Lal, Chief Conservator of Forests, Ministry of Environment and Forests, Regional Office (Central), Lucknow, U.P. for constantly encouraging for writing this manuscript. Technical assistance of Dr.(Mrs.) Rubab Jaffer and Shri Ashok Kumar, Technical Assistants is also thankfully acknowledged.

REFERENCES

1. MOEF, GOI, 1989. Gazette Notification S. O. No. 102 (E) dated February 1, 1989, pp 1-7.

2. MOEF, GOI, 1992. Gazette Notification S.O. No. 319 (E) dated May 7, 1992, pp 1-5.

3. MOEF, GOI, 1994. The Environment Impacts Assessment Notification S.O. No. 60 (E) dated January 27, 1994 (as amended on May 5, 1994) pp 1-7.

4. Singh, H; Duraisamy, A; Subramanium, U and Debabrate, D, 1994.
5. **Handbook of Environmental Procedures and Guidelines**, pp 4-10 (MOEF, New Delhi, India).

Rehabilitation of an Asbestos-contaminated Mine Site: A Case Study from Cornwall, South West England

M.J. Heath[1] and D.P. Roche[2]

[1]*Camborne School of Mines, University of Exeter, Redruth TR15 3SE, UK*
[2]*David Roche Geo Consulting, 19 Richmond Road, Exeter EX4 4JA, UK*

INTRODUCTION

This paper describes the investigation, remediation and monitoring of an asbestos-contaminated mine site at South Wheal Francis near Camborne, Cornwall, South West England. The work was commissioned by the site owner, Kerrier District Council, which is also the local authority with environmental responsibility in the area in which the mine is situated, and was undertaken by the Earth Resources Centre, University of Exeter, working with Frank Graham Consulting Engineers (now WSP).

South Wheal Francis is an important heritage site forming part of the Mineral Tramways Project, a major tourism initiative intended to attract many visitors. The site is also very close to the small settlement of Piece, with its primary school and popular public house. The presence of asbestos on site was, therefore, considered to have wide implications for public health and was the subject of detailed investigation leading to extensive remediation works.

SITE DESCRIPTION

South Wheal Francis Mine is an abandoned copper-tin mine situated immediately east of Treskillard, approximately 3 km south east of Camborne, Cornwall (National Grid Reference SW 681394). The site extends to some 8 hectares and features many important mine buildings including several Scheduled Ancient Monuments which make it of high archaeological value (Smith, 1992).

The mine is situated at an elevation of about 170 metres above Ordnance Datum on the northern boundary of the Carnmenellis granite at its contact with the surrounding killas (Devonian slates and sandstones). The area is heavily mineralised. The recorded history of the mine dates from 1824 (1). Originally worked for copper, production had switched to the deeper tin by 1877. The mine became part of Basset Mines Ltd. in 1896 and closed in 1918. Steam to power the mine machinery was generated in the Boiler House using a bank of six Lancashire boilers (2) and, as the main asbestos contamination occurs within and around this building, it appears to be the boiler insulation that is the main source of asbestos on site.

South Wheal Francis is of special importance owing to the wide range of buildings that have survived at the site. These buildings are clustered mainly around Marriott's Shaft, with two additional buildings around Pascoe's Shaft at the western end of the site (Figure 1). Asbestos contamination was first observed at the surface in and around the Boiler House, Pumping Engine House and Winder House during early work to consolidate buildings. The detailed investigation of the nature and extent of this contamination and the subsequent cleanup operation are described in this paper.

Mine sites are characterised by a range of potential hazards, each of which has been assessed and addressed during the development of South Wheal Francis as a heritage site. These hazards include voids (shafts, stopes and other excavations), derelict buildings (some initially in a dangerous condition), heavy metal contamination and, of particular importance at this site, asbestos contamination. Geotechnical and contaminant studies were carried out at South Wheal Francis by Frederick Sherrell

Consulting Engineering Geologists of Tavistock, Devon, in 1992/93. These investigations included trial pitting and drilling and an assessment of chemical contamination at the site. Remediation of the hazards associated with voids, derelict buildings and heavy metal contamination are beyond the scope of this paper but the presence of high levels of heavy metals as well as asbestos in contaminated soils has implications for their disposal, and some heavy metal determinations (not reported here) were carried out on asbestos-contaminated material prior to removal for disposal.

SITE INVESTIGATION
The site investigation described here focuses only on the asbestos contamination but, having determined the distribution of this principal hazard, further investigations were carried out to determine the level of heavy metals in the asbestos-contaminated soils, as this affected the cost of disposal. The asbestos survey was carried out in three phases.

Visual inspection
The first phase of site investigation consisted of a walkover survey with visual inspection of material exposed on the ground and on other surfaces. Blue fibrous material was observed at the ground surface in and around the Boiler House and in the Winder House and Pumping Engine House. Samples of this blue fibrous material were collected and analysed at the Earth Resources Centre, University of Exeter, by X-ray diffraction (XRD) techniques. This confirmed that the blue fibrous material exposed on site was the mineral *riebeckite* (an asbestiform amphibole also known as *crocidolite*), referred to here as *blue asbestos*. Having confirmed the presence of this highly hazardous material at the surface, the next stage of investigation entailed a soil survey to determine the distribution of buried asbestos around the site.

Soil surveys
The purpose of the soil survey was to obtain representative soil material in order to determine the possible presence of buried asbestos. Samples were collected by augering to depths of up to 0.60 m, the sampling depth being limited by the stony nature of the ground. Special attention was paid to mounds and other bodies of dumped material.

As the auger survey was essentially of a 'walkover' type, precise sampling locations were not recorded except where samples were taken for analysis. Augering density varied across the site according to the nature of the ground, previous knowledge of asbestos distribution around the buildings, and accessibility. In the areas surrounding Marriott's Shaft (Zone I, Figure 1) and Pascoe's Shaft (Zone II), augering took place at about 5 - 10 m intervals (subject to accessibility). In the landscaped area (Zone III) and the north eastern extension of the site (Zone IV), the sampling interval was approximately 20 - 25 m; in the thickly vegetated Zone V (where access was difficult), the sampling density was lower.

From a total of some 220 auger points, 18 soil samples were selected for analysis. At each auger point, soil material was examined visually and any fibrous material present was noted. Samples were selected for analysis by X-ray diffraction on the basis of this visual examination and to provide a reasonable cover of samples across the site.

Of the 18 auger samples selected for analysis, only one (from the Boiler House floor) was shown positively to contain riebeckite. As the cleanup operation later revealed, auger sampling has its limitations where the asbestos occurs as discrete *objects* (ropes, bags, sheeting etc.) rather than as a dispersed contamination. There was, therefore, some overlap of the investigation and cleanup phases of the work, as the full extent of the asbestos contamination could not be determined until excavation commenced (as discussed further below).

Surface (deposit) sampling
In addition to the soil sampling, deposit (dust) sampling was also carried out from surfaces in buildings to determine the extent of asbestos contamination in accumulated aeolian material. Sampling of these surface deposits took place under damp conditions with subsequent analysis by X-ray diffraction (XRD). Selected surface material was sampled by hand-picking and scooping with a knife. These samples included fine dust collected from surfaces within and around buildings, representative samples of the widely distributed blue fibrous material, and some white fibrous material collected from the Miner's Dry. A total of 17 surface samples were collected. Riebeckite (blue asbestos) was identified by XRD in six of these samples, all located within the main group of buildings around Marriott's Shaft. In these samples, the riebeckite was found to occur either in the form of small woven mats or pads (up to a few cm in dimensions) or, in one case, as asbestos rope. Blue fibrous material was also identified visually in Pascoe's 80-inch engine house and on the surface of a small dump adjacent to Pascoe's Whim engine

house. No asbestos was identified in the fine dust samples collected on ledges and other surfaces. The white fibrous material observed in the Miner's Dry was identified as glassy in structure and appeared to be glass fibre.

The surface contamination appeared to be due largely to the dispersion of asbestos by the wind and, importantly, by nesting birds, notably jackdaws (*Corvus monedula*). In fact, much of the blue asbestos observed at the surface occurred beneath nesting sites.

RISK ASSESSMENT

The investigation revealed the presence of considerable quantities of blue asbestos at the site, and the subsequent review and assessment of the findings confirmed the need for remediation of the asbestos contamination.

Asbestos is a fibrous silicate mineral which is chemically inert, heat resistant and mechanically strong, and has in the past been used extensively for industrial purposes, including fire protection and thermal insulation. Asbestos materials are prone to break down into fibrous particles or dust, especially when dry, exposed or disturbed. Inhalation by humans and animals by breathing in dust is therefore the greatest risk, and long term harm to health is a consequence in various forms including asbestosis, bronchial carcinoma and mesothelioma, which can be fatal. Ingestion by humans and animals is another possible route for asbestos fibres to enter the body, either by direct ingestion or in drinking water.

Blue asbestos (crocidolite/ricbcckite) is considered to be the most harmful form of asbestos to human health. Various control limits and action levels have been set for exposure by humans to asbestos inhalation. The most commonly quoted is the control limit for blue asbestos exposure of 0.6 fibres per millilitre of air over 10 minutes (or 0.2 fibres per millilitre over 4 hours) (3).

Risk assessment methodologies were used to evaluate the risks and the need for remediation. In the existing conditions, there was ample opportunity for direct inhalation by humans or animals when walking or playing on the site, with the potential to transport the particles home attached to clothing or hair, and there was also the likelihood that airborne particles could be windblown off site onto adjoining agricultural land and nearby habitation. Under existing conditions, the risks and consequences of asbestos contamination were considered to be high. During the remedial works, there were similar risks to workers, and also an enhanced level of risk from site disturbance allowing more widespread contamination unless the works were properly controlled and implemented. Following completion of the remediation, which was carried out under proper management and control, the remaining risks from asbestos should be reduced to low and acceptable levels, with all asbestos either safely removed from the site or safely buried and covered.

REMEDIATION PLAN

Before any remediation works are undertaken, it is important that the proposed actions are carefully considered and planned. A Remediation Plan was prepared to identify the methods and procedures, to allocate responsibilities and to obtain approvals. It formed the basis of the employer's requirements from the remediation works contractor. The contractor appointed was required to demonstrate proper accreditation and competence to undertake asbestos removal works, and to produce a detailed method statement describing the implementation and management of the works. Technical competence as well as contract price were, therefore, emphasised as the basis for selection and award.

REMEDIATION WORKS

Having determined the distribution of asbestos contamination at the site, the remediation programme was implemented during the period March - May 1997 by licensed contractors R. I. & H. J. Bartlett Ltd., of Sturminster Newton, Dorset, with the assistance of specialist asbestos contractors Southern Counties Asbestos Removal Ltd. (SCI). During the removal work, air quality was monitored for health and safety purposes by Argus Laboratories Ltd., Brighton. The work included the bulk excavation and removal of the contaminated material, the installation of a protective capping layer and the reinstatement of the surface for safe public access. The excavation work focused on the buildings and their surroundings in Zone I; in other parts of the site, where no buried asbestos had been identified, surface material was carefully collected by hand and removed along with the buried asbestos. The total cost of the work was in excess of £250,000, funding being provided by English Partnerships under the Leasehold Reform, Housing and Urban Development Act 1993.

Excavation and capping (Zone I)

Following careful excavation of the surface soils, with careful damping of the contaminated materials during dry periods, and bulk excavation and removal of all asbestos materials, the reduced ground surface was carefully inspected and checked to be clean and ready to receive a new surface capping layer. The capping layer (shown in Figure 2) comprised:

- a geosynthetic separator membrane - Terram 900 - to act as a physical barrier to prevent possible upward migration of any undiscovered asbestos fibres residual in the underlying soils;
- a galvanised wire mesh (or "rabbit net"), with 20 mm mesh size and 1 mm gauge wire, to act as a physical barrier to rabbits or other burrowing rodents which might otherwise compromise the capping system;
- a granular aggregate surface layer of well-graded crushed rock material derived locally from a client-specified source of mine waste, with a 500 mm minimum depth, to achieve final surface levels.

As the buildings present at the site were of historical importance, their protection was part of the excavation plan and great care was taken to ensure that the fabric of the buildings themselves was not damaged. It was, therefore, necessary to carry out much of the excavation work by hand or with small mechanical excavators. There was also close liaison with the district archaeologist.

Until excavation commenced, the full extent of the asbestos contamination could not be fully determined. Thus there was some overlap between the site investigation and cleanup. This created some difficulty as far as the work schedule was concerned, the cleanup period being unavoidably extended as more asbestos was discovered. Excavation continued until all observed asbestos was removed, even where it extended beyond its originally-identified boundaries. Prior to implementation of the remediation works, the volume of asbestos-contaminated materials to be excavated and removed was estimated to be in the order of 50m^3, but approximately double this volume was actually removed.

Forensic walkover (Zones II - V)

In addition to the excavation of buried asbestos in Zone I, surface contamination was removed from Zones II - V through the implementation of a 'forensic walkover' undertaken by a team of workers in which all observed surface contamination was collected by hand during a systematic sweep on foot of the entire site.

Nesting sites

In light of the observed occurrence of asbestos contamination beneath nesting sites, and its presence in fallen nests, the cleanup programme included removal of asbestos from the upper parts of buildings used as nesting sites. This was carried out manually at the end of the nesting season to minimise disruption to the life cycle of the birds themselves, access being gained through the use of a hydraulic lift.

UK LEGISLATION GOVERNING THE HANDLING AND DISPOSAL OF ASBESTOS WASTES

The excavated asbestos-contaminated material was handled, transported and disposed of in accordance with UK regulations governing the management of hazardous wastes. These regulations prescribe the correct procedures that ensure the safe management of hazardous materials.

Waste containing asbestos is classified as 'controlled waste' under the Environmental Protection Act, 1990 (EPA) and as 'special waste' under the Special Waste Regulations 1996. Under these regulations, movements of asbestos waste have to be traced by means of a 'consignment note' system until they reach a disposal site (a landfill) licensed under the Waste Management Licensing Regulations 1994 to accept asbestos waste. Licensed sites operate special procedures to ensure the safe disposal of asbestos. Carriers of asbestos waste must also be licensed and are subject under the EPA to a 'Duty of Care', which places a responsibility on everyone handling or controlling the waste to ensure it is managed safely and transferred only to persons authorised to deal with it.

The transport of asbestos waste must also be in accordance with a number of regulations, Statutory Instruments and codes of practice. The Health and Safety Executive (HSE) also issues guidance on the transport of dangerous materials.

Asbestos waste should be kept separate from other waste, should be double-bagged in heavy-duty polythene bags and clearly marked with the label prescribed for asbestos prior to transport to the disposal site. The asbestos waste should also be carried in a suitable container, such as an enclosed skip, which should also be labelled. Carriers who transport waste must be registered with the appropriate regulatory authority: the Environment Agency (in England and Wales) or the Scottish Environment

Protection Agency.

In the case of South Wheal Francis, asbestos was double-bagged, the outer bag red in colour and suitably labelled, and was transported in sealed skips by licensed road carrier to a licensed landfill approximately 10 km from the site. The requirement for such handling of asbestos waste imposes costs on the cleanup of sites like South Wheal Francis, but wastes resulting from the cleanup of historically contaminated land are exempt from the Landfill Tax established under the Finance Act of 1996.

Safe working with asbestos was a central consideration at all aspects of the works, which were undertaken with reference to current UK regulations and guidance as published by HSE and the Construction Industry Research and Information Association (CIRIA), including *The Control of Asbestos at Work (Amendment) Regulations 1992* and the accompanying *Approved Code of Practice* (3) and the *Guide to Safe Working Practices for Contaminated Sites* (4).

HEALTH & SAFETY ASPECTS
As asbestos is a hazardous material, great care was taken to ensure the health and safety of site workers during all stages of the site investigation and cleanup and a risk assessment was carried out before any work commenced.

Site investigation
During the site investigation, special care was taken during sampling. As the soil environment in Cornwall is almost always damp (and sampling was only carried out under these conditions), there was no asbestos dust hazard associated with soil sampling and no special protective measures were needed. Protective suits, filter masks, goggles and gloves were, however, available at all times in case site conditions changed. Surface sampling was considered to present greater risk requiring some protective measures to be taken. Protective suits (with hoods), filter masks and wellington boots were worn at all times during sampling, though conditions remained generally wet during the sampling period.

Remediation
During the remediation phase, the working area remained fenced off at all times with access permitted only to authorised and properly protected personnel. Within this area, workers wore disposable red protective paper suits with hoods, filter masks or respirators (depending on site conditions), goggles and protective boots. Access to the work area was restricted to a single controlled entrance/exit where, on leaving the work area, protective suits were removed and bagged for disposal and where boot washing facilities were available.

Health and safety monitoring was carried out throughout the cleanup operation to ensure the protection of site workers. The results of this monitoring, carried out by specialist health and safety contractors, Argus Laboratories Ltd., demonstrated that maximum permitted levels of airborne fibres were not exceeded at any time during the asbestos removal operation.

ENVIRONMENTAL MONITORING PROGRAMME
In addition to health and safety monitoring to ensure the safety of site workers, environmental monitoring was also carried out throughout the 8 week period of asbestos removal in order to determine the extent of any asbestos dispersion that might have accompanied disturbance of the buried asbestos. The primary asbestos contamination was confined to the buildings around Marriott's Shaft, notably the Boiler House, Winder House, Pumping Engine House and Compressor House, and the monitoring was focused on asbestos removal from these buildings.

Two approaches to air quality monitoring were adopted:

(A) **Passive sampling**: 'frisbee' dust samplers were installed at four different points on site, their locations being chosen so that sampling would allow for changing wind direction (Figure 1). Using the 'frisbee' technique, dust collected in the sampler is washed into a collection bottle by rainfall or with water and is later separated by filtration. The filtrate is then analysed for blue asbestos (riebeckite) by X-ray diffraction. Three of the four samples collected using the 'frisbee' samplers did not show any detectable asbestos and there was only a very small trace in the fourth sample (F3, Figure 1). It is worth noting that sample F3 was collected at a site immediately north of the heavily contaminated area to the east of the Boiler House (see Figure 1), this was the scene of extensive excavation works during the cleanup operation and represented one of the major potential sources of airborne contamination. The absence of any serious contamination from sample F3 indicates a good level of environmental control

441

during the works. The results suggest the absence of any widespread airborne asbestos dispersion during the asbestos removal operation.

(B) **Active sampling**: thirteen active (pumped) samples were collected, each over a period of one working day using a Negretti personal sampler operating at a pumping rate of approximately 2.0 litres per minute, some 960 litres of air being sampled over an eight hour period. Each sample was collected at a height of approximately 1 metre above ground level, at a distance of between 5 and 10 metres from the working area in a downwind direction. Using the personal samplers, dust is collected on a filter which is then examined for asbestos by scanning electron microscopy (SEM). The samples collected by this active sampling did not show any significant asbestos dispersion during the cleanup operation. Some fibrous material was observed in some samples but most of these fibres were shown by scanning electron microscopy with secondary electron analysis to consist of paper (probably from the protective suits worn by site workers) and organic material (from vegetation).

A small number of fibres identified as 'probably asbestos' were observed in four samples but represented very low levels of contamination. This suggested that very little dispersion of asbestos took place during the cleanup. This is consistent with the results of the health and safety monitoring carried out by Argus Laboratories Ltd. which showed that the maximum permitted level of airborne asbestos fibres was never exceeded during the asbestos removal operation.

POST-REMEDIATION INSPECTION
Following completion of the asbestos removal operation, a visual inspection of surfaces, walls and exposed ground was carried out to ensure that no visible asbestos remained on site. Special attention was paid to areas previously seen to be contaminated with asbestos and to areas used for the handling and loading of asbestos-contaminated material. The inspection included both the Marriott's Shaft and Pascoe's Shaft areas. The visual inspection did not reveal any remaining asbestos contamination at the surface, even at the previously contaminated parts of the site. Furthermore, none of 10 additional surface samples collected immediately after asbestos removal showed any trace of blue asbestos contamination, even though many of these samples were collected from locations where dispersed or residual asbestos contamination might have been suspected. The results indicate that no asbestos remained at the surface and that no significant asbestos dispersion occurred during the cleanup operation.

CONCLUSIONS
Blue asbestos (riebeckite, variety crocidolite) was found to contaminate the site of the former South Wheal Francis copper-tin mine. The primary asbestos contamination was confined to the main buildings around Marriot's Shaft, especially the Boiler House where it was used for boiler insulation. Secondary surface contamination was also present due mainly to dispersion by the wind and by nesting birds.

The blue asbestos contamination was investigated and removed successfully and the site made safe for public access. The programme was not entirely straightforward, however, as the nature of the asbestos contamination (as discrete objects such as ropes, bags etc.), made identification of the extent of the contamination problematic (individual objects are difficult to locate in an auger survey). There was, therefore, some overlap between the site investigation and cleanup phases of the work. This had implications for scheduling the cleanup operation, delays being introduced as additional asbestos contamination was discovered as the work progressed.

The results of the passive and active air quality monitoring show that no significant airborne dispersion of asbestos occurred during the asbestos removal operation. This was confirmed by the analysis of surface deposits collected after asbestos removal. This was ensured both by careful damping of soils during their removal and by the generally wet conditions of this Cornish site during the work period.

The results of the post-cleanup inspection show that no visible asbestos remained on site and that the surface environment appeared clear of asbestos. The overall results show that the impact of the asbestos removal on environmental air quality was negligible and that its removal was accomplished successfully without any significant environmental impact.

REFERENCES
1. Dines, H. G. (1956). *The metalliferous mining region of South-West England*. Memoirs of the Geological Survey of Great Britain. HMSO, London.
2. Smith, J. R. (1992). *Kerrier Land Reclamation Scheme. Marriott's and Pascoe's Shafts South Wheal Francis: Archaeological Assessment*. Cornwall Archaeological Unit Report.
3. Health and Safety Executive (1992). *The Control of Asbestos at Work (Amendment) Regulations 1992*

with accompanying Approved Code of Practice.
4. CIRIA (1992). *Guide to Safe Working Practices for Contaminated Sites.* CIRIA, London.

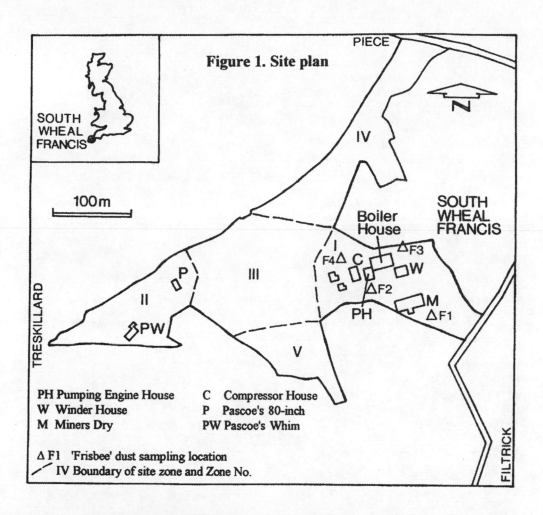

Figure 1. Site plan

SOUTH WHEAL FRANCIS

100m

PH Pumping Engine House   C   Compressor House
W Winder House   P   Pascoe's 80-inch
M Miners Dry   PW Pascoe's Whim

△ F1   'Frisbee' dust sampling location
IV Boundary of site zone and Zone No.

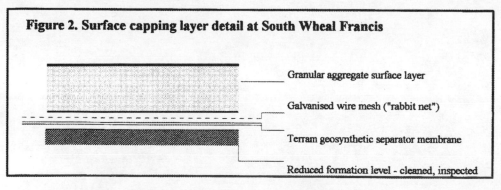

Figure 2. Surface capping layer detail at South Wheal Francis

Granular aggregate surface layer

Galvanised wire mesh ("rabbit net")

Terram geosynthetic separator membrane

Reduced formation level - cleaned, inspected

Sustainable Development of India and Coal Sector

J.L. Jethwa

Scientist-in-charge, Central Mining Research Institute Regional Centre, 54-B, Shankar Nagar, Nagpur 440 010, India

1.0 INTRODUCTION

Mining industry is an integral part of the very human life. Coal , oil and natural gas accounts for 30%, 31% and 20% respectively of the world's total energy needs. In India, coal based energy meets about 70% of the total energy requirement. Despite all efforts to raise energy through unconventional sources, coal continues to remain our prime energy source at the present and at least for the next decade. Mining industry, thus occupies the center stage and is indispensable for the overall industrial growth.

Coal mining shall remain essential for sustainable growth of the country till alternate energy sources are tapped in a big way or till India can afford to import coal. Both the alternatives do not seem to emerge in a decade or so. The need therefore is to take a pragmatic view of the problem considering the following issues:

1. What is the share of the coal industry in environmental and ecological degradation of the country?

2. What the mining industry needs to contribute to be acceptable from environmental considerations?.

3. What could be the long-term solution to the problem?

2.0 COAL SECTOR & ENVIRONMENT

In the process of mining, particularly in open cast mining, environmental equilibrium is disturbed on many fronts e.g. land degradation, water, air and noise pollution etc. Over the last few years, environmental considerations are taking precedence and many a times put constraints in the project commissioning and in the routine execution of activities.

2.1 Land Degradation

Following types of land degradation occurs due to mining :

1. Destruction of vegetation-forest and meadow land
2. Soil removal and burial
3. Disturbance to water regime by deep craters, cutting across aquifers breaking the continuity of recharging zone.
4. Formation of dumps with undulating, unfertile and salty debris.
5. Subsidence and mine fire

The land degradation due to factors induced by mining in different regions and due to deforestation on account of mining and other regions are shown in Tables I & II.

Table I : Causes and extent of land degradation in different coalfields (after Sachdev, 1995)

Coalfields	Land area degradation due to mining causes (in Ha)				
	Fire	Subsidence	Excavation	Dumps	Total
Raniganj	600	5094	138	370	6202
Jharia	1732	3497	1268	630	7127
East & West Bokaro	-	526	940	100	1566
Other coalfields	-	3394	NA	NA	3394
	2332	12511	2346	1100	18289

Table II– Deforestation and land degradation due to mining and other reasons – a comparison

	Reasons		Area (lac. ha)
Deforestation	1.	River valley projects	4.01
	2.	Agriculture	24.33
	3.	Roads & transportation	0.55
	4.	Mining & related industries	1.25
	5.	Miscellaneous	3.8
	Reasons		Area/sq.km.
Land Erosion in Damodar Valley	1.	Deforestation	2,200
	2.	Agriculture	3,350
	3.	Mining	110

2.2 Environmental Degradation

Environmental degradation due to mining occurs in the form of production of SPM, SO_2, NO_x Noise, Fly rock and Ground vibration on account of blasting activity and deterioration in the water quality. Air pollution due to mining and other industries is shown in Table III

Table III – Air pollution due to mining & other industries (CMRI studies, 1994-96)

Sl.No.	Industries	S.P.M.	SO_2	NO_x
	Standard limit	500	120	120
a)	Steel plants			
	1. Bokaro	15,007.56	1936.78	370.60
	2. Durgapur	3,045.81	563.07	552.40
b)	Thermal Power Station			
	3. Chandrapur	263.00	1315.07	1479.45
c)	Fertilizer Industry			
	4. F.C.I., Sindri	310.00	N.A.	70.00
d)	Hard Coke Oven Plants			
	5. Bhowra	19.31	0.86	1.39
	6. Barari	25.33	0.60	0.58
e)	Bee Hive Coke Oven			
	7. Govindpur	1899.54	474.89	534.25
	8. Chirkunda	2009.14	502.28	565.07
f)	Cement Plants			
	9. Kheladi	1577.00	N.A.	N.A.
	10. Bokaro	788.50	N.A.	N.A.
g)	Mining Industry			
	11. Pokharia	50.55	-	-
	12. Underground mine	47.90	-	-

Based on the facts, it is found that environmental pollution on account of mining is very less as compared to other sources and industries (Tables – II & III). The following are some of the facts which reveal the contribution of mining towards environmental degradation.

• Deforestation due to mining is 5% of the total deforestation
• Mining industry contributes only 2% in the total land erosion
• Mining induced land degradation is 10% of the total land degradation
• Mining does induce water pollution but it is very less as compared to the pollution caused by other

industries
- Air pollution due to mining is very less as compared to other industries like steel, coke, cement etc.
- Other industries are also equally responsible for noise pollution as mining industry

3.0   REMEDIAL MEASURES
3.1 Short-term Strategy
- Environmental management policy for sustainable development of India
3.2 Long-term Strategy
- Develop cleaner alternate energy sources like hydro, solar, tidel and wind
- Develop cleaner process to produce energy from coal

4.0   CONCLUSIONS
- Coal will continue to the primary energy source atleast for the next decade.
- Follow environment management plan religiously to lessen the geo-environmental effects.
- Develop cleaner energy generation process from coal
- Trap alternate cleaner energy source like hydro, solar, tidal, wind, etc.

5.0   REFERENCE
Sachdev R.K. (1995). Environmental issues in coal mining in India, WOMEC, New Delhi, Dec. 11-14, pp. 45-58.

Table I – Deforestation and land degradation due to mining and other reasons – a   comparison

	Reasons	Area (lac. Ha)
Deforestation	6. River valley projects	4.01
	7. Agriculture	24.33
	8. Roads & transportation	0.55
	9. Mining & related industries	1.25
	10. Miscellaneous	3.8
	Reasons	Area/sq.km.
Land Erosion in Damodar Valley	4. Deforestation	2,200
	5. Agriculture	3,350
	6. Mining	110

Table II – Air pollution due to mining & other industries (CMRI studies, 1994-96)

Sl.No.	Industries	S.P.M.	SO$_2$	NO$_x$
	Standard limit	500	120	120
a)	Steel plants			
	3. Bokaro	15,007.56	1936.78	370.60
	4. Durgapur	3,045.81	563.07	552.40
b)	Thermal Power Station			
	3. Chandrapur	263.00	1315.07	1479.45
c)	Fertilizer Industry			
	4. F.C.I., Sindri	310.00	N.A.	70.00
d)	Hard Coke Oven Plants			
	5. Bhowra	19.31	0.86	1.39
	6. Barari	25.33	0.60	0.58
e)	Bee Hive Coke Oven			
	7. Govindpur	1899.54	474.89	534.25
	8. Chirkunda	2009.14	502.28	565.07
f)	Cement Plants			
	6. Kheladi	1577.00	N A	N.A.
	7. Bokaro	788.50	N.A.	N.A.
g)	Mining Industry			
	8. Pokharia	50.55	-	-
	9. Underground mine	47.90	-	-

Table III - Forest Area affected due to various reasons between 1951 and 1971 (Haque, 1992)

Sr. No.	Reasons	Area (Lac. Ha)
1.	River valley projects	4.01
2.	Agriculture	24.33
3.	Road and other communication	0.55
4.	Industrial projects & mining	1.25
5.	Others	3.8

Coal Bed Methane Harnessing an Environmental Risk for a Fuel Source

S.D. Prasad[1] and B. Rai[2]

[1]*CMD, MECL, Nagpur, India*
[2]*General Manager (Mining), MECL, Nagpur, India*

Introduction :

The energy requirements of the globe are poised to multiply day by day to meet the anticipated rapid growth of the industry. As a result, the conventional indigeneous resources are likely to be depleted at a faster rate. This situation can be averted, provided an alternate source of energy is identified.

Coal Bed Methane, a hydrocarbon (CH_4) in gaseous form originates during the formation of coal out of accumulated plant material. The process under humid condition starts with bacterial decay called humification. Later bacterially decomposed plant material gets converted into peat. . Subsequent sedimentation over it, its positioning at depth and tectonic forces convert it gradually into lignite, sub-bituminous, bituminous and finally anthracite coal during the geological time period of carboniferous and Permian. This transformation is the process of attainment of maturity of coal where in the proportion of carbon keeps increasing and that of oxygen and other materials decrease. Coal acquires crystalline structure. Its increasing vitrinite constituent increases its reflectivity. Its reflectivity in oil is taken as an index of maturity and ranges approximately from 0.4 to 1.7 for Indian Bituminous coals. As coal matures, methane gets trapped in fractures and fine pore spaces. The pressure exerted by naturally formed water keeps methane adsorbed on internal surfaces of coal. This methane is in mono molecular state and is not as free as natural gas in oil or in gas fields. Thus all coals have methane in it. Mature coals are likely to have higher content of methane.

Methane an added risk in coal mining :

Association of methane with coal adds a serious fire risk in underground mining of coal.. Methane forms an explosive mixture with about 9% oxygen. Coal mine explosions are largely due to methane gas explosions.

Size of investment ,in exploiting gassy coal from underground mines, increases substantially due to costlier design of electrical equipment. The operating cost further increases due to increased quantum of ventilation and safety measures needed to reduce the risk of methane concentration in mine air, reduce source of fire and contain fire from spreading into larger area, if at all methane in mine catches fire.

Methane as green house & clean fuel gas :

While mining gassy coal, associated methane gas gets released into atmosphere along with mine ventilating air adding to the concentration of green house gases. Thus use of coal adds one more green house gas in the form of methane due to its shorter residence (life) and higher potency in atmosphere than that of CO_2. However, methane is remarkably a clean fuel when burnt as it does not produce SO_2 or dust particles.

Methane's contribution in global warming is 18%, making it the second most significant green house gas after carbon dioxide (66%), N_2O-5%, and CFC(11%] (Kruger 1995)

Methane Emission from different sources:

Sl. No.	Source	Methane Emission (1000 tons)
1.	Live Stock	65 – 100
2.	Rice	60 – 100
3.	Natural Gas & Oil	32 – 68
4.	Biomass Burning	28 – 51
5.	Liquid Wastes	29 – 40
6.	Coal	24 – 40
7.	Landfills	20 – 28
8.	Manure	8 – 18
9.	Minor Industries	4

Emission of methane from mined coal range from 3 m^3 per tonne to as much as 85 m^3 per tonne.

Estimated Methane Emissions from Coal Mining in Ten Largest Coal Producing Countries in 1990 :

Country	1990 Coal Production (million tons)		Estimated Methane Emission (in 1000 Tons)	
	Underground	Surface	Low	High
People's Republic of China	1023	43	8.4	13.0
United States	385	548	3.3	5.2
CIS	393	309	4.9	7.1
Poland	154	58	0.7	1.6
South Africa	112	63	0.8	2.3
India	80	120	0.8	2.3
Germany	77	359	1.0	1.6
United Kingdom	75	14	0.7	0.9
Australia	52	154	0.5	0.8
Czechoslovakia	22	85	0.3	0.5
Sub Total Top 10	**2402**	**2120**	**21.4**	**35.3**
Global Total	4740		24	39

Tapping methane in advance of mining coal

Methane drainage for making gassy mines safer has been practiced for a long time. One of the most efficient drainage has been from the region of caved rocks after extraction of coal through long wall caving technology. The percentage of methane in such drainage from mines vary from 30% to 70%. Where as such drainage may be useful for its use in heating or electricity generation for the coal mine.

450

such concentrations are not good enough for a commercially viable project in particular for chemical feed stock.

For obtaining higher concentration of methane attempts have been made to obtain methane from virgin gassy coal seams. Advance drainage of coal seam in virgin condition makes subsequent coal mining easier if the depth of mining and geological condition keep the cost of coal within marketable range.

The idea of draining coal bed methane for fuel even before coal is extracted has the potency to improve coal mining environment in underground mining by reducing risk associated with fiery methane and reduce content of methane being released to surface environment by isolating methane from getting mixed with mine ventilating air.

Extracting CBM in advance yields higher purity of methane having better prospects for its commercial utility. Mining coal may not have commercial viability from coal beds thinner than 1.2m in India and also from deep seated coal beds all over the world but CBM from thin seams and deep seated seams can be extracted.

Understanding Methane in coal bed

As already mentioned during the transformation of plant material into coal in the laboratory of earth's crust, as coal matures, methane becomes trapped in the fractures of coal. The pressure exerted by naturally formed water keeps the methane "adsorbed" on internal surfaces of coal. However this coal bed methane is in monomolecular state and is not a free gas, as in natural oil/gas fields. Therefore, all coal fields of the world
have coal bed methane, the only difference being the quantity of gas in individual coal seams.

Unlike oil and natural gas, which is generated in shales and clays, but migrates to some other rocks – like sands and limestones – the coal bed happens to be both the source of gas and its storage reservoir.

Porosity plays an important role in building up methane gas reserves in the coal bed (bastia and Shankar, 1994). Unlike the conventional reservoir, methane is not compressed in the pore spaces of coal, instead it is physically attached to coal at molecular level through microporosity of coal. Microporosity makes up about 70% of the total porosity in coal bed and is equivalent to a conventional reservoir having 20% porosity, saturated with 100% gas. On account of this difference coal has higher gas storage capacity than sands containing petroleum gas. It is, therefore, necessary that before undertaking production, a detailed exploration has to be taken up to (i) identify the probable coal basins rich in CBM & (ii) to ascertain quantum of CBM.

Releasing CBM for its drainage :

Since coal beds contain methane gas in an adsorbed state, to produce this trapped methane, the water in coal bed must be removed to reduce pressure on methane molecules. The process involves drilling wells from surface into coal bed and injecting fluids, water, and fine sand, into the bore well to create cracks and fractures in the coal bed. These openings provide channels through which water and gas can flow into the bore well and then removed to surface. However, unlike conventional gas wells, which produce gas immediately on completion of drilling, the coal bed methane wells pump out only water for considerable length of time month,) before methane is produced. In short, gas desorption is the main production mechanism. This is accomplished by hydraulic fracturing of coal bed through bore wells, draining water, which is always present in the pore structure of coal; and reducing pressure to allow desorption process to take place.

The construction of full scale conventional gas production well is very costly for conducting CBM potential investigation. In order to reduce such investigation cost, drill cores of coal is taken out by slim hole wire line drilling machine and a number of tests are conducted to assess the CBM potential.

The oil crises of the seventies, all over the world, forced even the most advanced nation like USA to search for alternative source of fuel. USA has 28% of world reserves of coal. Extensive research in the coal field established that CBM locked up in the coal beds have been effectively exploited to supplement the energy needs of the country. Today, USA is the most advanced nation in utilisation of CBM which amounts to 7% of total energy needs of the country and about 70% of the total natural fuel consumption. Production of coal bed methane has now taken the shape of an industry, mainly due to sustained efforts in USA. USA has huge world coal reserves and in 1994 it was estimated that they contain 675 trillion cubic feet of methane gas (Hamilton et al 1995). Total assessment of reserves is yet to be competed and, therefore, these reserves will increase in future. It is not known, as to how much of this can be produced economically under the present technology, The situation is very similar to petroleum, where oilfields previously considered uneconomical, beonme profitable with use of new extraction technologies.

451

List of Exploration Data Needed by CBM Industry :

1.	Total area of the coal field.
2.	Number of Seams and their thickness
3.	Proximate and Ultimate analyses
4.	Rank of Coal
5.	Geological structure
6.	Hydrogeology
7.	Geothermal gradient
8.	Reflectance
9.	Maceral Composition
10.	Mineral Matter
11.	Dispersed Organic matter
12.	Gas composition and percentage
13.	Permeability and porosity
14.	Cleats and fracture pattern
15.	Adsorption isotherms
16.	Quality of adsorbed water
17.	Reservoir Engineering and modelling
18.	Open hole drill stem testing
19.	Side wall pressure core testing, etc.

Lead Work in USA :

US Govt. support to CBM :

Although under the current know-how and gas prices in USA, which are abundant, it is difficult for coal bed methane to survive as fuel source, the American government is allowing considerable tax benefits and rebates to the companies engaged in such work.. US Government is interested in promoting coal bed methane as it has the following merits:

1.	Methane is environmentally compatible,
2.	Methane is a new clean source of energy,
3.	Harnessing Coal Bed Methane transforms a hazard into resource,
4.	CBM is capable of power generation in 2 to 3 years
5.	Draining CBM in advance makes future coal mining safe
(a)	Additional biogenic gas can be generated
(b)	Pore plugging waxes can be removed, and permeability enhancement may take place as cleat aperture width is enlarged during biogassification.

On the basis of American experience of coal bed methane production, coals having 6-7 m^3/tonne of coal gas or high rank (Ro-0.7 to 1.7),buried at depth should be made the target for CBM extraction (Scott, 1995).

Indian CBM experience :

India is the third largest producer of coal but its mining is limited to shallow depth. In future deeper mining shall have to be undertaken which have higher methane content. The exploitation of CBM from

deeper levels have further advantage of having additional resources of energy besides making coal mining safe. The extensive research has ascertained that there is good future for CBM in our country to make it an additional source of energy. The heat value of CBM is 8500 K.cal. compared to 9800 K.cal. of natural gas and can be utilised either at site or it can be transported through pipe line to distant places.

The awareness of CBM came to India in Nineties after Economic liberalisation policy in India attracted several international and national companies to undertake initiatives for exploitation of CBM. However the Indian Government were not ready to receive the offers. The existing law proved to be inadequate to support the entrepreneurs. Besides the following reasons can also be quoted for a poor response towards CBM in India-

- India not updated in CBM exploration technology

- Inadequate expertise and experience in CBM production

- Methane during mining viewed as a nuisance and not a valuable resource and hence drained off to be wasted in atmosphere for safety reasons

Attempts made by **Modi-Mckenjee** to undertake CBM exploitation in Coal India command area proved to be waste of time and money. **AMOCO** (USA) after initial attempts seem to abandon the CBM projects. However Reliance Industries Ltd., Great Eastern Energy Corporation Ltd., ONGC, and GAIL are still interested to pursue CBM projects.

Although, under Indian Coal Mines Act (sub regulation (5) of regulation 116 of coal mines regulation, 1957) it is mandatory to carry out gas emission measurements in mines, its basic approach is to facilitate better ventilation design.

CMRI has developed technology to estimate CBM by measuring measuring methane oozing out of coal cores over a period of time. Gas released from the time core was detached from seam to its sealing in the laboratory is estimated and the residual gas is measured after grinding core to fine size. Three components put togather give total CBM. However only half of this gas is available for extraction. CMRI estimated CBM for Chano –Rikba dip side block of North Karnpura Coal field.

Gas Authority of India Ltd. engaged MECL which collected all available necessary data from published sources/literature from all the major and minor coalfields belonging to Lower Gondwana Group, located in the five intracretonic basins of peninsular India. In all, 42 coalfields were studied with respect to geographical information, general geological set up, geological structure, number of coal seams, variation in the thickness of coal seams, quality characteristics and their variation, resource potential and depth-wise disposition of coal reserves.

MECL report was submitted in two stages. In the first stage, coal resources quantity and its characteristic were compiled. In the second stage, report information concerning likely occurrence of Coal Bed Methane was made which is summerised below in detail. The rough estimate of coal bed methane availability was kept at 30 trillion cu. ft. which comes to approximately 850 billion cu.mt.

C B M resources of potential coalfields :

Sl.No.	Name of the Coalfields	Total coal Resources (in billion tonnes	Resource + 300m depth (in billion tonne)	CBM poten (in billion cu.m.)
1.	Raniganj Coalfield	23.13	10.09	109.0
2.	Jharia Coalfield	19.42	5.21	145.0
3.	West Bokaro Coalfield	4.50	0.41	32.0
4.	East Bokaro Coalfield	5.81	2.74	40.0
5.	N. Karanpura Coalfield	13.65	3.94	67.0
6.	S. Karanpura Coalfield	5.99	2.68	28.0
7.	Godavary Valley Coalfield	13.02	7.64	50.0
	TOTAL	**85.61**	**32.71**	**471.0**

Great Eastern Energy Corporation Limited had undertaken study of Coal Bed Methane in Raniganj coal field in West Bengal. 2 nos. of deep bore hole were made out of which one was located in Surajnagar in the district of Bardwan and other was located at Poradiha in district of Bardwan during May 1996 to May 1997. Coal Bed Methane content in the coal seam was found to be sufficient for exploitation and company is very soon taking up pilot drilling for production.

Oil and Natural Gas Commission selected MECL to undertake Coal Bed Methane exploration in Raniganj and Jharia coal field. 10-12 Bore holes each about 1000 m depth were identified for Coal Bed Methane exploration in Raniganj and Jharia coalfields The drilling work was scheduled to start from 1st October 2000. The entire work for drilling is likely to be completed in a years time using high capacity hydrostatic drill rigs and accessories. MECL shall undertake geophysical logging work as detailed below:

Geophysical logging of bore holes will be done by using MECL's Digital RG (Robertson Geologging, U.K.) Logging Systems which digitize data right at the point of measurement in a probe (Sonde). This facility enables accurate stable data measurements with no significant signal distortion during cable transmission.. The unit is fitted on 3 Tonne mini truck with 4 wheel drive for convenience of logging operation in all types of terrains. Logging shall be done for 17 parameters over a maximum depth coverage upto 1700 m. the parameters are Self potential, Single point resistance, Short normal, Long normal, Temperature, High resolution density, Long spaced density, Natural gamma, Caliper, Focused resistivity. Short spaced neutron, TA, TB, TC, TD, (transit times with two transmitters and two receivers), and Sonic velocity.

Coal India CBM Demonstration Project

Recently a CBM project has been undertaken with assistance from United Nations Mine Development Programme and General Environment Fund with the main purpose to demonstrate the project so as to provide the latest technology, expertise and experience. The project would develop and coordinate multinational investment efforts in development and replications of CBM technology. The project will be executed by CMPDIL at Sudamdih and moonidih coal mines of BCCL

Coal methane shall be tapped by-
1. Advance vertical bore hole from surface into virgin coal areas.
2. Bore hole from surface into mined out areas.
3. Long bore hole from on going under ground coal mine workings along the seam, upwards and downwards into into coal measure strata.

Methane gas shall be used to power dump trucks upto 50 tons pay load as fuel in place of diesel. The project shall be completed in 5 years time at a total capital cost of Rs.80 crores.

CBM from lignite fields in India

The Director General, Hydro Carbons, Ministry of Petroleum Gas is actively associated for search for CBM in lignite field in Tamil Nadu and Rajasthan. In the state of Tamil Nadu one bore hole has been done upto a depth 500m by MECL at Maranguri and huge deposits have been established. Another bore hole of 90mts. has been done and handed over to CMRI for further testing. Similarly in the state of Rajasthan, Research & Development Scheme is under execution by MECL on behalf of Government of Rajasthan where 2 fields i.e. Barmer and Bikaner have been identified as potential reservoirs for production. Drilling is likely to start shortly. Lignite has been found upto 600-800 mtrs. depth in this area and holds a big future for huge source of energy and fuel in Rajasthan.

ONGC

ONGC has already drilled 3 bore holes in Jharia field south of Damodar river in Partabpur block and gas production has started from one of the boreholes for further successful production. Studies are being conducted for utilisation of gas at nearby Bokaro Steel Plant.

Conclusion :

It may be concluded that Indian Coal Reserve have a prospect for obtaining about 850 BCM of coal bed methane which has a potency as an important non-conventional source of energy. Harnessing this CBM shall also reduce the ecological problems by reduction of release of methane into atmosphere. It may also be concluded that CBM exploration demands import of technology which shall be undertaken for through the projects of ONGC undertaken with the help of MECL. The future prospects for this non-conventional clean energy source may depend upon the results of scientific exploration of coal bed methane undertaken by ONGC and other organisations who may choose to invest on updated technology for CBM exploitation. An increased Govt. support and incentive may increase the pace of development of CBM which has potency to reduce Indian oil import foreign exchange drainage.

Kolar River Water Pollution due to Ash Bund Effluent of Koradi Thermal Power Station

D.M. Vairagade[1], H.D. Juneja[1], R.T. Jadhav[2] and L.J. Paliwal[3]

[1]*Department of Chemistry, Nagpur University Campus, Nagpur 440 010, India*
[2]*Department of Chemistry, Ramdeo Baba Kamla Nehru College of Engineering, Nagpur, India*
[3]*Department of Chemistry, Hislop College, Nagpur 440 001, India*

INTRODUCTION

Ground water is used for various purposes such as domestic supply, power generation in thermal power stations, industries and agriculture in most parts of the world as it is a replenishable resource and has inherent advantages over surface water. There has been a substantial increase in the demand for fresh water due to growth in population. The rapid growth of urban areas has affected the ground water quality due to over exploitation of resources and improper waste disposal practices. Consequently there is always a need and concern for the protection and management of ground water quality.

Ramaswami and Rajaguru[1] have conducted studies on ground water quality at Tiruppur town in Coimbatore district of Tamil Nadu (India) and reported that several parameters exceeded the permissible limits for various uses pointing out the necessity of proper treatment and disposal of wastes in the area. Ravichandran and Pundarikanthan[2] have studied ground water quality in Chennai with the context of polluted water way of the city and confirmed the increasing concentration of chloride towards the coast due to saline intrusion. Vijayaram[3] carried out the pollution studies of ground water in Sembattu, Tiruchirapalli in relation to waste disposal and reported that most of the ground water is not suitable for drinking and industrial utilization. Murali Krishan and Sumalatha[4] have conducted some preliminary studies on the quality of ground water of Kakinanda town and recommended that any water source must be thoroughly analyzed and studied before being used for domestic purposes.
In the present study, investigations have been carried out to evaluate the physicochemical characteristics of ground water in this central region of the country.

THE STUDY AREA

Kolar river, originally a fresh water stream originating form chicholi-wad district Chhindwada (Madhya Pradesh); passes through the Suradevi village and joins the river Kanhan and Pench. It receives ash bund disposal waste water from Koradi Thermal Power Station (KTPS) during its flow over a distance of 3 km. About 30 percent of raw sewage is directly discharged into the Kolar river, which becomes most unhygienic during its last stretch a few kilometers before joining the river Kanhan and Pench. The present investigation was undertaken to assess the physico-chemical characteristics of ash bund disposal waste water and evaluate the impact on the river water quality.

MATERIALS AND METHODS

Six water samples from Kolar river were collected in plastic cans. Once every month from October 1979 to September 1998. Water sample No. 1 (WS1) was collected from the water fall of ash bund disposal waste water. Water sample No. 2 (WS2) was collected form the confluence point of effluent discharge into Kolar river. Water sample No. 3 (WS3) was collected from a distance of 1 km upstream from the confluence point while water sample No. 4 (WS4) was collected from 1 km down stream from the confluence point. Water sample No. 5 (WS5) was collected from the agriculture use intake point. Water sample No. 6 (WS6) was collected from domestic water use point. The physico-chemical characteristics of the samples were analysed as per standard procedures[5]. For other parameters, samples were preserved by adding an appropriate preservative[6].

RESULTS AND DISCUSSION

Data on the quality of ash bound disposal waste water is presented in Table 1. The D. O. (Dissolved

Oxygen) levels in (WS1–WS6) varied between 6.4 mg/l to 6.7 mg/l; which was further reduced to 6.4 mg/l after the confluence of sewage into the river. The pH values were always alkaline in all the samples in river. However, it was reduced to 7.97 at WS4 and WS5. The BOD level of the river varied between 16 mg/l to 69 mg/l. The values of alkalinity, total dissolved solid (TDS) and total hardness were always higher in the river water. In water sample No. 3 blue green colour results from contact of the river water with organic debris such as leaves and other plant material in various stages of decomposition, WS1 and WS2 were milky white due to the discharge of ash bund waste water consisting mainly of fly ash slurry.

The data indicates that the ash bund disposal waste water brought in higher amounts of suspended solids apart from chemical effluent load. The river is not likely to carry considerable amount of suspended solids originating naturally. The WS1 and WS2 brought in higher concentration of suspended solids which were mainly composed of fly ash particles which renders the Kolar river unfit as a source of drinking water.

The BOD levels of the river varied between 16 mg/l to 69 mg/l. The BOD level in WS3 was high due to discharge of sewage water into Kolar river from Binagaon. The BOD level of WS6 is reduced due to the dilution by Kolar river water.

COD levels of Kolar river varied between 33 mg/l to 150 mg/l. WS3 recorded the highest level; while it was minimum in WS1. The disposal of sewage water into adjacent natural stream leads to depletion of DO and increase in biological oxygen demand of water. In addition the unhygienic conditions may lead to the spread of water borne diseases in the surrounding population.

Values of the nutrients like calcium, magnesium, chloride, sulphates, phosphates, sodium and potassium at WS1 to WS6 show the extent of contamination in the Kolar river (Table No. 2).

CONCLUSION
The results of the analysis of the various parameters indicate a considerable pollutional load being incorporated on the Kolar river due to ash bund disposal and discharge from the KTPS. Pollutional loads show substantial increase with respect to suspended solids and dissolved solids. The latter are likely to affect adversely the Kolar river basin by percolation. The high pollutional load is likely to be carried forward with successive decrease into the Kanhan river system. Thus the pollutional load generated by KTPS shall have wide spread implications on a large area as these are carried by the river system through long distances.

ACKNOWLEDGEMENT
The authors are thankful to Head of the Department of Chemistry, Nagpur University, Nagpur for his encouragement. DMV thanks Mr. A. R. Deshpande, and D. K. Banergee of Khaperkheda Thermal Power Station for providing laboratory facilities to carry out this work. The authors are also thankful to the Principal R. K. N. Engineering College, Nagpur.

REFERENCES :
- Ramaswami V. and Rajaguru P; Ground water quality of Tiruppur; Indian J. Environ. Hlth. 33, 187–197 (1991).
- Ravichandran S. and Pundarikanthan N. V.; Studies on ground water quality of Madras; Indian J. Environ. Hlth 33, 481–487 (1991).
- Vijayram K.; Pollution studies of ground water quality in coastal regions of south Madras; Indian J. Environ. Hlth. 34, 318–325 (1992).
- Murali Krishan K.V.S.G. and Sumalatha V.; Ground water quality in Kakinada, Porc of the All India Seminar on Ground Water Management in coastal Areas, 5–6th June, Visakhapatnam (1983).
- Apha, Standard Methods for the Examination of water and waste water, 16th ed : American Public Health Association, New York (1985).
- Jain C. K. and Bhatia K. K. S.; Physico-chemical analysis of water and waste water user's manual UM–26, National Institute of Hydrology, Roorkee (1987).

Table 2 : Pollutants in the Kolar river water (AV ± SD) AV – Average Value   SD – Standard Deviation

	Calcium Mg/l	Magnesium mg/l	Chloride Mg/l	Sulphate mg/l	Phosphate mg/l	Sodium mg/l	Potassium Mg/l
WS1	49.54±5.68	16.50±4.06	55.22±21.11	94±6.86	3.0±0.3357	69±10.189	6.5±0.4532
WS2	42.69±10.22	15.62±3.10	55.05±5.49	88±7.29	2.8±0.3162	60±3.233	6.2±0.6578
WS3	40.47±4.25	27.63±6.09	80.22±50.74	28±6.53	3.0±0.2480	44±3.801	4.8±0.6331
WS4	45.94±2.65	20.44±3.54	58.53±21.55	58±6.45	2.38±0.4117	49±4.134	5.4±0.4177
WS5	45.68±2.65	20.24±2.65	57.21±20.67	50±4.76	2.1±0.2713	44±3.060	5.5±0.3554
Ws6	45.71±3.36	19.86±2.76	59.72±18.97	42±5.59	2.0±0.440	42±6.775	5.1±0.337

Table 1 : Monthly variations in the physico-chemisal characteristics of Kolar river samples (AV ± SD)

Date Sample No.	Colour	Temp. °C	pH mg/l	Total suspended solids mg/l	Total dissolved solids mg/l	Turbidity NTU	Dissolved Oxygen Mg/l	BOD Mg/l	COD mg/l	Total Alkalinity mg/l	Total hardness mg/l	Conductivity µmhos/cm
16-10-97 to 16-9-98 WS1	Milky white	24.33±2.19	8.05±0.25	243±200	449±91.24	277±78.07	6.7±1.10	16±5.49	33±11.22	135±30.80	199±15.4	525±199.48
16-10-97 to 16-9-98 WS2	Milky white	23.9±2.32	8.02±0.25	278±196	441±139.36	280±93.09	6.7±1.11	30±8.25	65±18.36	135±31.07	198±19.60	528±97.68
16-10-97 to 16-9-98 WS3	Blue green	24±2.41	8.04±0.21	77.25±96.65	386±116.82	15.54±24.15	6.4±1.43	69±6.61	150±11.44	180±41.78	231±23.79	597±222.51
16-10-97 to 16-9-98 WS4	Slightly Milky white	24.3±2.29	7.97±0.20	73.33±130	366±94.30	78.33±62.14	6.45±1.30	47±11.98	119±11.84	154±29.51	209±19.31	530±127.49
16-10-97 to 16-9-98 WS5	Slightly Milky white	24.2±2.20	7.97±0.25	51.33±48.25	406±85.96	68±46.90	6.5±1.19	46±9.0	107±14.87	156±44.00	208±16.34	534±136.27
16-10-97 to 16-9-98 WS6	Slightly Milky white	24.5±2.18	8.03±0.18	45.51±47.61	361±84.93	65±37.43	6.4±1.07	30±5.13	77±10.77	158±28.61	206±16.84	485±140.61

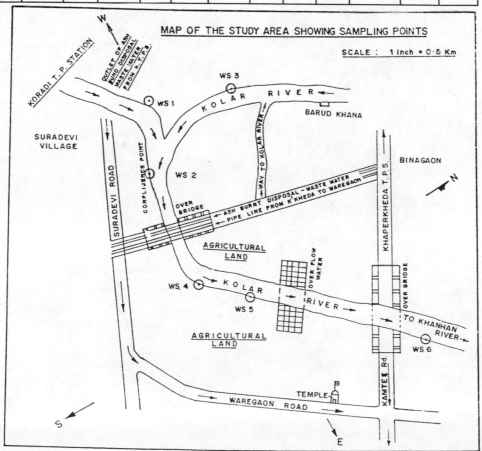

MAP OF THE STUDY AREA SHOWING SAMPLING POINTS

SCALE : 1 Inch = 0·5 Km